POLITICAL ESSAY ON
THE KINGDOM OF NEW SPAIN

VOLUME 1

ALEXANDER
von HUMBOLDT
IN ENGLISH

A series edited by Vera M. Kutzinski and Ottmar Ette

POLITICAL ESSAY ON THE KINGDOM OF NEW SPAIN

VOLUME 1

A Critical Edition

ALEXANDER VON HUMBOLDT

*Edited with an Introduction
by Vera M. Kutzinski and Ottmar Ette*

Translated by J. Ryan Poynter, Kenneth Berri, and Vera M. Kutzinski

*With Annotations by Giorleny Altamirano Rayo,
Tobias Kraft, and Vera M. Kutzinski*

THE UNIVERSITY OF CHICAGO PRESS   CHICAGO AND LONDON

The University of Chicago Press, Chicago 60637
The University of Chicago Press, Ltd., London
© 2019 by The University of Chicago
Published 2019
Printed in the United States of America

28 27 26 25 24 23 22 21 20 19   1 2 3 4 5

ISBN-13: 978-0-226-65138-5 (cloth)
ISBN-13: 978-0-226-65141-5 (e-book)
DOI: https://doi.org/10.7208/chicago/9780226651415.001.0001

Library of Congress Cataloging-in-Publication Data
Names: Humboldt, Alexander von, 1769–1859, author. | Kutzinski, Vera M., 1956– editor, translator,
writer of introduction, writer of added commentary. | Ette, Ottmar, editor, writer of introduction. |
Poynter, J. Ryan, translator. | Berri, Kenneth, translator. | Altamirano Rayo, Giorleny D., writer of
added commentary. | Kraft, Tobias, 1978– writer of added commentary. | Humboldt, Alexander von,
1769–1859. Works. English.
Title: Political essay on the Kingdom of New Spain : a critical edition / Alexander von Humboldt ;
edited with an introduction by Vera M. Kutzinski and Ottmar Ette ; translated by J. Ryan Poynter,
Kenneth Berri, and Vera M. Kutzinski ; with annotations by Giorleny Altamirano Rayo, Tobias Kraft,
and Vera M. Kutzinski.
Other titles: Essai politique sur le royaume de la Nouvelle-Espagne. English (Kutzinski and Ette)
Description: Chicago ; London : The University of Chicago Press, 2019. | Includes bibliographical
references and index.
Identifiers: LCCN 2019005351 | ISBN 9780226651385 (cloth : alk. paper) | ISBN 9780226651415
(e-book)
Subjects: LCSH: Mexico—Description and travel. | Mexico—Geography. | Mexico—Population. |
Mexico—Statistics. | Mines and mineral resources—Mexico. | Mexico—Social conditions—To 1810. |
Agriculture—Mexico. | Mexico—Politics and government—1540–1810.
Classification: LCC F1211 .H92613 2019 | DDC 917.2—dc23
LC record available at https://lccn.loc.gov/2019005351

♾ This paper meets the requirements of ANSI/NISO Z39.48-1992 (Permanence of Paper).

Contents

VOLUME 2

Note on the Text

The translation in this edition is based upon the full text of the second, revised French edition of Alexander von Humboldt's two-volume *Essai politique sur le royaume de de Nouvelle Espagne* from 1825 to 1827; it includes all footnotes, tables, and maps. The page numbers in the outer margins of each page refer back to that edition, which has never been translated into English before. English-speaking readers previously had to rely on John Black's *Political Essay on the Kingdom of New Spain* (1811), which was based on the first edition and has long been out of print. Now more than two hundred years old, Black's translation was notorious even in its time for its omissions, inaccuracies, capricious alterations, and misrepresentations; nonetheless, it was frequently reprinted and excerpted during the nineteenth century. Since then, there has only been one attempt at producing a new or even a corrected English translation (see Editorial Note). The only prior English translation of this text used its earlier, significantly shorter version from 1808 to 1811 (see Editorial Note).

We firmly believe that maintaining the original structure of this resolutely nonlinear text keeps the reader engaged more directly and more fully with Humboldt's conceptual and narrative process. Throughout our work, we have retained his paragraphing and the placement of tables. Humboldt's tables are part of his overall narrative. There are, however, a few instances when decisions about design made it necessary to relocate tables slightly from their original positions. Brief bracketed comments inserted into the text alert the reader to such changes.

Humboldt typically supplied indexes for his work, as he did in this case. We have augmented his own index to provide readers with additional navigational assistance to a text that has an often baffling amount of detailed information on many topics. The index appears in the second volume of this edition.

We have placed any additions and corrections in [square brackets], reserving the use of parentheses entirely for Humboldt himself. We have also kept intact Humboldt's seemingly whimsical use of capitalization and italics and all words, phrases, and quotations originally written in Spanish and other languages (such as Latin and Greek). Only where meanings are not evident from context do we add our own translations of foreign words and phrases in [square brackets] to signify additions to the text. Humboldt himself rarely converts currencies and other units of measure, illustrating their variety across different parts of the world. We have followed his practice, offering, at the beginning of our annotations, explanations of currencies and units of measure no longer in use. To avoid encumbering this already densely textured work with endnote numbers, we have used the symbol ▼ to signal that annotations related to a name, concept, or historical event follow the translation. The annotations are ordered in accordance with the translation's pagination. To help readers trace at least part of the vast scientific network that Humboldt created, we provide a briefly annotated bibliography—Humboldt's Library—of the sources he acknowledges in this work. The annotations and Humboldt's Library are available on the Press's website at press.uchicago.edu/books/humboldt.

The HiE site also includes a link to the digital version of Humboldt's maps prepared by the David Rumsey Collection from Volume V of the French edition. The Editorial Note affords more information about the thinking behind this translation and about the historical and scholarly contexts for *The Political Essay on the Kingdom of New Spain.*

The HiE Team

All the Bumps in the Road:
Alexander von Humboldt's Mexican Tableau

AN INTRODUCTION BY

VERA M. KUTZINSKI AND OTTMAR ETTE

The Prussian natural scientist and humanist Alexander von Humboldt (1769–1859) traveled in the Americas between 1799 and 1804. This five-year journey would make him one of the most famous men of his time. Some regarded him as second in importance only to Napoleon Bonaparte; others even compared him to Aristotle. By all accounts, Humboldt was the most celebrated modern chronicler of North and South America and the Caribbean. Prior to setting out for the New World from the Spanish port of La Coruña on June 5, 1799, Humboldt and his travel companion, the French botanist and physician Aimé Bonpland, had hoped to join Nicolas Baudin's expedition to Australia, which had to be canceled because of financial problems. After Napoleon's invasion of Egypt foiled Humboldt's alternative plan to visit North Africa, Humboldt opted to go to Spain and organize an expedition to the Americas. Phillip von Forell, Saxony's ambassador in Madrid, helped the Prussian to succeed in what no foreigner, especially a non-Catholic, had been able to do: be granted a passport and unlimited access to Spain's American colonies. That Humboldt did not need any money from King Carlos IV likely helped. Unlike most other scientific explorers before and after him, Humboldt had a considerable inheritance from his mother with which to underwrite his own expeditions. In this way, he could ensure his freedom from any country's commercial and political interests.

A brief sketch of Humboldt and Bonpland's itinerary after their delayed departure from blockaded La Coruña will help situate their visit to

New Spain, one of four viceroyalties that the Spanish Crown had created in order to administer its overseas possessions. Humboldt's transatlantic voyage followed Columbus's route, stopping first in the Canary Islands (June 19–25, 1799). Tenerife, the largest island of the Canaries, was the first non-European island that Humboldt had ever visited. It was there that he learned to describe a place in all its different dimensions: anthropological, botanical, economical, geographical, geological, historical, political, and sociocultural. The island of Tenerife, where he climbed the first of many volcanoes, became the theoretical and practical model that Humboldt would apply elsewhere on his travels. He had expected to cross from the Canaries directly to Cuba, but an outbreak of "fever" thwarted these plans when the captain of the *Pizarro* decided to divert to Cumaná in today's Venezuela. In Cumaná, Humboldt first became acquainted with the "equinoctial regions," the tropics he had dreamed of ever since developing an interest in plants as a boy. Explorations on the mosquito-infested Orinoco and Casiquiare rivers followed in short order. These adventurous explorations would become the best-known part of Humboldt and Bonpland's American voyage. After returning to the Venezuelan coast, Humboldt proceeded to the island of Cuba, where he stayed for nearly three months, from December 19, 1800, to March 15, 1801. He then visited the regions of South America that later became Colombia, Ecuador, and Peru, where he climbed (or attempted to climb) some of the world's highest volcanoes, most famously the Chimborazo. From Lima, Peru, Humboldt sailed north to the Ecuadorian port of Guayaquil on the waters now known as the Humboldt Current. He continued on to Acapulco in New Spain and then to Mexico City. From the capital, he made many expeditions to the colony's different regions and climbed more volcanoes, this time the Jorullo and the Nevado de Toluca, though not the Popocatépetl, as some have claimed (see figure 1). On his inland excursions, Humboldt also spent much time studying the working conditions of miners, most of whom were descended from the Mexica, the same pre-Columbian peoples whose languages and cultures he researched in the archives of New Spain. This research became the foundation for two major works: his *Views of the Cordilleras and of Monuments of the Indigenous Peoples of the Americas* and his *Political Essay on the Kingdom of New Spain*. Both are part of the vast corpus of Humboldt's writings on the Americas, brought together in his *Voyage to the Equinoctial Regions to the New Continent* (1805–1839) in no fewer than thirty volumes. Although a comparatively small part of his entire opus, the original five volumes of the

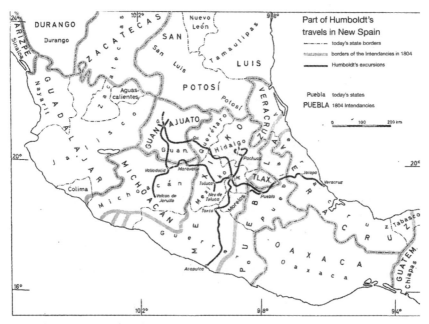

Humboldt's Expeditions in New Spain

Political Essay on the Kingdom of New Spain have a special place: they made Alexander von Humboldt famous among his contemporaries.

Science in the Field and in the Archive

Alexander von Humboldt arrived in Acapulco, then a major port for the trade with China, on Tuesday, March 22, 1803. He would stay in New Spain for close to a year, departing for a second visit to Havana on March 7, 1804. As the travelers—they also included Humboldt's close friend Carlos Montú-far, who had joined the team in Quito—disembarked the Spanish frigate in Acapulco after an almost thirty-day journey at sea, a curious spectacle unfolded in front of the local bystanders. It was the same spectacle wherever Humboldt and Bonpland went. A seemingly endless stream of crates and packages emerged from the ship's hold, filled with specimens of plants and animals, books, and, most importantly, the most advanced scientific instruments of the times. There were barometers, hygrometers, cyanometers, thermometers, sextants, repeating circles, theodolites, and different kinds of field glasses. A pioneer of scientific fieldwork, Humboldt cherished these instruments almost as much as his human travel companions.

Like the books he carried with him, the instruments were what made his fieldwork possible in the first place. It is not surprising, then, that we find representations of these instruments, along with botanical dictionaries and guides, in the countless paintings and images that—as in the work of Friedrich Georg Weitsch and Eduard Ender, among others—visualized Humboldt's voyage to the Americas emblematically and epistemologically. In these paintings, the instruments always take center stage.

One cannot, however, reduce field research to the examination and measuring of natural-scientific phenomena with the help of instruments. As his *Political Essay on the Kingdom of New Spain* and *Views of the Cordilleras and Monuments of the Indigenous Peoples of the Americas* show very clearly, Humboldtian field research was not limited to reading the Book of Nature, as it were, but also extended to reading the Book of Culture. His science-in-the-field included natural-scientific and cultural-scientific objects alike, combining observational works with research in local libraries and archives, in which he often made major discoveries. This was especially the case in the viceregal and ecclesiastic archives of New Spain, where Humboldt had access to an extensive collection of data: maps, geographical descriptions, production statistics, and the like. Above all, he was interested in census data, the papers of the Royal Mining Board, and ancient Mexica codices—"fragments of Aztec manuscripts written on maguey paper" (570). He supplemented what he found in the official archives with information that he received either in person or later through letters from experts such as José María Fagoaga, Fausto Delhuyar, Andrés del Rio, and Ludwig von Eschwege, as well as from specialists in botany such as Vicente Cervantes and José Antonio Pichardo, among many others.

That Humboldt's *Voyage to the Equinoctial Regions of the New Continent* has long been an indispensable compendium for historians of the natural and social sciences follows from the narrower definition of fieldwork, as does its neglect by cultural and especially literary historians. Humboldt's voluminous and innovative writings on the New World made him the undisputed father of modern geography, early American studies, transatlantic cultural history, and environmental studies. Countless natural and social scientists followed in Humboldt's footsteps, among them anthropologists, archeologists, astronomers, botanists, ecologists, economists, engineers, epidemiologists, geologists, geographers, mineralogists, political scientists, and volcanologists—all fields to which Humboldt made signature contributions. Along with not a few politicians in Europe and the Americas, notably Carlos IV of Spain and Thomas Jefferson, they mined Humboldt's writing for his data and built on his scientific insights. By contrast, humanities scholars have, until very recently, neglected and underappreciated Humboldt's work, including distinctive features of Humboldtian writing that are very much of interest to humanists, whether they specialize in literary studies, art history, history, or philosophy.

To begin with, Alexander von Humboldt's works are not mere scientific reports; they were neither so for their times, nor are they by any of today's standards. They are, in fact, works of art that crisscross many different literary and nonliterary genres in an attempt to do justice to the comparative global framework Humboldt always used for his analyses. To show how people, plants, animals, goods, and ideas moved across the Earth's varied regions, he availed himself of the forms of the essay (with the political essay as a subgenre), the travelogue, and the tableau. Because they are also works of art, Humboldt's writings do what academic science writing typically does not: incorporate multiple perspectives and integrate information from diverse fields of knowledge in often surprising, sometimes even irritating, ways. Humboldt's writings are not easy reads, and they are so by design. Bending and reshaping familiar generic schemas and other writerly conventions keep readers attentive to new inputs. For, above all, Humboldt wanted his readers to keep an open mind when confronted with cultural and other differences in the Americas. Humboldt himself was not one to presume anything: "How could a traveler stranded on an island, or who had lived in a distant country for some time, presume to judge the different faculties of the soul and the prevalence of reason, wit, and imagination of other peoples?" (249). Open-mindedness marks his thinking and writing; it makes Humboldt's works more than just the stuff of history. His writing's intellectual and emotional energy can influence both our present and our future. His discourse on the tropics, which revolutionized the ways in which Europeans thought and still think about the New World, is our present case in point.

As it crosses genres and media, Humboldtian writing, in *New Spain* and elsewhere, settles into distinct patterns that organize myriad minute details into larger pictures—he called them *tableaux*—that attempt to show how everything, absolutely everything, is interrelated. Rendering these relations intelligible was as important to Humboldt as fieldwork and archival research. His goal was to demonstrate how differing and seemingly incompatible objects and phenomena come together in patterns and networks on both a hemispheric and a global scale. The imaginative dimensions with which his desire for "full impressions" infuses his writing are perhaps the most important and lasting part of his intellectual legacy. Humboldt also has bequeathed us steadfast democratic beliefs inspired by the French Revolution. Fueled by his comparative global perspective on politics, economics, and science, these convictions, which we will call Humboldtian ethics, come to light most clearly when he passionately condemns slavery and other forms of colonial exploitation throughout his writing—in *New Spain* notably the massive deforestation on the Central Mexican Plateau and labor practices reminiscent of today's sweatshops. All these elements come together in Humboldt's unorthodox tableau-travelogues in ways that earlier English translations have all but buried. John Black's 1811 translation of the *Political Essay on the Kingdom of New Spain* was no exception (more on this translation in our Editorial Note).

Untold Riches: Of Mining and Men

Contrary to popular belief, it was not his travel narrative, the three-volume *Relation historique* (*Personal Narrative*), but his monumental *Political Essay on the Kingdom of New Spain* that established Alexander von Humboldt's popular reputation in Europe. Combining natural sciences and economics with cultural and political history, the *Political Essay* not only explores New Spain's legendary mineral and agricultural resources but also chronicles the fate of its indigenous peoples. The five volumes, first published in French between 1808 and 1810, consist of a methodological introduction followed by six books (which include close to two hundred tables), notes, a supplement, and a detailed index. Almost half of Volume I is taken up by Humboldt's lengthy "Reasoned Analysis of the Atlas of New Spain," in which he discusses the intricacies and difficulties of cartographic representation at a time when most maps were still exceedingly inaccurate. Using local sources, such as Indian muleteers, Humboldt measured distances and used his state-of-the-art instruments to determine elevation, temperature, geological composition, and the exact geographical positions of towns and cities, along with what we now know as their time zones. He explains in detail the twenty-one physical and geographical maps in the accompanying atlas (Volume V) and gives his readers a structure within which to understand the astronomical observations that appear throughout all the volumes. These maps of New Spain are the most precise maps ever to be drawn of the region, barring satellite imaging. They provided a wealth of information about geography, national borders, and natural re-sources, both to the Spanish King and to then-US president Thomas Jeffer-son, with whom Humboldt had shared his maps of New Spain even prior to their publication, much to the consternation of the Spanish Crown. Hum-boldt's dedication to Carlos IV might have been an attempt at mitigating the political fallout from this undiplomatic gaffe. What Humboldt could not have known was that the dedication would be suppressed in all Spanish-language versions of *New Spain*.

Humboldt's introductory discourse on maps opens onto discussions of the shape of the land, soil and climate (and thus its potential for agriculture), trade with Europe and the United States, diseases such as yellow fever, and military defenses. As always in Humboldt's work, comparisons with other parts of the world abound: among others, his far-flung geography encom-passes the British colonies, Europe, the United States, Canada, and the Asian parts of the Russian Empire (Humboldt would visit the Urals and Siberia in 1829). He adds to this comparative global mix frequent excur-sions into New Spain's pre-Columbian past, motivated in part by Humboldt's interest in indigenous languages such as Nahuatl and in the origins of local nomenclature. For him, "[k]nowledge of languages [was] not only necessary for drawing from sources and for collecting a large amount of information that would otherwise be lost to science" (17). Such knowledge was even more crucial, he insisted, "for facilitating the type of philological inquiry to which the learned geographer must submit the names of rivers, lakes, mountains, and local people in order to discover their identity on a large number of maps" (17). Humboldt's concern with this sort of inquiry permeates all of *New Spain* and is most visible in his precise attention to geographical nomenclature; it might also explain why he would have main-tained variations in spelling throughout. He also collected linguistic ma-terials for his brother Wilhelm von Humboldt, whose work on indigenous American languages unfortunately remained unfinished.

What attracted so many of his contemporaries to these rather unwieldy volumes was not so different perhaps from what stirred the imaginations of the early sixteenth-century conquistadors: the prospect of untold riches that were to be found in the New World. Even if, by the beginning of the nine-teenth century, the myths of El Dorado and the Seven Cities of Cíbola had been summarily discredited, the effect that Humboldt's reports about Mex-ico's silver mining had on many Europeans was somewhat similar. Hum-boldt's *Political Essay* quite literally provided the map for a new treasure hunt. Instead of "the vague and oft-repeated statements about 'enormous amounts of gold' or 'immense treasures'" (II.248), he offered meticulously researched data to formulate answers to the following questions:

> What is the geographic location of the mines that provide the vast amount of silver that the Veracruz trade sends to Europe annually? Was this amount of silver produced by a large number of small, scattered mining operations, or can it be considered as having been furnished almost in its entirety by three or four metalliferous veins of extraordinary wealth and *power*? What is the total amount of precious metals mined yearly in Mexico? What is the relationship between this amount and the total production of all of the mines of Spanish America? What is the average estimated wealth, in ounces per quintal, of Mexico's silver ore? [. . .] Is it possible to know the exact amount of precious metals that have passed from the Kingdom of New Spain to Europe and Asia since the conquest of Tenochtitlan? Considering both the current state of mining and the geological composition of the country, is it likely that the annual production of Mexico's mines might increase, or should one assume, as have several famous writers, that silver exports from the Americas have already reached their *peak*? (II.63)

Each of these questions is a vector pointing in a different analytical direction: toward geography, toward geology, toward economics, and toward politics. These were also precisely the questions to which many, including the King of Spain, wanted answers at a time when the Spanish Empire was about to unravel. Flatly contradicting those who regarded New Spain's silver mines as nearly exhausted, Humboldt concluded that "[f]or the last one hundred and thirteen years, mining output has been steadily on the rise" (II.174): "The output of the mines in Mexico has tripled in fifty-two years and sextupled in one hundred years," he wrote, and he predicted that "it will increase even more as the country becomes more populated and as knowledge contributes to progress" (II.462). Humboldt backed up his claims with solid evidence derived in part from his excursions to the famous mining district of Guanajuato, among many other locations (see figure 1, above).

Equipped, then, with a list of the richest mining districts in New Spain and their annual outputs over time (see II.80–82), along with accurate locations for most of the famous mines, European investors fell into a veritable frenzy. The majority of them were British, owing to the fact that the *New Spain* was first translated into English almost instantaneously after its first appearance (see Editorial Note). Millions of British pounds and other European capital were invested in Mexican silver mining, with fairly predictable results: the market crashed, and investors blamed Humboldt for their losses.

What went wrong? For one, the changing political landscape in a Mexico that was moving toward independence no doubt had a significant effect on how the country's commodity markets behaved around 1810, several years after Humboldt's visit. For another, one ought not to derive singular conclusions from Humboldt's writings. Humboldt visited New Spain in 1803–1804. In other words, he was there at a time when the reorganization of the Bourbon state—under royal emissary José de Gálvez, Viceroy Croix, and finally the second Count of Revillagigedo—had made New Spain the most prosperous colony on the planet. Free trade became the norm, marking the end of various royal trade monopolies. The latter created an incentive for previous traders who had flourished under such monopolies to enter other fields, notably mining, agriculture, and industry. Reorganization also meant better systems for taxation and revenue collection, mining regulations, and the founding of a mining school in Mexico City. These resources quickly turned the capital into a modern city, with better roads, streetlights, new buildings, and improved drainage. Humboldt, in sum, experienced a Golden Age of the Spanish Empire which, as he himself cautioned, could not and did not last.

Although the *Political Essay on the Kingdom of New Spain* was the first systematic chronicle of the origins of the "enormous amount of amassed capital among the mine owners" (II.7), Humboldt conspicuously deferred accounts of the country's fabled silver mines, including the exact location of the mines, until Volume III. Prior to that, his narrative alternates between detailed population statistics and demographic analyses broken down into the New Spain's administrative units, or intendancies (Volumes I and II), and close-ups of its agriculture and the trade in colonial commodities. In these close-ups, he hastened to remind his readers "that the value of the gold and silver of the mines in Mexico is almost a quarter less than the value of agricultural output" (II.58) and "that mining coal, iron, or lead can become as profitable an undertaking as mining a vein of silver" (II.62–63). Humboldt's elaborate and quite enthusiastic discussions of New Spain's agricultural output and potential constitute a notable counterpoint to his discussion of silver mining. Trained as a mining engineer and inspector at the School of Mines in Freiberg, Saxony, Humboldt regarded precious metal mining as a source of morally corrupting wealth that distracted from Mexico's more solid possibilities for public prosperity through agricultural production, manufacturing, and commerce (Volume IV). In the short term at least, his cautionary notes went unheeded.

Fractal Spaces: "All the Curves in the Road"

Although the *Political Essay of the Kingdom of New Spain* may be the first modern regional economic geography, Alexander von Humboldt himself called it a "political *tableau* of New Spain" (II.460). While the word may at times be translated simply as "table," a tableau is a hybrid genre that allows for the combination of numbers and narrative with figures of visual imagination. (Appropriately, Tableaux is also the name of interactive visualization software developed in 2003.) Given his penchant for visualization, it is no coincidence that Humboldt begins *New Spain* with maps and mapmaking to introduce readers to his core idea that *everything* is interrelated. For instance, when he explains how he used chronometers in determining astronomical positions, he is also quick to emphasize that each measurement affects the entire process of mapmaking: "Any corrections that more recent and more precise observations oblige us to make to one of these positions must necessarily affect the entire system. Failure to recognize this interdependence has led to the most bizarre distortions in the location of mineral deposits and in the determination of the relative distance between sites" (17). As *New Spain* unfolds, inundating us with deliberations that range from local labor conditions to floods on the Mexican Plateau to yellow fever, we come to understand that the same is true when it comes to observing and comparing social practices and cultural phenomena domestically, nationally, and globally.

Humboldt's approach to representing these practices and phenomena frequently anticipates the French mathematician Benoît Mandelbrot's influential study, *The Fractal Geometry of Nature* (1977). Like Mandelbrot nearly two hundred years later, Humboldt sought to describe the irregular and fragmented patterns that exist in the world around us. Mandelbrot used the famous example of the length of the British coastline to present a fractal, post-Euclidean geometry of nature. He started with the assumption that nature had a far higher degree of complexity than the mathematician Euclid of Alexandria (d. 366 BCE) could admit, having left certain forms aside for being supposedly *formless*. One of Humboldt's own illustrations concerns the distance between Mexico City and Puebla, which resonates with Mandelbrot's remarks about coastlines. When commenting on his drawing of the eastern slope of New Spain, he notes that "the absolute distance from Mexico City to Puebla is only twenty-seven leagues, but it appears to be two leagues longer on the profile drawing, which presents, so to speak, *all the curves in the road*" (103; emphasis added). Measuring distances between places accurately is no less important to making the maps for overland travel as astronomical determinations are to navigation at sea. Humboldt looked for ways to articulate in context the "full effect" (*Totaleindruck* in German) of diverse natural and cultural phenomena. To do so, he used individual, or local, measurements and observations as structural components that contain a whole structure in a smaller, reduced form. If such microcosmic patterns repeated themselves in macrocosmic structures—be they natural or cultural—then to understand a larger structure as a whole, that is, the interrelations it contained within itself, one only had to increase the scale.

Humboldt experimented most with visualizing patterns and relations between word and image in his *Views of the Cordilleras and Monuments of the Indigenous Peoples of the Americas*. That he worked on the first edition of *New Spain* around the same time that he wrote *Views of the Cordilleras*, published just a few years later (1810–1813) explains some of the shared characteristics of otherwise seemingly rather different books. *New Spain* focuses in on that part of the Americas represented in *Views of the Cordilleras* with more than forty images and texts. In both texts, Mexico City serves as a focal point, as the key component that contains a whole structure in reduced form. In *Views*, the city's main plaza, the Zócalo, already appears on the third engraving, a clear signal of its special significance. Mexico City is no less important in the *Political Essay*, where Humboldt writes the following:

No city on the new continent, including those of the United States, has such exceptional and solid scientific establishments as does the capital of Mexico. I limit myself here to mentioning the School of Mines, whose director is the learned Delhuyar (we shall return to it when we discuss the mining of metals), the botanical garden, and the Academy of Painting and Sculpture. This academy is called the *Academía de los Nobles Artes de México*. It owes its existence to the patriotism of several Mexican individuals and the protection of minister Gálvez. The government granted it a spacious building where one finds a finer and more complete collection of plaster casts than is to be found in any part of Germany. One is amazed to see that the Apollo Belvédère [the Pythian Apollo], the Laudon group, and even more colossal statues were able to travel over mountain roads that are at least as narrow as those of the Saint Gotthard. It is surprising to find these masterpieces of antiquity assembled here in the Torrid Zone, where the plateau is at a higher elevation than the Great Saint Bernard monastery. It cost the king almost two hundred thousand francs to transport the collection to Mexico City. The other remnants of Mexica sculpture, colossal basalt and porphyry statues that are covered with Aztec hieroglyphics and often show similarities with the Egyptian and Hindu style, should be collected in one of the courtyards of the Academy building, if not in the building itself. It would be strange to place these monuments of the early civilization of our species, the work of a semi-barbarous people living in the Mexican Andes, beside the handsome shapes created beneath the skies of Greece and Italy. (274)

In this passage, Humboldt mentions the arts and the sciences, the mountainous landscapes and climates of Europe and the Americas, and European and non-European works of arts, all as if in the same breath. Crossing diverse areas and objects of knowledge, the traveler oscillates here between nature and culture, Old and New Worlds, even if, when it comes to the arts, Humboldt still favors Europe. Humboldt creates here in prose something like an imaginary museum of the world's cultures, which invites visitors to find their own way in the spaces that open up between the worlds, between different times and different spaces. At first, the order of the world's cultures seems to settle into a Western hierarchy that resembles Goethe's view of world literature. At second glance, however, we also can appreciate something new: an unusual sensitivity for the hybrid, for a relational perspective that brings together that which does not, at first glance, seem to belong together—recall the genre of the tableau, which attempts to do the same. The routes by which the European plaster casts, these "masterpieces of antiquity," arrived in New Spain, represent displacements, whose spatial extraordinariness Humboldt emphasizes. It is these displaced objects that work against familiar hierarchies and facilitate entirely new webs of relations and meanings when seen together with the remnants of Mexica sculpture. In this way, the above passage functions as a microcosm in which we learn how to read the entire *Political Essay on the Kingdom of New Spain*.

At the level of narrative, Alexander von Humboldt's desire to articulate an alternative modernity by connecting the Old Worlds with the New, and the natural sciences with both the social sciences and the humanities, produced an inimitable voice. Scrupulous statistical details—always the backbone of his theories—alternate with anecdotal observations, tirades against African slavery and landowners' and missionaries' abuses of the indigenous peoples, and painstaking descriptions of landscapes, plants, animals, and cultural artifacts. Humboldt moves rapidly between close-ups and broad, at times startling, comparisons, using the process of scaling up and down in dizzying succession. He draws readers in by showing them absolute distances *and* all the twists and turns in all the roads he traveled, whether in actuality or in his imagination. He always helps readers understand local phenomena as parts of a far-flung web of interrelations.

A Historian of the Americas

An important thread, or vector, running through all of Alexander von Humboldt's writings is the history of the Spanish discovery and conquest of the New World. Most frequently, Humboldt pulls in the figure of Columbus, as he does most conspicuously in his *Examen critique de l'histoire de la géographie du Nouveau Continent* (Critical examination of the historical development of the geographical knowledge of the New World). Often dubbed "the second discoverer of the New World," Humboldt had a certain fascination with Columbus—fed no doubt by his Huguenot mother's maiden name, Colomb, which would be Colón in Spanish—and in whose imaginative footsteps he tried to follow when he set out on his journey to the Americas. In *New Spain*, however, the infamous conquistador Hernán (Fernando) Cortés steps into the foreground—Cortés "the hero whose eyes were set on all branches of the national industry in the midst of a bloody war" (533). With Cortés, Mexico's foremost Spanish conqueror, Humboldt ineluctably pulls the history of the Spanish conquest forward into eighteenth- and nineteenth-century New Spain, so he can make the past pierce the present at appropriate moments. Effectively, the two temporal orders run parallel throughout the entire text, something we begin to notice at those very moments of rupture. Volume IV, and thus *New Spain* as a whole, concludes with an "Extract from Fernando Cortés's Will and Testament" from 1537, a copy of which Humboldt had found in one of Mexico City's archives, where he spent long hours when not out in the field. With this document in New Spain's second edition, Humboldt lends special emphasis to being a "historian of the Americas," as he calls himself in his *Political Essay on the Island of Cuba*. This is quite different from being a historian of the conquest. As a historian of the Americas, Humboldt also draws frequently and extensively on the writings of indigenous chroniclers, such as the Nahua intellectuals Chimalpahin Quauhtlehuanitzin, Alva Ixtlilxochitl, Alvarado Tezozomoc, and Juan Bautista Pomar, as well as the Peruvian Garcilaso de la Vega, known as El Inca, for their alternative perspectives on the history of the conquest. Here and in *Views of the Cordilleras*, Humboldt comments extensively on ancient Mexica codices. Here, we can see a far more nuanced approach than those who have viewed Humboldt's gaze as exclusively imperial have been willing to grant.

Similarly underestimated has been Humboldt's discourse on the tropics, to him the very heart of the Americas. The "Torrid Zone," or the "equinoctial region," as he also called the tropics, were Humboldt's "real element," the emotional center of all his scientific studies. They would forever remain connected with his person and his popular image. Already as a young man, Humboldt had been acutely aware of the so-called Berlin Debate on the New World, which had reached its first peak after Cornelius de Pauw published the initial volume of his *Réflections philosophiques sur les Américains* (Philosophical reflections about the Americans) in 1768. Building on the writings of the Count of Buffon a step further, De Pauw insisted on a categorical divide between the Old and the New World, portraying America (and Americans) as weak, unformed, and simply incapable of independent progress. For him, the Europeans were humanity's true, and indeed only, representatives. Other thinkers, notably Guillaume-Thomas Raynal and, later, Georg Wilhelm Friedrich Hegel, sided with de Pauw, setting off a heated worldwide controversy. Humboldt fervently opposed the ideas of Buffon, Raynal, De Pauw, and their supporters, launching a bitter polemic against all those armchair philosophers who had never once set foot in the Americas. We see traces of this polemic in the *New Spain* when Humboldt writes: "It would be unnecessary to refute here Mr. de Buffon's baseless claims regarding the supposed degeneration of the domesticated animals introduced to the New Continent. These ideas spread easily, not only because they flattered the vanity of the Europeans but also because they were linked to brilliant theories about the primitive state of our planet. Now that the facts are being examined with greater care, physicists are able to acknowledge the harmony that exists in the very places where that eloquent natural historian thought to find only contrasts" (II.33–34). Humboldt's own discourse on the topic emphasized concord where others assumed discord, strength where others saw only weakness, civilization where others supposed primitivism. This emphasis is as characteristic a part of his American travel writings as are his metaphoric combinations of natural and cultural imagery and his insistent use of different genres and even languages. Languages, he insisted, needed to be submitted to close scrutiny "in the same way one consults historical monuments" (II.36). It is the only way, he argued, to distinguish what is older, that is, attributable to the ancient Mexica populations, from what is newer, such as the sociolinguistic construction of race in a country whose diverse population consisted of Indians, Creoles, Mestizos, Europeans, and Africans.

The *Political Essay on the Kingdom of New Spain* is "respectfully" dedicated to "Your Catholic Highness," Carlos IV of Spain, whom Alexander von Humboldt had met in person in 1799. What Humboldt wrote at the end of his dedication, however, was not a simple expression of gratitude, which may well be the reason why the dedication was suppressed in all Spanish-language editions of *New Spain*: "How could one not like a good King when He speaks of national interest, of perfecting social institutions, and of the eternal principles on which rest peoples' prosperity?" Traveling in the Spanish colonies prior to their political independence allowed Humboldt not only to gather the scientific specimens that he deposited in collections throughout Europe but also to register disjunctions between Enlightenment philosophy, socioeconomic practices, and political institutions. Unrestricted access to local government archives provided him with the opportunity to observe that scientific progress had failed to foster human equality in many parts of the Americas. Because he was financially independent, Humboldt was in a far better position than most other European explorers before him—notably James Cook, Alessandro Malaspina, and Louis Antoine Bougainville, to name but a few—to voice critical opinions about nineteenth-century colonial practices, and he did so quite unabashedly. That most of the mining workers were indigenous peoples added fuel to Humboldt's humanitarian fire, which also targeted the Catholic missionaries in New Spain and their maltreatment of the Indians. The following passage shows clearly that he did not mince his words: "To go '*a la conquista*' and 'to conquer' (*conquistar*) are the technical terms that missionaries use in America to signify that they have planted crosses around which the Indians have built a few huts; unfortunately for the indigenous peoples, however, the words *to conquer* and *to civilize* are not synonymous" (429n). As a result of such thinly veiled criticisms, neither the *Political Essay on the Kingdom of New Spain* nor the *Political Essay on the Island of Cuba* (1825–1826) received a warm reception in Spain. To Humboldt's chagrin, his reputation for being critical of colonial powers would also prevent him from being able to visit British India.

But unlike his *Political Essay on the Island of Cuba*, which was immediately banned on that island for its abolitionist sentiments, Humboldt's work on New Spain was such a treasure trove of coveted information that it proved an instant success. What further contributed to the work's popularity, in Latin America and the United States, was that independence movements were springing up all over the Spanish colonies at the time of its first publication, eroding Spain's control of a significant part of the Western Hemisphere. *New Spain*, especially in its revised second edition, shows Humboldt as an ardent, though perhaps belated, supporter of Latin American independence; in fact, his volumes delivered a blueprint for socioeconomic development in postcolonial Mexico. By the time Humboldt decided to publish a second edition in the mid-1820s, independence had already become a reality in most former Spanish colonies in the Americas, with the exception of Cuba and Puerto Rico. In the decades leading up to and including the Mexican War, *New Spain* came to play a fundamental role in shaping the relations between Mexico and its closest neighbor, the United States of America. It bolstered Humboldt's influence on US politics and letters, which had begun with his 1804 meeting with Thomas Jefferson and extended, via correspondence with Albert Gallatin and others, into the post–Civil War period.

Humboldtian Ethics: Knowledge for Living Together

The unabridged (re)translation in this critical edition is a fundamental departure from the image of Alexander von Humboldt that John Black's *Political Essay on the Kingdom of New Spain* created. We hope to show in these volumes a Humboldt for whom knowledge incurred moral responsibilities. Building on the foundations of Enlightenment thought—not only Kant's "Physical Geography" but also his ethics—in ways that were not always easily compatible with nineteenth-century European colonial endeavors and emerging nationalist beliefs, Humboldt forged a highly unusual vision of modernity. Its curiosity and its openness toward other parts of the world were significantly at variance with then-dominant beliefs in the desirability of the economic and cultural predominance of "old Europe" over the rest of the world.

In *New Spain*, Alexander von Humboldt presents himself as an explorer who navigated geographical, cultural-historical, and socioeconomic spaces: geographical spaces that would very soon become the United States of Mexico; the historical spaces of pre-Columbian Anáhuac and its diverse cultures; and the socioeconomic hierarchies of a New Spain. In this context of analyzing the latter's complexities, Humboldt resembles the protagonist of the first novel written by a Spanish American—a *criollo*—in Spanish America: José Joaquín Fernández de Lizardi's *El Periquillo Sarniento* (1816; *The Mangy Parrot* [2005]). Like the novel's *pícaro*, Humboldt puts on display the different strata of New Spanish society, from the viceroy and other high colonial administrators, the Catholic bishops, and the wealthy miners and businessmen at the top of the ladder, right down to the bottom rungs occupied by indigenous laborers exploited by a dying colonial regime. Humboldt readily identified New Spain as "the land of inequality. Nowhere is there a more alarming imbalance in the distribution of wealth, civilization, agriculture, and population" (254). Humboldt's vision in *New Spain* also goes significantly beyond that of Fernández de Lizardi's narrator to imagine an independent country able to determine its own future. In so doing, Humboldt was confident that a free Mexico would leave behind a past and a present of inequality.

Humboldt tellingly concludes his *Political Essay on the Kingdom of New Spain* by insisting "that the well-being of the whites is intimately linked with that of the coppery race, and that it cannot find any durable happiness in the two Americas, unless that humiliated race, which has not been debased by being oppressed for so long, participates in all the benefits that derive from the progress of civilization and the perfecting of the social order" (II.436). This unequivocal call for racial and economic equality was then—and regrettably still is—an unpopular sentiment. Humboldtian ethics are not only a way of thinking about the world in which everything is interrelated; they are also a vision of the future that is much more than just a promise. For Humboldt, the future is a practice of politics as *cosmopolitics* that points toward what one might call conviviality, a practice of living together in peace and difference on a national and a global scale. Humboldtian ethics envision New Spain and the New World as the place where the possibilities might truly exist for a *really* new and better world.

Nashville / Potsdam, March 2017

VOLUME 1

To His Catholic Majesty,

Charles IV,

King of Spain and the Indies.

Sire,

Having enjoyed for many years the protection and benevolence of Your Majesty's Scepter in the distant lands subject to Him, I am merely carrying out a sacred duty in depositing at His Throne the testament of my profound and respectful gratitude.

In 1799, I was fortunate enough to have a private audience with Your Majesty at Aranjuez. He deigned to applaud the zeal of a humble individual whose love for the sciences led him to the banks of the Orinoco and the peaks of the Andes.

The confidence inspired in me by Your Majesty's favors has emboldened me to place His august name at the head of this work. It sketches the portrait of a vast kingdom whose prosperity, Sire, is dear to Your heart.

None of the Monarchs who have occupied the Castilian Throne has so widely spread the precious knowledge of this beautiful part of the world, which, in both hemispheres, is subject to Spanish laws. The coasts of the Americas have been charted by skilled astronomers, and with the munificence worthy of a great Sovereign. Exact maps of these coasts, as well as detailed maps of several military ports, have been published at Your Majesty's expense. He has commanded that the state of the population, of trade, and of finance be published annually in a Peruvian journal in Lima.

What was lacking was a statistical essay on the kingdom of New Spain. I have gathered a large number of materials that were in my possession into a work whose first sketch was honored with the attention of the Viceroy of Mexico. I would be fortunate indeed if I could flatter myself that my humble work, in a new format and drafted with greater care, will not be found unworthy of being presented to Your Majesty.

It is pervaded by the sense of gratitude that I owe to a Government whose protection I enjoyed, and to that noble and loyal Nation that welcomed me not as a traveler but as a fellow countryman. How could one fail to please a good King, when one speaks to Him of the national interest, of the development of the social institutions and the eternal principles on which the prosperity of peoples rests?

I am, respectfully,

SIRE,

The very humble and obedient servant
Of Your Catholic Majesty,
Baron A. von Humboldt
Paris, March 8, 1808.

▼ Editor's Preface

Spanish America today is one of the greatest spectacles that, in the history of human civilization, have ever attracted the attention of philosophers. Just by considering the peoples of the Mexican race—those who inhabit Guatemala and New Spain—we see eight and a half million inhabitants creating new social institutions; establishing a great federal state; and exploiting freely the immense riches of their territory. Situated between the coasts of the seas of Asia and Europe, concentrated on the back and the slopes of the Cordilleras where varied climates follow each other as if they were stories, we see these peoples reach toward a great future.

When matters so deserving of the most serious contemplation occupy the mind, one believes that reprinting a work that contains very precise material about the questions to which Spanish America gives rise is something that would please the public. The *Political Essay about the Kingdom of New Spain* brings together in its own representations [tableau] the physical and moral state of a country five times as large as France. It was first published   xii at a time when the metropole still exerted complete influence, when Europe, torn by interminable resurgent wars, had little concern for the affairs of the New Continent. Nonetheless, these volumes, upon their initial appearance, created a lively sensation. And, to prove that public interest has only grown, instead of lessening as time passed, it suffices to recall that, since that time, in Europe as in America, people have continued to reprint, translate, excerpt, and copy this work, and seize upon the geographical maps it includes. All those who were in charge of colonial administration recognized the necessity of consulting Mr. Humboldt's works. The Spanish administration derived from them information about population growth, domestic consumption, and trade balances. Since the Mexican Revolution, the new government of that country has frequently cited the works of Mr. Humboldt in official writings that reported about anything from the calculation of the territorial area and the wealth of the mines to the spread of different human races across the plains and on the highlands of the Cordillera. Because of this book, because of the conceptual clarity and the precision of its results, it was easy for the owners of the Mexican mines to find, in England, several million pounds sterling with which either to revive mining operations that had been languishing for a long time, or to begin to explore lodes that had barely been touched.

xiii   Most recently, on a solemn occasion on July 21, 1824, the *executive branch of the Mexican government* declared that "Mr. Humboldt's *Political Essay* offers the most complete and most exact representation [tableau] of the country's natural riches, and that reading this great work has significantly contributed to resuscitating the nation's industrial operations and has given it confidence in its own strength."

The statistics of a country consist of varied elements, some of which, like the laws of physical nature, are fixed while others are mutable, dependent on the vicissitudes of the people's moral and political situation. The *Political Essay on the Kingdom of New Spain*—and this is a characteristic of all works held in broad esteem—focuses on the unchanging fundamentals of public prosperity. In it, the author discusses the effects that the composition of the soil, the climate, and a more or less vigorous vegetation have on agriculture, mining, industrial technologies, and trade relations with foreign nations. These effects and these interrelations are the same no matter what the form of government under which a people lives. When it comes to statistical variables, Mr. Humboldt, upon our request, has been willing to adjust the most essential ones. In additional notes, he has commented on the progress that the population has made since the time of his travels.

The *Political Essay* does not limit itself to painting a tableau of agriculture, mineral wealth, manufacture, trade, finances, and military defense in the Mexican context. It also deals with the other parts of the Spanish

xiv   Americas, for which he analyzes with equal care the principal foundations of public wealth. In this respect, the book holds more than its title seems to suggest. Besides statistical observations about Mexico, one finds, if not information about all of the Spanish possessions in the Americas, at least very precise summaries about their populations, the profits of their mines, their exports, and their public revenues before the great revolution that disconnected them from the metropole. This *comparative statistical research* extends also to the federation of the United States and to Great Britain's Asian possessions.

Publishing here for the first time, according to official documents from the archives, the value of the amounts of silver (149,350,721 marks) extracted from the mines of Mexico during the long interval from 1690 to 1800, Mr. Humboldt examines the amount of cash (5,706 million piasters) that flowed back from one continent to the other from the end of the fifteenth century to the beginning of the nineteenth. In his analysis—one of the most important among the research that the study of political economy offers—he has often delved into the amounts of gold and silver, and he has corrected some errors that the famous author of the *Wealth of Nations* could not help but commit, mistakes based on [the lack of] empirical information. This examination has even been respectfully cited in the parliamentary debates in Great Britain. To the topics that pertain to the progressive accumulation of monies in Europe and in Asia, and of the movement of metals from west to east, Mr. Humboldt has added other reflections that, because of the wide-ranging interest in them, might well have been the subjects of separate books. We will only mention here the research on the characteristics of *yellow fever* in the Torrid Zone, studies that were reprinted countless times and in different languages during the periods of the epidemics in Cádiz and Barcelona; the reflections on the *Oceanic Canals* imagined between the Atlantic and the South Sea; summaries of the feats of hydraulic engineering (in canals and tunnels) undertaken to prevent the flooding of Mexico City; and finally the geognostic descriptions of the Cordilleras of Anahuac and their manifold ramifications. xv

In the *Political Essay*, as in the author's *Views of the Cordilleras and Monuments of the Indigenous Peoples of the Americas*, one finds descriptions of the remnants of buildings and of Toltec and Aztec sculptures that testify to an advanced civilization. The first travelers, monks, and warriors had talked about them since the time of the conquest. But the qualms that had motivated their reports at a time of philosophical uncertainty, of often exaggerated relations, had consigned to oblivion the traces of an ancient culture of our species. Recently, an eager voyager, ▼ Mr. Bullock, brought back from Mexico to London small effigies and plaster models of the tiered pyramids, of the sacrificial stone, of the zodiac and the many Aztec deities, which excited vivid interest. Many of them are imaged in the *Picturesque Atlas* of the *Voyage to the Equinoctial Regions*.

In reissuing the *Political Essay*, we found it important not to change its title, to disfigure the work through heedless changes, or to overburden it with accessories of purely passing interest. Mr. Humboldt has described the state of the colonies at the end of their European domination, and it was deemed desirable to preserve the original simplicity of his representations [tableau], with a sparkle of color always born from having the objects he described right in front of him. Each page recalls the time when he wrote it. We see there the missteps in public administration; but those are shown with a moderation without which all hope that they may be remedied would vanish. If the author had attempted, in this new edition, to bring everything into the present, he would have stripped from his book its physiognomy and individual character. Moreover, in writing from such a great distance, he would have risked to fall into grave errors. Internal wars have reduced mining and slowed down the commercial use of these vast lands. The quantities of gold and silver coined at the mint of Mexico City, and the exports from ports such as Veracruz, Tampico, Alvarado, Acapulco, and San Blas, have been so erratic that even the most careful statistical facts about Mexico during the past few years are but of slight interest to those who would like to find out what this beautiful country might become once its institutions are fully established and the public peace is not threatened from the outside. To predict with any degree of accuracy the future situation of Mexico, to get a precise idea about the Mexican population's consumption of agricultural products and of European manufactured goods, it would be necessary, for a long time yet, to consult Mr. Humboldt's *Political Essay*. This traveler caught the country at the moment of the greatest glory that it could have reached amidst the restrictions imposed by the metropole's power, a time when the annual production of the the mines was at 1,600 kilograms of gold and 537,000 kilograms of silver, that is, 23 million piasters; when the value of native manufactured goods was at seven to eight million; and when the consumption of foreign merchandise amounted to more than twenty million piasters. It is to be hoped that indepedendence, with its new social institutions and investment capital flowing from Europe, will enliven the business and lift Mexico to a level of prosperity higher than what it had enjoyed up to now. But before achieving this happy result, the country must once again go through *the state of production and consumption* that immediately preceded the political troubles.

We will recall that the numerous additions to and improvements of this new edition are, to a large extent, interpolated into the text. They pertain, above all, to the account of astronomic geography that comprises the introduction to the work; to the discussion of the extent to which oceanic canals would be useful for the trade that both Europe and North America conduct with India and China, with Peru, with Guatemala and with the northwestern coast; to population increases among indigenous peoples; to the independent Indian tribes that inhabit the northern regions; to the census of the inhabitants of Mexico City; to the production of currency during the public troubles; to the always decreasing export of gold from Brazilian *washes*; to the Veracruz trade, where the totals from 1795 to 1820 rose to 538,640,163 piasters; to the yearly consumption of canvas in the Mexican interior; to the revenue the government derived from three kinds of taxation (Alcabalas as well as taxes on Pulque and on spirits made from sugarcane); and finally to funds that the *Tribunal de Mineria* [mining board] spent between 1777 and 1813 to boost mining activites. The constant relations that the author has maintained with the Mexican government and with persons who, at different times, occupied distinguished positions in the administration of this country, gave him access to a great amount of entirely new information which appears in this edition for the first time. We also felt that the public would appreciate the addition of Hernán Cortés's testament, which Mr. Humboldt found in the archives of the Monte-Leone family in Mexico City and which carries the imprint of the personality and great qualities of this extraordinary man.

xviii

Alexander von Humboldt's Preface
to the First Edition

I reached Mexico by the South Sea in March 1803 and resided in this vast kingdom for a year. Having carried out research in the province of Caracas, on the banks of the Orinoco, the Río Negro, and the Amazon, in New Granada, Quito, and on the coast of Peru, where I had journeyed to observe the passage of Mercury over the sun in the southern hemisphere on November 9, 1802, I was struck by the contrast between the civilization of New Spain and the lack of culture in the parts of South America where I had just traveled. Captivated by this contrast, I resolved not only to conduct a focused study of the statistics of Mexico but also to research which factors had the greatest influence on the advances of the population and the national industry.

My personal situation provided me with all the means to reach the goal that I had set for myself. No printed sources could supply me with useful material, but I had at my disposal a large number of unpublished accounts; active curiosity about this material was such that copies had multiplied and were circulating in the remotest parts of the Spanish colonies. I compared the results of my own research with the information contained in the official documents that I had been collecting for several years. An interesting, though very brief, stay in Philadelphia and Washington in 1804 prompted me to establish parallels between the present state of the United States and that of Peru and Mexico, which I had visited shortly before.

It is thus that my statistical and geographical materials grew to the point that it became impossible to include their results in my *Personal Narrative of the Voyage to the Equinoctial Regions of the New Continent*. I entertained

I.2

the hope that a distinctive work, published under the title of *Political Essay on the Kingdom of New Spain,* would be well received at a time when, more than ever, America stirs the interest of Europeans. ▼ Several copies of the first version of this work, published in Spanish, exist in Mexico City and on the Peninsula. Believing that it could be of use to those who are called to the administration of New Spain and who, often after a long stay, still I.3   lack a precise view of the state of these beautiful and vast regions, I made my manuscript available to anyone who wished to study it. The ensuing exchanges prompted important corrections. Even the Spanish government honored the manuscript with special attention. My work provided material for several official documents devoted to the trade interests and the manufacturing industry of the Americas.

The *Political Essay on New Spain* that I am publishing here is divided into six large sections. The first book offers general considerations on the country's area and physical appearance. Without going into details of descriptive natural history (details that I am saving for another part of my work), I have examined the influence of the geological configuration and the unevenness of the terrain on climate, agriculture, trade, and coastal defense. The second book deals with the population in general, its progressive increase, and the division into castes. The third presents the particular statistics of the ▼ intendancies, their population and their area, calculated according to the maps I drew on the basis of my astronomical observations. In the fourth book, I discuss the state of the agriculture and mines; in the fifth, the progress of factories and of commerce. The sixth book contains research on state revenues and on the country's military defense.

I.4   In spite of the extreme care that I took in verifying the results I used, I have probably committed errors that will become apparent once my work inspires the inhabitants of New Spain to study their own country. I can count on the indulgence of those who know the difficulties of this sort of research and who have themselves compared the statistical tables published annually in Europe's most civilized regions.

A Reasoned Analysis of the Atlas
of New Spain

In publishing maps of ▼ New Spain and drawings that illustrate the uneven-   I.5
ness of the Mexican terrain in vertical projections, I must give an account to
astronomers and naturalists of the materials that I have used in this work.
When an author limits himself to compiling information, when he draws
from little-known sources and merely gathers together in one place what has
been previously published in the form of printed works and engraved maps,
then a simple list of sources can suffice. This is not at all the case when an
atlas is based on the author's own astronomical observations and measures,
or when, in order to draw up new maps, he makes use of sections of outlines
and handwritten notes preserved in archives or hidden away in monasteries. In the latter case (in which I find myself), geographers have the right   I.6
to expect a reasoned presentation of the means employed so that they can
confirm the position of the most important points. In presenting this work
to the public, I shall carefully distinguish between what results from simple
convergences, which may only suggest greater or lesser probabilities, and
information that has been directly deduced from astronomical observations and geodesic or barometric measurements conducted on site. I shall
attempt to give a succinct analysis of the materials at my disposal, while
reserving the purely astronomical details for the *Recueil d'observations et de
mesures* that I am publishing with ▼ Mr. Oltmanns.[1] This approach will,

---

1. This astronomical work, which includes both the geographic leveling of the Cordilleras and a table with 700 positions, of which 235 are based uniquely on Mr. Humboldt's
own observations, appeared in two in-quarto volumes completed in 1811. The second volume

I hope, enable each the different parts of my work—the Statistics of Mexico, the *Personal Narrative of the Voyage to the Equinoctial Regions*, and the volumes on astronomy—to demonstrate that the desire for accuracy and the love of truth have guided me throughout the course of my travels. May my modest efforts contribute in some small way to dispel the darkness that for centuries has enshrouded one of the most beautiful regions on Earth!

---

includes (on pp. 466–564) a detailed account of astronomical observations made in New Spain.

I

A Condensed Map of the Kingdom of New Spain

I compiled and drew up this map at the Royal School of Mines (*Real* I.7
Seminario de Minería) in 1803, shortly before my departure from Mexico
City. ▼ Mr. Delhuyar, the school's learned director, had long collected notes
on the locations of the mines of New Spain, and on the thirty-seven districts
across which the mines are distributed, under the name *Deputaciones de
Minas*. He wanted to have a detailed map drawn up of the most interesting
mines for the use of the high college, the *Tribunal de Minería*. Such a project was indeed necessary both for the administration of the country and
for those interested in its national industry. On most of the maps printed
in Europe, one looks in vain for the town of Guanajuato with its 70,000
inhabitants; nor does one find the names of the famous mines of Bolaños,
Sombrerete, Batopilas, and Zimapán. No map to date shows the position of
the Real de Catorce in the intendancy of San Luis Potosí, the mine where
silver ore valued at twenty million francs was extracted annually. Because
of its proximity to the Río del Norte [Rio Grande], this mine seems to have
already aroused the greed of colonists who have recently settled in Louisiana. I.8
Since I had already calculated most of my astronomical data while still in
Mexico City in order to establish set points to which other points could
then be linked, and since I had at my disposal a large number of materials
and unpublished maps, I decided to expand my original plan. Rather than
putting on the map only the three hundred sites known to be considerable
mining ventures, I set myself the task of assembling all the references that
I could procure, and of examining the differences in position that so much
varied material suggested in each case. The imprecision that abounds in

the geography of Mexico should not be surprising when one considers the obstacles that halted the advances of civilization not only in the colonies but also in the mother country. This is especially true when one considers the long period of peace that these countries have enjoyed since the beginning of the sixteenth century. The wars with ▼ Hyder Ali and Tippu Sultan in Hindustan, the continuous march of armies, and the need to find the shortest route of communication each contributed to the expansion of geographical knowledge. It has only been in the past thirty or forty years, however, that we have attained more precise knowledge of Hindustan, a country that has been explored by the most adventurous of European travelers. Although I worked diligently on my map of Mexico for three or four months, I.9   I foresaw that it would still be quite inaccurate in comparison with the maps of those parts of Europe that first became civilized. This thought, however, did not discourage me. Considering the advantages of my personal situation and position, I would like to think that, despite the serious flaws that would inevitably mar it, my own work would be preferable by far to any of the other attempts to date to introduce the geography of New Spain.

The atlases that accompany this *Political Essay* and my *Personal Narrative* will be judged favorably if readers bear in mind the breadth of information that geographical studies nowadays require. The fundamental bases of these studies are a discussion of measurements (astronomical observations, geodesic research, and itineraries) and a critical comparison of descriptive works (journeys, statistics, the history of wars, and missionaries' reports). If all countries were charted trigonometrically, all triangles were properly angled, and the extremities of their chains were fixed by equally precise astronomical observations, then the assembling of maps would be a simple graphic and manual exercise. But our current state of knowledge falls woefully short of these conditions. For the foreseeable future, the wisdom of geographers will be brought to bear on dubious information. At present, a judicious critique must be based on two completely separate branches of I.10   knowledge: (1) on a discussion of the relative value of the astronomical methods used to determine the position of places, and (2) on the geographer's necessary study of the descriptive works that contain detailed information about the lengths of itineraries, the tributaries of rivers, and irregularities in the terrain. The first branch of knowledge assumes both an existing set of astronomical observations consistently calculated according to the most recent tables and methods and an intuitive skill that can only be acquired through the actual practice of astronomy and that leads one to settle on a definitive result from a long series of occultations, eclipses of satellites, and lunar distances, taking into account the margin of error in each type of observation and the circumstances under which the observations were made. Astronomy is the field of geography that has been most successfully sustained at present. This advantage is due to the precision of instruments, the skill of the observers, and, above all, the perfecting of astronomical tables. Astronomy provides the most basic foundation for drawing maps: elements that are, so to speak, immutable. In countries for which we lack geodesic research and concatenated triangles, we must increase the number of astronomical positions and carefully link the points determined by *absolute methods* (such as star occultations, solar eclipses, and lunar distances) using ▼ *chronometric lines*, that is, a series of points whose longitudes are based not only on the movement of time but whose end points coincide with the results yielded by the *absolute methods*. The use of chronometers, often dangerous when not carefully planned, puts a certain number of positions into a relationship of mutual dependence. Any corrections that more recent and more precise observations oblige us to make to one of these positions must necessarily affect the entire system. Failure to recognize this interdependence has led to the most bizarre distortions in the location of mineral deposits and in the determination of the relative distance between sites.

    I.11

    While it is indispensable in studies of astronomical geography to refer to sources, to the very works in which scholars have presented their findings, it is even more so in the second branch of geographical research; that is, in discussions of itineraries, voyages, and descriptive narratives in general. Translations are most often either truncated or flawed, both when it comes to names and to converting units of measure. Travelers themselves rarely draw the maps that accompany the accounts of their travels, and these maps do not contain all possible materials. As ▼ D'Anville has already remarked, in essential points, these maps most often directly contradict the descriptive accounts for which they are intended. In that illustrious geographer's time, any discussion of astronomical fundamentals could only have been quite incomplete. It was through the exceptional care with which he collected itineraries, studied travelers' accounts, and combined distances and mineral deposits that he was able to give many of his works the degree of perfection we admire to this day. Knowledge of languages is not only necessary for drawing from sources and for collecting a large amount of information that would otherwise be lost to science; it is even more crucial for facilitating the type of philological inquiry to which the learned geographer must submit the names of rivers, lakes, mountains, and local peoples in order to discover their identity on a large number of maps. Our atlases are laden

    I.12

with names for which rivers were invented, just as in the catalogue of organized beings, commonly called the *Systema naturae*, plants or animals for which multiple names exist are indicated as being two or three distinct species. This desire to compile information without a critical eye, filling gaps, and combining varied material, gives rise to a veneer of precision in maps, especially those of less-visited countries, whose inaccuracy one recognizes when one is actually on site. As ▼ Mr. de La Condamine has astutely written: "Most of our atlases are brimming with details that are as false as they are circumstantial." Generally, insofar as it is detectable in maps, progress in the field of geography is much slower than it should be, because, at any given time, enormous amounts of valuable results are scattered in works by different nations. Astronomical observations and topographical information may accumulate over a long period of time without even being used; due to an otherwise highly commendable principle of stability and preservation, geographers frequently prefer not to add or change anything rather than sacrifice a lake, a mountain range, or the tributary of a river that has appeared on maps for centuries.

I.13

One may say that it is premature to draw up general maps of a vast kingdom about which we lack precise data. But following this same logic, with the exception of the province of Quito and the United States, we should not publish any map of either the interior of the continental Americas or several parts of Europe (for instance, Spain or Poland) where, on surfaces of over 1,600 square leagues, it is impossible to find a single point whose position was established by astronomical means. Less than fifteen years ago, there were scarcely twenty places in central Germany whose longitude had been determined within a sixth or an eighth of a degree.

In the part of New Spain situated north of the twenty-fourth parallel, in the provinces called *internas* (in New Mexico, the province of Coahuila, and in the intendancy of Nueva Vizcaya) , the geographer is reduced to drawing conclusions on the basis of travel journals. Since the ocean is located far from the most inhabited part of the country, the geographer lacks the means of connecting places in the interior of this vast continent with slightly better-known points on the coast. Thus, outside the city of Durango, the traveler finds himself wandering, as it were, in a desert. Despite the assistance of a few hand-drawn maps that generally contradict each other, there are no more resources to be found there than were available to ▼ Major Rennell when he drew up the maps of the interior of Africa. This is not at all the case for the section of Mexico located between the meridians of Acapulco and Veracruz, and between the capital of Mexico City and the

I.14

Real[1] de Guanajuato. This region across which I traveled from March of 1803 to February of 1804 is the most cultivated and inhabited part of the kingdom; since my journey, the position of a large number of points in the region has been determined astronomically.

It would be desirable for a traveler experienced in making observations to traverse the northern portion of the Kingdom of New Spain in three directions, armed with a ▼ sextant—or a reflection-based repeating circle—a time-keeper, an achromatic telescope, and a portable barometer, for measuring elevations. He would set his route (1) from the town of Guanajuato to the Presidio of Santa Fe or to the village of Taos in New Mexico; (2) from   I.15 the mouth of the Río del Norte [Rio Grande], which empties into the Gulf of Mexico, to the Sea of Cortés [Gulf of California], especially at the junction of the Río Colorado and the Río Gila; and (3) from the town of Mazatlán in the province of Sinaloa to the town of Altamira on the left bank of the Río Pánuco.

The first of these three journeys would be the most important and the easiest to carry out: here, the chronometer would be subjected to the smallest changes in temperature. One should not, however, rely exclusively on the passage of time; rather, to determine longitudes, one should also use *absolute astronomical observations*—namely, of Jupiter's satellites, star occultations, and above all, the distances between the moon and the sun—methods that, since the publication of ▼ Delambre's, Zach's, Bürg's, and Burckhardt's excellent tables, deserve the greatest degree of confidence. In the course of an astronomical journey from Mexico City to Taos, the positions that I assigned to San Juan del Río, Querétaro, Zelaya, Salamanca, and Guanajuato could be verified; the longitude and latitude of San Luis Potosí, Charcas, Zacatecas, Fresnillo, and Sombrerete—five sites famous for the wealth of their mines—could also be determined; one would pass via the city of Durango and the Parral to Chihuahua, the residence of the governor of the *Provincias internas*. Traveling along the Río Bravo, one would reach the capital of New Mexico via the Paso del Norte, and, from there, the   I.16 village of Taos, the northernmost point of that province.

The second journey, during the course of which the observer would face a scorching climate, would be the most difficult; but it would yield fixed points in the New Kingdom of León, the province of Coahuila, Nueva Vizcaya, and Sonora. Operations should be directed from the mouth of the Río Bravo del Norte via the episcopal seat of Monterey, to the Presidio of

1. The word *Real* indicates a place where metal deposits are mined.

Monclova. Following the route by which the ▼ chevalier de Croix, vice-roy of Mexico, arrived in the province of Texas in 1778, one would reach Chihuahua, and the second journey would thus connect to the first. From Chihuahua, one would proceed through the military establishment *(Presidio)* of San Buenaventura in the town of Arispe [Arizpe], and from there to the mouth of the Río Gila, either via the Presidio of Tubac [Arizona?] or the missions of ▼ Pimería Alta [southern Arizona], or the savannahs where the Tonto Apache roam.

The third excursion, which involves crossing the kingdom from Altamira to the port of Mazatlán, would join to the first excursion at the town of Sombrerete. Through a detour to the north, it would serve the purpose of setting the position of the famous mines of Catorce, Guarisamey, Rosario, and Copala. I have limited this list to the most dubious locations, the positions of which left me with the greatest uncertainty when I drew up my map of New Spain. Besides, it is known that very few days are needed to determine the latitude and longitude of each of the locations I just listed. Only the largest towns, such as Zacatecas, San Luis Potosí, Monterrey, Durango, Chihuahua, Arispe, and Santa Fe in New Mexico, would require a stopover of a few weeks. Even if the observer were not exceptionally skilled, the astronomical means indicated would easily provide an accuracy of within twelve to fifteen seconds[1] for latitude and one-third of a minute for absolute longitude. How many large towns are there in Spain and in the easternmost and northernmost parts of Europe whose geographic determination is still far from being this precise?

I.17

I.18

1. Margins of error differ depending on whether one observes the circum-meridional height of the sun or stars with a reflecting instrument whose images, weakened and deformed by the mirrors of the artificial horizons, appear as indeterminate. In the first case, we can attain an accuracy of within six to eight seconds of arc; in the second, errors often range as high as within twenty and twenty-five seconds of arc (see the introduction to my *Recueil d'observations astronomiques*, vol. 1, pp. vi, xiv, and xxvii). We must keep in mind, however, that even on our most detailed maps of the New Continent, we can rarely distinguish one arcminute or 950 toises. A famous astronomer has rightly affirmed that nowadays, even since the introduction of repeating circles, there are not even three places on the face of the earth whose latitude can be ascertained *at an accuracy of within one second*. In 1770, the latitude of Dresden was inaccurate by nearly three minutes, and the latitude of the Berlin observatory was uncertain by nearly twenty-five seconds until 1806. In 1790, prior to ▼ Mr. Barry and Mr. Henry's observations, the position of the Mannheim observatory was inaccurate by one minute and twenty-one seconds in latitude. Nonetheless, the Jesuit priest ▼ Christian Mayer had made observations there using a Bird quarter circle with a radius of eight feet (*Ephémérides de Berlin*, 1784, p. 158, and 1795, p. 96). Prior to ▼ Le Monnier's observations, the correct latitude of Paris had been misconstrued by nearly fifteen seconds.

The three journeys that we are proposing to an enlightened government are easy to carry out and would transform our understanding of the geography of New Spain. The positions of Acapulco, Veracruz, and Mexico City have been confirmed at various times by the work of ▼ Galiano, Espinosa, Bauzá, Cevallos, Gama, and de Ferrer, as well as by my own. It would require only one excursion by officers of the Royal Navy stationed at the port of San Blas to determine the important positions of the Bolaños mines and the town of Guadalajara. The astronomical expedition that the government entrusted to Mr. De Cevallos and ▼ Mr. Herrera for the purpose of charting the coastline of the Gulf of Mexico would determine the mouth of the Río Huasacualco southeast of Veracruz. It would be a simple task for these skilled astronomers who possess superb English instruments to sail up this river, which has become famous because of the projected canal that would connect the Antillean Sea with the equinoctial Great Ocean. They would be able to measure the width of the Mexican isthmus by setting the position of the port of Tehuantepec and the sandbar of San Francisco at the mouth of the Río Chimalapa.

Few countries on Earth present as many advantages for trigonometric observations as New Spain. The great Valley of Mexico and the vast plains of Zelaya and Salamanca, as smooth as the surface of the waters that must have covered their terrain for centuries, have many elevated plateaus from 1,700 to 2,000 meters above sea level, bordered by mountains that are visible from great distances. These same plateaus invite the astronomer to measure a few degrees of latitude toward the northernmost limits of the Torrid Zone. In the intendancy of Durango, in one section of San Luis Potosí, triangles of extraordinary size could be measured on terrain covered with grass and devoid of forests. In discussing these advantages, however, we must distinguish between the needs of science and those of the administration. The measure of an arc of the meridian between the nineteenth to twenty-fourth parallels, together with precise observations of the length of the pendulum, would likely be of great interest to the perfecting of our knowledge of the shape of the Earth. However (and this is worth mentioning here), this interest is subordinate to another that is more materially linked to the advancement of national prosperity. In order to govern New Spain well, to open lines of communication via roads and canals, it is necessary to find means for quick and easy implementation. It would be a brilliant undertaking for the government to cast a network of triangles over a country bristling with mountains and covering 118,000 square leagues, at twenty-five per degree, to establish a series of delicate operations across a

I.19

I.20 terrain five times the size of France, and publish a map of Mexico at a scale of 1:80,000. But it would also be too ambitious a project for anyone to hope that it could be completed in less than a century and a half. The beautiful trigonometric map of Sweden, which is being elaborated as I write, has succeeded only in charting nine hundred square leagues per year.[1] People have criticized the scrupulous precision with which ▼ Mr. Fidalgo and Mr. Churruca, both officers of the Spanish Navy, have examined even the slightest

I.21 curves of South America's coastline.[2] This work was painstaking and very costly. In my opinion, however, it would be wrong to censure the men, who presented the court of Madrid with such a fine hydrographic charting project. A nautical map can never be detailed enough. Safe navigation, easy recognition of one's bearings when setting anchor, and the necessary means of defense against an enemy threatening to land—these all depend on the most in-depth knowledge possible of the coasts and the bottom of the sea. It is sometimes of no consequence whether the determined latitude of a city in the hinterlands is exact to within two minutes; but on the coast, it is of the utmost importance to know the position of a cape with all the precision that astronomical methods can provide. Each point on a hydrographic map must be determined with equal precision, since any one of them may serve as a point of departure or reconnaissance for a navigator. By contrast, maps of the interior of a large region are invaluable even when they show just a few locations whose position has been astronomically determined.

1. For the large map of France that is currently being created, an officer and his aide determine an average of twelve to sixteen primary points during a five-month period; the seven other months of the year are devoted to geodesic calculations. Each triangle in this survey covers approximately twenty-seven square leagues (twenty-five per degree) and costs 340 francs, not counting the salary of an officer—chief of operations—and an aide. After eight months in the field and four months of desk work, an officer working only with secondary triangles based on the primary points and tracing broad ▼ geodesic lines of 200,000 by 200,000 meters could observe and calculate between 120 and 150 points, in the direction of the meridian and of the parallels. At a scale of 1:40,000, these calculations would cover a sheet measuring eight by five decimeters, or a surface area of thirty-two leagues and four-tenths (twenty-five per degree). Not counting the officer's salary, each point would cost twenty francs. Each sheet (or one eighty-thousandths) contains approximately one thousand altitude measurements. Thirteen-inch repeating circles are used for primary triangles, and eight-inch repeating ▼ theodolites are used for the secondary triangles. The zenithal distances of each observed point are measured at each field station.

2. Mr. Rennell, one of our century's most learned geographers, has observed that whereas the English have very precise maps of the moorings along the Bengal coast, for a very long time there was no acceptable map of the channel that separates England and Ireland (*Description de l'Indostan*, vol. I, *Preface*).

If it is preferable to refrain from plotting the location of the Spanish possessions in the interior of the Americas with the meticulous precision used for drawing the coastline; if, given the current state of affairs, it is more useful to limit the project to *astronomical bearings and the drawing of chronometric lines*—in other words, to conduct a preliminary study using reflecting instruments and chronometers, lunar distances, and the observation of satellites and stellar occultations—then it would be no less important to the success of this project to add to such precise and purely astronomical tools other, secondary means provided by the nature of the region with its high, solitary peaks. Once the absolute height of these peaks is known—calculated by means of either a barometer or geometric operations—vertical angles and azimuths taken either at sunset or at sunrise can connect these mountains to points whose latitude and longitude have been reliably confirmed. This method is based on the use of perpendicular bases. By evaluating how many meters in error one might be in measuring each base and admitting different hypotheses, one can easily determine how much this margin of error might affect the astronomical position either of the mountain itself or of other related points. Often, precise knowledge of the lower limit of perpetual snows can offer the same advantages as do measurements of a solitary peak. I have used this method to determine the longitudinal difference between the capital, Mexico City, and the port of Veracruz. Two large volcanoes—Puebla, known as Popocatepetl, and the Orizaba Peak, both visible from the platform of ▼ the ancient pyramid of Cholula—served to connect two locations that are separated by 155,200 toises. Combining two geometrical measurements of mountains, azimuths, and the vertical angles Mr. Oltmanns calculated determined the location of the port of Veracruz at $0^h11'31''$ west of Mexico City, whereas astronomical observations yield a meridional difference of $0^h11'46''$. Modifying the initial result through some secondary observations made on the Cholula pyramid even yielded $0^h11'41.3''$, which demonstrates that, in this particular case, the azimuth method was at least correct by five seconds in time over a distance of three degrees.[1] Using this same ▼ *hypsometric method*, I was able to find the meridional difference between the Orizaba volcano and Veracruz at

I.22

I.23

1. "Mémoire astronomique sur la différence des méridiens entre Mexico et Vera-Cruz" by Mr. Oltmanns and Mr. Humboldt (Zach, *Monatliche Correspondenz*, November 1806, pp. 445, 454, 458). See also my *Recueil d'observations astronomiques*, vol. 1, pp. 133–38, and vol. II, pp. 537–46.

1°5′13″. According to Mr. Ferrer and ▼ Mr. Isasbirivil's trigonometric calculations, Veracruz is either at 1°4′57″ or at 1°6′30″.

If they are accessible, isolated volcanic peaks in the middle of a vast
plateau may offer a much more reliable and faster way of determining the
longitude of many neighboring locations within a few seconds. *Light sig-*
I.24   *nals* created by exploding a small quantity of gunpowder would be visible from a great distance by someone equipped with the means of finding
and recording the average time. ▼ Cassini de Thury and Lacaille were the
first to successfully use this light signal method. Mr. de Zach's recent field
work in Thuringia has demonstrated that, under favorable conditions and
in only a few minutes' time, this method can yield positions that are just as
precise as the results of several observations of satellites and solar eclipses.
In the Kingdom of New Spain, these signals could be shone from either the
Iztaccihuatl or the Sierra Nevada de Mexico, from the crag called the Monk
(the solitary peak of the Toluca volcano, where I arrived on September 29,
1803), from the Malinche near Tlaxcala, from the Cofre de Perote, and
from other mountains with accessible summits and heights of over 3,000 to
4,700 meters above sea level.

For the past twenty years, the Spanish government has made the greatest and the most extravagant sacrifices in order to perfect nautical astronomy and to obtain precise bearings of the coastline, and one may thus hope
that it will soon turn to the geography of its vast possessions in the [West]
Indies. This hope is bolstered by the fact that the Royal Navy has an excellent collection of instruments and is able to produce astronomers with
I.25   considerable experience in making observations. Providing its students
with a strong foundation in mathematics, the mining school in Mexico City
sends across this vast empire large numbers of enthusiastic young people
who are well trained to use the instruments placed in their hands. In a similar fashion, the British East India Company was able to obtain maps of its
own immense territory. The day has passed when governments sought their
security in secrecy, fearful of revealing their territorial treasures in the Indies to rival nations. The King of Spain has ordered that the bearings of
the coastline and its ports be published at the state's expense; he has no
fear that the most detailed maps of Havana, Veracruz, and the mouth of the
Río de la Plata might fall into the hands of nations which circumstances
have made enemies of Spain. One of the fine maps from the ▼ *Depósito hidrográfico* in Madrid displays the most valuable details of the interior of
Paraguay, details that are based on fieldwork by officers of the Royal Navy
enlisted to determine the borders between the Portuguese and the Spanish

colonies. With the exception of the maps of Egypt and some parts of the East Indies, the most precise work we have of any European continental possession outside Europe is ▼ Maldonado's map of the Kingdom of Quito. This all goes to show that, in the past fifteen years, the Spanish government has been far from apprehensive about advances in geography and has actually published all of the interesting materials it possesses about its colonies in the two Indies.

I.26

Having discussed the methods that seem best suited for quickly perfecting the maps of New Spain, I shall now give a succinct analysis of the material I was able to use for the geographical work I am presenting to the public.

Like all of the maps that I drew during my journey, the general map of the Kingdom of New Spain uses increasing latitudes (following ▼ Mercator's projection). The advantage of this projection is that it directly shows the real distance between two places. It is also the most appealing projection for sailors in the colonies who want to determine their bearings by setting the position of their ships in relation to two mountains visible from the open sea. If I had had to choose from among the various stereographic projections, I should have preferred ▼ Murdoch's, a method one should generally adopt. The scale of my map is thirty-two millimeters for each degree of the equator. The scale of increasing latitudes is based not on ▼ Don Jorge Juan's tables, but on those ▼ Mr. de Mendoza has calculated for the spheroid.[1]

To keep the map of Mexico to a more convenient size, its scale was limited to 15°–41° northern latitude, and from 96°–117° longitude. These limits did not allow for including on the same printing plate the intendancy of Mérida or the Yucatán peninsula, both of which belong to New Spain. In order to include the point that is the farthest east, Cabo Catoche, or, better yet, the island of Cozumel, one would have been obligated to add seven more longitudinal degrees to the east. This would have forced me to include on the same printing plate a portion of Guatemala (for which I have absolutely no information), all of Louisiana, all of western Florida, and parts of Tennessee and Ohio.

I.27

The Spanish settlements on the northwest coast of the Americas are not included on my general map of New Spain: these isolated settlements should be considered colonies, dependent on the Mexican metropole. To

---

1. [Mendoza y Ríos "Mémoire sur la méthode de trouver la latitude par le moyen de deux hauteurs du soleil,"] *Connaissance des temps*, 1793, p. 30[2].

show the missions of New California on the same printing plate, I would have been obligated to add eight longitudinal degrees to the west: the Presidio of San Francisco is the northernmost point. According to ▼ Vancouver, it is located at 37°48′30″ northern latitude, and 124°27′45″ western longitude.

These considerations suggest that a map of New Spain that deserves to be called a general map should include the vast extent of the region that I.28 lies between 89° and 125° longitude and 15° and 38° latitude. To avoid the inconvenience of representing on such a large scale regions that are not all of equal interest from the point of view of political economy, I have chosen to confine my work to narrower boundaries. I also had a second map drawn up in a much smaller format that not only allows the viewer to take in at a glance all the regions governed by the viceroyalty of Mexico, but also encompasses the Antilles and the United States of America.

Although, according to principles I have often stated, I continue to prefer new measurements to old ones, I have not used the new scale of centesimal degrees on my maps. The French Office of Longitudes has consistently used the former method of determining latitude, both in *Connaissance des temps* and in the new *Tables astronomiques* it has just published. A single individual would in vain oppose such deeply rooted practices by presenting latitudes only in terms of centesimal units. It is a pity that the introduction of the metric system, written into law by the decree of 13 brumaire, year IX [1799], has not become universally adopted, and that there is even more confusion in France now over the former units of measure. One is forced to conceive of a *metric foot* and of *inches*, which might be confused with *pouces de pied-de-roi* [king's foot inches]. The degrees of longitude that I pro-I.29 vide are counted to the west of the meridian that passes through the Royal Observatory in Paris. If the majority of the public were not so opposed to innovations, even when they are useful, I would prefer the universal meridian proposed by one of the foremost geometers[1] of our century to the Paris meridian. This universal meridian is based on the movement of the large axis of the solar ellipsis and is located at 185°30′ east of Paris, which corresponds to 166°46′12″ in the old sexagesimal system. Consequently, the meridian passes through the South Sea, twelve minutes of arc east of the island of ▼ Erromanga [New Hebrides], which is part of the archipelago

1. [Pierre Simon de] Laplace, *Exposition du système du monde*, p. 19. In the fourth edition of this work (p. 74), the author suggests measuring all terrestrial longitudes from Mont Blanc, "the mountain that dominates the immense, inalterable slope of the chain of the Alps."

of Saint-Esprit. Introducing a universal meridian based on nature, one that would not offend Europeans' national pride, would be all the more desirable since every day we see more and more primary meridians that are arbitrarily drawn on maps. In the past few years, Spain has used five: the meridian of Cádiz, the one most used by navigators; the meridians of Cartagena and the New Observatory on León Island; the meridian of the ▼ College of Nobles in Madrid (introduced by Mr. Antillón's beautiful maps); and finally the meridian of Galera Point on Trinidad Island. One could add two others that have been adopted by many geographers: the Tenerife and ▼ Ferro Island [Canary Islands] meridians. The latter gives rise to unavoidable confusion; D'Anville has it passing between Ferro's market town and the western cape of the island. Even without counting the meridian of Toledo, we have seven primary meridians within the possessions of the King of Spain alone.

I.30

In naming the seas that wash the Mexican coasts, I have followed the ideas that ▼ Mr. Fleurieu has proposed in his *Observations sur la division hydrographique du globe*, a work that unites broad perspectives with deep historical erudition. Spanish names have been included to make it easier to read travel accounts written in Spanish. In drawing my ▼ tableau of Mexico, I began by assembling all the points that had been fixed through astronomical observations. To emphasize the degree of reliability that these results deserve, I have organized them in a table that offers the type of observation and the observer's name. There are seventy-four such points, fifty of which are located in the interior of the country. Of this last group, only fifteen points were known before I arrived in Mexico in April of 1803. A discussion of the thirty-three points whose position I myself determined would be useful; all these points are located between 16°50′ and 20°0′ latitude and 98°29″ and 103°12′ longitude. As we fix these positions, we shall provide some historical details on the extraordinary errors that have proliferated in the most recent and most widely used maps.

I.31

Mexico City

Several meridional elevations of the sun and stars provided me with a reading of 19°25′45″ for the latitude of the capital at the monastery of San Agustín.[1] Mexico City's longitude is 6ʰ45′42″ or 101°25′30″; I deduced it

1. The main door of the cathedral in Mexico City is twelve seconds farther north and ten seconds (by arc) farther east than the monastery of San Agustín, near which I made my observations.

from eclipses of Jupiter's satellites, distances from the moon to the sun, the movement of time from Acapulco, and trigonometric calculations to evaluate the meridional difference between Mexico City and the port of Veracruz. I shall make clear, once and for all, that I use Mr. Oltmanns's figures exclusively. These figures are the result of extremely careful calculations by this distinguished geometer, who has calculated all the astronomical observations I made from the time I left Paris in 1798 until I returned to Bordeaux in 1804. The longitude of Mexico City ($6^h45'28''$) in the new *Astronomical Tables* published by the Office of Longitudes is based on a paper I presented to the most advanced class at the French Academy of Sciences on 4 pluviose, year XIII [January 24, 1805]. In that paper, the positions of the moon had not been calculated according to Mr. Bürg's tables. The previous year, I had decided on a result that was closer to the actual longitude; the mean of my observations published in Havana was $101°20'5''$.

I.32

According to Mr. Delambre's revised tables, three reappearances of Jupiter's first satellite, which I observed, suggested an average longitude of $6^h45'30''$; according to corresponding observations in Lancaster [Pennsylvania] and in Havana, it was $6^h45'21''$.

Thirty-two distances from the moon to the sun calculated by Mr. Oltmanns, according to the newest lunar tables, produced a longitude of $6^h45'50''$.

For the meridional difference between the port and the capital of Mexico, the movement of time from Acapulco gives a reading of $2'55.4''$: as a result, if we suppose that Acapulco is at $6^h48'38.2''$, then the longitude of Mexico City would be $6^h45'42.8''$.

The longitude of Guanajuato, determined by lunar distances, which I compared to that of Mexico City using my chronometer, indicated $6^h45'56''$ for the capital.

The result of the trigonometric calculation—or, more exactly, my attempt to connect the capital city with the port of Veracruz by means of azimuths and elevation angles taken on the two volcanoes of Orizaba and Popocatepetl (according to Mr. Oltmanns's calculations, and assuming Veracruz to be at $6^h33'56''$)—put the longitude of Mexico City at $6^h45'37.3''$.

I.33

All of these results, obtained in different ways and independently of each other, confirm the longitude we assign to the Mexican capital. This longitude differs by *one and a half degrees* from the longitude we have used up to now: in 1772, *Connaissance des temps* put Mexico City at $106°1'0''$, and again, in 1804, at $102°25'45''$. The map of the Gulf of Mexico published in 1799 by the *Depósito hidrográfico* in Madrid places the capital at

103°1′27″. Even before I was able to make my own observations in Mexico, however, the actual longitude was quite exactly known to three astronomers, two of whom were born in Mexico. Their work deserves to be revived. From 1778 onward, ▼ Mr. Velázquez and Mr. Gama had deduced a longitude of 101°30′ from their observations. But because there were no corresponding observations, and since they had made their calculations only in reference to ▼ Wargentin's old tables, they were uncertain (as they readily admit) up to nearly a quarter of a degree. This interesting result is found in a short pamphlet printed in Mexico City[1] that is little-known in Europe. Velázquez, the director of the highest tribunal of mines, set the longitude of the capital at 101°44′0″; this is proven by the invaluable unpublished accounts that ▼ Mr. Costansó has preserved in Veracruz. On a map of New Spain begun in 1772, Velázquez put Mexico City at a longitude of 278°9′, calculated from Ferro Island = 101°51′. In a note appended to the I.34 map, he states that "before his journey to California in 1768, all of Mexico was in the South Sea, that his map is the first to give the true position of the capital, and that he has confirmed this through several observations made in Santa Rosa, California; Temascaltepec; and Guanajuato." Don Dionisio Galiano, one of the most skilled astronomers in the Royal Navy, had also discovered the actual position of Mexico City when he crossed the kingdom to join ▼ Malaspina's expedition in 1791. It is true[2] that Mr. Antillón had deduced the longitude of 101°52′0″ from Galiano's observations; but his result still differs from my result by 1′48″ in time. I suspect that this difference is due to a slight error that may have slipped into the calculation. When I began my work on Mexico, the observations of Gama, Velázquez, and Galiano were entirely unknown to me. Moreover, Mr. Espinosa only sent me the details of Don Dionisio Galiano's observations in the winter of 1804, after I had returned to Europe. These results yielded a longitude that seems much more precise than the one Mr. Antillón had published. "During your stay in Spain in 1799," the learned director of the Hydrographic Office in Madrid wrote to me, "I was unaware of the observations I.35 of our mutual friend Mr. Galiano. They consist of two emersions of satellites during a stellar occultation and at the end of a lunar eclipse. Their result was 101°22′34″ = 6ʰ45′30″." The two emersions of the first satellite

---

1. Antonio de León y Gama, *Descripción orográfica universal del eclipse de sol del día 24 de Junio de 1778, dedicada al Sr Don Joaquín Velásquez de León*, 1778, p. iv.

2. Isidoro de Antillón, *Análisis de la carta de la América septentrional*, 1803, p. 34. This map puts Mexico City not at 101°25′, but at 101°52′, an error of twenty-seven minutes in arc.

provided Mr. Oltmanns with $6^h45'44.0''$, the occultation of one of Taurus's stars gave him $6^h45'35.6''$, and the lunar eclipse $6^h45'54.5''$. The average of Mr. Galiano's three observations, which were published only after I had returned to Europe, was $6^h45'44''$. As a result, the difference between the observations by the Spanish astronomer and my own—a difference assumed to be nearly half a degree—is reducible to less than two arcminutes. It is gratifying to find so much agreement between two observers who, without knowing each other, used different methods. On ▼ Thomas Jefferys's quite detailed maps, published in 1794, Mexico City is located at $20°2'$ latitude and $102°52'47''$ longitude, whereas in 1803, on his beautiful in-quarto map of the West Indies, ▼ Mr. Arrowsmith gives the longitude of Mexico City as $102°8'0''$ and the latitude as $19°57'$, an error of thirty-two minutes.

Some seventeenth-century Mexican astronomers came quite close to divining the capital's actual longitude. ▼ Father Diego Rodríguez of the order of Nuestra Señora de la Merced [Our Lady of Mercy] and a professor of mathematics at the imperial University of Mexico, together with the astronomer ▼ Gabriel López de Bonilla, adopted $7^h25'$ as the meridional difference between ▼ Uranienborg and the capital, hence the longitude of $101°37'45'' = 6^h46'29''$. In 1681, ▼ Don Carlos de Sigüenza,[1] the famous successor to Rodríguez's academic chair, was completely ignorant of the observations on which Bonilla had based this result. He published a short treatise[2] on the longitude "that should be attributed to Mexico City," citing the engineer Henri Martínez's observation of a lunar eclipse northwest of the capital city in Huehuetoca on December 20, 1619. This same Dutch engineer undertook the bold construction of the Desagüe de Huehuetoca canal designed to prevent the all-too-frequent flooding from Tenochtitlan valley. If we compare it to the longitude of Ingolstadt without making any modifications, Martínez's observation would give a longitude of $6^h32'16''$ for Mexico City. Compared to Lisbon, this same eclipse gives $6^h22'31''$. But since engineer Martínez did not use a ▼ telescope, Sigüenza assumes that the end of the eclipse was calculated fifteen minutes of arc earlier because of the twilight. The result of this rather arbitrary assumption is that Mexico City, compared to Ingolstadt, is at $6^h46'40''$ and, compared to Lisbon, at

I.36

I.37

---

1. Don Carlos de Sigüenza y Góngora, *Libra astronómica y filosófica escrita en 1681, Catedrático de Matemáticas de la Universidad de México, y impreso en la misma ciudad en*, 1690, §386.

2. See the above-cited work, §382–85. I owe my knowledge of Mr. Sigüenza's book, which is quite rare, to ▼ Mr. Oteiza, who was willing to recalculate several earlier observations by Mexican astronomers.

$6^{h}37'31''$. Mr. Oltmanns correctly notes that one of the corresponding observations must be incorrect by nine minutes of arc, since the actual meridional difference between Lisbon and Ingolstadt is only $1^{h}22'16''$, whereas the eclipse of December 20, 1619, would put it at $1^{h}13'0''$. Such old and careless observations cannot provide us with any certainty. Moreover, Rodríguez and Sigüenza, the two Mexican geometers whom we just mentioned, were themselves incapable of obtaining the results we have just stated. They were so unfamiliar with the meridional differences between Uranienborg, Lisbon, Ingolstadt, and the island of Palma that they concluded from the same information (specified in the *Libra astronómica y filosófica* [by Sigüenza y Góngora]) that Mexico City was located 283°38′ west of the first meridian of the island of Palma, or $96°40' = 6^{h}26'40''$. This longitude varies by 100 nautical leagues from the actual one and by 240 leagues from the longitude the geographer ▼ Jean Covens adopted in the middle of the last century.

In the Ephemerides of Vienna written by ▼ Father Hell in 1772 and in the Berlin astronomical tables for 1776, Mexico City is located at 106°0′. The assumption of an excessively western longitude is quite an old one. Mr. Oltmanns had already found it in the observations[1] of the Jesuit Father Bonaventura, who had stayed in the town of San Cosme y Damián in Paraguay. This little-known astronomer put Mexico City at $3^{h}13'$ west of his observatory and the observatory itself $3^{h}52'23$ west of Paris, which results in a longitude of $7^{h}5'23'' = 106°22'30''$ for Mexico City. On a Mexican map from 1755, the Jesuits of the Puebla de los Angeles put the capital city at a latitude of 19°10′ and a longitude of 113°0′—in other words, 240 leagues too far to the west.

▼ Chappe's voyage, edited by Mr. de Cassini, taught us nothing specific about the capital's position. Chappe's stay there was limited to four days, during which he could not make any astronomical observations. The observations ▼ Mr. Alzate provided for him did not help solve the problem. That Mexican clergyman, whom the Academy of Paris had made one of its corresponding members, was more zealous than precise in conducting his work. He tried to do too many things at once, and his knowledge was inferior to that of Velázquez and Gama, two Mexicans whose merits have been insufficiently recognized in Europe. On his map of New Spain published in Paris, Don José Antonio Alzate y Ramírez puts Mexico City at $104°9'0'' = 6^{h}56'36''$. Using the data from Venus's passage that Alzate observed in 1769, ▼ Lalande found $6^{h}50'1''$, and Pingré, $6^{h}49'43''$. The lunar

I.38

1. ▼ J.G. Triesnecker, *Ephemerides astronomicae*, 1803.

I.39   eclipse that Alzate observed in 1769 gave $6^h37'7''$; he only used the *old* lunar tables to calculate its end. Basing his calculation on two occultations of Jupiter's satellites that Alzate had observed in 1770 and comparing them to the *old* tables, Cassini[1] deduced a mean of $101°25' = 6^h45'9''$. In a paper that Alzate published on the geography of New Spain,[2] he assured his readers that, based on observations of satellites, the longitude of Mexico City was $6^h46'30''$. But in a note accompanying the map of the area surrounding the capital, drawn by Sigüenza and engraved in Mexico City in 1786, Alzate set the longitude at $100°30'0'' = 6^h42'0''$; he added that this last result, *the most certain of them all*, was based on more than twenty-five lunar eclipses, of which he had informed the academy in Paris.[3] Mr. Alzate's various observations show a difference of more than two degrees, even if one excludes the result deduced from the lunar eclipse of December 12, 1769. The observer was presumably very imprecise in measuring time. It is also possible that the longitude set by the satellites was too far to the east, because the eclipses of the first satellite were not separated from those of the third and the fourth.

I.40   The error about the position that has been attributed to the capital of New Spain for so long appeared in a most remarkable way during the solar eclipse of February 21, 1803. It was a total eclipse that alarmed the public, because the almanacs, having based their calculations on the assumption of a longitude of $6^h49'43''$, had announced that the eclipse would be barely visible. ▼ Don Antonio Robredo, Havana's learned astronomer, recalculated this eclipse according to my own observations of longitude[4] and found that the eclipse would not have been total if the longitude of Mexico City had been farther to the west than $6^h46'35.4'' = 101°38'49''$.

The latitude of the capital of Mexico has long been as problematic as its longitude. At the time of Cortés, Spanish navigators set it at $20°0'$, as ▼ Domingo de Castillo's map from 1541 confirms. This map was included in the Mexican edition of ▼ Cortés's letters.[5] D'Anville and other geographers kept this latitude. Jean Covens, who increased the longitude of Mexico City

1. [Chappe d'Auteroche,] *Voyage en Californie*, 1772, p. 104.

2. [Alzate y Ramírez, "Estado de la geografía de la Nueva España," *Asuntos Varios sobre Ciencias y Artes*,] 1772, no. 7, p. 56.

3. Carlos de Sigüenza, *Plano de las cercanías de México, reimpreso en 1786, con algunas adiciones de Don Josef Alzate (en la Impresa de Don Francisco Rangel.)*

4. [Robredo,] *Aurora, o Correo político económico de la Habana*, 1804, no. 219, p. 13.

5. Francisco Antonio de Lorenzana, *Historia de Nueva España escrita por Hernán Cortés, aumentada por el illustrado*, Mexico, 1770, p. 328.

by seven degrees, also assigned it a position that was 1°43′ too far north. Following Alzate, Chappe's expedition adopted a latitude of 19°54′. Using a quadrant, ▼ Don Vicente Doz, who is known for his fine observations in California, found[1] 19°21′2″, but since 1778, Velázquez and Gama have set the correct position. In February of 1790, Don José Espinosa and Don Ciriaco Cevallos found the latitude of 19°25′37″ for the cathedral of Mexico City, using a sextant with a radius of eight inches. By means of larger instruments, Mr. Galiano, in 1791, obtained 19°26′1.8″, which was eight or five seconds more than my own observations.

I.41

Veracruz

Latitude 19°11′52″. Longitude 6ʰ33′56″ = 98°29′0″. This longitude was deduced from stellar occultations observed by Mr. Ferrer and calculated by Mr. Oltmanns from three eclipses of the first satellite and from the longitude that my own calculations had assigned to Havana, using the movement of time to compare this port to that of Veracruz. We must bear in mind that I give the position of the northernmost part of Havana, whereas Mr. Ferrer's observatory was in ▼ Don José Ignacio de la Torre's house, which is situated thirty seconds of arc west of the fort of San Juan de Ulúa.

The longitude upon which I settled is almost identical to the one Don Mariano Isasbirivil and other officers of the Spanish Navy had found. It is only five arcminutes west of the position shown on the map of the Gulf of Mexico published in 1799 by the office of hydrographic works in Madrid. Mr. Antillón sets it at 98°23′5″; the *Connaissance des temps* for the year 1808 sets it at 98°21′45″. Using the movement of time, ▼ Don Tomás Ugarte, squadron leader in the service of the King of Spain, has compared Veracruz with Puerto Rico. He assigns 98°39′45″ to the first of these ports. In 1791 and 1792, Mr. Ferrer deduced the longitude of Veracruz from sixty series of distances from the moon to the sun and the stars; he obtained a mean of 98°18′15″. This adept astronomer should have published the data from his observations, so that they could be recalculated according to Bürg's tables, correcting the positions of the moon by observing that planet's [sic] passage through the Greenwich or Paris meridians. The same corrections should be applied to the results published in the account of Vancouver's expedition.

I.42

The position of the town of Veracruz also shared the same fate with Mexico City and the rest of the New Continent. All were believed to be

1. ▼ *Gazeta de México*, 1772, p. 56.

sixty or even one hundred forty leagues farther from Europe's shores than they actually are. Jean Covens put Veracruz at 104°45'0"; on his map of New Spain, Alzate put it at 101°30'. ▼ Mr. Bonne[1] rightly complained about the lack of agreement among the astronomical observations made in Veracruz.

I.43   After a long discussion, he settled on 99°37'. This is almost the same longitude that D'Anville and the author of *Le [petit] Neptune français* adopted; it is also the longitude that British astronomers have long preferred. ▼ Hamilton Moore settled on 99°49'47" and Mr. Arrowsmith (in his map of the Spanish possessions, 1803) on 98°40', whereas Thomas Jefferys, the King of England's geographer, put Veracruz at 100°23'47" in 1794.

Although the error of assigning excessively western longitudes to American ports is not new, the Abbot Chappe's finding errs in the opposite direction. He deduced a reading of 97°18'15" from his chronometer.[2] More eager than exact as an observer, Chappe neglected to measure the distances from the moon to the sun. These distances would have led him to see his error *of more than one degree*, an error he was induced to make because of his overconfidence in chronometric methods.

The earliest astronomical observation made in Veracruz (at the fortress of San Juan de Ulúa) was probably the viewing of the lunar eclipse in 1577. Comparing the end of this eclipse with his own corresponding observation in Madrid, Mr. Oltmanns found a meridional difference of $6^h26'$ and, as a result, a longitude for Veracruz of 102°30'.[3]

I.44   The Abbot Chappe set the latitude of the town at 19°9'38",[4] a position that was three minutes too far south. I have examined Chappe's small quadrant, which is still in Mexico, in the hands of the learned ▼ Father Pichardo. It is no surprise that such an imperfect instrument produced such imprecise observations. Other geographers put Veracruz twenty minutes of arc too far south, and Alzate's map of New Spain even gives a latitude of 18°50'0".

Since the publication of the first edition of this *Analyse de l'atlas de la Nouvelle-Espagne*, Mr. Oltmanns has had the opportunity to reverify the longitude of Veracruz ($6^h33'56"$), which is the same one on which I settled when I drew my maps. Using the occultation of Sagittarius ε (observed by Mr. Ferrer on August 25, 1799), Oltmanns found $6^h33'57.9"$. Use of the eclipses of Jupiter's satellites (compared to Mr. Delambre's tables) produced

1. [Rigobert Bonne,] *Atlas [de toutes les parties connues du globe terrestre]*, p. 11.
2. [Chappe d'Auteroche,] *Voyage en Californie*, p. 102.
3. [Académie des sciences, *Histoire] de l'Académie [royale des sciences]* for 1726.
4. [Chappe d'Auteroche,] *Voyage en Californie*, p. 103.

$6^h33'52.2''$. My own hypsometric observation, connecting the pyramid of Cholula and the Orizaba volcano with Mexico City and Veracruz, gave $6^h34'0.7''$. Their average would be $6^h33'57''$. I am appending other equally significant observations to these results. For the longitudinal difference between Veracruz and Havana, six chronometric observations gave Mr. Ferrer $0^h55'4''$, $0^h55'5''$ to Mr. Isasbirivil, and $0^h55'4''$ to the ▼ brigadier Montés. This same difference is $0^h55'2''$, according to the two satellites observed simultaneously by Mr. Churruca and Mr. Ferrer, in Veracruz and in Havana. According to the movement of time from Cumaná, which I measured with my chronometer during a slightly stormy passage, the latter port (Morro) is located at $5^h38'40''$. According to the satellites that I observed together with Mr. Galiano, it is at $5^h38'50''$; according to the fifteen stellar occultations that Mr. Ferrer observed from 1803 to 1811, it is $5^h38'49.3''$. If we agree with that capable Spanish astronomer's reading of $5^h38'51''$ for Havana, we find $63[I.°]\ 3'54''$ for Veracruz, which is two seconds of arc less than my result published in 1808. It is possible to infer from all these studies[1] that, even in Europe, there are few positions that offer the certainty we have for Veracruz, Havana, Puerto Rico, and Cumaná. By lowering Veracruz to Cap Français [Cap Haïtien] (the island of Saint-Domingue), we find $6^h33'53.7''$ because ▼ Borda's, Puységur's, Churruca's, Ferrer's, and Cevallos's chronometers have all given $1^h35'20''$ as the meridional difference between these two ports.

I.45

Acapulco

This is the most beautiful port of all on the Pacific Ocean coastline. According to my observations, made at the residence of the Contador ▼ Don Baltasar Álvarez Ordoño, Acapulco is located at $16°50'53''$ latitude and $6^h48'38'' = 102°9'33''$ longitude. Mr. Oltmanns deduced this position from two stellar occultations that the astronomers of the Malaspina expedition observed in 1791, and from twenty-eight distances from the moon to the sun that I myself measured. Those taken on March 27, 1803, (calculated according to Bürg's tables) produced $6^h48'33''$; the distances from March 28 gave $6^h48'23''$.

I.46

According to my own chronometer, the difference in time between the meridians of Mexico City and Acapulco is $2'54''$. Since we have found

1. Humboldt, *Recueil d'observations astronomiques*, vol. II, pp. 550–55. [Ferrer, "Occultations d'étoiles observées à la Havane,"] *Connaissance des temps* for 1817, p. 333.

Mexico City at 6ʰ45′42″ longitude, thanks to the mean of my lunar distances, it follows that if we exclude any other kind of observation, we find 6ʰ48′48″ for Acapulco. An uncertainty of fifteen seconds of arc is rather small when comparing two longitudes that were set by simple distances from the moon to the sun. In 1803, using ▼ Mason's lunar tables, I had found 102°8′9″.

The Atlas that accompanies the voyage of the Spanish navigators to the Fuca Strait assigns 102°0′30″ longitude and 16°50′0″ latitude to the port of Acapulco. This atlas is based on the observations of Malaspina's expedition. However, in the excellent book that I cited above, Mr. Antillón offers a result on the basis of these same observations that differs by nearly *one-third of a degree*. He assures us that the observations made in 1791 by the astronomers, who had embarked on the corvettes *La Descubierta* and *La Atrevida*, found Acapulco at 102°21′0″ longitude. This result seems less exact to me, although it conforms more to the unpublished documents these navigators left in Mexico. These same navigators themselves concluded 102°26′ from eight series of lunar distances, 102°20′40″ from an immersion of a first satellite, and 102°22′0″ from the time[1] of Guayaquil— all in an admirable agreement that is perhaps superficial, because of the errors in the old lunar tables. Moreover, I must point out that the longitude deduced from the observations made aboard the ▼ brigantine *Activa* in 1791 was farther west than Malaspina's longitude. After having examined the coast of Sonsonate and Soconusco, the *Activa* set the longitude of Acapulco at 102°25′30″, but we have no idea upon what kind of observations the longitude of 6ʰ48′23″ is based. Quite recently, two learned astronomers equipped with excellent instruments, ▼ Captain Basil Hall[2] and Mr. Henry Foster, found Acapulco chronometrically at 5°24′40″ east of San Blas; they therefore used a longitude of 102°14′2″ and assumed (based on a stellar occultation) 107°38′42″ for the port of San Blas. This result differs by only eighteen seconds from the longitude of Acapulco, which I determined via the distances of the moon to the sun; the difference is even slighter if, as we shall soon see, San Blas were a little farther east, as Mr. Hall assumes

I.47

I.48

---

1. The chronometric longitude of 102°22′ is also found on the detailed map of the port of Acapulco drawn by Malaspina's expedition and copied at the royal school of navigation in Lima. It seems that the astronomers of Malaspina's expedition had adopted positions for the entire coast of the South Sea which were much *farther west* than those set by the Office of Hydrographic Works in Madrid. The difference is (in arc) twenty minutes for Acapulco, sixteen minutes for Guayaquil, and eighteen minutes for Panama and Realexo.

2. [Basil] Hall, *[Memoir] on [The Navigation of] South America*, vol. II, p. 379.

it to be. Using two stellar occultations, the result of all these observations is 6^{h}48′40″ and 6^{h}49′0″, calculated by correcting the tables; 6^{h}48′58″ according to Jupiter's eight satellites; 6^{h}48′33″ according to my own lunar distances; 6^{h}48′48″ according to my chronometer; and 6^{h}48′56″ according to captain Basil Hall's chronometer. These absolute astronomical observations do not align as closely as one might have hoped. The result of the two stellar occultations differs by twenty seconds, which supposes unfavorable circumstances. My conclusion is that since Acapulco's location is less well defined than that of Veracruz, it still oscillates between 6^{h}48′38″ and 6^{h}48′56″. One may take these demarcations to be fairly close to each other, if one considers the general state of astronomical geography in Spanish America.

There is a note in the archives of the viceroyalty in Mexico City by one of the ▼ astronomers of Malaspina's expedition; it indicates that, at that time, it was believed possible to conclude a meridional difference of 2′21″ from observing a few eclipses of satellites that were observed simultaneously in the capital and in Acapulco. If one puts Mexico City at 6^{h}45′42″, then, according to all my observations, one gets 6^{h}48′3″ for the port of Acapulco. This is forty-seven seconds less than the reading that the two stellar occultations observed in Acapulco in 1791 produced, when calculated according to the most recent tables. The distance between the capital city and Acapulco is certainly greater than 2′21″, although it may also be less than the 2′54″ measured by my chronometer. The instrument was worn out by five years of use and had passed quickly, across mountainous terrain, between the extreme heat of the coast to the chilly weather of Guchilaque, from a temperature of thirty-six degrees to five degrees Centigrade. $\quad$ I.49

In the past, Acapulco was usually set four degrees farther west in the South Sea: even ▼ Jean Covens and Corneille Mortier, on their map of the Mexican archipelago, set the longitude of Acapulco at 106°10′0″. The older maps of the French Naval Office set it at 104°0′. It is interesting to note how this longitude has gradually shifted to the east. In the geographical paper appended to ▼ Raynal's work, Bonne sets it at 103°0′; in 1803, Arrowsmith put it at 102°44′.

The *Connaissance des temps* for the year 1808 sets a longitude (102°19′30″) for Acapulco that comes close to hitting the mark, but assigns to the port a latitude that is too far south by ten minutes of arc. This error is all the more striking in that, before Malaspina's expedition, the latitude was given as 17°20′ or 17°30′. D'Anville's maps, as well as those of the Naval Office, and the much older (1540) map drawn by the navigator Domingo de Castillo, $\quad$ I.50

all confirm this assumption. In Cortés's time, the capital of Mexico was believed to be three degrees west of Acapulco, almost on the same meridian as the port of Los Angeles. It is quite possible that the maps that the indigenous peoples themselves drew of their coast—maps that the emperor Moctezuma presented to the Spanish—gave rise to this opinion. Among the hieroglyphic maps in the ▼ Boturini collection, which are preserved in the viceroy's palace in Mexico City, I have not found the map of the western coastline but, rather, a very intriguing map of the outskirts of the capital. Lately, the individuals who have been involved in astronomy in Mexico City have allowed as a fact that the capital city and the port of Acapulco are on the same meridian.

The Road from Mexico City to Acapulco

As we have now set the position of the three major locations in the kingdom, let us briefly turn our attention to the two roads leading from the capital to the South Sea and to the Atlantic Ocean. One could even call the first one the Asia route, and the other, the Europe route; these titles indicate the main directions of seagoing commerce in New Spain. I have determined seventeen points, either in latitude or in longitude, on these two very busy roads.

I.51   The village of *Mescala*. Using the culmination [outer edge] of Antares, I found the village's latitude to be 17°56′4″; its longitude, determined by my chronometer, is 6ʰ47′30″, assuming Acapulco is at 6ʰ48′38″. According to angles measured in Mescala, the town of Chilpanzingo appears to be at 17°36′ latitude by 6ʰ47′7″ longitude.

Venta de *Estola*, a solitary house in the middle of a forest near a beautiful fountain. I took some heights of the sun there: the chronometer indicated a longitude of 6ʰ47′10″.

The village of *Tepecuacuilco*. Using ▼ Douwes's method, I found its latitude to be uncertain by nearly three minutes of arc at 18°20′0″; its longitude is 6ʰ47′26″.

The village of *Tehuilotepec*. Longitude 6ʰ47′26″. Double heights of the sun provided me with the latitude of 18°38′. This latitude, however, based on graphometric measures, is uncertain by several minutes. The position of Tehuilotepec is of interest because of its proximity to the large mines at Taxco.

The bridge of *Isla*, on the great San Gabriel plains. I found it to be located at 18°37′41″ latitude and 6ʰ46′33″ longitude.

The village of *San Agustín de las Cuevas*: longitude 6ʰ45′48″; latitude 19°18′37″. This village is at the western end of the great Valley of Mexico.

For a more detailed knowledge of this region, it would be helpful to include the distances that the locals, especially the mule drivers who lead their caravans to the large fair in Acapulco, calculate from village to village. Knowing the true distance from the capital to the port and assuming another third more for detours on a reasonably straight and easily accessible road, one would find the number of leagues traveled in these areas. This information is of interest to geographers who must otherwise rely on rudimentary travel journals when exploring less frequented regions. It is evident that people, the more difficult their roads become, tend to perceive distances in leagues as shorter. All things being equal, however, one may trust the mule drivers' judgment regarding comparative lengths; they do not know whether their pack animals walk two or three thousand meters in an hour's time, but they are aware of aliquot [fractal] proportions. Longtime habit has taught them to recognize one distance as a third, or a fourth, or twice that of another distance.

According to the Mexican mule drivers, the road from Acapulco to Mexico City is 110 leagues long. They count four leagues from Acapulco to Paso de Aguacatillo; to El Limón, three leagues; Los dos Arroyos, five; Alto de Camarón, four; la Garita [lodge] de los Dos Caminos, three; La Mojonera, half a league; Quaxiniquilapa, two and a half; Acaguisotla, four; Mazatlán, four; *Chilpanzingo*, four; *Sumpango*, three; Sopilote, four; Venta Vieja, four; *Mescala*, four; Estola, five; Palula, one and a half; La Tranca del Conejo, one and a half; Cuagolotal, one; Tuspa or Pueblo Nuevo, four; Los Amates, three; Tepetlapa, five; Puente de *Isla*, four; Alpuyeca, six; Xuchitepeque, two; Cuernavaca, two; S. María, three-fourths; *Guchilaque*, two and a half; Sacapisca, two; La Cruz del Márques, two; El Guarda, two; [San José] Axusco, two; *San Agustín de las Cuevas*, three; *Mexico City*, four. In this travel journal, the points where I made astronomical observations are printed in italics; the numbers indicate how many leagues there are from one place to the location immediately preceding it. Other guidebooks that are distributed to travelers arriving from the South Sea, either from the Philippines or from Peru, estimate the total distance at 104 or 106 leagues. But according to my observations, it is 151,766 toises as the crow flies. Increasing the distance by one-fourth to account for detours would give us 189,708 toises, or 1,725 toises for each Mexican mule driver's league.

I.52

I.53

The Route from Mexico City to Veracruz

I have determined thirteen points on this route, either by purely astronomical means or by geodesic operations, in particular by using azimuths or angles of elevation. Using my observations, Mr. Oltmanns decided on a position of 19°16′8″ for La Venta de Chalco at the eastern edge of the wide valley of Tenochtitlan; 19°0′15″ latitude, and 6ʰ41′31″ = 100°22′45″ for Puebla de los Angeles (near the cathedral); 19°26′30″ for La Venta de Soto; 19°33′37″ latitude and 6ʰ38′15″ longitude for the village of Perote, near the eponymous fortress; 19°37′36″ for the village of Las Vigas; and finally 19°30′8″ latitude and 6ʰ36′59.6″ = 99°14′54″ longitude for the po-

I.54 sition of the town of Jalapa. Don José Joaquín Ferrer, who, long before I did, had determined several points in the environs of Veracruz and Jalapa, found 19°31′10″ latitude and 99°15′5″ longitude for the latter. Both of us made our observations near the monastery of San Francisco, and the agreement between our findings could not be more satisfactory.

Four mountains in this fertile, cultivated region, three of which are always covered in snow, merit the greatest attention. Knowing their exact position would serve to connect several points of interest for the geography of New Spain. The positions of the two volcanoes known in Puebla or Mexico City (Popocatepetl and Iztaccihuatl) were confirmed by observations both in the capital city and in Cholula. For Popocatepetl, I found 18°59′47″ latitude and 6ʰ43′33″ = 100°53′15″ longitude; for the Sierra Nevada or Iztaccihuatl, 19°10′0″ latitude and 6ʰ43′40″ = 100°55′0″ longitude. Based on a series of geodesic measurements, Mr. Costansó had arrived at the latitude of 19°11′43″ for Iztaccihuatl and 19°1′54″ for Popocatepetl. Given that the engineer used a compass in making these calculations, and that the magnetic declination depends on several minor local causes, the precision of the results that were obtained is astonishing. These two colossal mountains, as well as the volcano or

I.55 Peak of Orizaba, are visible from the top of the Cholula pyramid whose position I set out to determine carefully. I found the chapel that crowns this ancient monument to be at 19°26′ latitude and 6ʰ42′14″ = 100°33′30″ longitude.

Mr. Ferrer concluded the position of the Cofre de Perote from several geodesic measurements made in El Encero and Jalapa: he found 19°29′14. Despite the inclement weather, I was able to carry instruments to the top of that mountain on February 7, 1804; the peak is 384 meters higher than the Peak of Tenerife [Teide]. From there, I observed the meridional height of the sun, which gave me the latitude of 19°28′57″ for the Alto de Cajón, which is located forty-three seconds (in arc) north of the Peña del Cofre [de Perote]. Using the angles that I measured between the Cofre and the Peak of Orizaba, Mr. Olt-manns found a longitude of 6ʰ37′54.6″ = 99°28′39″; this longitude differs by nearly twenty-six seconds from the one set by Mr. Ferrer. In the table of positions that this able astronomer recently sent to ▼ Mr. Arago, Mr. Ferrer decided on a reading of 19°28′54″ latitude and 99°26′55″ longitude for the Cofre; this agrees by within six seconds with the results of my own observations.

Precise knowledge of the position of the Peak of Orizaba is especially important for navigators when they land at Veracruz. The map of the Gulf of Mexico published by the Hydrographic Office in Madrid in 1799 puts this mountain one degree too far east, at a longitude of 100°29′45″. The elevation and azimuth angles that I took gave Mr. Oltmanns the latitude of 19°2′17″ and a longitude of 99°35′15″ = 6ʰ38′21″. Long before me, however, Spanish seafarers had already identified almost the correct position of the Peak of Orizaba. The error on the *Seno Mexicano*'s map, which was transferred to the French map,[1] must be attributed to a misunderstanding on the engraver's part. It was subsequently corrected in 1803 by the learned Mr. Bauzá's edition of the Spanish map. The name of the capital of Mexico was erased there, and Peak of Orizaba was located at a longitude of 99°47′30″. As the 1793 manuscripts in my possession prove, Mr. Ferrer set this mountain at a latitude of 19°2′1″ and a longitude of 99°35′35″. He later[2] settled on a reading of 99°33′5″. Mr. Isasbirivil decided on the same result; I had the opportunity to witness his exacting observations when we conducted observations together in Lima and Callao in 1802.

Surprisingly, the most recent map of the region of New Spain under discussion here, which bears the name of a highly regarded author, is also the most incorrect of all. I am referring to the great British map called *Chart of the West-Indies and Spanish Dominions in North-America*, by Arrowsmith, from June 1803. Between Mexico City and Veracruz, place names are assigned haphazardly. The position of the Peak of Orizaba is noted in a way that could be dangerous to sailors. The following table shows the positions of the most important points, as they are shown on this map. I include here the results of my own astronomical observations. In this table, longitudes have been calculated east of the town of Veracruz in order not to bring the absolute position of this port into the comparison.

I.56

I.57

---

1. [France,] *Carte des côtes du golfe du Mexique,* after the observations made by the Spanish, year IX [1800–01].

2. *Connaissance des temps* for 1817, p. 302, where the latitude of the peak is given as 19°12′17″, probably by mistake.

	ARROWSMITH'S MAP			RESULTS OF ASTRONOMICAL OBSERVATIONS	
	Latitude	Longitude		Latitude	Longitude
Mexico City	19°57′	3°38′	Mexico City	19°25′45″	2°56′30″
Mexico City's volcano	19°33′	3°0′	Popocatepetl	18°59′47″	2°24′15″
Puebla	19°33′	2°25′	Puebla	19°0′15″	1°53′45″
Peak of Orizaba	20°3′	1°50′	Pico de Orizaba	19°2′17″	1°6′15″
Tlaxcala Volcano	19°33′	1°54′			
Perote	19°48′	1°37′	Perote	19°33′37″	0°59′45″
False Orizaba	19°51′	1°12′			
Xalapa	19°36′	1°0′	Xalapa	19°30′8″	0°45′54″

I.58   The errors in *latitude* are therefore greater than *half a degree*. It is difficult to imagine what Arrowsmith's intentions were in placing the three mountains called Orizaba, False Orizaba, and Volcán de Tlaxcala on his 1803 map (since the 1805 version is but a copy of my own map). All three are shown *northwest* of the port of Veracruz, whereas the true Peak of Orizaba—and the Mexicans recognize only one, which, in the Aztec language, is called Citlaltepetl—is located *southwest* of Veracruz, between the town of Córdoba and the villages of San Andrés, San Antonio, Huatusco, and San Juan Coscomatepec. The note "visible from the high seas, at a distance of forty-five leagues" was appended to the *False Orizaba*. We know that Citlaltepetl is the first peak that sailors see as they near the coast of New Spain. For this reason, one might believe that this was why the learned British geographer called it the *False Orizaba*. In this case, however, the altitude of this problematic mountain would be incorrect by one degree, and Orizaba would be seven nautical leagues north of the town of Jalapa, whereas, in reality, it is located twelve nautical leagues south-southwest of there. Is it possible that Arrowsmith's Peak of Orizaba is the Cofre de Perote? But the Cofre is still southeast and not northwest of the village of Perote. Moreover, we also find the fable of the *two* mountains called Orizaba in Thomas Jefferys's atlas (*The West-Indian Atlas*, London, 1794), which in-

I.59   sists that it gives detailed information on the route from Veracruz to Mexico City. Its latitudes are incorrect by thirty-six minutes. It gives the longitudinal difference between the port and the capital as 2°29′ instead of 3°38′, as Arrowsmith's map does, and instead of 2°56′30″, which is the result of my own astronomical observations. It is also highly unlikely that the Volcán de Tlaxcala on the 1803 British map is the Sierra de Tlaxcala, which goes by the local name of Malinche. The latter is neither outstanding in elevation nor is it very far from Puebla. This confusion is all the more surprising when one considers that, in 1803, it was possible to study Don José Joaquín Ferrer's astute observations from 1798,[1] together with the maps by the *Depósito hidrográfico* in Madrid. I present these errors in detail in order to demonstrate the state of Mexican geography at the time when I began to devote my work to it. Even in Spain, the metropole of the colonies, Mr. Antillón, in 1802 located Puebla thirty-two minutes of arc farther south than its actual position on his map of North America.

I.60

Points Located between Mexico City, Guanajuato, and Valladolid

During the two expeditions I made—one to the ▼ Morán mines and to the porphyritic peaks of Actopan, the other to Guanajuato and the Jorullo volcano in the province of Michoacan—I determined the position of ten points whose longitudes are almost all based on the movement of time. These points have enabled me to introduce with some precision a large part of the three intendancies of Mexico City, Guanajuato, and Valladolid. Using distances from the moon to the sun, the longitude of the town of *Guanajuato* has been confirmed at $6^h53′7.5″$. Its latitude is 21°0′9″, as deduced by the observation of Grus's α [alpha]; 21°0′28″ by Fomalhaut; by the observation of Grus's β [beta], 21°0′8″. On their map engraved in Puebla in 1755, the Jesuits put Guanajuato at 22°50′ latitude and 112°30′ longitude, an error of 9°! Mr. Velázquez, who has observed the eclipses of Jupiter's satellites from

---

1. [Ferrer, "Geographische Ortsbestimmungen"] in Mr. de Zach's *Allgemeine geographische Ephemeriden*, 1798, vol. II, p. 393. I cite Mr. Ferrer's results according to this work. They do not always agree with the unpublished work (copies of which I own) that this excellent and tireless navigator wrote when he was actually at the sites; this work was most likely based on less careful calculations. I find it necessary to make this observation as a reminder to those who, often against my wishes, have procured copies of my own work. Only after having calculated all the observations that were made is it possible to settle upon a precise result. (Since the first edition of this *Political Essay*, Mr. Ferrer has published the following *latest results* of his Mexican observations in the *Connaissance des temps* for 1817: New Veracruz, latitude 19°11′52″, longitude 98°28′15″; Pico de Orizaba, latitude 19°2′17″, longitude 99°33′5″; Cofre de Perote, latitude 19°28′54″, longitude 99°26′55″; Jalapa, latitude 19°30′57″, longitude 99°12′55″; El Encero, latitude 19°28′8″, longitude 99°6′39″; Tampico, Barra, latitude 22°15′30″, longitude 100°12′15″; Nuevo Santander, Barra, latitude 23°45′18″, longitude 92°53′21″.)

I.61   Guanajuato, found this town at 1°48' east of Mexico City, but at 20°45'0" latitude, according to his unpublished map of New Spain. This latitudinal error of a quarter of a degree is even more astonishing if we know that the longitudinal error to which the Mexican astronomer admits is the same, to almost one arcminute, as the result of my own chronometric measures.

For the latitude of the town of Toluca, I found 19°16'24" using Grus's α [alpha], and 19°16'13" using Fomalhaut. Whenever possible, I have tried always to observe the same stars in the southern hemisphere in order to lessen the errors that would arise from the uncertainty of their declination.

The position of the Nevado de Toluca; the latitude of Pátzcuaro, a town located on the shore of the eponymous lake; and the position of Salamanca, San Juan del Río, and [San Salvador] Tisayuca are all based on less precise observations. Under certain circumstances, Douwes's method can only produce approximate results; but in a country that has so few fixed positions, one must often be content with less exact results. I believe that we may rest assured that the longitudes of Querétaro, Salamanca, and San Juan del Río are reliable. Using the movement of time, I positioned them at 102°30'30"; 103°16'0"; and 102°12'15". The latitudes of these three towns appear to be 20°36'39"; 20°40"; and 20°27".

I.62   In the valley of Mexico City itself, there are several very important points whose positions Velázquez, a very distinguished Mexican geometer, has determined. In 1773, this tireless man did a leveling together with trigonometric work. His goal was to prove that the waters of Lake Texcoco could be diverted to the Huehuetoca canal. I have in my possession Velázquez's unpublished triangles, which Mr. Oteiza calculated on site. Mr. Oltmanns has just repeated these calculations, subjecting the positions of the signals to the latitude and longitude that I have adopted for the monastery of San Agustín in the capital of Mexico. My own table of geographical positions contains the most recent results Mr. Oltmanns obtained. There is no longer any doubt concerning the oblique distances. However, the lack of observations of azimuths renders the reduction to perpendiculars, or differences in latitude and longitude, somewhat uncertain. We shall return to this subject when we analyze the maps of the environs of Mexico City.

Seventeen positions set by Mr. Ferrer on the outskirts of Veracruz depend on the port's longitude. Since I assume the longitude of this port to be 10'45" farther west than the position the Spanish astronomer gives, I felt that it was necessary to reduce the longitudes Mr. Ferrer has published to the Paris meridian by adding 8°47'15". According to the *Connaissance des temps*, the Spanish observer had calculated his lunar distances at a time

when Cádiz was thought to be 8°36′30″ west of Paris. Also according to   I.63
this same principle, I changed the absolute longitudes of Jalapa, the Cofre de Perote, and the Peak of Orizaba (all discussed earlier). For example, Mr. Ferrer puts the latter peak at 90°48′23″ longitude west of Cádiz,
whereas, using this same meridian, he fixes Veracruz at 89°41′45″.

Old and New California; Provincias Internas

The northwestern part of New Spain, the coastline of California, and the
section that the British call ▼ New Albion offer several points that have been
precisely determined by the geodesic measurements of ▼ Quadra, Galiano,
and Vancouver. Few maps of Europe are more detailed than the maps we
have of western America, from Cabo Mendocino to the Queen Charlotte
Strait.

After ordering the two reconnaissance voyages by ▼ Diego Hurtado
de Mendoza, Diego Becerra, and Hernando de Grijalva in 1532 and 1533,
Cortés himself saw the coastline of California and the gulf that, since then,
has been called the *Sea of Cortés.*[1] In 1542, the intrepid ▼ Juan Rodríguez
Cabrillo advanced north to 44° latitude; ▼ Juan Gaëtan discovered the
Sandwich Islands; in 1582, ▼ Francisco Gali discovered the northwest   I.64
coast of America around 57°30′ latitude. This information suggests that
Spanish navigators had visited these same regions long before ▼ Cook made
known to the world that part of the Great Ocean where he perished, a victim of his own ambition. Their names have not yet met with the fame that
they deserve. Narrow-minded policies have opposed their renown, and
the Spanish nation has not been able to enjoy the glory for which the brave
mariners of the sixteenth century had set the course. The reasons for the
mysteries that enshroud the Castilians' early discoveries are discussed in
the historical introduction to the ▼ *Voyage de Marchand* and in the introduction to the Summary of the Spanish expeditions to discover the Strait
of [Juan de] Fuca.

Venus's passage in 1769 occasioned the voyage of Mr. Chappe, Mr. Doz,
and Mr. Velázquez, three astronomers, the first of whom was French, the
second Spanish, and the third Mexican, and who, what is more, was raised
by a highly intelligent Indian from the village of ▼ Xaltocan. The expedition became important in terms of longitudes, for before these astronomers
arrived in California, the latitudes of Cabo San Lucas and Santa Rosa

---

1. Gómara, *Historia [General de las Indias]*, chap. 12.

had already been determined, with adequate precision, by Don Miguel Costansó, who is today a brigadier and head of the corps of engineers. A respectable officer, he has enthusiastically studied the geography of his

I.65 country. Using ▼ gnomons and perfectly constructed English octants, he has located San José at 23°2′0″ and Cape San Lucas at 22°48′10″. To that point, as Alzate's map demonstrates, San José was thought to be at 22°0′ latitude.

The details of the Abbot Chappe's observations, published by Cassini, inspire less than complete confidence. Equipped with a quadrant with a three-foot radius, Chappe, using Arcturus, found the altitude of San José to be 23°4′1″; using Antares, he determined 23°3′12″. The mean of all of these stellar observations differs by thirty-one seconds from the result taken from the solar passages across the meridian. Among the solar observations, some of the ranges are as wide as 1′19″. Nonetheless, Mr. Cassini refers to them as "very precise and very concordant."[1] I do not give these examples in order to discredit other astronomers, who have many other admirable qualities, but merely to demonstrate that a sextant with a five-inch radius would have been more useful to Abbot Chappe than a quadrant with a radius of three feet, an instrument that is difficult to set up and to verify. Don Vicente Doz puts San José at latitude 23°5′15″. The longitude of this village, which is famous in the annals of astronomy, has been deduced from Venus's passage and from some eclipses of Jupiter's satellites observed by Chappe and then compared to Wargentin's Tables. Mr. Cassini put it at an average of 7ʰ28′10″ or 112°2′30″; Father Hell at 7ʰ37′57″. The longitude

I.66 derived from Chappe's observations is 3°12′ farther east than the one on Alzate's map from 1768.[2] The Mexican astronomer Velázquez had a small observatory built in the village of Santa Ana, where he observed Venus's passage by himself and communicated the results of his observation to the Abbot Chappe and Don Vicente Doz. Mr. de Cassini published these results, which conform to the unpublished observations I was able to obtain in Mexico City; they could serve to determine the longitude of Santa Ana. Before Chappe arrived, Velázquez was aware of the enormous error in the longitude of California, having observed several eclipses of Jupiter's

1. [Jean Chappe d'Auteroche,] *Voyage en Californie*, p. 106.

2. José Antonio de Alzate et Ramírez, *Nuevo mapa geographico de la America septentrional: perteneciente al Virreynato de Mexico dedicado a los sabios miembros de la Academia Real de las Ciencias de Paris . . . Año de 1768.*

satellites from the Santa Rosa mission in 1768.[1] He communicated the true longitude of the Peninsula to European astronomers before they were able to conduct any of their own observations.

Spanish navigators had determined the position of Cabo San Lucas, which was called Cabo de Santiago in Cortés's day.[2] In unpublished works[3] ordered by the ▼ Chevalier de Azanza and preserved in the archives of the Viceroyalty of Mexico, I found that Mr. Quadra had located Cabo San Lucas at 22°52′ latitude and 4°40′ longitude west of the port of San Blas. If we agree with Malaspina in positioning S. Blas at 107°41′30″, this result gives 112°21′30″ for the southernmost cape of California. Malaspina's expedition (according to Mr. Antillón) also placed Cabo San Lucas at 22°52′ latitude, but at 112°16′47″ longitude. This chronometric position has been adopted in the atlas that accompanies the Voyage of the Spanish to the Strait of Juan de Fuca; but it is still 17′15″ farther west than the position that was published (I do not know by what authority) in the *Connaissance des temps* for 1818. I have adopted a meridional difference of 14′17″ between San José and the cape. It should be noted, however, that since these two points have not been compared to one another but simply determined by independent observations, there may be an error in their respective distance. According to information that I received from those who have visited these arid, uninhabited places, it would seem that the longitudinal difference is somewhat greater than I made it out to be. In Cortés's time, Cabo San Lucas was believed to be at 22°0′ latitude and 10°50′ west of the meridian of Acapulco, a relative longitude that is correct to nearly half a degree. Using the most recent astronomical tables, Mr. Oltmanns has recalculated the

I.67

I.68

1. Don José Antonio de Alzate, "Estado de la Geografía de la Nueva España y modo de perfeccionarla" ([▼ *Asuntos varios sobre Ciencias y Artes,*] December 1772, no.7, p. 55.)

2. Domingo de Castillo, *Mapa de California*, 1541.

3. Mr. de Azanza, Viceroy of Mexico, had ordered ▼ Mr. Casasola, a frigate lieutenant in the Royal Navy, to collect in four volumes all materials relating to navigations in the northern part of California under the viceroys ▼ Bucareli, Flores, and Revillagigedo. These works consist of: 1) the observations by Mr. Pérez, ▼ Mr. Cañisarez, Mr. Galiano, Mr. Quadra, and Mr. Malaspina, combined in an atlas of twenty-six maps; 2) a large folio volume entitled *Compendio histórico de las navigaciones sobre las costas septentrionales de California, ordenado en 1797[sic: 1799] en la ciudad de México*; 3) Don Juan Francisco de la Bodega y Quadra's voyage to the northwestern coast of California, commanding the frigates *Sta. Gertrudis*, *Aranzasa*, *Princesa*, and the schooner *Activa*, in 1792; and 4) *Reconocimiento de los cuatro establecimientos rusos al norte de la California* in 1788. Viceroy Flores ordered this expedition, which is described by ▼ Don Antonio Bonilla. Some of this invaluable material, which I was able to consult in the archives in Mexico City, has been published in [Espinosa y Tello's] *Relación del viage de las Goletas Sútil y Mexicana*, Madrid, 1802.

passage of Venus that the Abbot Chappe observed in San José. He found 7ʰ28′1.6″ for that village; the satellites had given 7ʰ28′7.5″. Since San José was also thirty-eight seconds east of Cabo San Lucas, we must put this cape at 112°10′38″ longitude and 22°52′28″ latitude. Based on the emersion of a single satellite and the end of a lunar eclipse (both of which were observed by the officers of Malaspina's expedition), Mr. Oltmanns[1] set the port of San Blas at 107°35′48″ longitude (and 21°31′15″ latitude). A stellar occultation recently gave a reading of 107°39′42″ to Captain Basil Hall.[2]

I.69   The coastline of California was explored in great detail by the Spanish expedition of the schooners Sútil and Mexicana in 1792. Vancouver's expedition covered the same coast from 30° latitude (from the San Domingo mission). Malaspina and the unfortunate ▼ Lapérouse both visited Monterrey, but the longitudes they give to this port differ by 1′16″! Although one may assume that the direction of the coasts and the longitudinal differences between various points have been determined perfectly, one is still somewhat at a loss when it comes to setting *absolute longitudes*. Vancouver's obser-

I.70   vations[3] of lunar distances put Nootka and almost the entire northwestern coast of the Americas at twenty-eight minutes (in arc) east of the longitudinal position that Cook, Marchand, and Valdés assign to it. It would be interesting to analyze the influence of Bürg and Burckhardt's lunar Tables on Vancouver's observations, which have unfortunately not been published in complete detail. I felt it necessary to give preference to the absolute longitude

1. Humboldt, *Recueil d'observations astronomiques*, vol. II, pp. 613–16.

2. Basil Hall, *Extracts from a Journal Written on the Coast of Chile, Peru, and Mexico, 1820–1822* (Edinburgh 1824), vol. II, p. 379. This navigator puts San Blas at the latitude of 21°32′24″; Cabo Corrientes at latitude 20°24′0″, longitude 108°2′41″; and a very high peak that was thought to be the volcano of Colima at latitude 19°36′20″, longitude 105°56′44″. Following Malaspina on my map, I put Cabo Corrientes at latitude 20°25′30″, longitude 107°55′51″, a position that corresponds closely to that of the famous British navigator's. I can make no judgment as to the identity of the mountain Captain Hall mentions as the volcano of Colima. Following the expedition itineraries, my map of Mexico puts this volcano at 19°3′ latitude and 105°30′ longitude. As I have already mentioned in the first edition of this work (vol. II, p. 257), this position is very dubious because of the complete absence of astronomical observations between Petatlán, Selagua, and the Playas de Jorullo. If Captain Hall's cross-bearings yield a result that is equally certain in terms of both latitude and longitude, then the Volcano of Colima would be north of the parallel of Puerto de Navidad and, as a result, quite far from the Punta de Colima. The latitudinal position of the Volcano of Colima is important to the geologist: one day it will be decisive in determining whether the Colima Peak, like the Volcano of Tuxtla, is located outside the *parallel of the Mexican volcanoes and nevados*, or if this mountain is aligned with the Peak of Orizaba, Popocatépetl, or the Puebla Volcano, the Nevado de Toluca, and the new Jorullo Volcano.

3. Vancouver, *Voyage autour du monde*, vol. II, p. 46.

of Monterrey, which was deduced from Malaspina's observations, not only because it is based on the occultation of stars and the eclipses of satellites, but above all, because the Spanish observations connect, as it were, New and Old California through the movement of time. Commanded by Don Alejandro Malaspina, the corvettes Descubierta and Atrevida chronometrically determined the longitudinal difference between Acapulco, San Blas, Cabo San Lucas, and Monterrey. If he adopted the easternmost position of this port, which is Vancouver's reference, the geographer would be unsure of the mineral deposits on the southernmost coastline. To avoid such problems, I have, following Malaspina, put Monterey at 36°35′45″ latitude and 124°23′45″ longitude.[1] Using lunar distances, Lapérouse[2] had placed it at 123°34′0″ and, using the chronometer, at 124°3′0″.[3] Vancouver deduced a longitude of 123°54′30″ from 1,200 distances from the moon to the sun. Since he had the leisure to locate the mineral deposits on the coastline with scrupulous precision, I felt inclined to respect the longitudinal differences he gives between Monterrey and the mission of San Diego, San Juan, San Buenaventura, Santa Barbara, and San Francisco. In this way, all these positions have been related to that of Monterrey. If, on the other hand, I had traced the entire northwest coast, using only Vancouver's observations, I would have been tempted to put the longitude of Cabo San Lucas farther to the east. One needs only point out that the striking difference[4] between the British and Spanish data, despite so much work, still persists. I am assuming that the absolute positions we have set for Acapulco, San Blas, and Cabo San Lucas are quite exact, and that the error of plus twenty-eight minutes (in arc) is found farther north. A false reading of the quotidian

I.71

I.72

---

1. [Antillón,] *Análisis de la carta [de la América septentrional]* 1803, p. 50. Mr. Oltmanns decided on 124°11′21″ for reasons that that skilled geographer has presented in the supplement to my [*Recueil d'*] *observations astronomiques*, vol. II, p. 613.

2. [Lapérouse,] *Voyage*, vol. III, p. 304.

3. Correcting Lapérouse's result by using Greenwich lunar observations, Mr. Triesnecker found a longitude of 123°42′12″, instead of 123°34′0″ (["XV. Geographische Längen aus Lapérouse's Entdeckungs-Reise durch gleichzeitige astronomische Beobachtungen berichtiget"], Zach, [*Monatliche*] *Correspondenz*, vol. 1 [3], p. 173).

4. I submit the following definitive results of observations by the most renowned navigators, based on the entirety of their work: Nootka, the Anse des Amis, according to Galiano and Valdés, 8ʰ35′40.2″; 8ʰ35′44.0″, according to Marchand; 8ʰ36′0″ according to Cook; 8ʰ36′55.1″ according to Vancouver; average, 8ʰ35′48″ = 128°57′1″. According to Lapérouse, Monterrey is at 8ʰ15′35.6″; according to Malaspina, at 8ʰ16′51.6″; according to Vancouver, 8ʰ16′35.0″; average 8ʰ16′20.7″ = 124°5′11″. It is hard to believe that there is still so much uncertainty about a coast that has been visited by so many scientific expeditions.

movement of a chronometer, together with the state of Mayer and Mason's old lunar tables, may have contributed much to this error.

Having discussed positions based on the astronomical observations of practiced observers, I shall now review those positions that may be considered suspect, either because of imperfect instruments, or because the observers' names do not inspire confidence, or finally because we cannot be sure that the results were not lifted from unpublished works that were incorrectly copied. Here is the information I have been able to glean from these early astronomical observations. It must be used with caution; but it is invaluable to the geography of a region that is so little-known to this day.

The Jesuit fathers have the honor of being the first to examine the Gulf of California or the Sea of Cortés. In 1701, ▼ Father Kino, a professor of mathematics at Ingolstadt and avowed enemy of the Mexican astronomer Sigüenza (against whom he has polemicized extensively in print), reached the junction of the Río Gila and the Colorado. Using an astronomical ring, he set the latitude of this *Junta* [conjunction] at 35°30′. When I consulted the unpublished map drawn in 1541 by Domingo de Castillo, which was discovered in the Cortés family archives, I found that the two rivers—the Río de Buena Guía and the Brazo de Miraflores—which appear to meet below 33°40′ latitude at the northernmost reach of the gulf, had already been identified by the middle of the sixteenth century. In 1538, using the meridional height of the sun, ▼ Father Nadal had found the conjunction of the Gila and the Colorado at 35°0′. ▼ Father Marcos de Niza put it at 34°30′. ▼ Delisle most likely based the position of 34°, which he adopted in his maps, on these figures. But a work printed in Mexico[1] cites the recent observations by ▼ two learned Franciscans, Father Juan Díaz and Father Pedro Font, who used an astronomical ring. These observations conform to each other and would seem to prove that the *Juntas* are much farther south than we have believed until now. At the mouth of the Gila in 1774, Father Díaz obtained a reading of 32°44′ on two consecutive days. In 1775, Father Font found 32°47′ there. The former also reassures us that the route he took—taking rhumbs and distances into consideration—reveals that the *Juntas* cannot be located at 35° latitude. In 1777, the positions that Father Font assigned to the missions of Monterrey, San Diego, and San Francisco (and which differ only by a few minutes from Vancouver's and Malaspina's observations) seem to support the precision of his work, but it is also possible that

I.73

---

1. [Arricivita,] *Crónica seráfica [y apostólica del Colegio de propaganda fide de la Santa Cruz] de Querétaro*, 1792, Prologue, p. 11 [iii].

the missionary simply copied the information that the navigators gave him. I.74
The precision of the observations made at the *Junta de los Ríos* is above
reproach, since an attentive and eager observer may often obtain satisfactory results by imperfect means. The latitudes that ▼ Bouguer measured on
the banks of the Río Magdalena, using a seven-to-eight-foot-high gnomon
and a few strips of reeds as a scale, differ by only four or five minutes from
the latitudes I found sixty years later with excellent reflection instruments
[sextants].

What seems more likely is that, by using his astronomical ring, Father
Font very inadequately determined the latitudes of the missions San Gabriel (32°37′), San Antonio de los Robles (36°2′), and San Luis Obispo
(35°17′). When comparing these positions to Vancouver's Atlas, I find that
the errors are sometimes +1°11′, and, at other times, -0°23′. It is true that
the British navigator did not visit the three missions personally but was able
to locate them in relation to the nearby coast, where he was examining mineral deposits. The lesson to be learned from this experience is that one must
be very wary of observations made with astronomical rings. Father Pedro
Font also visited the site of the ruins called *Las Casas grandes* [the Great
Houses], which he found at 33°30′. If it were exact, this position would
be of great importance because *Las Casas grandes* is the site of an older
civilization. We should not, however, confuse this second Aztec settlement
from which they traveled from Tarahumara to Calhuacan, with the *Casas* I.75
grandes or the third Aztec settlement south of the Llanos presidio in the intendancy of Nueva Vizcaya. I should like to acquaint myself with the observations of the ▼ Jesuit Father Juan Ugarte, who, according to Mr. Antillón,
found the errors on the maps of California in 1721. His is the glory of having
been the first to recognize that this vast region is a peninsula; but from the
sixteenth century onward, no one in Mexico would deny this fact, although
many years later, Europeans came to doubt it.[1]

Among the astronomical observations I consider a bit suspect are those
several Spanish officer engineers made during their frequently required visits to various small forts on the northern borders of New Spain. While in
Mexico City, I obtained brigadier ▼ Don Pedro de Rivera's travel journals
from 1724, those by Don Nicolás Lafora, who accompanied the Marquis

---

1. On an expedition funded by Cortés in 1539, ▼ Francisco de Ulloa recognized that the
Gulf of California extended to the mouth of the Río Colorado. It was not until the seventeenth century that California was thought to be an island (Antillón, *Análisis*, p. 47, number
55).

de Rubi on his 1765 expedition to research the military defense line of the *Provincias internas*, and Don Manuel Mascaró's unpublished journey from

I.76 Mexico City to Chihuahua and Arispe.[1] These esteemed travelers assure us that they observed the meridional height of the sun. I have no idea what instruments they used, and I fear that the manuscripts I have read were not always copied exactly. Having gone to the effort of calculating latitudes based on wind areas and given distances, I have found that the results do not conform well to the observed latitudes. In Madrid, Mr. Bauzá and Mr. Antillón have found the same discrepancy. I regret that none of the latitudinal observations the officer engineers made can be connected to a place whose position either Mr. Ferrer or I may have determined. It is true that Mr. Mascaró made observations in Querétaro. He and I differ by ten minutes of arc on the latitude of that town; but since my result is based on a method that is analogous to Douwes's, it is suspect by two minutes. Despite these uncertainties, the material on which I have just reported should not be slighted; on the contrary, it is of great assistance to anyone who wishes to draw a map of a part of the world that learned travelers have

I.77 so infrequently visited. We shall limit our discussion to some of the major points.

In his classic study of Virginia, ▼ Mr. Jefferson has attempted to fix the position of the Presidio of Santa Fe in New Mexico. He believes it to be at 38°10′ latitude; but we find 36°12′ by taking the average of Mr. Lafora's direct observations and those of ▼ Father Vélez y Escalante. Using a group of ingenious convergences, Mr. Bauzá and Mr. Antillón related Santa Fe to the Presidio del Altar and positioned the latter in relation to the coastline of Sonora; they found Santa Fe de Nuevo Mexico at 4°21′ west of the Mexican capital.[2] Mr. Antillón's map shows a difference of 5°. Taking a different approach but without knowledge of the work of these skilled Spanish astronomers, I was able to find an even greater longitudinal difference. I fixed

1. a) *Derrotero del Brigadier Don Pedro de Rivera en la visita que hizo de los Presidios de las Fronteras de Nueva España en 1724.* b) *Itinerario del mismo autor de Zacatecas a la Nueva Viscaya.* c) *Itinerario del mismo autor desde el Presidio del Paso del Norte hasta el de Janos.* d) *Diario de Don Nicolás de Lafora en su Viaje a las Provincias internas en 1766* [*Relación del viaje que hizo a los presidios internos*]. e) *Derrotero del mismo autor de la Villa de Chihuahua al Presidio del Paso del Norte.* f) *Derrotero de México a Chihuahua por el Ingeniero Don Manuel Mascaró en 1778.* g) *Derrotero del mismo autor desde Chihuahua a Arispe, Misión de Sonora.* h) *Derrotero del mismo autor desde Arispe a México en 1784.* The original versions of these eight unpublished works are preserved in the archives of the viceroyalty of Mexico.

2. [Antillón,] *Análisis de la Carta*, p. 44.

the longitude of Durango using a lunar eclipse observed by Dr. Oteiza; this position matches the one Mr. Antillón adopted. Supposing the latitude of Durango to be 24°30′, and Chihuahua, the capital of Nueva Vizcaya, where Mr. Mascaró observed at length, to be at 28°45′, I have estimated the value of the leagues that Brigadier Rivera discusses in his travel journal. The distances and rhumbs gave me fifty-three minutes of arc for the meridional difference between Durango and Chihuahua, using a geographical construction, from which derives a longitudinal difference of 5°48′ for Mexico City and Santa Fe. It is normal that this difference appears to be greater than the one on which Mr. Bauzá and Mr. Antillón decided, since those geographers put the capital of Mexico thirty-seven minutes (in arc) too far to the west. Nonetheless, the position they assign to Santa Fe depends more on the longitudes of San Blas and Acapulco than on that of Mexico City. I situate Santa Fe at 107°13′ absolute longitude; Mr. Bauzá and Mr. Antillón situate it at 107°2′, a highly probable result, and 5°28′ farther east than the longitude shown on the map of western Louisiana published in Philadelphia in 1803. This same map also shows the position of Cabo Mendocino as determined by the observations of Vancouver and Spanish navigators to be incorrect by 4°. Basing his work on several convergences, Mr. Costansó concluded that Santa Fe and Chihuahua were at 4°57′ and Arispe at 9°5′ west of Mexico City. On all the old unpublished maps I have seen, especially those that were completed before Velázquez returned from California, Durango is situated 3° east of Parral and of Chihuahua. Velázquez reduced this meridional difference to three minutes (in arc); but a graphic method based on the travel journals I have just cited provided me with fifty minutes.

I was happy to see that my convergences had led me to the same result that the learned astronomers in Madrid obtained about another geographical point in New Spain. The map I drew in Mexico City in the same year that Mr. Antillón published his analytical memoir[1] indicates (and this is proven by the copies preserved in Mexico) that the meridional difference of Tampico and Mazatlán (in other words, the width of the kingdom from the Atlantic Ocean to the South Sea), is 8°0′. Mr. Bauzá and Mr. Antillón found it at 8°20′, while Lafora's map gave 17°45′, and Arrowsmith's maps of the West Indies showed 9°1′. On my own map, I compared Tampico with Barra de Santander, whose longitude Mr. Ferrer observed; I assumed Tampico

I.78

I.79

1. [Antillón y Marzo,] *Análisis de los fundamentos de la Carta de la América septentrional.*

to be at ten minutes of arc east of Barra, in agreement with the maps of the Naval Office in Madrid. We shall return to the position of this port later in this essay.

The ▼ Count Santiago de la Laguna determined the latitude of the town of Zacatecas, famous for the wealth of its mines, not by using an astronomical ring or gnomons but by employing several quadrants with a radius of three to four feet, which were built in the country itself; the latitude deduced was 23°0′. ▼ Don Francisco Xavier de Sarría had deduced 22°5′6″ from several gnomonic observations! These observations have been recorded in the Chronicle published by the Franciscan fathers of Querétaro in Mexico, a work that is unknown in Europe. Zacatecas was formerly thought to be half a degree farther north; this is proven by a brief

I.80   latitude sheet which ▼ Don Diego Guadalajara published in Mexico City for those who wish to construct gnomons. The Count de la Laguna claims that he found the longitude of Zacatecas at 4°30′ west of Mexico City, but this result is probably highly suspect. Having fixed the position of Guanajuato, using the chronometer and lunar observations, I found the meridional difference for Zacatecas and Mexico City to be 2°32′, using rhumbs and the distances given in itineraries; Mr. Mascaró's route calculation gave me 3°45′. The Count de la Laguna has set the absolute longitude in a completely erroneous way. From a corresponding observation of a lunar eclipse in Bologna, he concluded that Zacatecas was at 7ʰ50′ east of that Italian city, which would give a 7ʰ13′59″ longitude for Zacatecas and consequently 7ʰ3′39″ (instead of 6ʰ45′42″) for Mexico City. Could some error have slipped in when the figures were copied? Could the meridional difference be 7ʰ30′ instead of 7ʰ50′?

I am assuming the longitude of *Durango* to be very close to 105°55′. Don Juan José Oteiza, a young Mexican geometer whose insights have often been a great help to me in my work, has observed (at Hacienda del Ojo, thirty-eight minutes in arc east of Durango) the end of a lunar eclipse that produced the result we have just given when compared to Mayer's old tables. Using the rhumbs and distances given in Rivera's and Mascaró's travel journals, ▼ Mr. Friesen concluded a position of 5°5′ east of Mexico

I.81   City, and consequently 106°30′. The latitude of Durango seems highly questionable. Rivera and his traveling companion, ▼ Don Francisco Álvarez Barreiro, confirm having found 24°38′ in 1724, using meridional heights of the sun; in 1766, Lafora settled on 24°9′. We have no idea what type of instruments these engineers used. If the latitude that the Count de la

Laguna, Mr. Sarría, and engineer Mascaró assign to the town of Zacatecas is correct, then Durango's latitude, based on rhumbs and distances marked in the itineraries, must be close to 24°25′.

There are a few places in the northern provinces of New Spain where the three engineers we have just mentioned made successive observations; this circumstance makes us somewhat more confident of the average result.

Chihuahua. Latitude, 29°11′, according to Rivera; 28°56′, according to Lafora; and 28°45′, according to Mascaró. Its longitude, deduced from rhumbs and distances, is 5°25′ west of Mexico City.

Santa Fe. Latitude, 36°28′, according to Rivera; 36°10′, according to Lafora. Its longitude by approximation is 107°13′, or 5°48′ by the meridian of Mexico City.

[The account of ▼ Pike's expedition might have led me to believe that the longitude I set for Santa Fe del Nuevo México was 6° too far east. But the result of the longitudes of gigantic mountains—Spanish Peak, 106°55′, James Peak, 107°52′—that have recently been determined by connecting the Rocky Mountains chronometrically with Council Bluffs on the Missouri, is that the longitude of Santa Fe is probably not more than 108° west of the Paris meridian. If we closely examine ▼ Major Long's work and Mr. Tanner's geographical discussions, we notice that Major Pike, a zealous and courageous explorer, was not very exact in his measurements, his evaluations of elevation, and his astronomical observations. I offer here the variations for Santa Fe since the first edition of the *Political Essay*:

I.82

HUMBOLDT	ROBINSON	PIKE	TANNER	LONG
1804		1810	1823	1824
Lat. 36°12′	36°25′	36°20′	36°15′	36°12′
Long. 107°15′	109°57′	113°55′	107°49′	107°15′

I am surprised to note that Mr. Tanner (*Geographical Memoir* on North America, 1823, p. 6) puts Santa Fe thirty-minutes of arc farther west than I do, although he, together with Major Long, offers 108°30′ for the longitude of *Highest Peak*. The position Long attributes to Santa Fe has been copied from the one I have published, following Lafora. The three peaks (Spanish, James, and Highest

Peak), have been brought into bearing with astronomically determined points on the shores of the Mississippi. But one will never be sure of the positions of Taos, Santa Fe, and the course of the Río del Norte until one has related these three peaks to Taos through the movement of time—or until we have made some good lunar observations in New Mexico itself. Major Pike gives 39°45′ latitude, 113°45′ longitude for James Peak, whereas Long finds 38°47′ latitude, 107°51′ longitude. This explorer has also pointed out (*[An Account of] Expedition[s,]* Vol. II, p. 354 [233]) the enormous errors that Major Pike committed in charting the course of the Arkansas and the Canadian River, as if they ran from NNW to SSE instead of from west to east, or at most from WNW to ESE. In various maps that were engraved based on Pike's information, the source of the Arkansas River is sometimes located at latitude 41°50′, longitude 115°55′, other

I.83   times at 40°30′ latitude and at 110°15′ longitude (Tanner, "Memoir," p. 7). Major Long's work has finally rectified some of these errors and changed the geography of the regions between the Mississippi, the Rocky Mountains, and the Missouri.

The following selective examples will suffice to prove the importance of these changes:

		LEWIS AND CLARK		LONG
Source of the Arkansas	lat.		41°47′	
	long.		114°25′	
United States Fort, at the mouth of the Saint Pierre River	lat.		43°55′	43°0′
	long.		95°49′	98°6′
Council Bluff	long.		99°10′	98°6′

These Bluffs are hills twenty-five toises high, which are found twenty-two minutes of arc north of the confluence of the Missouri and the Platte River, and three miles NW of Fort *Calhoun* or the *Engineers' Camp.* According to ▼ Graham and Long (*[Account of an] Expedition,* Vol. I, p. 153, vol. II, p. xxvi, table xlii), latitude 41°25′[4]″, longitude 98°4′8″ [corr. 95°43′53″]; the fort is an important military post located halfway between St. Louis and the village of Mandanes. It would be beneficial if Santa Fe and Taos were linked chronometrically with Council Bluffs.]

Presidio de Janos. Latitude 31°30', according to Rivera; 30°50', according to Mascaró. Somewhat suspect longitude of 7°40' west of Mexico City.

Arispe. Latitude 30°30', according to Rivera; 30°36', according to Mascaró. Approximate longitude, 9°53' (from Mexico City).

Geographic convergences based on travel journals lend credence to the following positions, whose latitude Mr. Mascaró and Mr. Rivera have determined. These results, which I have adopted in my map, match Mr. Bauzá's and Mr. Antillón's information. We differ, however, by nearly one degree on the absolute longitude of the town of Arispe, located in the province of Sonora, as well as on the longitude of the Paso del Norte in New Mexico. Some of these differences arise from the fact that Mr. Antillón's map puts Mexico City, Acapulco, and the mouth of the Río Gila farther east than I do.

I.84

PLACES	LATITUDE N	LONGITUDE WEST OF MEXICO CITY
Guadalajara	21°9'	3°37'
Real del Rosario	25°28'	7°1'
Presidio del Pasaje	26°30'	4°8'
Villa del Fuerte	26°30'	9°5'
Real de los Álamos	27°8'	9°58'
Presidio de Buenavista	27°45'	11°3'
Presidio del Altar	31°2'	2°41'
Paso del Norte	32°9'	5°38'

When the militia (*tropas de milicia*) was formed in the Kingdom of New Spain, a map of the province of *Oaxaca* was drawn up; that map shows eleven points (according to an author's comment) whose latitude was observed astronomically. While in Mexico City, I was unable to learn with any certainty whether (as we may rightly assume) these latitudes are based on meridional elevations measured with gnomons. The map of Oaxaca bears the name ▼ Don Pedro de Laguna, lieutenant colonel in the service of His Catholic Majesty. Some of the eleven points that deserve particular attention are located on the coast itself between the two ports of Acapulco and Tehuantepec; others are not far inland from the coast. Proceeding from west to east, we find:

I.85

PLACE	LATITUDE
Ometepec	16°37'
Xamiltepec	16°7'
Barra de Manialtepec	15°47'
Pochutla	15°30'
Puerto Guatulco	15°44'
Guiechapa	15°25'

In the Misteca Alta, the positions one found were:

San Antonio de la Cues	18°3' latitude
Teposcolula	17°18'
Nochistlán	17°16'

We could add to these positions that of the village of Acatlán in the intendancy of Puebla at 17°58', and the town of Oaxaca at 16°54' latitude. Although they are still somewhat inexact, all these positions are even more invaluable, since from the Puebla de los Angeles to the Isthmus of Panama, there is scarcely one point in the interior of these lands whose latitude has been determined astronomically until now. What does promote a certain degree of confidence in the positions in the province of Oaxaca is the agreement between the latitudes assigned to the town of Tehuantepec and Puerto Escondido in the maps of Don Pedro de Laguna and Mr. Antillón. The Spanish navigators put the first of these points at 16°22', and the second, which is near the village of Manialtepec, at 15°50' latitude.

I.86

I have candidly described the deplorable state of the geography of New Spain. I have pointed out the suspicions raised by my own observations and those of other explorers who preceded me. I have also demonstrated that a few positions in Mexico have been determined with the utmost precision in places where observatories have not yet been established. To the north and to the east, in the interior of these regions, errors in latitude can be greater than one degree. I fervently hope that my own maps may be revised soon, substituting more precise work for my own. The astronomical information they contain will last forever and support future geodesic fieldwork.

Having discussed those positions based on astronomical observations are more or less worthy of the geographer's trust, we must now consider the maps, almost all unpublished, that I used for different sections of my general Map of New Spain.

I.87

The mineral deposits and the curvature of the west coast on the shores of the Great Ocean, from Acapulco to the mouth of the Colorado River and to the volcanoes of the Virgins in California, were taken in large part from the map that accompanies the travel journal of the Spanish expeditions to the Strait of Juan de Fuca. Published in 1802 by the Naval Office in Madrid, this map is based on the route of Malaspina's corvette for the area north of Acapulco and San Blas. The coastline that extends southeast of Acapulco, however, has been very imperfectly studied. In order to trace it, I was obliged to consult Mr. Antillón's map of North America. The lack of precision with which the eastern coastline of Mexico north of Veracruz has been charted to date is also lamentable. The area between the mouth of the Río Bravo del Norte and that of the Mississippi is almost as unknown as the east coast of Africa between Orange River and Fish Bay. Mr. Cevallos and Mr. Herrera's expedition, equipped with excellent astronomical instruments, should produce precise maps of these arid, uninhabited regions. For the details of the eastern coastline, I followed the map[1] of the Gulf of Mexico published by order of the King of Spain in 1799 and perfected in 1803. Following Mr. Ferrer's observations, I have corrected several of the points I cited above. Since that able observer had put the port of Veracruz at 9'45" (in arc) less to the west than I have, I reduced the positions of the sites he determined in the environs of Veracruz to the longitude from Mr. Oltmanns's calculations. The error on the early maps had mainly to do with the longitude of the Barra de Santander that, according to Mr. Ferrer, was at 1°54'15" west of Veracruz, whereas the map in the Depósito put it at a difference of only 1°23' in longitude. I felt that I should follow Mr. Ferrer's[2] observations more closely by reducing the longitude of Tamiagua to that of Santander.

I.88

1. [Lángara,] *Carta esférica que comprehende las costas del Seno Mexicano, construida [de orden del Rey] en el Depósito hidrográfico de Madrid*, 1799.
2. According to Mr. Ferrer's last corrections (by this I mean his calculations and not his observations), he locates the Barra de Santander at 23°45'18" latitude, and 100°18'45" longitude. This position is eleven minutes (in arc) farther east than the one I used for my map in 1804 (*Connaissance des temps* for 1817, p. 303). Since Mr. Ferrer puts Veracruz only forty-five seconds (in arc) farther east than I do, the result of his report of 1817 ["Occultations d'étoiles observées à la Havane"] is that the meridional difference between Veracruz and Barra de Santander is 1°50'30". The *Depósito hidrográfico de Madrid* gives 99°48' for Barra, an error of half a degree to the east.

The terrain between the ports of Acapulco and Veracruz, between Mexico City, Guanajuato, the Santiago Valley, and Valladolid, between the Volcán de Jorullo and the Sierra de Toluca, is drawn according to several I.89 geodesic measurements that I took, either with a sextant or with an ▼ Adams graphometer. I base the area between Mexico City, Zacatecas, Fresnillo, Sombrerete, and Durango on the unpublished sketch that Mr. Oteiza so kindly made for me, based on material that he collected during his journey to Durango. Having marked the rhumbs very precisely and evaluated the distances based on the mules' gait, his map is not undeserving of our trust; I have used my own direct observations to correct the positions of Guanajuato and San Juan del Río. In this way, it was easy to convert time into distance and determine the length of the leagues in the region.

The journals of Mr. Rivera, Mr. Lafora, and Mr. Mascaró to which we have already referred were of great help for the *Provincias internas*, especially for the route from Durango to Chihuahua, and from there to Santa Fe and to Arispe, in the province of Sonora. We were only able to use this material, however, after long discussions, comparing it with the information Mr. Velázquez had collected during his expedition in California. Rivera's routes are often quite different from Mascaró's; the meridional difference between Mexico City and Zacatecas, or between Santa Fe and Chihuahua, is especially vexing, as we shall demonstrate below.

Mr. Costansó has adjusted the geography of the Sonora. This scholar, I.90 who is as modest as he is deeply learned, has collected everything that relates to the geographic knowledge of the vast Kingdom of New Spain over the last thirty years. He is the only officer-engineer who has devoted himself to detailed discussions of the longitudinal difference between the points that are the farthest from the capital. He himself has composed several interesting maps on which we can see how some ingenious convergences may replace astronomical observations to a certain degree. I am even happier to give Mr. Costansó his due because I have seen many unpublished maps in the archives in Mexico City where scales of longitude and latitude are little more than accidental decoration.

Here is a list of the maps and sketches that I consulted for the details of my own map. I believe that I have collected all the material that existed up to the year 1804.

Carta ó mapa geográfico de una gran parte del Reino de Nueva España by Mascaró and Costansó: Representing the vast space between 39° and 42° latitude, the map covers the area from Cabo Mendocino to the mouth of the Mississippi River. The work appears to have been done with great care;

I used it for the ▼ Moqui, for the environs of the Río Nabajoa [southern Utah], and for the Chevalier La Croix's route in 1778 from Chihuahua to Coahuila and Texas.

Mapa del Arzobispado de México [*Nuevo mapa geographico de la América septentrional*] by Don José Antonio de Alzate: an unpublished map drawn in 1768, revised by its maker in 1772; very poor quality, at least the part I re- I.91 viewed. Some mining locations shown on the map may be of interest to the geologist.

I made no use of the map of New Spain published in Paris in 1765 by ▼ Mr. *de Fer*, nor of governor *Pownall*'s map published in 1777, nor of *Sigüenza*'s map, which the Academy of Paris had engraved under the name of Alzate, and *which has been considered the best map of Mexico to this day.*

Carte générale de la Nouvelle-Espagne, from 14° to 27° latitude, drawn by Mr. *Costansó*. This unpublished map offers valuable knowledge of the coastline of Sonora. I also consulted it for the segment that extends from Acapulco to Tehuantepec.

Carta esférica que comprehende el trozo de costa entre el puerto de Acapulco y el surgidero de Sonsonate, provided by the brigantine *Activo* in 1794.

Carta geográfica de las provincias de Nueva Galicia, Nueva Vizcaya, Sinaloa, Sonora y California, drawn by Mr. Velázquez in 1772: Includes the regions located between 19° and 34° latitude, from the mouth of the Río Colorado to the Cholula meridian. It was drawn up in order to establish the location of the most remarkable mines in New Spain, especially those in Sonora.

Manuscript map of a part of New Spain, from the parallel of Tehuantepec to that of Durango, drawn up by ▼ *Don Carlos de Urrutia*, by order of the viceroy Revillagigedo. It is the only map of the country that shows the I.92 division into intendancies; for this reason, it was very useful to me.

Mapa de la provincia de la Compañía de Jesús de Nueva España, engraved in Mexico City in 1765. Is it merely by chance that this map, which is very poor, puts Mexico City at 278°26′ longitude, while the same capital city is located at 270° longitude on the map bearing the title *Mapa de distancias de los lugares principales de Nueva España* [Coromina, *Mapa, y tabla geographica de leguas communes, que ai de unos à otros lugares, y ciudades principales de la America Septentionale*] that the Jesuit fathers had engraved in Puebla de los Angeles in 1755?

In Rome, I discovered the *Provincia Mexicana apud Indos ordinis Carmelitarum* (*erecta* 1588), *Rome*, 1738 [by Franceschini]. On that map, Mexico City is located at 20°28′ latitude.

Father Pichardo de San Felipe Neri, a very enlightened clergyman who owns the Abbot Chappe's small quadrant, was kind enough to provide me with two unpublished maps of New Spain, one by *Velázquez* and the other by *Alzate*. Both maps diverge from the one engraved by the Academy of Paris. They show the locations of many remarkable mines, whose positions I sought in vain on other maps.

Plano de las cercanías de México: Sigüenza's map, republished by Alzate in 1786. Another map of the valley of Mexico City appears annually in the almanac called *Guía de Forasteros*; this is Mr. Mascaró's map ["Mapa de las cercanías de México," 1794]. Neither of these two maps, nor the one published by ▼ López in 1785, shows the lakes in their actual location. López's map indicates longitudinal degrees marked on the meridian, a rather odd mistake for a geographer to the King!

I.93

Detailed map of the surroundings of the Doctor, of the Río Moctezuma, which receives its waters from the canal of Huehuetoca, and of Zimapán, by M. Mascaró: The environs of *Durango, Toluca*, and *Temascaltepec* are represented with great care on the unpublished maps drawn for my use by Mr. Juan José Oteiza.

Carte manuscrite de tout le royaume de la Nouvelle-Espagne, from 16° to 40° latitude, by ▼ Don Antonio Forcada y la Plaza, 1787. This map appears to have been drawn with great care. Those familiar with these places think the same about Mr. Forcada's unpublished map of the *Audiencia de Guadalajara* [*Mapa geográfico del territorio que comprende la audiencia de Guadalajara*], 1790.

Plano en el que se representa la dirección de los dos caminos que bajan de México para Veracruz drawn by ▼ Don Diego García Conde, lieutenant colonel and superintendant of roads. This unpublished map, a series of triangles measured with a graphometer and a compass, is based on observations by Mr. *Costansó* together with Mr. García Conde. The work was done with the utmost care and above all represents in great detail the area around the Cordillera's slope from Jalapa and Orizaba to Veracruz.

Maps of the routes that run from Mexico City to Puebla, north and south of the Sierra Nevada [likely the same as *Carta geográfica que conprehende en su extensión mucha parte del Arzobispado de México y alguna de los Obispados de Puebla, Valladolid de Michoacan, Guadalajara y Durango*] drawn by Don Miguel de Costansó on the order of the viceroy, the ▼ Marquis de Branciforte.

I.94

[Corral] *Mapa de una Porcion de Costa del Seno Mexicano*: This map reaches as far as Perote and also shows the difference between the projected roads from Jalapa to Veracruz.

Map of the terrain between Verzcruz and the Río Xalmapa: 1796 [*Mapa chorografico que manifiesta el terreno comprehendido entre la presa del rio de Jamapa y la ciudad de Vera Cruz*].

Manuscript map of the province of Jalapa, with detailed surroundings of Antigua and Nueva Veracruz [Costanzó and García Conde. *Mapa general de los terrenos que se comprenden entre el río de La Antigua y la Barra de Alvarado, hasta la Sierra de Orizaba y Jalapa*].

Manuscipt map of the province of Oaxaca and of the entire coastline from Acapulco to Tehuantepec, drawn by Don Pedro de Laguna. This map is based on eleven positions whose latitudes are claimed to have been determined by direct observations. As for the course of the Río Huascualco, made famous by the projected canal that is to connect the South Sea with the Atlantic Ocean, I found it drawn on the maps of two officer engineers, ▼ *Don Augustin Cramer* and *Don Miguel del Corral*. These maps [Corral, *Descripcion Geografrica [sic]*] are preserved in the archives of the viceroyalty of Mexico.

Anonymous mapa of the Sierra Gorda in the province of Nuevo Santander from 21° to 29° latitude. [José de Escandón, 1747. *Mapa de la Sierra Gorda y Costa del Seno Mexicano desde la Cuidad de Querétaro, c.1747*]. This is a hand-drawn map painted on velum, decorated with primitive Indian figures. It is very exact about the environs of Sotto la Marina and of Camargo.

The course of the rivers between the Río del Norte and the mouth of the Río Sabine has been copied from an unpublished map that ▼ General Wilkinson generously shared with me in Washington, upon his return from Louisiana.

Mapa de la Nueva Galicia. Unpublished map drawn by ▼ Mr. Pagaza [urtundua] in 1794, based on his own observations and on Mr. Forcada's map.

Nuevo mapa geografica de las provincias de la Sonora y Nueva-Vizcaya de la America septentrional dedicated to Mr. de Asanza and drawn by the engineer *Don Juan de Pagazaurtundua* in Cádiz. This hand-drawn map is four feet long and shows in great detail the sites in the mountains where the Indian savages hide to make their forays and attack travelers. It also presents in great detail the environs of Paso del Norte, especially the uninhabited area called the *Bolsón de Mapimi* [Comarca Lagunera].

Mapa geográfico de la Provincia de Sonora from 27° to 36° latitude, dedicated to colonel ▼ *Don José Tienda de Cuervo*. The author of this map appears to have been a German Jesuit priest who lived in the *Pimería Alta,* the northernmost part of the province of Sonora.

I.95

Manuscript map of the Pimería Alta. The map extends as far as the Río Gila. The famous ruins of Casas Grandes are shown on the map at 36°20′ latitude, with an error of three degrees!

Mapa de la California, unpublished map by Fathers Francisco Garcés and Pedro Font, 1777. This map was engraved in Mexico City; it has an error of at least three minutes for all latitudes. It is of interest for its representation of the Pimería Alta and the Río Colorado.

Carta geográfica de la Costa occidental de la California que se descubrió en los años 1769 y 1775, por Don Francisco de Bodega y Quadra and Don José Canizares, desde los 17 hasta los 58 grados [Carta geographica que contiene la costa ocidental de la California]. This small map, engraved by ▼ Manuel Villavicencio in Mexico City in 1788, is drawn on the San Blas meridian. It will be of interest to those who study the history of the discoveries in the Great Ocean.[1]

The Gulf of Cortés appears in detail on the map of California that accompanies the *Noticia de la California, del* ▼ *Padre Fr. Miguel Venegas [Mapa de la California, su golfo y provincias fronteras en el continente de Nueva España]* 1757; but the actual positions of the missions that are now found on that peninsula are shown on the map that has been appended to the ▼ *Life of Father Fray Junípero Serra*, printed in Mexico City in 1787.

Manuscript map of the province of Nueva Vizcaya from 24° to 35° latitude drawn by the engineer Don Juan de Pagazaurtundua in 1792, based on data collected in Chihuahua. An interesting piece of work, done by order of ▼ Mr. de Nava, captain-general of the *Provincia internas*; I used it for the entire intendancy of Durango. The environs of the town of Durango appear to be less precise.

Mapa manuscrito de las fronteras septentrionales de la Nueva España by the engineer *Don Nicolás Lafora*. It develops the Marquis de Rubi's defense project and assisted me in verifying the position of the small forts called *Presidios*. I have seen a copy of this map, which is three meters long, in the archives of the viceroyalty.

Mapa del Nuevo México desde 29° hasta 42° de lat. This unpublished map [by Arrowsmith?] represents the regions below the forty-first parallel in great detail. It also contains details on *Lake* (there is some doubt here) *Timpanogos* [Lake Utah] and on the sources of the *Río Colorado* and the *Río del Norte*.

1. There are some interesting details on the *Map of New California, by Order of the Captain-general of the Internal Provinces*.

Map of New Spain engraved in 1795 by López [de Vargas Machuca, *Mapa Geográfico del Gobierno de la Nueva Granada ó Nuevo México con las provincias de Nabajo y Moqui*]. I made no use of it. It seems to be quite flawed concerning the sources of the Río del Norte. The regions between these sources and those of the Missouri are shown in better detail on the *Map of Louisiana published in Philadelphia in* 1803 [Arrowsmith, *Chart of the West Indies and Spanish Dominions in North America*?].[1]

Despite its considerable imperfections, I dare say that my general map of New Spain has two essential advantages over all other maps that have appeared to date: it presents the locations of 312 mines , as well as the re-   I.98
cent division of the country into intendancies. The mining developments are represented in accordance with the catalogue of those places ordered by the High Court of Mines throughout this vast empire. I have used special symbols to designate the seats of the *Deputaciones de Minas* and the mining sites associated with them. The catalogue that I was given usually indicated the rhumb and the distance from a larger town. I combined these notes with what was available on older unpublished maps, among which Velázquez's were most useful. This work was as meticulous as it was painstaking. When no map indicated the name of the mine, I simply located the mine in relation to the mineral deposits shown in the catalogue, by reducing the itinerary distances, or *local leagues*, to absolute distances, according to the convergences presented by analogous cases. Since the population of New Spain is concentrated on the large interior plateau of the central mountain range, the map of Mexico is very unevenly dotted with place names. We should not, however, assume that there are completely uninhabited locales wherever the map indicates neither village nor hamlet. I chose to locate only those places whose position was the same on *several* of the unpublished maps with which I worked. Most maps of America are produced in Europe and are full of the names of places whose existence is unknown in   I.99
the country itself. These errors persist, and their sources are often difficult to surmise. On my own map, I prefer to leave ample blank space rather than getting carried to hazardous guesses.

1. The *Map of the Internal Provinces of New Spain from the Sketches of Mr. Pike* [1807] is today regarded as a mere copy of Mr. von Humboldt's map, with a few modifications in the sources of the Arkansas River (Tanner, *American Atlas*, 1823, p. 9). Like the map that goes with Taylor's *Selections* [*from the Works of the Baron de Humboldt*], the map that accompanies the work published under the modest title of *Notes on Mexico Made in the Autumn of 1822, by a Citizen of the United States* [Joel Roberts Poinsett] is based on Mr. von Humboldt's map. However, it also includes the roads from Tampico to San Zimapán.

Only those who have attempted to draw geographic maps will be able to relate to the great difficulties I encountered in indicating mountain ranges. I was obliged to give preference to the use of hatching in orthographic projection over showing mountains in profile. The latter method, which is both the most inaccurate and the oldest, produces a mixture of two types of very heterogeneous projections. I cannot conceal, however, that a real advantage almost compensates for this inconvenience. The earlier method offered signs that simply announced "that the terrain was mountainous, and that there were mountains in such and such province." The vaguer this hieroglyphic language, the fewer errors it exposes. On the contrary, the hatching method forces the mapmaker to tell more than he knows about the terrain, even more than it is possible to know about the geological constitution of a vast area of land. Considering the most recently published maps of Asia Minor and Persia, one would think that the learned geologists had identified the relative elevations, as well as the confines and orientation of the mountains. In that part of the world, we find snake-like mountains chains that interweave like streams; the Alps or the Pyrenees would seem less familiar to us than those faraway regions. Nonetheless, educated persons who have traveled throughout Persia and Asia Minor claim that the way mountains cluster together there is entirely different from the type of mountainous formations shown on Arrowsmith's large map of Asia, which has been profusely copied in France and Germany.

I.100

The shape of a country is largely determined by water; but the courses of rivers only show that different levels are present in the terrain through which their waters flow. Knowledge of broad valleys or basins and the study of dividing ridges are of great interest to the hydrographic engineer. It is, however, a false application of the principles of hydrography for geographers to attempt to determine, from the comfort of their studies, the direction of mountain ranges in countries where they assume they have precise knowledge of the course of the rivers. They imagine that two large basins of water can only be separated by high elevations, or that a sizable river can only change its course whenever a group of mountains stands in its way. But they forget that—either because of the type of rock or because of the gradient of the strata—rivers are rarely born on the highest plateaus, while the mouths of the greatest rivers are located far from high mountain ranges. Attempts that have heretofore been made to draw physical maps based on theoretical assumptions have not, therefore, been very felicitous. The complete disappearance of deep-sea currents and the majority of rivers that have altered the surface of the earth make it all the more difficult to *guess*

I.101

the true configuration of a given terrain. The most accurate knowledge of rivers that existed in the past and of those that exist today might teach us something about the *slope of valleys*, but it would be of no use to us in regard to either the absolute elevation of mountains or the positions of their ranges.

On my Map of New Spain, I have traced the direction of the Cordilleras not according to vague assumptions or hypothetical combinations but, rather, on the basis of a great deal of information from individuals who have visited the Mexican mines. The highest group of mountains is on the outskirts of the capital city, at 19° latitude. I myself have climbed that part of the Cordilleras of Anahuac, between the parallels of 16°50′ and 21°0′, an expanse of over 140 leagues. It was in that region that I made most of my barometric and trigonometric measurements, whose results contributed to the geological profiles in my Mexican atlas. Mr. Velázquez's unpublished maps, as well as those by Mr. Costansó and Mr. Pagaza[urtundua] helped me greatly with the northern provinces. Mr. Velázquez, the director of the *Tribunal de Minería*, has traversed most of New Spain. On the map to which I refer above, on page 91, he traced the two branches of the *Sierra Madre de Anáhuac*: the *eastern branch* that extends from Zimapán toward Charcas and Monterrey in the Kingdom of León, and the *western branch* that extends from Bolaños to the Presidio de Fronteras. Unpublished narratives by ▼ Mr. Sonneschmidt, a learned Saxon mineralogist who visited the mines at Guanajuato, Zacatecas, Chihuahua, and Catorce, the work of Mr. del Río (a professor at the mining school in Mexico City), and the work of ▼ Don Vicente Valencia, a resident of Zacatecas, also provided me with very useful insights. I owe other clarifications to the gracious advice I received from the celebrated chemist and director of mining, Mr. Don Fausto Delhuyar in Mexico City; ▼ Mr. Chovell in Villalpando; Mr. Abad y Queipo in Valladolid; Mr. Anza in Tasco; Colonel Obregón in Catorce; and many rich mine owners and missionary monks who took an interest in my work. Despite the care I took to learn about the direction of the Cordilleras while I was on site, I consider that aspect of my work to be far from perfect. Having spent the past twenty years wandering around mountains and collecting material for a geographic atlas, I know the great risk involved in attempting to draw mountains on a terrain that extends for 118,000 square leagues!

I had hoped to be able to draw two maps of New Spain on a large scale—one physical and the other completely geographical; but I feared that this would make the Mexican atlas too bulky. The hatching, which indicates the lines of the steepest slope and the shifts in the terrain, also overburdens

I.102

I.103

a map that is already laden with several place names. These names often become illegible when the engraver uses chiaroscuro to heighten the map's visual effect. These considerations suggest that the geographer who has carefully analyzed the astronomical positions[1] of the sites is unsure of what his preference should be: should he preserve the sharpness of line and *letter*, or should he illuminate the relative elevation of the mountains? One of the most beautiful maps published in France, produced by the War Office in 1804, aptly demonstrates how difficult it is to reconcile these two opposing interests: the geologist's interest and the astronomer's. Fearing that the present work would become too long, the difficulties of publishing an atlas without the financial participation from any government forced me to abandon my original project, which was to append a horizontally projected physical map to each segment of terrain.[2]

1. In my eighth chapter, I discuss the extraordinary regularity of the positions of the Mexican volcanoes. I am somewhat unsure of the latitude of the *Colima* volcano and the longitude of the Cerro *Tancitaro*, which was taken twice from a distance. I fear that an error was introduced while my angles were being copied, but I am certain of the latitude of the peak of *Tancitaro* to within eight minutes of arc.

2. There are several more or less incomplete copies of Mr. von Humboldt's large map of New Spain. We cite here only those by Mr. Arrowsmith, Major Pike, and ▼ Mr. A. F. Tardieu, Sr. (*Map of Louisiana and Mexico*, 1820). ▼ Here is our traveler's opinion (*Relation historique*, vol. I; *Introduction*, p. 21) of the first two of these copies: "My *Carte générale du Royaume de la Nouvelle-Espagne, dressée sur des observations astronomiques et sur l'ensemble des matériaux qui existaient à Mexico en* 1804, was copied by Mr. Arrowsmith, who appropriated and published it in a larger scale in 1805 (before the English translation of my work by *Longman, Hurst,* and *Orme* appeared in London) under the title *New Map of Mexico, compiled from original documents by Arrowsmith*. This map is easily recognizable for its many chalcographic errors, the explanation of symbols that were not translated from French to English, and the word *Ocean*, which is inserted in the middle of mountains, in a place where, on the original, is written "*The height of the Toluca plateau is 1,400 toises above the level of the Ocean*." Mr. Arrowsmith's approach is even more blameworthy in that ▼ Mr. Dalrymple, Mr. Rennell, Mr. d'Arcy de La Rochette, and many other excellent British geographers offer no precedent for what he has done either on their own maps or in their companion analyses. A traveler's complaints must be valid when mere copies of his work are distributed under others' names." In his *New American Atlas*, Mr. Tanner also benefited from Mr. von Humboldt's work. Not content simply to reveal his sources, however, Mr. Tanner pays a stirring homage to the candid and truthful character of the *Policial Essay*: "That portion of the map of Mexico for which Mr. von Humboldt is solely responsible," according to Mr. Tanner (*New American Atlas*, 1823, p. 6), "bears the mark of precision that twenty years of testing have not refuted. This map continues to be what it was when first published: the standard for any new map of Mexico, that is, until the entire country is subjected to actual geodesic surveys."

II

Map of New Spain and Its Bordering Countries to the North and to the East

I have explained above the factors that led me to restrict my large Map of New Spain to an area too narrow to cover the kingdom in its entirety, from New California to the intendancy of Mérida, on the same plate. The second map in the Mexican atlas is intended to remedy this flaw. It shows not only all the provinces governed by the viceroy of Mexico and the two commanders of the *Provincias internas*, but also the island of Cuba, whose capital may be considered the military port of New Spain, Louisiana, and the Atlantic portion of the United States. ▼ Mr. Poirson, an able Parisian engineer, drew up the map from material with which Mr. Oltmanns and I provided him. It covers the vast area between 15° and 42° latitude and 75° and 130° longitude. I had originally planned to extend this map as far south as the mouth of the Río San Juan, in order to indicate the various canals that the court of Madrid has proposed to build and that would serve to establish a channel of communication between the two oceans. I shall discuss this subject in the second chapter of this work. But, while carrying out this project, having noticed that if I used a smaller scale, the Yucatán peninsula and the coastline of Monterrey would not be represented in as much detail as they seemed to deserve, I chose to keep a larger scale and to extend my map southward only as far as the Gulf of Honduras.

The main section of the map, which includes the Kingdom of New Spain, is a faithful copy of my large map, which I have just analyzed. The addition of the Yucatán is based on the map of the Gulf of Mexico published by the *Depósito hidrográfico* in Madrid. New California was drawn from the atlas that accompanied the logbook of the corvettes *Sútil* and *Mexicana* and from Mr. Espinosa's paper, published in 1806 and entitled *Memoria sobre las observaciones astronómicas que han servido de fundamento a las cartas de la costa N.-O. de América, publicadas por la Dirección de los trabajos hidrográficos*. Whenever this account presented results that differed from those found in the *Relación del viaje al Estrecho de Fuca*, the former were chosen, on the assumption that they were based on a more solid foundation.[1] Mr. Espinosa's work was also useful for the small group of islands that ▼ Mr. Collnett called the Archipelago of Revillagigedo, in honor of the Mexican viceroy who reorganized financial affairs and ordered important statistical reports.

I.107

The islands of San Benedicto, Socorro, Roca Partida, and Santa Rosa, located between 18° and 20° latitude, were discovered by Spanish navigators at the beginning of the sixteenth century. Hernando de Grijalva discovered the island of Santo Tomás in 1533; it is now called the island of Socorro. In 1542, ▼ Ruy López de Villalobos landed on a small island that he called Nublada, clearly marking its distance from Santo Tomás. Villalobos's Nublada is now called San Benedicto. It is less certain that what that navigator called the Roca Partida corresponds to modern hydrographers' Santa Rosa because there is great confusion about the position of this reef. Juan Gaetan[2] even places it two hundred leagues west of the island of Santo Tomás. On Domingo de Castillo's map from 1541, this island appears at 19°45′ latitude as a shoal thirty-six miles long; this map was long buried in the archives of the Cortés family in Mexico City. In recent years, the Revillagigedo Islands group has only been sighted three times: by the pilot ▼ Don José Camacho in 1779, while sailing from San Blas to New California; by the ▼ ship's captain Don Alonzo de Torres in 1792, during a voyage from Acapulco to San Blas; and finally

I.108

1. I put Monterrey at 36°35′45″ latitude and 124°12′23″ longitude and Cabo San Lucas at 22°52′33″ lat. and 112°14′30″ long. The longitude of Monterrey, which I fixed definitively with Mr. Espinosa while compiling my map of Mexico, differs less from Mr. Vancouver's numbers than the result that Mr. Antillón has published. There is a difference of eighteen minutes in arc between the Spanish and British navigators. See above, p. 70 [p. 48 in this edition]. (It is important to keep in mind here that the beginning of this geographical Introduction from p. 1 to p. 97 was written in Berlin during the month of September 1807, and that all that followed was published in the spring of 1806 [1809]. Additions that appear in this second edition of the *Political Essay* in 1825 will be easily recognized.)

2. [Giovanni Battista] Ramusio, [*Delle navigationi et viaggi,*] vol. 1, p. 375 (Venice edition, 1613).

by Mr. Collnett[1] in 1793. These three navigators' observations diverge sharply. It seems, however, that Mr. Collnett has more or less correctly fixed the position of the island of Socorro, by taking several series of distances from the moon to the sun. The location of the entire group of small islands was determined according to these distances, which were calculated using Mason's tables.

On this map, I drew the *Rocky Mountains* as far as the forty-second parallel from material that was available in 1804. The agreement between Major Long's astronomical observations and the longitude I assigned to Santa Fe and Taos is such that, even today (in 1825), I would not modify the longitude of the eastern part of these mountains. The three peaks are located at:

Spanish Peak	37°20′ lat.	106°55′ long.
James Peak (believed to have a height of 1,798 toises)	38°38′ lat.	107°52′ long.
Big Horn or Major Long's Highest Peak or Tanner's Long Peak:	40°13′ lat.	108°30′ long.

I.109

Given the current state of geographical knowledge about these regions, I think that if one were to continue the Mexican Cordilleras northward, one would have to put their eastern edge at

38° latitude by	107°20′ longitude
40°	108°30′
45°	113°0′
63°	124°40′
68°	130°30′

I am setting the longitude of the northernmost extremity of the Andes range in the Rocky Mountains, in accordance with the corrections that ▼ Captain Franklin has recently made to Mr. Mackenzie's map. The

---

1. Collnett, *Voyage to the South [Atlantic]*, p. 107. Mr. Collnett found Cape San Lucas at 22°45′ latitude and 112°20′15″ longitude. This latitude seems incorrect by nearly seven minutes. Mount San Lázaro, which Mr. Collnett fixed at a position of 25°15′ latitude and 114°40′15″ [longitude] (pp. 92 and 94) is certainly not the same mountain that Ulloa called Cape San Abad in 1539. I (with Mr. Espinosa) put that peak at 24°47′ latitude and 114°45′30″ longitude.

errors at 67° and 69° latitude appear to be from 4° to 6° longitude, but at the parallel of the Slave Lake, they are virtually non-existent (the mouth of the Mackenzie River is at 128°, according to Franklin, 135° according to Mackenzie; the mouth of the Copper Mine River is at 115°35′, according to Franklin, at 111°, according to ▼ Mackenzie and Hearne; the mouth of the Slave River in the eponymous lake is at 112°45′, according to Franklin and at 113° west of Greenwich, according to Mackenzie). These data suggest that (1) the Rocky Mountains are located at 60° and 65° latitude by 127° and 128° longitude west of the Paris meridian; (2) the northern extremity of the mountain chain, to the west of the mouth of the Mackenzie River, is I.110 at 130°20′ longitude; (3) the Copper Mountains group is at 118° and 119° longitude and at 67° and 68° latitude. Mr. Tanner's fine maps are still affected by the previous error of 6° to 7° at the mouth of the Mackenzie River. It seems to me that that geographer puts the Rocky Mountains too far west by 2° to 3° at 60° to 65° latitude, and by 0°30′ at 55° latitude, although he fears "having placed the mountains 3° farther east than they are generally placed at the forty-fifth and forty-eighth parallels."

One must admit, moreover, that all of the longitudes given for the central mountain chain north of 50° are highly uncertain. Since there have been no observations of lunar distances in these regions, one can rely only on the positions of Council Bluffs and the sources of the Arkansas River, which are 10° to 12° farther south.

For the countries that border on New Spain, the engineer ▼ Lafond's fine map of Louisiana has been used, as has (for the United States) Arrowsmith's map, corrected by ▼ Rittenhouse, Ferrer, and Ellicott's observations. Mr. Oltmanns has discussed the positions of New York and Lancaster in his scholarly paper, included in the second volume of my *Recueil d'observations astronomiques* [Collection of Astronomical Observations], p. 92. This work contains the materials used to draw the map of Cuba. It is unnecessary to go into great detail here about an area that is only incidental to this map.

I.111 Several points in the interior of the island of Cuba and on its southern coast, between the ports of Batabanó and Trinidad, were fixed by the astronomical observations I made in 1801 in the *King's Gardens* before my departure for Cartagena de Indias.

On the *Carte du Mexique et des pays limitrophes* [Map 3] the following points are based on the astronomical observations I made while sailing from Cumaná to Havana, crossing the Vibora shoals, and also while navigating from Batabanó to the Gulf of Darién.

PLACE NAME	LONGITUDE						LATITUDE		
	in hours			in arc					
	h	′	″	°	′	″	°	′	″
Havana, the *Morro*	5	38	52.5	84	43	8	23	9	27
Trinidad, Cuba	5	29	24.5	82	21	7	21	48	20
Cabo San Antonio, NW	5	49	9.5	87	17	22	21	55	0
Punta de Mata-Hambre	5	38	31.0	84	37	45	"	"	"
Boca de Jagua	5	31	37.5	82	54	22	"	"	"
Cayo Flamingo	5	36	14.1	84	3	32	22	0	0
Cayo de Piedras	5	34	28.8	83	37	12	21	56	40
Cayman Grande, E. point	5	31	56.3	82	59	4	19	19	0
Cayman Brac, E. point	5	28	30.5	82	7	37	19	40	0
Cape Portland	5	17	14.3	79	18	35	"	"	"
Las Ranas	5	13	34.4	78	23	35	17	28	0
Little-known reefs on the shores of Vibora	5	22	55.4	80	43	49	16	50	0

These positions are discussed in the *Recueil d'observations astronomiques* that I published together with Mr. Oltmanns, vol. II, pp. 7, 11, 13, 56, 66, 68, 109, and 112. According to Mr. de Puységur, Cape Morant, located at 17°57′45″ latitude and 78°35′23″ longitude, was set 5° in arc farther east by Mr. Poirson. This more easterly position is documented by several Spanish maps.

I.112

As for the position of the city of Washington, we have not chosen to adopt the longitude assigned to it by the *Connaissance des temps* for 1812, namely, 78°57′30″, or half a degree too far east. If this position were correct, the geographers of the United States would be hard-pressed to determine the location of Baltimore or Cape Hatteras. The occultation of Aldebaran of January 21, 1793, observed in Washington, had been calculated by Lalande, who, in fact, deduced a longitude of 5ʰ15′51″ from it; but ▼ Mr. Wurm[1] redid the calculation and found 5ʰ17′16″ or 79°19′0″.

1. [Wurm in] Zach, ["Beytrag zu geographischen Längenbestimmungen,"] *Monatliche Correspondenz*, [vol. 7,] Nov. 1803, p. 382. ▼ Mr. William Lambert decided on 79°15′45″ longitude, perhaps too far east, since Ferrer finds Georgetown at 70°25′7″.

This latter result is in agreement with Mr. Ellicot's observation of a solar eclipse in Georgetown just west of Washington in 1791 that gave $5^h17'40''$ or $79°25'9''$. Although for the eastern part of the United States, one has generally used Arrowsmith's map; a few minor changes were made based on ▼ Mr. Ebeling's research and on the information that ▼ Mr. de Volney gathered during his expedition west of the Allegheny Mountains.

I.113   The northwest coast of North America, from Cabo San Lucas to Cabo San Sebastian, was drawn according to Mr. Oltmanns's learned research in his work on the Geography of the New Continent.[1] The findings of both Vancouver and Alessandro Malaspina have been taken into consideration: the longitude of the island of Guadalupe seems somewhat dubious. An *ukase* proclaimed in 1799, during the reign of emperor Paul I, declared that the entire coast north of the fifty-fifth parallel belonged to the Russian government. In this *ukase*, the northwest coast is consistently referred to as the *northeast coast of America*, an extraordinary title that is justified circumstantially, "since from Kamchatka, one must sail east to find America" (▼ Storch, *Russland*, Vol. I, pp. 145, 163, 265, and 297).

    Although Mr. Oltmanns's results, in the large table of positions placed at the beginning of my *Recueil d'observations astronomiques*, do not significantly differ from those I have presented above on pages eighty-five to ninety, it would nonetheless be useful to record here the longitudes on the

I.114 western coastline that have been corrected by eight points.

PLACE NAMES	LONGITUDE		
	°	′	″
Acapulco	102	9	33
San Blas	107	13	48
San José	112	1	8
Cabo San Lucas	112	10	38
Cabo Mendocino	126	49	30
Punta del Año Nuevo	124	43	53
Monterey	124	11	21
Nootka	128	57	1

1. [Alexander von Humboldt and Jabbo] Oltmanns, *Untersuchungen über die Geographie des Neuen Continents* (Paris: F. Schoell), part II, p. 407. *Recueil d'observations astronomiques*, vol. II, pp. 592–619.

III

Map of the Valley of Mexico City, Formerly Known as Tenochtitlan

Few regions spark such a wide range of interests as does the valley of Tenochtitlan, the site of an early civilization of American peoples. Not only Mexico City but also, in particular, the valley's older monuments, the pyramids of Teotihuacán, which were dedicated to the sun and the moon, are steeped in memories. They are described in the third book of this work. Students of the history of the conquest will enjoy seeking out on my map both Cortés's military positions and those of the Tlaxcalteca army. The geographer will reflect with interest on the prodigious elevation of the Mexican terrain, its rivers that do not empty into the sea, and the extraordinary shape of a chain of trachytic and basaltic mountains surrounding the valley like a circular wall. He recognizes that this entire valley is the bottom of a dried-up lake. To his eyes, the basins of fresh and salt water that fill the middle of the plateau and the five lagoons—Sumpango, San Cristóbal, Texcoco, Xochimilco, and Chalco—are merely the remains of a large body of water that once covered the entire valley of Tenochtitlan. To engineers and hydraulic architects, the projects undertaken to protect the capital city from the dangers of flooding are, if not models to be emulated, then at the very least subjects worthy of serious study.[1]

I.115

---

1. See below, in my *Statistical Analysis*, the studies on the position of the predecessor of Mexico City, the pyramids of Teotihuacán, the position of the lakes, the artificial canal (*Desagüe*), through which the waters of the valley flow toward the Gulf of Mexico, and the two plateaus of Cholula and Toluca, a part of which is included in my Map of the Valley of Tenochtitlán.

Despite the threefold connection of this area to history, geology, and hydraulic architecture, no map exists from which one might get a sense of the valley's true shape. The sole basis for both the map of the environs of Mexico City that ▼ López [de Vargas Machuca] published in Madrid in 1785 and the map that accompanies the *Guía de Forasteros de México* is Sigüenza's old map, drawn up in the seventeenth century. These sketches are certainly not worthy of being called topographical maps, since they represent neither the current location of the capital nor the state of its lakes in Moctezuma's time.

I.116

Sigüenza's map, which measures only twenty-one by sixteen centimeters, is titled *Mapa de las aguas que per [sic] el círculo de noventa leguas vienen a la laguna de Tezcuco [Texcoco], delineado por Don Carlos de Sigüenza y Góngora, reimpreso en México con algunas adiciones, en 1786, por Don José Alzate.* The scale of latitudes and longitudes that Mr. Alzate appended to Sigüenza's map has structural flaws of over three minutes in arc. The absolute longitude of the capital—which the learned Mexican claims to be the result of twenty-one observations of Jupiter's satellites and to have been *approved* and *verified* by the Academy of Sciences in Paris—is off by one degree. Mr. Alzate's map has been slavishly copied by all those geographers who have attempted to publish maps of the Valley of Mexico. In terms of direct distances, it gives:

a) 1°1′ in equatorial arc from the summit of the Popocatepetl volcano to the village of Tisayuca, located at the northern end of the valley. (The actual distance is 0°53′.)

b) 0°32′ from the center of Mexico City to Huehuetoca, where the canal that drains the lakes begins (actual distance 0°23′).

c) 0°20′ from Mexico City to Chiconautla (actual distance 0°15′).

d) 0°32′ from the rock (Peñol) of Los Baños to Sumpango (actual distance 0°21′).

I.117

e) 0°13′ from the Peñol de los Baños to San Cristóbal (actual distance 0°8′).

f) 0°29′ from the village of Tehuiloyuca to Texcoco (actual distance 0°21′).

Here are errors of 16,000 up to 20,000 meters for distances that Mr. Velázquez had measured with great accuracy during a geodesic survey in 1773, and for which there is perhaps not even a deviation of one hundred meters. Nonetheless, Mr. Alzate had access to Velázquez's triangulation, and he might have used it as we did— ▼ Don Luis Martín, Mr. Oltmanns, and me—in compiling the map in the Mexican atlas. I did not make any astronomical observations in Pachuca, but I did do so in Real de Morán, whose latitude is greater than that of Pachuca. I found Morán to be at 20°10′4″ latitude; yet, Mr. Alzate puts Pachuca at 20°14′. On his map, the old town of Tula is placed nearly a quarter of a degree too far north.

Mascaró's map from the *Guía de México (Mapa de las cercanías de México)* is only fourteen by ten centimeters. It is thus twelve times smaller than the map appended to the present work. It can be considered a copy of Sigüenza and Alzate's map, although its representation of the northern part of the valley is somewhat narrow. According to Father Alzate, the summit of the volcano of Popocatepetl is separated from Huehuetoca by a distance of 1°14′; according to Mascaró, this distance is 1°11′. But the actual distance is 1°1′. One obtains this distance by connecting Huehuetoca with the rock of   I.118 Los Baños, using Velázquez's triangles and then connecting this rock with the volcano of Popocatepetl and the pyramid of Cholula using my own astronomical observations and several azimuths.

There are certain maps, according to which the waters of the lakes near Mexico City do not flow northeastward toward the Gulf of Mexico, as they actually do, but, rather, northwest toward the South Sea. This error is also found, along with many others, on the map of North America published in London by ▼ Mr. Bowen, geographer to the King.

When I arrived in Mexico City in the spring of 1800, I had thought of drawing up a map of the valley of Tenochtitlan. Through astronomical observations, I had proposed to demarcate the boundaries of this long oval-shaped valley. I had also taken several positional angles from on top of the tower of the cathedral in the capital, from on top of the hills of Chapultepec and the Peñol de los Baños, from the Venta de Chalco, and from the summit of the Chicle Mountain, from Huehuetoca, and from Tisayuca. The position of the two volcanoes of La Puebla and the Peak of Ajusco had been determined using a special *hypsometric method*, namely through angles of elevation and azimuths. Since I had only a short time to devote to this work, I could not hope to assemble on my map all of the small Indian villages that dot the lake shores. My main goal was to determine the general shape of the   I.119 valley carefully and to draw up a physical map of the region, where I had just measured several elevations with the help of a barometer.

Circumstances enabled me to publish a topographical map from more precise material. ▼ Don José María Fagoaga, a respectable person who (by a rare conjunction in any country) possesses both a large fortune and a love of science [knowledge] and the common good, sought to leave me with a precious keepsake of his country by giving me a sketch of a map of the valley upon my departure from Mexico City. At Mr. de Fagoaga's invitation, one of my friends, Don Luis Martín, a mineralogist and able engineer, drew up a map based on geodesic operations that had been carried out at different periods between Mexico City and the village of Huehuetoca, during the digging of the Texcoco, San Cristóbal, and Sumpango canals. Mr. Martín used some of my measurements, checking them all against the astronomical observations that I had made at the extremities of the valley. Numerous geological excursions that he made on the outskirts of the capital and at the Volcanoes of La Puebla enabled him to render faithfully the shape and the relative height of the mountains that separate the plateau of Mexico City from the plateaus of Toluca, Tula, Puebla, and Cuernavaca.

I.120   The map that I owe to Mr. de Fagoaga's friendship is not, however, the one included in my Mexican atlas. When I examined this map closely and compared it both to Mr. Velázquez's triangulation (of which I have a detailed original manuscript) and to the table of astronomical positions fixed by my own observations, I observed that the eastern shore of Lake Texcoco and the entire northern part of the valley required significant alterations. Mr. Martín himself had recognized the inaccuracy of his first attempt, and I thought it would please him to know that I enlisted Mr. Oltmanns to supervise the redrawing of the map of the valley on the basis of the combined materials that I had gathered. Each point was discussed separately, and averages were adopted whenever several incompatible measurements were obtained.

Here is the concatenation of triangles that Mr. Velázquez measured in 1773, from the bathing rock (the *Peñol de los Baños*) near Mexico City to the Cerro del Sincoque north of Huehuetoca. The angles were measured with an excellent English theodolite measuring ten inches in diameter and
I.121   equipped with two twenty-eight-inch telescopes.

NUMBER OF TRIANGLES		NAMES OF STATIONS	OBSERVED ANGLES		DISTANCES CONVERTED (INTO MEXICAN VARAS, AT 2.32258 PER TOISE)	
I.	A.	Garita de Guadalupe .	57°	42′	from A to B	4,474
	B.	Garita de Peralvillo	84°	57′	from B to C	6,233
	C.	Cumbre del Peñol	37°	21′	from A to C	7,346

NUMBER OF TRIANGLES		NAMES OF STATIONS	OBSERVED ANGLES		DISTANCES CONVERTED (INTO MEXICAN VARAS, AT 2.32258 PER TOISE)	
II.	A.	Garita del Peralvillo	81°	27′	from A to C	4,806
	B.	Cumbre del Peñol	40°	44′	from B to C	7,283
	C.	San Miguel de Guadalupe	57°	49′		
III.	A.	San Miguel de Guadalupe	62°	25′	from A to C	29,136
	B.	Cumbre del Peñol	103°	31′	from B to C	26,560
	C.	Texcuco	14°	4′		
IV.	A.	Cumbre del Peñol	61°	35′	from A to C	20,229
	B.	Texcuco	46°	25′	from B to C	24,562
	C.	Cruzes del Cerro de San Cristóbal.	72°	0′		
V.	A.	Texcuco	35°	1′	from A to C	20,694
	B.	Cruzes del Cerro de San Cristóbal	57°	19′	from B to C	14,100
	C.	Crestón de Chiconautla	87°	40′		
VI.	A.	Crestón de Chiconautla	76°	35′	from A to C	14,631
	B.	Crestón de Chiconautla	53°	3′	from B to C	17,809
	C.	Xaltocan	50°	22′		
VII.	A.	Crestón de Chiconautla	59°	47′	from A to C	19,677
	B.	Cruzes del Cerro de S. Cristóbal	76°	8′	from B to C	17,513
	C.	Hacienda de Santa Inés	44°	5′		
VIII.	A.	Cruzes del Cerro de San Cristóbal	23°	5′	from A to C	17,800
	B.	Hacienda de Santa Inés	80°	46′	from B to C	7,072
	C.	Xaltocan	76°	9′		
IX.	A.	Xaltocan	65°	19′	from A to C	11,738
	B.	Hacienda de Santa Iñes	71°	30′	from B to C	10,884
	C.	Sumpango	36°	11′		
X.	A.	Sumpango	49°	34′	from A to C	12,718
	B.	Hacienda de Santa Iñes	74°	46′	from B to C	10,033
	C.	Tehuiloyuca	33°	40′		

(continued)

NUMBER OF TRIANGLES		NAMES OF STATIONS	OBSERVED ANGLES		DISTANCES CONVERTED (INTO MEXICAN VARAS, AT 2.32258 PER TOISE)	
XI.	A.	Sumpango	57°	12'	from A to C	20,927
	B.	Tehuiloyuca	83°	30'	from B to C	17,647
	C.	Sincoque (Cerro de)	37°	17'		
XII.	A.	Tehuiloyuca	24°	30'	from A to C	10,783
	B.	Sincoque	29°	43'	from B to C	9,020
	C.	Hacienda de Jalpa	125°	47'		
XIII.	A.	Hacienda de Jalpa	32°	19'	from A to C	12,288
	B.	Sincoque	101°	44'	from B to C	6,709
	C.	Loma del Potrero	47°	57'		
XIV.	A.	Loma del Potrero	113°	30'	from A to C	8,672
	B.	Sincoque	37°	50'	from B to C	
	C.	Puente del Salto	28°	20'		

I.122   Mr. Velázquez measured two bases: one of 3,702 and one-half Mexican varas, on the plain separating the village of San Cristóbal and the Loma de Chiconautla, which is often flooded; the other base of 4,474 varas, on the causeway that leads from the capital to the sanctuary of San Miguel de la Guadalupe. The second base was even measured twice. Successively resolving the series of triangles according to these values, one finds the direct distance between the cross on Cerro San Cristóbal and the crest (*crestón*) of the Loma de Chiconautla. One of the bases gives this distance at 14,099 varas; the other at 14,101. The third triangle and the three last ones each have an obtuse angle; but for these same triangles, even an error of one minute in the most acute angle would only produce a difference of three or four varas in the length of the sides. In other words, this is a highly valuable calculation for discerning the topography of the Valley of Tenochtitlan.

Specific marks on my map indicate the positions that are based on Mr. Velázquez's triangulation and those that I determined astronomically. The results of my barometric measurements, calculated according to ▼ Mr. Ramond's coefficient, were added. To make the map more accessible to students of the history of the conquest, I have placed the former

Mexican names next to those used today. I have tried to be very exact in transcribing Aztec spellings, taking my lead only from Mexican authors and not relying on the work of ▼ Solís, Robertson, or Raynal and [de] Pauw, who mangle the names of the cities, provinces, and kings of Anahuac in the most bizarre ways.

I.123

IV

Map of the Projected Points of Communication between the Atlantic Ocean and the South Sea

This map was drawn up to show the reader in a single tableau nine points that could provide means of communication between the two Oceans. It serves to clarify the information given in the second chapter of the first book. In a set of nine sketches, I have represented the shared points between the Ounigigah and the Tacoutche-Tesse rivers, and between the Río Colorado and the Río del Norte; the isthmuses of Tehuantepec, Nicaragua, Panama, and Cúpica; the Guallaga [Huallaga] River and the Gulf of St. George, and, finally, the ravine from the Raspadura to the Chocó, which ships have used since 1788 to sail from the Pacific Ocean to the Antillean Sea. The most interesting sketches are the ones of the small by-wash channel of the Raspadura and the Isthmus of Tehuantepec. I have drawn the course of the Huasacualco (Guasacualco) and the Chimalapa Rivers from material I found in the archives of the viceroyalty of Mexico, especially the plans drawn by the engineers Don Miguel de Corral and Don Agustin Cramer whom the viceroy Revillagigedo had sent to those sites. The distances have been corrected through itineraries drawn up very recently, when indigo from Guatemala began passing through the forest of Tarifa, in other words, via a new road opened to the Veracruz trade.

On the *Map of the Dividing Range*,[1] the isthmus of Panama was largely drawn according to Mr. Fidalgo, ▼ Mr. Noguera, and Mr. Ciscar's astronomical and trigonometric calculations. I refer my readers to the fine Map published by the office of longitudes in Madrid, the *Carta esférica del Mar*

1. *Mexican Atlas*, plate IV, No. VII, *Introduction*, p. lx.

de las Antillas y de las costas de Tierra firme, desde la isla de la Trinidad hasta el golfo de Honduras, 1805 [by Miguel Moreno et al.]. According to the studies conducted by Fidalgo's expedition, Mandinga Bay reaches southward to 9°9′ north latitude, and the town of Panama lies seven minutes (in arc) east of the town of Portobelo. Don Jorge Juan concluded from the measurements that he took at the Chagre River that Panama was located thirty-one minutes in arc west of Portobelo.[1] According to the Depósito's 1805 map, the isthmus would measure only fifteen minutes (in arc) south of Mandinga Bay, or 14,258 toises in width, whereas, according to ▼ La Cruz's map, its width is fifty-five minutes in arc, or 52,277 toises. Despite the high degree of confidence that Mr. Fidalgo's measurements of the coastline merit, one should bear in mind that his operations cover only the *northern* coastline, and that, at present, this portion of the coast has only been connected to the southern coastline by a concatenation of triangles or by the movement of time. It is only by these means or by using several corresponding observations of satellites and the occultations of stars that one may solve the significant problem of the longitudinal difference between Panama and Portobelo. I call this problem significant because the longitude of Panama affects that of the mouth of the Río Chepo and, as a result, the position of the part of the Gulf of Panama that corresponds to the meridian of Punta San Blas and the fort of San Raphaël in Mandinga. When one looks at the shape of the northern and southern coastlines, one can easily see that, although their average direction is more or less from east to west, the width of the isthmus is not merely a function of their latitudes. According to the map of the *Depósito hidrográfico* of 1817 (*Cuarta Hoja* de la *Prov[incia] de Cartagena* [Moreno, *Carta Esférica en quatro hojas de las Costas de Tierra Firme*]) the *minimum* width of the isthmus of Panama is once again twenty-five minutes (in arc) or 23,775 toises, almost eight nautical leagues. The depth of Mandinga Bay is not given as 9°9′, but rather as 9°28′ latitude.

What is the height of the mountains at the point where the isthmus is the narrowest? How wide is the isthmus at the point where the mountain range is at its lowest? These are the two main questions that an enlightened government must attempt to answer, using experienced observers, who need only be equipped with a sextant, two chronometers, and a barometer. No measurement of elevation, nor any leveling of the terrain, has ever been done in the isthmus of Panama: neither the Simancas archives nor those of the Council of the Indies contain any important document that might

I.125

I.126

1. [Juan and Ulloa,] *Voyage [historique] dans l'Amérique méridionale*, vol. I, p. 99.

shed light on the possibility of constructing canals between the two seas. It would be wrong to accuse the ministry in Madrid of withholding things with which it was less well acquainted than were geographers in London or Paris.

On the small Map of Chocó[1] that shows the canal that was dug, by order of the priest of Novita, through an area called *Bocachica*, I have indicated as uncertain the direction of the coastline that stretches from Punta San Francisco Solano to the Gulf of San Miguel. I hope that we shall soon ascertain a more precise position for Cúpica (or Cupique), where ▼ Mr. Goyeneche, the Spanish helmsman, made his camp.

1. *Mexican Atlas*, plate IV, No. VIII, Chapter II.

V

A Condensed Map of the Road from Acapulco to Mexico City

I compiled and drew this itinerary map as I traveled from the coasts of the South Sea to Mexico City from March 28 to April 11. A précis of the astronomical observations that form the basis for this map is given above (p. 50 [p. 38 in this edition]); it presents the results of a barometric leveling,[1] and shows the unevenness of the Anahuac terrain and the *agricultural lines*, the direction of which changes with the elevation of the terrain.

---

1. See my *Recueil d'observations astronomiques*, vol. I, pp. 318–20.

VI

Map of the Road from Mexico City to Durango

Since the plateau of New Spain, which stretches across the ridge of the Cordilleras is the most inhabited part of the kingdom, I thought it would be of interest to present in detail, on three itinerary maps, the road that leads from Mexico City through Zacatecas, Durango, and Chihuahua to Santa Fe in New Mexico. This route is suitable for travel by coach and runs at an elevation of 2,000 meters above sea level, at least as far as the town of Durango and perhaps even farther. Since I used other materials for these maps than those that formed the basis of my general map of Mexico, I must now explain the reason for the noticeable differences between the different parts of the Mexican atlas. Following the example of D'Anville, Rennell, and other famous geographers, on my general map I presented the results that seemed the most likely according to several combinations. The absence of any direct observations must be compensated for by combinations and by the tricks of rational analysis. Averages taken from observations among which there are quite radical extremes may provide useful approximations. In D'Anville's day, there were hardly any places in Hindustan whose position had been determined astronomically. Nonetheless, according to Mr. Rennell's own account, that excellent geographer who had only a few vague itineraries of the interior of India at his disposal, was able to compile maps of astonishing accuracy.

Since I was drawing maps of the Mexican plateau using simple travel journals, it would have been hazardous to modify the intermediate points. The main goal of these maps is to represent the topography to a degree of detail that could not be shown on the large map: it seemed more advantageous

86

not to change anything about the winds and distances marked by the engineers. Since the latitudes of the most extreme points were already known, the calculation of the sine and cosine of the observed rhumbs produced the longitudinal difference and the value of the *local leagues*. These results are trustworthy if several latitudes on the same route have been corrected by astronomical means; this is the case for the road from Mexico City to Durango. In this instance, the navigators' method was used; that is, *the estimate* was corrected by *the observed latitude*. Mr. Friesen, whose talent as a distinguished draughtsman goes hand in hand with his solid knowledge of mathematics, was kind enough to undertake these calculations. Using Mercator's projection, he also drew the three roadmaps found in the Mexican atlas. They bear no noticeable difference from the large map, with the exception of the longitude of Santa Fe (which, according to [Pedro de] Rivera, is at 107°58′ instead of 107°13′) and the latitude of the Presidio del Paso, which my large map places eight minutes of arc farther south. The latter map also gives positions that, in the current state of our geographical knowledge, seem to me to be the least incorrect, rather than the most exact. The scale of the three itinerary maps is the same as that of the general map: three to two.

I.129

The map showing the road from Mexico City to Durango via Zacatecas is based on my own astronomical observations and on Mr. Oteiza's travel journals. Next to each place name between Mexico City and Guanajuato has been added the number of toises above sea level (according to my barometric leveling) of the plateau terrain.

This calculation put the capital of Mexico east of Zacatecas at 3°45′, by Mr. Mascaró's route, and at 1°58′ by Rivera's. This enormous difference proves the uncertainty of the rhumbs in a mountainous region, and especially over twisting roadways. Along with Mr. Oteiza, we have adopted 2°35′, which is almost the average of the results obtained by the two engineers. According to Rivera, Durango is 1°20′ west of Zacatecas, 1°57′ according to Oteiza. Mr. Friesen found that the rhumbs in Lafora's journal put the town of Querétaro 1°33′ east of Zacatecas and forty-seven minutes of arc east of Mexico City. This latter difference is off by eighteen minutes of arc, since, according to my chronometer, Querétaro is located at 102°30′30″ longitude.

I.130

VII

Map of the Road from Durango to Chihuahua

This road crosses a large part of the province of Nueva Vizcaya. Mr. Rivera and Mr. Mascaró traveled this route, the former proceeding directly from Durango to the capital of the *Provincias internas*, the latter traveling through Zacatecas, Fresnillo, La Laborcilla, and Abinito. Basing his findings on Rivera, Mr. Friesen found 1°10′ as the meridional difference between Chihuahua and Durango. According to the same traveler, Zacatecas would be 2°3′ east of Chihuahua; according to Mr. Mascaró, it would be at 2°53′. Given that this method of estimation is inherently inaccurate, this is a satisfactory degree of similarity. The two engineers differ widely, however, on the longitude of some of the intermediate points. Both men crossed the Río Florido. Based on the rhumbs and the distances that he reports, Mr. Mascaró puts this point at 3°22′, Rivera puts it at 2°12′ west of Zacatecas. Compiled according to Rivera's information, our road map offers several points of interest, such as the Parral mines and the military forts of Paso del Gallo, Mapimi, Cerro Gordo, and Conchos. It would be desirable for someone to determine the elevation of the plateau that extends from Durango to Chihuahua or to the Paso del Norte. I deduced the elevation of Durango from a series of Mr. Oteiza's barometric observations. I believe that the central plateau of New Spain descends rapidly from Durango toward the Bolsón de Mapimi. Assuming that the incline of the Río del Norte is no greater than that of the Río de la Magdalena in New Granada, the Presidio del Paso and the land that is south of this military post can be no more than six hundred meters above sea level.

I.131

VIII

Map of the Road from Chihuahua
to Santa Fe in New Mexico

There is an abundance of material for this part of the country. Since distances are considerable, and because such an uninhabited region offers few hamlets that can be sighted from great distances, the reporting of rhumbs becomes subject to serious errors. Using trigonometric tables, Mr. Friesen has calculated Rivera's and Lafora's routes with great precision. According to Rivera, Santa Fe is located fifty-three minutes of arc west of Chihuahua, but, for Lafora, it is ten minutes of arc east of that town. Comparing some of the intermediary points, one finds that both journals put the Paso del Norte and Ojo Caliente (near the Presidio of I.132 Carizal) on the same meridian. But according to Lafora, the longitudinal difference between Paso del Norte and Chihuahua is thirty-five minutes of arc less, the difference between Muerto and Paso is sixteen minutes of arc less, and between Santa Fe and Muerto is twelve minutes of arc less than Rivera's measurements. On his map of North America, Mr. Antillón put Santa Fe forty-five minutes of arc west of Chihuahua. I found it necessary to decrease this distance on my general map, reducing it to twenty-three minutes of arc. Mr. Costansó even ventures to assume that these two places share more or less the same meridian. Since my combined observations determined that the position of the capital of Quito was incorrect by almost one degree of longitude, one should not be surprised by a few other such discrepancies in the northern part of New Spain. Otherwise, I preferred to follow Rivera's journal without modifying his result for the longitude of Santa Fe, which is most likely too far to the west. According to that traveler, the results are:

Mexico City east of Durango	3°18′
Durango east of Chihuahua	1°20′
Chihuahua east of Santa Fe	0°53′
Hence: Mexico City east of Santa Fe	5°21′

I.133

This last result differs by twenty-seven minutes of arc from the one on which I settled for my large map, because Rivera puts Durango too far to the east, to the same degree that he puts Santa Fe too far west. Mr. Antillón sets the latitude of the Presidio de Paso at 33°12′, whereas Rivera insists that he found it at 32°9′ through direct observation. This latitude is perhaps even smaller, for the distances and rhumbs that Mr. Rivera indicates put it at 31°42′. I did not want to change this result at all, because amid so many uncertainties, this small road map was to be based exclusively on Mr. Rivera's journals. The unpublished work that the engineer Lafora left in Mexico City gives 33°6′, a latitude that is fairly close to the one shown on Mr. Antillón's map. But the position of Santa Fe and the number of leagues that Lafora allows between that town and Paso makes one suspect that this agreement is purely accidental.

I.134

The sources of the rivers that originate between 33° and 42° latitude on the eastern slope of the central Cordillera in the *Provincias internas* have long been the object of uncertainty and vague geographic conjectures. Some (the Río de la Nueces, the Río Colorado de Texas, and the Río de los Brazos de Dios) flow directly into the Gulf of Mexico, while others (the Canadian River, the Arkansas River, and the Platte River) flow either into the Mississippi or the Missouri. Formerly (as Alzate's map proves), the Río Colorado and the Río Rojo were believed to originate 6° and one-half east of the central mountain range. Formerly, the river that runs near the Presidio de San Antonio de Bejar was the only one that was assumed to be linked to this range; called the Río de Medina, at 36° and 37° latitude, it was apparently confused with the Río Mora and the Canadian River, two tributaries of the Arkansas. The expeditions of Pike and, especially, Major Long, have cleared up some points that were in doubt when the first edition of my work was published. According to ▼ Mr. [Edwin] James (the scholarly editor of Long's travel journal), "[w]e are as yet ignorant of the true position of the sources of Red river [(Major Long's map puts it at 35° latitude, or ten leagues east of the meridian of Santa Fe)]; but we are well assured the long received opinion, that its principal branch rises 'about thirty or forty miles east of Santa Fe' is erroneous. Several persons have recently arrived at St. Louis in Missouri, from Santa Fe, and, among others, the brother of ▼ Captain Shreeves [Shreve], who gives information of a large and frequented road, which runs nearly due east from that place, and strikes one of the branches of the Canadian, [and] that at a considerable distance to the south of this point in the high plain is the principal source of Red river. His account confirms an opinion we had previously formed, namely, that the branch of the Canadian explored by Major Long's party, in August 1820, has its sources near those of some stream which descends toward the west into the Rio Del Norte, and consequently that some other region must contain the head of Red river. From a careful comparison of all the information we have been able to collect [(lat. 37°3′, 103°32′ long. west of Greenwich)], we are satisfied that the stream on which we encamped on the 31st of August [July] is the Río Raijo [Rojo] of Humboldt, long mistaken for the source of the Red river of Natchitoches, and that our camp of September [August] 2nd was within forty or fifty miles east from Santa Fe. In a region of red clay and sand, where all the streams have nearly the color of arterial blood, it is not surprising that several rivers should have received the same name; nor is it surprising that so accurate a topographer as the Baron Humboldt, having learned that a Red river rises forty or fifty miles east of Santa Fe, and runs to the east, should conjecture it might be the source of the Red river of Natchitoches. This conjecture (for it is no more) we believe to have been adopted by our geographers, who have with much confidence made their delineations and their accounts correspond to it" (Long, *[Account of an] Expedition*, vol. III, p. 175–77).

I.135

According to Major Long's work, the Río Rojo on my Map of New Spain of 1804 is the northern branch of the Canadian River into which the Río de Mora flows. The Río Rojo de Natchitotches originates from two branches at 34° and 35° latitude; as a result, the true sources of the Río Colorado in Texas are probably at 33° and one-half. One should also keep in mind that only the origins of the ▼ Paducah (or the South Fork of the Platte River), the Arkansas River, and the Canadian River are known with any certainty. Any information about the sources of the Río Rojo and the Río Colorado is still vague. The mistaken theory that the Río Mora is the same as the Río Rojo was repeated on my Map of New Spain, following an indication on a large unpublished map that I own, which is called the *Mapa geográfica de una [gran] parte de la América septentrional comprehendida entre los 19° y 41° de latitud.* As for the Río Napestle [Upper Arkansas River] and its tributaries (the Río del Sacramento and the Río Dolores), there is still doubt that one should believe (as I did) that it is a tributary of the Arkansas and, consequently, identical to the Paducah River, which is the southern branch of Platte River. The distance from Taos to the sources of the Napestle make the second assumption more likely, especially if the Sierra de Almagre [Almagre Mountain] (lat. 39°36′) on my map, and, on the unpublished maps I saw in Mexico City, is Longs Highest Peak (latitude 40°13′) and not James Peak (latitude 38°38′). The Río Napestle begins slightly south of the Sierra de Almagre, whose latitude on the unpublished Mexican maps alternates between the latitude of Taos and Santa Fe in New Mexico.

I.136

IX

Map of the Eastern Part of New Spain from the Plateau of Mexico City to the Shores of Veracruz

This map, which encompasses the area between 18°40′ to 19°45′ latitude and between 98°0′ to 101°35′ longitude, includes the roads from Veracruz to Mexico City, by way of Orizaba or Jalapa, which is the most interesting part of New Spain. Here, one can make out the interior plateau and the eastern slope of the Anahuac Cordillera, which is directly across from the arid coasts of the Gulf of Mexico. Cleverly distributing the shading on his map, Mr. Friesen has indicated the unevenness of the terrain and the relative elevation of the mountains; he drew this map according to another one that I had sketched in the Americas. The scale is three millimeters per minute of equatorial degree. Consequently, this scale corresponds to the scale of maps VI, VII, and VIII by a ratio of approximately four to one; it corresponds to the scale of map I by a ratio of six to one.

The material used to compile the map of the eastern part of the Anahuac plateau has been sufficiently discussed in the preceding pages. A map drawn by Mr. García Conde and the geodesic survey that this learned officer conducted in 1797 together with Mr. Costansó, the colonel of the corps of engineers, may be taken as the foundation of my work on map IX. No detail of the contours of the terrain has been changed, but its entirety has been corrected, in keeping with the results of my astronomical observations. When I had determined the positions of the four great peaks of the Cordillera (Popocatepetl, Iztaccihuatl, Citlaltepetl, and Naucampatepetl[1]) using

I.137

---

1. The Indians also call the Cofre de Perote *Nappateuctli, Nauvpavewizi*, or *Tepetlkaliatl.*

azimuths and observations of the firmament, and the positions of Mexico City, Cholula, Puebla, and Jalapa, it was easy to determine the rest, using partial reductions. The coastline of the Gulf of Mexico, from the mouth of the Alvarado River to Punta María Andrea, has been corrected according to Mr. Ferrer's chronometric observations. I added the results of my barometric leveling to map IX, as I did for all the other maps of my Mexican atlas.

I.138

X

Map of Incorrect Positions

This drawing shows the incorrect positions attributed to the ports of Veracruz and Acapulco and to the capital of New Spain. It demonstrates how flawed the maps of Mexico are that have been published until now. I made this drawing following the *Mapa crítica Germaniæ* [1753] that was drawn by the famous astronomer ▼ Tobie Mayer.

In this *Map of Incorrect Positions,*[1] the result Mr. de Cassini obtained from the observations of longitude found in the Voyage of the Abbot Chappe (which was recorded in the *Connaissance des temps* for 1784) has been distinguished from the result adopted by the members of the Academy of Sciences who were responsible for publishing Alzate's map in 1772. One finds the following note there:

"Mr. Chappe's Voyage in California provided corrections of the position of various places, which it may be of interest to present here."

	LONGITUDE OF EL HIERRO ISLAND			NORTHERN LATITUDE		
Nueva Veracruz*	285°	35′	15″	19°	9′	30″
Mexico City	278°	16′	30″	"	"	"
San José	267°	52′	30″	22°	1′	0″

* Probably a typographical error: 285° instead of 282°.

---

1. *Mexican Atlas*, plate X, *Introduction*.

The following question was recently put forth: "To what degree does the result of my observations, which were undertaken to determine the position of Mexico City, differ from Mr. Chappe's observations?" In this regard, I would remind my reader that that astronomer made his observations in Veracruz and Saint Joseph but not in Mexico City proper, and that Mr. Alzate's observations, knowledge of which we owe to the Abbot Chappe, differ from one another by more than two degrees of longitude.

I.139

XI

Map of the Port of Veracruz

This Atlas of New Spain would indeed be incomplete if it did not contain a map of the port from which all Mexican riches flow toward Europe. To this day, Veracruz is the only port that can berth European warships. The map that I am publishing here is an exact copy of the one drawn up in 1798 by ▼ Mr. Orta, captain of the port of Veracruz. I have reduced it to half of its original scale and appended a few notes on longitude, winds, atmospheric tides, and the annual amount of rain. One glance at this map proves the difficulty of waging a military attack against a country that offers no more shelter to vessels on its eastern coastline than a treacherous mooring between shoals.

The double lines on the map of the port show the direction that ships must take to set anchor. As soon as the helmsman sees the buildings of the I.140 town of Veracruz, he must steer the ship such that the tower of the church of St. Francis blocks the tower of the cathedral. He must follow this course until the jutting angle of the bastion of San Crispín appears behind the bastion of St. Peter. He should then tack portside by pointing the prow toward the Isle of Sacrifices [Isla de Sacrificios]. Near Soldado point [Punta Soldado?], the Gallega shoal has several buoys (*palos de marca*) that warn vessels entering Veracruz of two quite dangerous rocks, called the *Laja de Fuera* and the *Laja de Dentro*.

XII

Physical Tableau of the Eastern Slope
of the Anahuac Plateau

The horizontal projections that are commonly called geographical maps only give an imperfect sense of the irregularities of a country's terrain and physiognomy. The undulations of the terrain, the shape of the mountains and their relative height, and the steepness of the slopes can only be completely represented in a drawing by using the method of *leveling by cross section* and by orienting the hatching very precisely according to the *lines of the steepest slopes*. A map drawn according to these principles[1] compensates, to a certain extent, for a relief map. Lines drawn on a two-dimensional map can have the same effect as a relief model if the extent of the terrain is small, and if all its aspects are perfectly well-known. The difficulties of this type of work become almost insurmountable, however, if the horizontal projection includes a mountainous region with a surface area of several thousand square leagues.

I.141

In the most populated region of Europe—for example, France, Germany, or England—agricultural plains generally vary in elevation by only one or two hundred meters. Their absolute height is too inconsiderable to have a noticeable effect on the local climate.[2] As a result, it is not of equal

1. ▼ Mr. Clerc, the director of topography at the École Polytechnique, who has an extraordinary talent for configuring terrain, is preparing to publish a work on map drawing and on the design of relief maps that will be a landmark in the history of topography.

2. The interior of Spain is a striking example: the absolute elevation of Castile on the outskirts of Madrid is over 600 meters. See my paper on the configuration of the terrain of Spain in the *Itinéraire [descriptif de l'Espagne]* by ▼ Mr. Alexandre de Laborde, vol. I, pp. cxlvii–clvi; and more recently, the "Perfiles de la Península Española según las dos

importance to the farmer and to the geographer to know the exact height of these plains. On maps of Europe, one has only to indicate the highest mountain ranges and the foothills that extend toward the plains.   I.142

In the equinoctial zone of the New Continent, especially in the kingdoms of New Granada, Quito, and Mexico, the atmospheric temperature, the aridity or humidity, and the type of agriculture that the inhabitants practice all depend on the extreme elevation of the plains that form the ridge of the Cordilleras. The geological makeup of these areas is an object of study of equal importance to the statesman and to the naturalist traveler. The inaccuracy of our graphic methods becomes even more evident when we use them on highly elevated plateaus: it is much more striking on a map of New Spain than on a map of France. To give a complete understanding of these countries whose terrain has such an unusual configuration, I had to avail myself of methods that geographers had not yet begun to implement. The simplest ideas are usually the ones that come about last.

I have represented entire countries and vast areas of terrain in vertical projections, as has long been the practice for drawing sections of a mine or a canal.[1] In my ▼ *Essai de pasigraphie géologique*, I shall explain in detail the principles according to which these physical views are to be constructed.   I.143 Since the sites whose absolute elevation must be determined are rarely on the same line, each section is composed of several planes oriented in different directions; alternatively, there is only one plane that is placed off the path on which perpendicular lines fall. In this case, the distances represented on the physical map are different from the absolute distances, especially when the average direction of the points whose height and position have been determined deviates considerably from the direction of the projection plane.

The application of *graphic methods* to various objects of physical geography offers the advantage of instilling the inner conviction that always accompanies ideas received immediately and quickly by the senses. These ideas are not merely *imitative*, representing shapes spatially according to sections made by planes, as in the geometric configuration of a terrain

direcciones SE.-NO y SO.-NE por el Baron de Humboldt" on the large *Carte d'Espagne de MM. Donnet et Malo*, 1823 [*Mapa civil y militar de España y Portugal*]. The small geological map that is attached to the [*Notice historique*] *sur une importation de [six cents] mérinos* by ▼ Mr. Poyféré de Cère in 1809 was also drawn according to my measurements. Unfortunately, not all parts of this map were drawn to the same scale of elevation.

1. My first attempt in this genre was the physical map of the course of the Río Magdalena and the road from Honda to Santa Fé de Bogotá, which was engraved against my wishes in Madrid in 1801. See my *Recueil d'observations astronomiques*, vol. I, p. 370.

that results from different modes of projection; they may also, by extension, serve to demonstrate all relations of size and quantity—anything that can be numerically increased or decreased. One may thus draw—and not without benefit to natural philosophy—curves of average monthly temperatures, atmospheric pressure, and humidity by using the division of time as one of the coordinates. Examinations of the distribution of heat and the direction and intensity of magnetic forces on the surface of the globe have produced *isothermal bands*, curves of equal magnetic inclination and declination; that is, the *isodynamic lines* on which a single needle makes the same number of oscillations in a given period of time. Physical geography is limited to *imitative graphic methods* that use projections to show the relative positions of points whose various systems constitute large areas of land on the surface of the earth. It is normal that a need for *maps* in the strict sense—in other words, for the geometral plane of a country and of the respective locations of places projected on a horizontal plane—was felt first, rather than a need for *cross sections* or *vertical sections* that represent points on the surface of the earth at their elevation above sea level. The first type of projection represents planed or curved *surfaces*, whereas the second type only represents *lines*.

I.144

In emergent societies, surveys measuring the appearance of physical properties must have preceded leveling operations. Thus, for a long time, maps *strictu sensu* only revealed the outline of the spaces they represented, the curvature of coastlines, the course of rivers, and, like ancient Roman tableaux of roads (*itineraria picta*), the positions of places relative to each other in certain directions. The physical aspect of the country was completely neglected. Only in the past half-century has geometric representation achieved the degree of accuracy that enables it to represent the polyhedral shape of the surface of the globe whenever irregularities are considerable and comparable. This is not the case when plateaus are connected by gentle slopes. Differences in level arise simultaneously from the degree of inclination of partial slopes and the length or continuity of this inclination; in other words, from the more or less considerable extent to which the same slope continues. No matter how large the scale used for a chorographic map of several thousand square leagues, no drawing that employs hatching could represent a slope of just one degree. If, however, this same slope were protracted extensively, it could lead to considerable heights. Where the vast surface of South America stretches six hundred leagues wide from east to west is the most striking example of the continuity of one slope covered with profitable terrain.

I.145

When scales of height and distance are carefully combined, vertical sections may represent—simultaneously and equally well—low plains and high plains, the slopes that connect them, and all the undulations of a terrain that disappear almost entirely on a geometral map. Vertical projections applied to entire countries are consequently not only of great interest to the study of the shape of a terrain; their proliferation is also useful for correcting the *representation of relief* on our ordinary maps. The lack of such material is what made ▼ Mr. Dupain-Triel's ingenious attempt at representing all of France through highly developed level curves so imprecise.

I.146

The need for cross sections, or vertical projections, has long been felt in mining work, or subterranean geometry, and in the drawing of canals and roads. To show the comparative height of the mountains of the world, mountains were grouped in the bizarre shape of slender peaks, with no regard for the geographical position of their location. The Abbot Chappe published [a map of] the road from Petersburg to Tobolsk less from actual measurements than from vague information, mixing his sketch of the landscape and the effects of an aerial perspective with the drawing of a cross section. This type of projection, however, should have been subjected to set rules and then applied to the representation of entire countries. The vertical sections of New Spain that I drew in 1803, several copies of which remain in the Americas, were, I believe, the first attempt to do orographic work.

For profiles of entire countries and canals, the scale of distances cannot be equal to the scale of elevation. If one were to try to make both scales the same size, one would either be forced to make impossibly large drawings, or to adopt a scale of elevation so small that the most remarkable irregularities of the terrain would be undetectable. On the twelfth plate, I used two arrows to indicate what the height of Chimborazo and Mexico City would be if the physical tableau were subjected to the same scale in all its dimensions. In this case, an elevation of five hundred meters would occupy only one millimeter of space on the drawing. By contrast, if the scale of elevation given for plates XII, XIII, and XIV (which is approximately 270 meters per centimeter) were also used for itinerary distances, one would need a fifteen-meter-long copper plate to represent the extent of the terrain between the meridians of Mexico City and Veracruz! Because of these varying scales, my physical maps do not show the actual sloping of the terrain any better than do the profiles of canals and roads drawn up by geographical engineers. Depending on the type of projection used, these slopes seem steeper in drawings than they actually are.[1] This disadvantage is exacerbated when plateaus are very high but cover only a small area or when they are separated

I.147

1. See my *Essai sur la géographie des plantes*, p. 53.

by deep and narrow valleys. The visual effect of the profile of a country depends mainly on the relationship between the scales of distance and of height. I shall not discuss the principles that I applied to this kind of map. Any graphic method must follow rules, and it seemed all the more necessary

I.148   to me to report a few of them here since several recently published imitations of my tableaux transgress against both the style and the laws of orthogonal projection. They are partial cross sections shaded like landscapes and projected onto several planes at the same time, with no indication of the direction of the planes in relation to the large circles of the sphere.

Physical maps using vertical projections should not be constructed unless one knows the three coordinates of longitude, latitude, and the elevation above sea level of the points through which the projection plane passes. Only by combining barometric measurements with the results of astronomical observations can one draw the cross section of a country. This type of vertical projection will become more frequent as travelers devote themselves more diligently to barometric observations. Even today, there are very few parts of Europe for which we have the materials that are indispensable for drawing up tableaux similar to those of equinoctial America.

The construction of profiles for plates XII, XIII, and XIV is absolutely uniform. The scale of all three tableaux is the same; the ratio of the scale of distance to the scale of height is approximately one to twenty-four. The three maps also show the types of rock that compose the surface of the ground. Such geognostic knowledge is of interest to farmers; it is especially

I.149   useful to engineers responsible for planning canals and roads.

I have been criticized for not showing on these cross sections either the superposition or depositing of the secondary or primary layers, or their incline and direction. Specific reasons prevented me from indicating these phenomena. In my travel journals, I have all the necessary geological material for drawing what are usually called mineralogical maps. I have published much of this material in my work on the leveling of the Andean Cordillera and in the *Essai géognostique sur le gisement des roches dans les deux Hémispheres*. After a lengthy examination, however, I decided to separate entirely the geological profiles that reveal the superposition of rocks from the tableaux that show the irregularities of the terrain. It is very difficult—I would say almost impossible—to draw a geological cross section of a large area, if this cross section is to be governed by a scale of height. A layer of gypsum that is one meter thick is often as interesting to the geologist as an enormous mass of gneiss, mica schist, or porphyry because the existence of very thin layers and the way their deposits were formed shed light on the

relative age of the formation. How is one to draw the profile of entire provinces, if the size of the scale must be such that it is possible to distinguish
such minor masses? In a narrow valley, for instance the Papagayo (plate
XIII), how would one indicate in a space of one or two millimeters that this   I.150
valley on the map is composed of different formations one on top of the
other? Those who have given thought to graphic methods and have tried
to perfect them will feel, as I do, that these methods cannot combine all
possible advantages. A map that is covered with too many signs becomes
bewildering and thereby loses its greatest advantage, which is to show a
great many different relations at once. Different types of rock and their
mutual superposition are of more interest to the geologist than is the absolute elevation of formations and the thickness of their layers. A geological
profile need only show the general appearance of a country; it is only by
removing scales of height and distance that it can clearly convey important information about mineral deposits or stratification. Taking the narrowest definition that the word geography once had—that is, something
concerned only with surface area and shape—one could make a distinction
between *geographical cross sections* and *geological cross sections*. The latter
indicates both compositional relationships and the series of superimposed
rocks.

The tableau of the eastern slope of New Spain presents three partial
profiles. It simultaneously shows the astronomical positions of the points
of intersection, their respective distances, and the angle that each partial
secant plane makes with the meridians. The three cross sections of the map
are represented in different colors. Mexico City, Puebla de los Angeles, and   I.151
the hamlet of Cruz Blanca, located between Perote and Las Vigas, are the
points where the three secant planes intersect. The longitude and latitude of
these three points, based on my own observations, have been added, as has
the average direction of each cross section, its length expressed in French
leagues, which are common leagues of twenty-five to a degree. The scale of
distances for this profile (plate XII) is exactly the same as the one on which
the geographical map is drawn (plate IX). The vertical projection occupies
more space than does the horizontal projection, because the former retains
the itinerary distances from one place to another. For instance, the absolute
distance from Mexico City to Puebla is only twenty-seven leagues, but it
appears to be two leagues longer on the profile drawing, which presents,
so to speak, all the curves in the road. It also shows the number of leagues
that one travels from Mexico City to Puebla via the Venta de Chalco, via Río
Frío and Ocotlán.

The two large volcanoes east of the valley of Tenochtitlan—the Orizaba Peak and the Cofre de Perote—have been placed on the profile according to their true longitudes. They are represented as they appear during a *bright spell*, when heavy mist covers their bases and their peaks appear above the clouds. Despite the enormity of these colossal mountains, I have not attempted to represent their full contours because the scales of height and distance are so disproportionate. If they had been connected to the plateau, these volcanoes would have disfigured the tableau, appearing as two columns rising above the plateau. I have attempted to carefully render their strange shape, dare I say the special physiognomy of the four great mountains of the Anahuac Cordillera. I hope that those who, while traveling from Veracruz to Mexico City, have been struck by the imposing presence of these majestic peaks, will acknowledge that the contours have been faithfully drawn on this plate, as well as on plates XVI and XVII.

I.152

To lodge a few important facts of physical geography firmly in readers' minds, the respective elevations of the Chimborazo and several mountains in the Alps and the Pyrenees have been inserted on either side of the tableaux, near the scales of height. These tableaux also show the elevation of the limit of perpetual snows on the equator, at the parallel of Quito, and at 45° latitude,[1] the average air temperature at the foot and on the slope of the Cordilleras, and, lastly, the elevation at which some Mexican plants begin to appear or stop growing in the mountainous part of the country. Some of these phenomena are indicated repeatedly on several of my maps; this repetition is similar to that presented on the thermometric scales, which provided, albeit inexactly, the maximum and minimum temperatures observed in a given zone. I thought that the profiles in the *Mexican Atlas* that are somewhat analogous to the large tableau in my *Geography of Plants* might help generate interest in the study of natural history considered in terms of the mutual influence of climate and elevation.

I.153

1. According to Mr. von Humboldt's most recent studies ("Mémoires sur les neiges de l'Himalaya" in *Annales de chimie et de physique*, 1820, vol. XIV, p. 56), the line of the perpetual snows in the Andes of Quito (latitude 1° to 1°30′) is at a height of 2,460 toises; in Mexico City (latitude 19° to 19°12′), it is at 2,350 toises; in the Himalayas (latitude 30°40′ to 31°4′) on the southern slope at 1,950 toises; on the northern slope, it is probably at 2,605 toises; in the Caucasus (latitude 42° to 43°) at 1,650 toises; in the Pyrenees (latitude 42° and one half to 43°) at 1,400 toises; in the Swiss Alps (latitude 45° and three-fourths to 46° and one half) at 1,370 toises; in the Carpathian Mountains (lat. 49°10′) at 1,330 toises; in Norway (latitude 61°–62°) at 850 toises; (latitude 67°) at 600 toises; (latitude 70°) at 550 toises; (and at latitude 70° and one half), when affected by misty summers on the coast, at 366 toises.

XIII

Physical Tableau of the Western Slope of the Plateau of New Spain

This tableau, of the central section of the Valley of Tenochtitlan, and its cross section (plate XVI) have been drawn according to the principles we have just presented for the profile of the eastern slope of the Cordillera. The extent of the country, whose vertical projection is represented on the thirteenth plate, is drawn in horizontal projection on the fifth plate. The profile and the plane do not share the same scale, because the same number of leagues, when con- I.154 sidered an itinerary distance on this plane, takes up one-quarter less space than on the profile. By contrast, plates XIII and XIV were drawn to uniform scale so that they might be combined into a single cross section that would extend from the Atlantic Ocean to the South Sea; such a cross section would show the geologist the extraordinary contours of the entire country. I developed the map of the road from Mexico City to Acapulco (plate V) somewhat less than the large scale of the ninth map would have demanded. In order to use the sketches that were made of an area of terrain covering three degrees, moving up the western coastline toward the capital of New Spain, it was necessary to subject this drawing to a smaller scale. Its ratio to the scale of the ninth plate is the same as three to four.

Those who wish to combine profiles XIII and XIV should know that by dividing up the two vertical scales (on which the heights of Puy-de-Dôme [Auvergne region in France] and Vesuvius are indicated), the projection planes of these profiles intersect each other at nearly a right angle in the center of Mexico City. The mean direction of the first cross section, which comprises several planes, is from east to west; the mean direction of the second cross section, showing the road from Mexico City to Acapulco, is

I.155 from SSW to NNE.[1] The extension of the first cross section would reach nearly through Pátzcuaro and Zapotlán toward the Villa de la Purificación. Extended westward, this plane would end on the South Sea coast between Cabo Corrientes and the port of La Navidad. Since New Spain widens oddly in this western direction, the descent of the Cordillera, from the valley of Tenochtitlan to the plains of the intendancy of Guadalajara, would consequently be twice as long as the road from Mexico City to Acapulco drawn on plate XIII. Furthermore, the barometric measurements I made between Valladolid, Pátzcuaro, Ario, and Acámbaro prove that, if one drew this transversal cross section in the direction of the parallels of nineteen or twenty degrees, one would see the central plateau maintain the great height of 2,000 meters over more than sixty leagues west of Mexico City. In the direction of cross section XIII, however, the plateau no longer reaches this elevation as soon as one leaves the valley of Tenochtitlan toward the SSW.

I doubt that a cross section running from east to west, from Veracruz to the small port of La Navidad, could give a more specific idea of the geological composition of New Spain than the combination of my two profiles, XIII and XIV. One has only to consider the direction of the Anahuac Cordillera to understand what I am proposing here. The central mountain range runs from the province of Oaxaca to the province of Durango, from SE to NW. Consequently, for the projection plane to be perpendicular to the longitudinal axis of the Cordillera, it must not be placed parallel to the equator: it should run from NE to SW. Considering the specific structure and the limits of the group of mountains near the capital of Mexico, one will realize that the combination of the two cross sections XIII and XIV present the geological composition of the country less imperfectly than purely theoretical ideas might lead one to believe. In the mountainous region between 19° and 20° latitude, there is no sign of a longitudinal crest. Even those parallel secondary mountain ranges, which are much rarer in nature than in the work of geologists (where they appear most arbitrarily, like rows of dikes and rocky ridges) do not exist there. The Anahuac Cordillera is wider to the north; as a result of this widening, the inclined planes formed by the eastern and western slopes of the Cordillera do not, as a rule, run in a parallel direction. This direction is almost north and south along the coastline of the Gulf of Mexico, and southeast and northwest on the slope facing the Great Ocean. For the cross sections to be perpendicular to the directions of the slopes, they cannot be on the same plane of projection.

1. Precisely N.14° E.

XIV

Physical Tableau of the Central Plateau of the Cordillera of New Spain

The profile of the road that leads from Mexico City to the mines of Guanajuato, the richest in the known world, was drawn under my guidance in Mexico City by ▼ Mr. Rafaël Davalos,[1] a young man with a keen interest in science and a student at the School of Mining. This drawing shows the prodigious height of the Anahuac plateau as it extends to the north, well beyond the Torrid Zone. The extraordinary configuration of the Mexican terrain recalls the high plains of central Asia. It would be very interesting to see my profile continued from Guanajuato to Durango and Chihuahua, and especially as far as Santa Fe in New Mexico. As we shall later demonstrate, the Anahuac plateau maintains toward the north an absolute elevation of over two thousand meters across an expanse of two hundred leagues, and of over eight hundred meters across an expanse of five hundred leagues.

1. Mr. Davalos and ▼ Mr. Juan José Rodríguez (born in Parral, in the *Provincias internas*) were most willing to assist me for several months in composing several geological maps that I plan to add to my work on rock deposits. I am delighted to proclaim publicly my gratitude to these two distinguished persons of outstanding talent and application.

XV

Profile of the Canal of Huehuetoca

The canal of Huehuetoca, or Nochistongo, was dug in the seventeenth century, in the mountain range that borders the valley of Tenochtitlan toward the north. Its purpose is to protect the capital from floods. The profile that I am presenting to the public was drawn by Mr. Friesen in Berlin, on the basis of sketches by ▼ Don Ignacio Castera, an architect of hydraulic structures in Mexico City. It explains everything that is mentioned in the third book about the celebrated cut in the mountain through which the artificial river called *el Río de Desagüe* flows. Comparing this plate, number XV, with map number III, one sees that the four planes of projection, combined into a single profile, pass through the villages of Carpio, San Mateo, and Huehuetoca, whose height above sea level I determined by using barometric measurements. I had to use an extremely large scale for this profile, to show the slight difference in level that exists between the great plaza in Mexico City and Lake Texcoco. Since the drawing covers an expanse of terrain that is almost twenty common leagues, it was necessary to introduce a much larger disparity between the scales of distance and height than in

the three preceding cross sections. As a result, there appears to be a great drop in the canal, but the basins of the three lakes—one on top of the other, as if they were tiered—also seem much closer to their actual form. One can see how, if they were to overflow, these lakes would flood Mexico City.

Profile number XV is the only one of my physical tableaux that simultaneously shows several different parallel projections planes, which are identifiable by different colors. This method, which does not break any of the rules of projection, has long been adopted in drawing important roads and canals. If one wanted to represent the profile of a valley, for example, the valley of Quito, bordered by high mountains to the east and to the west, one could make the secant plane pass through the longitudinal axis of the valley and, using perpendiculars, project the contours of the eastern and western peaks onto the same plane. A profile whose construction followed this method would not present confusing ideas to the viewer, if one identified the highest peaks of the two Cordilleras using different colors, and if these solitary peaks were positioned such that they did not overlap one another.

The small sketches I–IV that have been added at the bottom of the plate are drawn to a different scale. They represent the old bridge of Huehuetoca and the different cross sections of the canal of Nochistongo. One will recognize (Number IV) the vestiges of the old gallery of ▼ Enrico Martínez. Drawing Number II shows the deplorable state of the trench, brought about   I.160 by the continuous erosion of rainwater. Drawing number III shows the embankment currently under construction on the lateral slopes of the canal and intended to decrease the danger of collapse. On the large profile, three white lines indicate the points of incision in the mountain, the height of which corresponds to the level of the three lakes of Sumpango, San Cristóbal, and Texcoco.

XVI

A Picturesque View of the Volcanoes of Mexico City and Puebla

This plate and the one that immediately follows were originally supposed to appear in the picturesque Atlas that accompanies the *Personal Narrative* of my voyage to the equinoctial regions, because this atlas includes sketches suitable for introducing the physiognomy of the colossal summits that crown the ridge of the Cordilleras. I thought that the contours of the Andes, compared with those offered in ▼ Mr. Ebel's excellent travelogue and by Mr. Osterwald's beautiful drawings, would be of keen interest to geologists who wish to conduct a comparative study of the Swiss Alps and the Cordilleras of Mexico and Peru. Although the goal of this work is to describe the territorial wealth rather than the geological composition of New Spain, I felt I should append some picturesque views (Numbers XVI and XVII) to the Mexican atlas as a supplement to the map of the valley (plate III) and to give a better impression of the beauty of Mexico City's setting. These two summits, Popocatepetl and Citlaltepetl—the first is visible from Mexico City and Cholula, the second from Cholula and Veracruz—helped me verify the meridional difference between Mexico City and the port of Veracruz, utilizing perpendicular bases, azimuths, and angles of elevation, a (hypsometric) method that has had little application until now.[1]

I.161

Mexico City is half as far from the two *Nevados de la Puebla* as the cities of Bern and Milan are from the central range of the Alps. This great proximity contributes much to lending an imposing and majestic aspect to the

---

1. See above, p. 32 [p. 28 in this edition], and my *Recueil d'observations astronomiques*, vol. I, p. 373. [I.Q]

Mexican volcanoes. The contours of their summits, covered with perpetual snow, appear even more clearly defined, because the air through which the sunlight reaches the eye is finer and more transparent. The snow glistens with extraordinary brightness, especially when it stands out against a sky that is always a shade darker than the celestial blue we see above our plains in the temperate zone. In Mexico City, the observer is in a stratum of air whose barometric pressure is only 585 millimeters. It is easy to imagine that the loss of light is very slight in such an uncondensed atmosphere, and that the summits of the Chimborazo and Popocatepetl, as seen from the plateaus of Riobamba or Mexico City, must present contours that are more distinct and sharply etched than if they were seen at the same distance from the Ocean coast.

I.162

Iztaccihuatl and Popocatepetl, the latter of which has the same conical shape as Cotopaxi and the Peak of Orizaba, are referred to indiscriminately by the locals as the Volcanoes of either La Puebla or Mexico City, because they are visible almost equally well from these two cities. I have no doubt that Iztaccihuatl, which ▼ Cardinal Lorenzana calls *Zihualtepec*, is an extinct volcano; there is no Indian legend, however, that dates back to the time when this mountain (whose contours are similar to the Pichincha volcano) belched fire. The same goes for the *Nevado de Toluca*. Since the earliest days of the conquest, the Spanish have customarily called *Volcán* any isolated peak that penetrates into the region of perpetual snows. The words *Nevado* and *Volcán* are often confused; in Quito, I have even heard such strange expressions as *Volcán de Nieve* and *Volcán de Fuego*. The Cotopaxi, for example, is reputedly a *fire volcano*, because its periodic eruptions are known, whereas the Corazón and the Chimborazo are called *snow volcanoes* because the indigenous peoples suppose that they hide no fire in their bosom.

In the kingdom of Guatemala[1] and in the Philippine islands, the volcanoes that flood their surroundings are called *water volcanoes* (*volcanes de*

I.163

1. Lorenzana in a note to the Letters of Cortés: "En Guatemala hay dos volcanes, uno de fuego y otro de agua" [In Guatemala, there are two volcanoes, one of fire and another of water.] This *Volcán de Agua* lies between the Volcán de Pacaya and the Volcán de Guatemala, called the *Volcán de Fuego*. It is covered with snow for several months of the year, and on September 11, 1541, it belched "a torrent of water and stones" that destroyed the *Ciudad Vieja*, or Almolonga, the former capital of the Kingdom of Guatemala, which should not be confused with *Antigua Guatimala*. See ▼ Remesal, *Histoire de la provincia de San Vicente*, book IV, chap. 5, and ▼ Juarros [y Montúfar], *Compendio de la Historia de Guatemala*, vol. I, pp. 72, 85.

agua). The examples I have just given show that on Spanish maps, the word Volcán is often understood quite differently from the meaning other European nations give it.

Don Luis Martín has drawn the volcanoes of la Puebla as they appear in clear weather from the terrace of the School of Mining (*Seminario Real de Minería*). During my stay in Rome, ▼ Mr. Gmelin, a famous artist who honors me with a special friendship, has embellished both Mr. Martín's drawing and a sketch that I did of the Orizaba Peak. The contours have not been altered, and the shading has made the rocky masses more imposing.

The Volcanoes of Puebla were drawn in the month of January, in a season when the lower level of perpetual snow dropped almost to the height of

I.164 the spike of the Peak of Tenerife [Teide], or 3,800 meters of absolute elevation. During my stay in Mexico City, I saw such heavy snowfalls on the mountains that the two volcanoes were almost joined by the same layer of snow. The *maximum* height[1] of the snow level, as I found it in the month of November 1803, was nearly 4,560 meters.

The Sierra Nevada, or Iztaccihuatl, is only a few meters higher than Mont Blanc; Popocatepetl surpasses that mountain in height by 625 meters. Furthermore, the plain that stretches from Mexico City to the foot of the volcanoes is already higher than the summit of Mont d'Or and the famous mountain passes of Petit Saint Bernard, Mont Cenis, Simplon, Gavarnie, and Cavarere.

Cortés passed between the peaks of these two volcanoes with his army and six thousand Tlaxcalteca on his first expedition against the city of Mexico. To prove his courage to the indigenous peoples during this arduous march, ▼ Don Diego Ordaz tried to reach the peak of Popocatepetl. Although he did not succeed in his endeavor,[2] the emperor Charles V allowed him to place a volcano on his coat-of-arms. After the capital was taken in 1522, it is not known if ▼ Francisco Montaño drew the sulfur that was used

I.165 to make gunpowder from the crater of Popocatepetl or—what seems more likely to me—from some lateral crevice.

1. See chap. II.
2. Cortés, [*Historia de Nueva-España*, Lorenzana ed.], pp. 318 and 380; Clavijero, [*Storia antica del Messico*,] III, pp. 68 and 162.

XVII

Picturesque View of the Peak of Orizaba

The Peak of Orizaba, whose position Mr. Arrowsmith[1] and other geographers have muddled on their maps, enjoys the same fame among navigators as the Peak of Tenerife [Teide], the *Silla de Caracas*, the *Tafelberg* [Table Mountain, South Africa], and the *Peak of Saint Elias [Mount St. Elias]*. I have drawn it as it appears from the road that leads from Jalapa to the village of Oatepec (Huatepec) near the Barrio de Santiago. Only the part that is covered with perpetual snow is visible from this station. The first map of my drawing is a dense forest of *Liquidambar styraciflua*, melastomes, arbutus trees, and pepper plants. It is noteworthy that the craters of the two great Mexican volcanoes, Popocatepetl and Citlaltepetl, are both inclined toward the southeast. In the equinoctial region of New Spain, the mountains generally slope more rapidly toward Gulf of Mexico, and the rocky shoals there go most often from NW to SE. To distinguish better the active volcanoes from the extinct ones, I have taken the liberty of inserting a I.166 small column of smoke in the drawings of the Peak of Orizaba and the great volcano of Puebla, although I have not seen this smoke either in Jalapa or in Mexico City itself.[2] On January 24, 1804, ▼ Mr. Bonpland and I saw a large cloud of ashes and very dense vapors issue from the mouth of Popocatepetl.

---

1. See above, pp. 41–42 [in this edition].

2. Mr. Bullock overlooked this passage when he criticized Mr. von Humboldt (*Six Months Residence in Mexico*, 1824, p. 121) for having drawn the smoke above the crater of the Peak of Orizaba. We should also like to remind this worthy traveler of ▼ Mr. Visconti's letter (*Vues des Cordillères*, vol. II) and the explanation *of the position of the hands and feet* of the kneeling Mexican figures, in response to his remark, p. 531.

We were on the Tetimpa plain at that time, near San Nicolás de los Ranchos, where we took a geodesic measurement of the volcano. The strongest eruptions of the Peak of Orizaba, which the Indians also call *Pojauhtecatl* or *Zeuctepetl*, were from 1545 to 1566.

Eight years before I arrived in Mexico, Mr. Ferrer had already measured Citlaltepetl, or the Orizaba volcano, by taking vertical angles from a great distance of the summit of the peak, near the Encero. In a paper in the *Transactions of the [American] Philosophical Society [held in] Philadelphia*, he assigns a height of 5,450 meters to the volcano.[1] My own measurement, which gives it 155 meters less, was taken on the small plain near Jalapa, where the vertical angle was also only 3°43′48″. Despite the extraordinary consistency of refractions in the tropics and despite my precautions, I do not believe that during the course of my expeditions I was able to ascertain the height of a single mountain in America as precisely as the geodesic works of General Roi, ▼ Tralles, Delambre, Zach, and Oriani have established the height of some European mountains. Making these minute calculations is similar to the chemical analysis of minerals: they are only done with great precision when one enjoys the perfect tranquility and leisure that a traveler can seldom find in distant climes.

Plate number XVII and the preceding one were engraved by my compatriot ▼ Mr. Arnold, an exceptionally talented young artist who was taken from the arts in the flower of his youth. I found the azimuth[2] of the Orizaba Peak in Jalapa by successively measuring the distances from the edge of the sun to the summit of the peak, south 33°35′30″ west. Mr. Ferrer finds south 33°36′30″ west.

I.167

---

1. Also see ▼ Purdy, *Columbian Navigator*, 1824, p. 198. If Mr. Ferrer's measurement is correct, then the Peak of Orizaba would be higher than Popocatépetl.

2. For the details of these observations, see my *Recueil astronomique*, vol. II, p. 530.

XVIII
Map of the Port of Acapulco

The commerce of New Spain has only two outlets, the ports of Veracruz and Acapulco. The first port engages in trade with Europe and with the coast of Caracas, Havana, the United States, and Jamaica. The second port is the central point for South Sea and Asian trade. It receives ships from the Philippine Islands, Peru, Guayaquil, Panama, and the northwest coast of North America.

After having presented itinerary maps in the greatest detail of the road-ways of Europe and Asia, I felt it was important to publish precise maps of the ports of Veracruz and Acapulco. It would be difficult to find two harbors that offer a greater contrast. The port of Acapulco looks like an enormous man-made basin, while the port of Veracruz barely deserves the name of harbor: it is a rather unfortunate anchorage among the shoals.

The map of the port of Acapulco that I am presenting has never been published, although there are several copies of it in the Americas. It was made in 1791 by the officers who had embarked under Malaspina's orders in the corvettes Descubierta and Atrevida. I owe it to the kindness of Mr. Bauzá, the director of the Hydrographic Office in Madrid. The drawing completely corresponds to another map of Malaspina's, nearly one meter in length, which I studied in Acapulco during my stay there in 1803.

The longitude I assigned to the port of Acapulco (109°9′33″) at the residence of Contador Don Balthazar Álvarez Ordoño, is greater than the one adopted in the *Voyage de la Sútil et Mexicana au Détroit de Fuca*. But a paper in the Almanac of Cádiz indicates that the members of the Hydrographic Office now fix a position farther west than mine and identical to the one on my chronometer[1] by moving Acapulco to the longitude of Mexico City and neglecting the lunar distances that were observed on March 27 and 28, 1803.

Mr. Espinosa found Acapulco's location to the west of Paris by measuring the movement of time from the port of San Blas[2] 102°17′21″; using Jupiter's two satellites, observed simultaneously in Acapulco, Greenwich, and Paris at 102°24′15″, and using eight satellites compared with corrected tables at 102°15′47″, or at a mean of 102°19′8″. This is the same longitude that Mr. Antillón adopted in the analysis of his Map of the Americas. Furthermore, during the Malaspina expedition's stay in Acapulco in 1791, two stellar occultations were observed for which there were, however, no corresponding observations in Europe. The frigate captain Don Juan Ciscar calculated them using Bürg's tables. Using the occultation of February 19, he found Acapulco at 102°9′45″, and using the occultation of April 15, at 102°35′45″. Distances from the moon to the sun taken on February 12, but calculated by groups without correcting the location of the moon by observing a passage through the median, gave 102°24′37″.

Here are several determinations made by very different methods. All give a *slightly more western* longitude than the result of my own observations, which I adopted in my atlas before I was aware of Mr. Espinosa's interesting paper. Stellar occultations are certainly preferable to any other kind of observation, if they are done under favorable circumstances. But the results of the occultations of two of Leo's stars that were observed in Acapulco vary, according the Mr. Ciscar's calculation, by twenty-six minutes, and by five minutes (in arc) according Mr. Oltmanns. The Spanish astronomers also admit an extremely serious error into the tables for the first satellite. They fix it at thirty-five seconds of arc in time, whereas Mr. Oltmanns, comparing Mr. Delambre's tables with the observations made from January to May of 1791, found the error in the tables only -7.6″ for the immersions

---

1. See above, p. 35 in this edition.

2. It is significant that the longitude of San Blas is only based on two celestial observations: on a satellite that was compared to the tables and on a lunar eclipse. The results of these two observations vary by 5′45″ in arc. Captain Hall reports 107°39′42″ for San Blas, based on a stellar occultation. See *Extracts from a Journal, etc.* 1824, vol. II, p. 279.) Mr. Espinosa's paper provides an instructive example of the extreme caution that the use of chronometers demands, if one does not verify chronometric longitudes by other purely celestial observations. On Malaspina's expedition, *four of Arnold's chronometers* gave Port Mulgrave the same longitude of 142°38′57″ to nearly nine minutes of arc; however, lunar distances have proven that the actual longitude was 142°0′27″. The four clocks had all changed their diurnal movement at the same time.

and -14" for the emersions. According to the calculations published in the second volume of our *Recueil d'observations astronomiques*, he believes that the actual mean taken from the observations of the Malaspina expedition is 102°14'30", and that by allowing only half-value to our observations, the longitude of Acapulco could be fixed at 102°9'33"; in other words, it would be three and a half minutes farther west than the longitude given in my Mexican atlas. One should not be astonished that such doubts remain about the position of a South Sea port when one considers that, only a few years ago, the longitude of Amsterdam was uncertain not by three or four minutes but by one-third of a degree. Examining the details of my observations,[1] one finds 102°9'57" = 6ʰ48'39.8", using my chronometer or the movement of time in Guayaquil; by fourteen distances from the moon to the sun, taken on March 27 (and correcting the error of the tables using Greenwich observations), 6ʰ48'34"; by fifteen distances taken on March 28, 6ʰ48'23". At the small fort of San Carlos, Captain Basil Hall reported, by the movement of time from San Blas, a longitude of sixteen seconds of arc in time farther west than the longitude given by my own chronometer; he decided on 102°14'2". By simply presenting these facts, I believe I have proven that the longitude of Acapulco is already circumscribed within narrow margins of error, so that observations of occultations alone might fix it with more precision.

I.172

1. *Observations astronomiques*, vol. e II, pp. 439, 456, and 464.

XIX

Map of the Various Routes by Which Precious Metals Flow from One Continent to Another

The amount of gold and silver that the New Continent sends annually to Europe is more than nine-tenths of the total production of all mines in the known world. The Spanish colonies, for example, produce almost three and a half million *marks of silver* per year, while the annual yield[1] of all the European states, including Asiatic Russia, barely exceeds the sum of three hundred thousand marks. An extended stay in Spanish America gave me the opportunity to obtain clearer ideas about the wealth of precious metals of Mexico, Peru, New Granada, and the viceroyalty of Buenos Aires than the information found in the work of Adam Smith, Robertson, and Raynal. With this background, I was able to conduct research on the historical accumulation of precious metals over a long period of time in southern and southeast Asia. I presented the main results of my hypotheses in a small map I sketched on the high seas in 1804, during the crossing from Philadelphia to the coast of France. In its own way, this map shows the flow and outflow of precious metals. Their movement is generally from west to east, the opposite of the movement of the Ocean, the atmosphere, and the civilization of our species.

I.173

---

1. For the mines of Europe, see the excellent statistical table of mineral wealth that is appended to the *Mémoire générale sur les mines [De la richesse minérale]* by ▼ Héron de Villefosse, p. 240.

XX

Figures Representing the Surface Area of
New Spain and Its Intendancies,
Advances in the Mining of Metals, and
Other Subjects Relating to the European Colonies
in the Two Indies

In the figure that represents advances in gold and silver mining in New   I.174
Spain,[1] using the method from ▼ Mr. William Playfair's [*Tablaux*] *Arith-*
métique linéaire, I have marked the year 1742 as uncertain. According to
the table given to me at the Mint of Mexico City, 16,677,000 piasters were
coined in that year. There is a great difference between this quantity and
the mass of precious metals struck into coin in 1741 and 1743. The comparison with the table that presents only the exploitation of silver alone,
leads me to believe that the sum of 16,677,000 is inexact.

    The figures collected on plate XX explain what is written below[2] about
the extraordinary disproportion between the expanse of the colonies and
the surface (*area*) of the European metropoles. Representing the intendancies as squares inscribed within each other has emphasized the inequality
of the territorial division of New Spain. This graphic method is analogous
to the one Mr. Playfair was the first to use, quite ingeniously, in both his
commercial and political atlas and his statistical maps of Europe. Without
attaching much importance to this type of sketch, I cannot regard them as
mere displays of wit that are foreign to science. It has been said that the
map on which Mr. Playfair has charted the progress of the national debt of
England resembles the profile of the Peak of Tenerife. But we recall here
that physicists using quite similar illustrations have long demonstrated   I.175
the workings of the barometer and the hygrometer, and average monthly

---

1. *Mexican Atlas*, plate XIX.
2. Chapters I and VIII.

temperatures. It would hardly be suitable to use graphs to explain moral ideas, the prosperity of nations, the progress of their constitutional trajectory, or the more or less rapid decay of literature. Nevertheless, it is appropriate to use geometric figures to represent whatever relates to extent and quantity. Projections applied to elements of political economy speak to the senses without tiring the mind; above all, they have the advantage of fixing our attention on several important facts and facilitating numerical comparisons.

Tableau of Geographical Positions in the Kingdom of New Spain, Determined by Astronomical Observation

(The positions that are marked with an asterisk have been established either by triangulations or by vertical angles and azimuths)

PLACE NAME	NORTHERN LATITUDE			LONGITUDE WEST OF PARIS						OBSERVER'S NAME AND NOTES
				In Degrees			In Time			
Interior of New Spain	°	′	″	°	′	″	h	′	″	
Mexico City	19	25	101	101	25	30	6	45	42	Humboldt at San Agustín monastery
San Agustín de las Cuevas (village)	19	18	37	101	27	0	6	43	48	*Idem.*
Cerro Ajusco* (mountains)	19	18	37	101	27	0	6	46	11	*Idem.*
Venta de Chalco, (farm)	19	16	8	"	"	"	"	"	"	*Idem.*
Morán (mine)	20	10	4	100	46	0	6	43	4	*Idem.*
Actopan (village)	20	17	28	101	9	15	6	44	37	*Idem.*
Totonilco el Grande (village)	20	17	55	100	53	0	6	43	32	*Idem.*
Tisajuca (village)	"	"	"	101	11	30	6	44	46	*Idem.*
Toluca (village)	19	16	19	101	41	45	6	46	47	*Idem.*

(continued)

PLACE NAME	NORTHERN LATITUDE			LONGITUDE WEST OF PARIS						OBSERVER'S NAME AND NOTES
				In Degrees			In Time			
I.177 Nevado de Toluca	19	11	33	101	45	38	6	47	2 ½	*Idem.*
San Juan del Río (city)	"	"	"	102	12	30	6	48	50	Humboldt at San Agustín monastery.
Querétaro (city)	20	36	39	102	30	30	6	50	2	*Idem.*
Salamanca (city)	20	40	103	15	0	6	53	0	*Idem.*	
Guanajuato (city)	21	0	15	103	15	0	6	53	0	*Idem,* at the home of Don Diego Rul.
Valladolid, (city)	19	42	0	103	12	15	6	52	49	*Idem,* at the Bishop's palace.
Pátzcuaro (city)	"	"	"	103	40	0	6	54	40	*Idem.*
Las Playas de Jorullo (farm)	"	"	"	103	50	33	6	55	22	*Idem.*
Jorullo volcano*	"	"	"	103	51	48	6	55	27	*Idem.*
Puente de Ixtla (farm)	18	37	41	101	34	45	6	46	19	*Idem.*
Tehuilotepec (city)	"	"	"	101	48	0	6	47	12	*Idem,* near the steam engine.
Tasco (city)	18	35	0	101	49	0	6	47	16	*Idem.*
Tepecuacuilco (village)	18	20	0	101	48	0	6	47	12	*Idem.*
Fuente de Estola (inn)	"	"	"	101	44	0	6	46	56	*Idem.*
Mexcala (village)	17	56	4	101	49	0	6	47	16	*Idem.*
Popocatépetl* (volcano)	18	59	47	100	53	15	6	43	33	*Idem,* mountain top.
San Nicolás de los Ranchos (village)	19	2	0	100	41	0	6	42	44	*Idem.*
Ixtaccihuatl* (mountain)	19	10	0	100	55	0	6	43	40	*Idem.*
Cholula Pyramid (ancient monument)	19	2	6	100	33	30	6	42	14	*Idem,* top of the pyramid.

PLACE NAME	NORTHERN LATITUDE			LONGITUDE WEST OF PARIS						OBSERVER'S NAME AND NOTES	
				In Degrees			In Time				
La Puebla de los Angeles (city)	19	0	15	100	22	45	6	41	31	*Idem.*	
Venta de Soto (farm)	19	26	30	"	"	"	"	"	"	*Idem.*	
Cofre de Perote (mountain)	19	28	57	99	28	45	6	37	55	*Idem.*	
Las Vigas (village)	19	37	37	"	"	"	"	"	"	*Idem.*	
Xalapa (city)	19	30	8	99	15	0	6	37	0	*Idem.*	I.178
Cerro de Macultepec (mountain)	19	31	49	99	14	35	6	36	58 ½	HUMBOLDT.	
Pico de Orizaba* (volcano)	19	2	17	96	35	15[1]	6	38	21	HUMBOLDT and FERRER, top.	
El Encero (farm)	19	28	25	99	8	32	6	36	34	FERRER.	
Texcoco* (city)	19	30	40	101	11	15	6	44	45	VELASQUEZ.	
Zumpango* (village)	19	46	52	101	24	0	6	45	36	*Idem.*	
El Peñol* (hill)	19	26	4	101	22	30	6	45	30	*Idem.*	
Xaltocan* (village)	19	42	47	101	21	15	6	45	25	*Idem.*	
Tehuiloyuca* (village)	19	43	17	101	28	5	6	45	54	*Idem.*	
Hacienda de Xalapa (farm)	19	47	58	101	29	45	6	45	54	*Idem.*	
Cerro de Chiconautla* (hill)	19	38	39	101	16	0	6	45	4	*Idem.*	
San Miguel de Guadalupe* (convent)	19	28	48	101	24	45	6	45	39	*Idem.*	
Huehuetoca* (village)	19	48	38	101	32	45	6	46	11	*Idem.*	

1. Instead of 96°35′15″, read 99°35′15″ [Humboldt's correction from Errata in IV.311].

(continued)

PLACE NAME	NORTHERN LATITUDE			LONGITUDE WEST OF PARIS						OBSERVER'S NAME AND NOTES
				In Degrees			In Time			
Garita de Guadalupe* (gates of Mexico City)	19	28	38	101	24	45	6	45	39	*Idem.*
Cerro de Sincoque* (hill)	19	49	28	101	33	30	6	46	14	*Idem.*
Hacienda de Santa Iñes* (farm)	19	42	25	101	24	15	6	45	37	*Idem.*
Cerro de San Cristóbal* (mountain)	19	35	5	101	21	30	6	45	26	*Idem.*
Puente del Salto* (bridge)	19	54	30	101	36	0	6	46	24	*Idem.*

Eastern Shores of New Spain.

PLACE NAME	NORTHERN LATITUDE			LONGITUDE WEST OF PARIS						OBSERVER'S NAME AND NOTES
Campeche, (city)	19	50	45	92	50	45	6	11	23	Ferrer and Cevallos.
Punta de la Desconocida	20	49	45	92	44	30	6	10	58	Cevallos and Herrera.
Castillo del Sisal	21	10	0	92	19	45	6	9	19	*Idem.*
Alacrán (western point)	22	27	50	92	7	40	6	8	30	*Idem.*
Alacrán (northern reaches)	22	35	15	92	0	45	6	8	3	Cevallos and Herrera.
Mouth of Río de los Lagartos	21	34	0	90	30	15	6	2	1	*Idem.*
Punta SO del Puerto	22	21	30	91	58	15	6	7	57	*Idem.*
Northern point of Conboy	21	33	30	89	5	0	6	56	20	*Idem.*
Southern point of Conboy	21	28	50	89	4	0	6	56	45	*Idem.*

I.179

PLACE NAME	NORTHERN LATITUDE			LONGITUDE WEST OF PARIS						OBSERVER'S NAME AND NOTES
				In Degrees			In Time			
Bajo del Alerta	21	33	0	89	11	15	6	56	45	*Idem.*
Shoals of Diez Brazas	20	32	10	94	14	5	6	15	56	*Idem.*
Small island SW of triangle	20	55	50	94	31	52	6	18	7 ½	*Idem.*
Bajo del Obispo	20	30	14	94	30	23	6	18	1 ½	*Idem.*
Veracruz (port)	19	11	52	98	29	0	6	33	56	HUMBOLDT and FERRER.
Island of Sacrifices (center)	19	10	10	98	26	40	6	33	47	FERRER.
Shoals of the Pájaro	19	10	55	98	26	10	6	33	45	*Idem.*
Isla Verde	19	11	16	98	25	26	6	33	42	*Idem.*
Islote Blanquil-las, (center)	19	12	55	98	26	45	6	33	47	*Idem.*
Anegada de Fuera, south-ern point	19	12	12	98	24	35	6	33	38	*Idem.*
Northern [point]	19	12	55	98	25	5	6	33	40	*Idem.*
Shoals of the Gallega	19	13	20	98	28	22	6	33	53 ½	*Idem.*
Punta Gorda	19	14	30	98	31	20	6	34	5	*Idem.*
Mouths of Rio Antigua	19	18	41	98	37	17	6	34	29	*Idem.*
Bernal Chico	19	37	45	98	46	5	6	35	4	*Idem.*
Bernal Grande	19	39	42	98	45	43	6	35	3	*Idem.*
Punta Mari Andrea	19	43	15	98	45	43	6	35	3	*Idem.*
Barra de Tamiahua	21	15	48	"	"	"	"	"	"	*Idem.*
Santander[1] (city)	23	45	18	100	32	23	6	42	9 ½	*Idem.* I.180

1. According to the final calculations of Mr. Ferrer:

(continued)

PLACE NAME	NORTHERN LATITUDE			LONGITUDE WEST OF PARIS						OBSERVER'S NAME AND NOTES
				In Degrees			In Time			
Lago de San Fernando, or La Carbonera	24	36	0	100	18	40	6	41	15	FERRER.
Mouth of the Río Bravo del Norte	25	55	0	99	51	10	6	39	25	*Idem.*

Western Coasts of New Spain.

Acapulco (port)	16	50	29	102	6	0	6	48	24	HUMBOLDT at the governor's mansion.
Western reaches of the Playas de Cujuca	17	15	0	103	5	15	6	52	21	MALASPINA expedition.
Morro Petatlan (hill)	17	32	0	103	48	45	6	55	15	*Idem.*
Port de Selagua, (uncertain)	19	6	0	106	53	5	7	7	32	*Idem.*
Cabo Corrientes	20	25	30	107	59	0	7	11	56	*Idem.*
Small island NNW of Cabo Corrientes	20	45	0	108	7	15	7	12	29	*Idem.*
Cerro del Valle (hill)	21	1	30	109	35	0	7	18	20	*Idem.*
Marías Islands (southern cape of the easternmost)	21	16	0	108	37	45	7	14	31	*Idem.*
San Juan Mountain	21	26	15	107	23	0	7	9	32	*Idem.*
San Blas (port)[1]	21	32	48	107	37	45	7	10	31	*Idem.*
Piedra Blanca	21	33	0	107	47	45	7	11	11	*Idem.*
San Juanico Island	21	45	30	109	1	35	7	16	6	*Idem.*
Isabela Islet	21	50	30	108	17	5	7	13	8	*Idem.*

1. According to Captain Basil Hall: lat. 21°32′24″, long. 107°39′42″.

PLACE NAME	NORTHERN LATITUDE			LONGITUDE WEST OF PARIS						OBSERVER'S NAME AND NOTES	
				In Degrees			In Time				
Cabo San Lucas	22	52	23	112	10	38	7	28	42	CHAPPE, DOZ, and MEDINA.	
San José Mission (village)	23	3	25	112	1	8	7	28	4	*Idem.*	
Todos los Santos Mission	23	26	0	112	38	15	7	30	33	MALASPINA'S expedition.	
San Lázaro Mountain	24	47	0	114	41	15	7	38	5	*Idem.*	I.181
Mountain north of Abreojos	26	59	30	116	8	15	7	44	33	*Idem.*	
Cedar Island, (southern point)	28	2	10	117	43	15	7	50	33	*Idem.*	
Isla de San Benito, (highest section)	28	18	22	118	6	15	7	52	25	*Idem.*	
Isla Guadalupe, (southern cape)	28	53	0	120	37	15	8	2	29	*Idem.*	
Isla de San Bernardo	29	40	40	118	17	15	7	53	9	*Idem.*	
Isla de San Martin, or de los Coronados largest and easternmost island)	32	25	10	119	38	55	7	58	36	*Idem.*	
San Diego (port)	32	39	30	119	38	15	7	58	33	VANCOUVER and MALASPINA.	
Isla San Salvador (southern point)	32	43	0	120	50	15	8	3	21	MALASPINA'S expedition.	

(continued)

PLACE NAME	NORTHERN LATITUDE			LONGITUDE WEST OF PARIS						OBSERVER'S NAME AND NOTES
				In Degrees			In Time			
Isla S. Nicolás (western cape)	33	16	30	121	56	15	8	7	45	*Idem.*
San Juan (mission)	33	29	0	120	13	30	8	0	54	VANCOUVER and MALASPINA.
Isla de Juan Rodríguez Cabrillo, (western cape)	34	0	0	122	51	15	8	11	25	MALASPINA expedition.
Santa Buenaventura	34	17	0	121	45	30	8	7	2	VANCOUVER.
Presidio de Santa Bárbara (mission)	34	26	0	122	5	30	8	8	22	VANCOUVER and MALASPINA.
Monterey (presidio)	36	36	0	124	11	8	8	16	44 ½	MALASPINA expedition.
Punta del Año Nuevo	37	9	15	124	42	53	8	18	51 ½	*Idem.*
Farallones (rocks)	37	48	10	125	21	15	8	21	25	*Idem.*
San Francisco (port)	37	48	30	134	57	0	8	19	48	VANCOUVER and MALASPINA.
Cabo Mendocino	40	29	0	126	48	45	8	27	15	MALASPINA EXPEDITION.
I.182 Nootka (port)	49	35	13	128	55	15	8	35	41	*Idem.* (Both this position and the previous one are outside of New Spain.)

Revillagigedo Islands

Isla de Santa Rosa (center)	18	37	0	116	23	45	7	54	33	COLLINET, CAMACHO and TORRES (Mr. Espinosa's paper).

PLACE NAME	NORTHERN LATITUDE			LONGITUDE WEST OF PARIS						OBSERVER'S NAME AND NOTES
				In Degrees			In Time			
Isla del Socorro, (mountain top, which is over 1,115 meters high)	18	48	0	112	29	15	7	29	57	*Idem.*
Roca Partida	19	4	0	113	25	45	7	33	43	*Idem.*
Isla de San Benedito, (southern cape)	19	15	40	113	13	45	7	28	55	*Idem.*

More Uncertain Positions

	°	′	″	°	′	″	h	′	″	
Guatulco (port)	15	44	0	"	"	"	"	"	"	PEDRO DE LAGUNA.
Barra de Manialtepec	15	47	0	"	"	"	"	"	"	*Idem.*
Pachutla (village)	15	50	0	"	"	"	"	"	"	*Idem.*
Xamiltepec (village)	16	7	0	"	"	"	"	"	"	*Idem.*
Guiechapa (village)	15	25	0	"	"	"	"	"	"	*Idem.*
Ometepec (village)	16	37	0	"	"	"	"	"	"	*Idem.*
Nochistlán (village)	17	16	0	"	"	"	"	"	"	*Idem.*
Teposcolula	17	18	0	"	"	"	"	"	"	*Idem.*
San Antonio de los Cues (village)	18	3	0	"	"	"	"	"	"	*Idem.*
Guadalajara (city)	21	9	0	105	22	30	7	1	30	MASCARÓ and RIVERA.
Zacatecas (city)	23	0	0	103	55	0	6	55	40	Count DE LA LAGUNA.
Real del Rosario (mine)	23	30	0	108	26	30	7	13	46	MASCARÓ and RIVERA.

(continued)

PLACE NAME	NORTHERN LATITUDE			LONGITUDE WEST OF PARIS						OBSERVER'S NAME AND NOTES
				In Degrees			In Time			
Durango (town)	24	25	0	103	55	0	7	3	40	Oteiza.
Presidio del Pasaje	25	28	0	105	33	30	7	2	14	Mascaró and Rivera.
Villa del Fuerte	26	50	0	110	333	30	7	22	14	*Idem.*
Real de los Álamos, (mine)	27	8	0	111	23	30	7	25	34	*Idem.*
I.183 Presidio de Buenavista	27	45	0	112	28	30	7	29	45	*Idem.*
Chihuahua (city)	28	50	0	106	50	0	7	7	40	Mascaró and Lafora.
Arispe (town)	30	36	0	111	18	30	7	25	14	Mascaró and Rivera.
Presidio de Janos	"	"	"	109	5	30	7	16	22	Mascaró.
Presidio del Altar	31	2	0	114	6	0	7	36	24	Mascaró and Rivera.
Paso del Norte (Presidio)	32	9	0	107	3	0	7	8	12	Mascaró.
Junction of the Río Gila and the Colorado	32	45	0	"	"	"	"	"	"	Fathers Díaz and Font.
Las Casas Grandes (near Río Gila)	33	30	0	"	"	"	"	"	"	Father Font.
Santa Fe (city)	36	12	0	107	13	0	7	8	52	Lafora.

Barra de Santander	lat.	23°	45′	18″	long.	100°	18′	45″
Barra de Tampico		22°	15′	30″		100°	12′	15″
Alvarado		18°	34′	16″		98°	28′	15″
Assuming Veracruz to be at						98°	28′	15″

Tableau of the Most Remarkable Elevations Measured in the Interior of New Spain

The work published under the title *Nivellement barométrique fait dans les régions équinoxiales du Nouveau-Continent, en* 1799–1804, contains almost two hundred points in the interior of New Spain, whose elevation above sea level I determined either by using a barometer or by trigonometric methods. In the following table, I have merely assembled the absolute heights of the most remarkable mountains and cities. The points that are marked with an asterisk are dubious. The page number near each point refers to the text of the special Statistics of Mexico. One may also consult my *Recueil d'observations astronomiques et de mesures barométriques* (vol. I, pp. 318–34) compiled by Mr. Oltmanns.

NAME OF THE OBSERVATION SITE	ELEVATION ABOVE SEA LEVEL ACCORDING TO MR. LAPLACE'S FORMULA	
	In Meters	In Toises
POPOCATÉPETL VOLCANO, Volcán Grande de México or de Puebla	5,400	2,771
PICO DE ORIZABA OR CITLALTEPETL	5,295	2,717
NEVADO DE IZTACCÍHUATL, Sierra Nevada de México	4,786	2,456
NEVADO OF TOLUCA, on the Frailes rock	4,621	2,372
COFRE DE PEROTE OR NAUHCAMPATEPETL	4,089	2,098

(*continued*)

NAME OF THE OBSERVATION SITE	ELEVATION ABOVE SEA LEVEL ACCORDING TO MR. LAPLACE'S FORMULA	
	In Meters	In Toises
CERRO DE AXUSCO, six leagues SSW of Mexico City	3,674*	1,885*
TANCITARO PEAK	3,200*	1,642
EL JACAL, top of the Cerro de las Navajas	3,124	1,603
MAMANCHOTA or ÓRGANOS DE ACTOPAN, NE of Mexico City	2,977	1,527
COLIMA VOLCANO	2,800*	1,437
JORULLO VOLCANO, in the intendancy of Valladolid	1,301	667
MEXICO CITY, at the San Agustín convent	2,277	1,168
PACHUCA	2,484	1,274
MORBAN, mine near Real del Monte	2,595	1,331
REAL DEL MONTE, mine	2,781	1,427
TULA, city	2,053	1,053
TOLUCA, city	2,688	1,379
CUERNAVACA, city	1,656	849
TAXCO, city	1,784	915
CHILPANCINGO, city	1,380	708
PUEBLA DE LOS ANGELES, city	2,194	1,126
PEROTE, small settlement	2,354	1,208
XALAPA, city	1,321	678
VALLADOLID, city	1,952	1,001
PATZCUARO, city	2,202	1,130
CHARO, city	1,907	978
VILLA DE ISLAHUACA, in the intendancy of Valladolid	2,585	1,326
SAN JUAN DEL RÍO, little settlement	1,978	1,015
QUERÉTARO, city	1,940	995
CELAYA, city	1,835	941
SALAMANCA, city	1,757	902
GUANAJUATO, city	2,084	1,069
VALENCIANA MINE	2,328	1,194
DURANGO, city	2,087*	1,071

The following elevations, taken from Mr. Sonneschmidt's *Mineralogische*   I.186
Beschreibung [Mineralogical description], may be added to the height of the
two hundred points that I have measured in the Kingdom of New Spain. This
scholar has only shown the barometric elevations, but Oltmanns calculated
them according to ▼ Laplace's formula, assuming that the mercury column in
Mr. Sonneschmidt's barometer was 1.9 lines too short,[1] and the temperature
of the instrument 2° R. higher than that of the exterior air.

PLACE NAME	BAROMETRICAL ELEVATION		AIR TEMPERATURE	ABSOLUTE ELEVATION		OBSERVATIONS
				in toises	in meters	
	po	li	° R.			
Cardonal	22	1.9	18	1,076	2,097	Intendancy of Mexico City, NE part
Real del Doctor	20	5.9	16	1,419	2,767	*Id.*
Zimapán	22	11.9	18	900	1,755	*Id.*
Valley between Zimapán and El Doctor	24	10.9	24	564	1,099	*Id.*
Mecameca	21	0.9	14.5	1,286	2,507	On the Mexican road at the Puebla volcanoes.
Fraile Peak	15	5.9	1 ½	2,567	5,004	Part of Popocatepetl.
Upper pine tree limit, on Popocatepetl	18	4.9	9.5	1,867	3,639	At the Cofre near Perote, I found this limit to be at an elevation of 2,022 toises.

1. This result is based on the comparison of barometric elevations given by Mr. Sonneschmidt for four sites where I took my instruments. The difference between our observations is

For	Mexico City	2.7	
	Real del Monte	1.9	1.9[lines]
	Pachuca	2.0	
	Guanajuato	0.9	

I.187   The height of 2,456 toises that I assigned to the *Sierra Nevada de Puebla* (Iztaccihuatl) is not based on a direct measurement but on elevation angles, azimuths, and distances. Mr. Sonneschmidt was more fortunate than I: he brought his barometer to the summit of Iztaccihuatl and saw the mercury level stabilize there at 16^{po} 6.4^{li}. Assuming a temperature of 6.5°R., this reading, according to Mr. Oltmanns's hypsometric tables, gives only 2,317 toises, or 4,516 meters. I do not know, however, if Mr. Sonneschmidt measured the same part of the Sierra Nevada where I took elevation angles from the terrace of the School of Mining in Mexico City and from the pyramid of Cholula.[1]

The farm at Pátzcuaro, near Zipaquirá[2] is, according to ▼ Mr. Ontiveros, at 880 toises (1,670 meters) above sea level, with the barometer stabilizing there at 23^{po} 2^{li} and the thermometer at 19° R.

Mr. Alzate confirms[3] that he saw the barometer stabilize at 18^{po} 3^{li} on the summit of the Picacho de San Tomás, which is part of the Cerro de Ajusco, and that "as a result, the Picacho is 4,300 *varas* above sea level." According to Mr. Laplace's formula and assuming the air temperature to be 9° R., Mr. Oltmanns reported 1,899 toises, or 3,702 meters.

END OF THE GEOGRAPHICAL INTRODUCTION

1. *Recueil d'observations astronomiques*, vol. II, p. 574.
2. Intendancy of Valladolid.
3. Sigüenza [y Góngora,], *Plano de las cercanías de México*.

Political Essay on
the Kingdom of New Spain

BOOK I

General Remarks on the Total Area and
the Physical Aspect of the Country—
The Influence of the Uneveness of
the Terrain on Climate, Agriculture,
Commerce, and Military Defense

CHAPTER I

The Extent of the Spanish Possessions in the Americas—
A Comparison of These Possessions with the British Colonies and
with the Asian Part of the Russian Empire—The Naming of New Spain
and Anahuac—The Boundary of the Aztec Kings' Empire.

Before drawing up the political tableau of the Kingdom of *New Spain*, it is important to consider the area and population of the Spanish possessions in both Americas. By generalizing ideas and considering the relationship between each colony, its neighboring colonies, and the metropole, one is sure to obtain accurate results and to give the country that one is describing the merit that it deserves for its territorial wealth.

I.190

The Spanish possessions on the New Continent occupy the vast expanse of land between 41°43′ southern latitude and 37°48′ northern latitude. This space of seventy-nine degrees is not only equivalent to the entire length of Africa, but also surpasses by far the breadth of the Russian Empire, which encompasses 167 degrees of longitude below a parallel whose degrees are no more than half the equatorial degrees.

The southernmost point of the New Continent that is inhabited by the Spanish is ▼ Fort *Maullin*, near the tiny village of *Carelmapu*[1] on the coast of *Chile*, opposite the northernmost reaches of the island of ▼ *Chiloé*. A road from *Valdivia* to *Fort Maullin* has been started; it is a bold but eminently useful undertaking, since constantly rough seas prevent mariners from approaching this dangerous coast for the better part of the year. There are no Spanish settlements to the south and southeast of Fort *Maullin*, in the Gulf of *Ancud* and the Gulf of *Reloncavi*, by which one reaches the great lakes of *Nahuelhapi* and *Todos los Santos*. There are, however, Spanish settlements on the islands near the east coast of *Chiloé*, up to 43°34′ southern latitude,

I.191

1. See note A at the end of this book.

where the island of *Caylin* (opposite the high peak of *Corcobado*) is inhabited by a few families of Spanish origin.

The northernmost point of the Spanish colonies is the Mission of *San Francisco* on the coast of *New California*, seven leagues northwest of *Santa Cruz*. As a result, the Spanish language has spread over an expanse of more than 1,900 leagues. Under the wise administration of ▼ Count *Floridablanca*, regular mail communication was established from *Paraguay* to the northwest coast of North America. A monk assigned to the *Guaraní* Indian Mission can correspond regularly with another missionary living in *New Mexico* or the nearby lands of *Cabo Mendocino*, without their letters ever straying far from the continent of Spanish America.

The American domains of the King of Spain are twice the size of the United States from the Atlantic Ocean to the South Sea; they are four times as large as the entire British Empire in India. Their total area is only one-quarter less than either Asiatic Russia or—to use an even more striking comparison—the half-surface of the moon. I thought it would be interesting to draw up a tableau that would show these differences and the striking disproportion between the *area* and population of the mother country compared to those of the colonies. To make this disproportion more obvious, I sketched the drawing on the last plate according to exact scales. A red parallelogram that serves as a base represents the surface area of the metropoles; a blue parallelogram that sits on this base shows the Spanish and British possessions in America and Asia. These tables, which are analogous to Mr. *Playfair*'s, have something frightening and ominous about them, especially when one beholds the great catastrophe represented by the fourth figure, which has become the source of the prosperity of the United States. This plate alone can give rise to important considerations for those who are called upon to watch over the happiness and the tranquility of the colonies. The fear of a future evil is certainly not in itself a very noble cause for action, but it should be a powerful one for great political bodies, as it is for simple individuals.

The Spanish possessions in the Americas are divided into nine large governments that may be considered independent of each other. Five of these nine governments—the viceroyalties of *Peru* and *New Granada*, the *capitanías generales* of *Guatemala*, *Puerto Rico*, and *Caracas*—are entirely located in the *Torrid Zone*. The four other divisions—the viceroyalties of *Mexico*, *Buenos Aires*, the *Capitanía general* of *Chile*, and that of *Havana*, which includes the Floridas—encompasses countries that are largely outside the two tropics, that is, in the temperate zone. We shall see in the course

I.192

I.193

of this work that this position alone does not determine what these regions produce. The confluence of several physical causes, such as the great height of the Cordilleras, their enormous masses, and the number of elevated plateaus more than two or three thousand meters above sea level, give some of these equinoctial regions a temperature that is suitable for growing wheat and European fruit trees. Geographical latitude has little effect on the productivity of a country where nature has gathered every climate on the ridges and slopes of its mountains.

Among the colonies subject to the domination of the King of Spain, Mexico at present occupies the prime position, both because of its territorial riches and because of its favorable location for trade with Europe and Asia. We shall only discuss the political value of the country here, considering its present stage of civilization, which is far superior to that of the other Spanish possessions. Several branches of agriculture have probably achieved a greater degree of perfection in the province of *Caracas* than in New Spain. The fewer mines a colony has, the more its inhabitants tend to focus on agricultural production. The soil is more fertile in the provinces of *Cumaná*, *Nueva Barcelona*, and *Venezuela*; it is more fertile on the banks of the *Lower Orinoco* and in the northern part of *New Granada* than in the Kingdom of Mexico, many regions of which are barren, arid, and devoid of vegetation. But considering the size of the Mexican population, the number of good-sized cities that are located in close proximity to one another, the prodigious value of its metal mining, and the influence of this activity on European and Asian trade; and recalling the low levels of civilization found in the rest of Spanish America, one is tempted to lend credence to the preference that the court of Madrid has granted to Mexico over the rest of its colonies.

I.194

The name *New Spain* refers generally to the vast expanse of land controlled by the viceroy of Mexico. If one takes the term in this sense, one must consider the thirty-eighth and tenth parallels as the northern and southern limits of the country. Nevertheless, the *captain-general of Guatemala*, who is considered an administrator, is only somewhat dependent on the viceroy of Mexico. According to its political division, the Kingdom of *Guatemala* encompasses the governments of *Costa Rica* and *Nicaragua*. It borders on the Kingdom of *New Granada*, which includes *Darién*, the Isthmus of *Panama*, and the province of *Veragua*.[1] Throughout this work, whenever

I.195

1. The northwest boundary of New Granada passes through Punta Careta (latitude 9°36′, longitude 84°43′) on the Antillean Sea coast and through Cabo Burica (latitude 8°5′,

we use the names *New Spain* and *Mexico*, we are excluding the *Capitanía general de Guatemala*, an extremely fertile country, densely populated in comparison with the rest of the Spanish possessions, and much better cultivated, since the terrain, stressed by volcanic action, contains almost no metal mines. I consider the intendancies of *Mérida* and *Oaxaca* as the southernmost and, at the same time, easternmost parts of New Spain. The boundary separating Mexico from the Kingdom of *Guatemala* meets the Great Ocean to the east of the port of *Tehuantepec*, near *Barra de Tonala*; it hits the coastline of the Sea of the Antilles near the bay of Honduras.

The name *New Spain* was first given, in 1518, exclusively to the province of *Yucatán*, where ▼ *Grijalva*'s soldiers had found carefully tended fields, buildings several stories high, and populated cities. In his first letter to ▼ Emperor Charles V in 1520, *Cortés* already extended the name New Spain to include all of *Moctezuma*'s empire. If one believes *Solís*, this empire extended from *Panama* to *Nueva California*. But the scholarly research of the Abbot *Clavijero*,[1] a Mexican historian, informs us that *Moctezuma*, the *Sultan of Tenochtitlan*, ruled over a much smaller territory. To the east, his kingdom was bordered by the *Guasacualco* and the *Tuxpan* rivers, and to the west, by the plains of *Soconusco* and the port of *Zacatula*. A glance at my general map of New Spain, divided into intendancies, reveals that, according to the boundaries I have just traced, Moctezuma's empire only included the intendancies of *Veracruz*, *Oaxaca*, *Puebla*, *Mexico*, and *Valladolid*. I would estimate its area to be between eighteen or twenty thousand square leagues.

At the beginning of the sixteenth century, the *Santiago* River separated the agrarian peoples of *Mexico* and *Michoacán* from the barbarian and nomadic hordes called *Otomí* and *Chichimeca*. These savages often made forays as far as *Tula*, a town near the northern edge of the valley of *Tenochtitlan*. They occupied the plains of *Zelaya* and *Salamanca*, whose fine cultivated fields and many scattered farms we now admire.

Nor should the name *Anahuac* be confused with that of *New Spain*. Before the conquest, Anahuac designated only the country between 14° and 21° of latitude. Besides Montezuma's Aztec empire, the small republics of *Tlaxcala* and *Cholollan*, the kingdom of *Texcoco* (or *Acolhoacan*) and that of

I.196

I.197

---

longitude 85°7′). Humboldt, *Relation historique*, vol. III, p. 78. For the boundaries of Guatemala, see p. 76. [I.Q?]

1. [Clavijero,] *Dissertazione sopra i confini di Anahuac.* See *Storia antica del Messico*, vol. IV, p. 265.

Michoacán, which included a part of the intendancy of Valladolid, belonged to what was known as *Anahuac*.

Even the name *Mexico* itself is of Indian origin. In the *Aztec* language, it signifies the home of the war god, whose name was *Mexitli* or *Huitzilopochtli*. But it seems that before the year 1530, the city was more commonly called *Tenochtitlan* than *Mexico*. *Cortés*,[1] who had made but little progress in the local languages, corrupted the name of the capital into *Temixtitlan*. In a work that is exclusively devoted to the Kingdom of Mexico, such etymological observations should not be seen as too meticulous. Furthermore, the audacious man who overthrew the *Aztec* monarchy considered it vast enough to advise[2] Charles V to add the title of *Emperor of New Spain* to that of Roman Emperor.

One is tempted to compare both the total area and the population of Mexico with those of the two empires with which this beautiful colony enjoys relations of union and rivalry. Spain is five times smaller than Mexico. Barring any unforeseen disasters, one may estimate that in less than a century, the population of New Spain will equal that of the metropole. The United States of North America, since the ▼ cession of Louisiana and since it does not *wish* to recognize any other border than the *Río Bravo del Norte*, now comprises 260,000 square leagues (at twenty-five to the equinoctial degree). Its population is not much larger than that of Mexico, as we shall see later, when we carefully examine the population and surface area of New Spain.[3] I.198

If the political strength of two states depended solely on the area of the globe that they occupy and the number of their inhabitants; if the nature of the terrain, the contour of the coastline, the climate, the people's vitality, and, above all, the degree of perfection of its social institutions were not the principal aspects of this great dynamic calculation, the Kingdom of New Spain could, at the present time, rival the confederation of American republics. Both feel the inconvenience of a population that is too unevenly distributed. Although its terrain and climate are less favored by nature, the I.199

1. [Hernán Cortés], *Historia de Nueva España* (Mexico, 1770), p. 1.
2. In his first letter, dated Villa Segura de la Frontera, October 30, 1520, Cortés wrote, "Las cosas de esta tierra son tantas y tales que Vuestra Alteza se puede intitular de nuevo Emperador de ella, y con título y no menos mérito, que él de Alemaña, que por la gracia de Dios, Vuestra Sacra Magestad posee" [The things of that land are such that Your Highness can install Himself as its new Emperor, and with a title and no less merit than that of Germany, which, by the grace of God, is in Your Holy Majesty's possession] (Lorenzana [1770 edition], p. 38).
3. In 1824, the population of New Spain (without Guatemala) was estimated at 6,800,000 and the population of the United States at 10,220,000. In 1800, the latter had only been 5,306,000; in 1810, 7,240,000 (Humboldt, *Relation historique*, vol. III, p. 70 [I.Q?]).

population of the United States is growing at an infinitely greater rate, for the simple reason that, unlike the Mexican population, it does not include almost two and a half million aborigines.[1] These Indians have been brutalized by the despotism of the ancient Aztec sovereigns and by the humiliations of the early conquistadores. Although they are protected by Spanish laws, which are, in general wise and humane, they benefit very little from this protection, since they are so distant from the supreme authority. The Kingdom of New Spain has one marked advantage over the United States: there are almost no slaves, neither African nor of mixed race. The European colonists have only rightly begun to appreciate this advantage since the tragic events of the revolution in Saint-Domingue. For it is so true that the fear of physical violence has a more powerful effect than moral considerations on the true interests of society, or on the principles of philanthropy and justice that are so often proclaimed in parliament, the constituent assembly, and the writings of philosophers!

The number of African slaves in the United States has reached over one million:[2] they represent one-sixth of the entire population. The southern states, whose political influence has increased since the acquisition of Louisiana, have imprudently increased the number of slaves. The black slave trade [traite des nègres] has at least been abolished, through a national act equally motivated by justice and prudence. This would have been done long ago, had the law allowed the President of the United States (a magistrate[3] whose name is dear to the true friends of humanity) to oppose the introduction of slaves and thereby spare future generations great misfortunes.

I.200

To facilitate the comparison of the large political divisions of Spanish America, we shall insert the following table (which Mr. Humboldt has just published in the third volume of his *Relation historique* [*Personal Narrative of the Voyage to the Equinoctial Regions*], p. 64) at the end of this chapter.

---

1. We shall see below that, in 1810, ▼ Mr. Navarro estimated the number of pure-blooded Indians living in New Spain at 3,676,000, a figure that represents more than half of all the Indians in Spanish America.

2. In 1824, they numbered 1,620,000, or one-quarter of all the free and enslaved blacks on the New Continent.

3. Mr. *Thomas Jefferson*, author of the excellent *Notes on [the State of] Virginia*.

MAJOR POLITICAL DIVISIONS	SURFACE AREA IN SQUARE LEAGUES, 20 PER EQUINOCTIAL DEGREE	POPULATION (1823)
I. SPANISH-AMERICAN POSSESSIONS	371,380	16,785,000
Mexico or New Spain	75,830	6,800,000
Guatemala	16,740	1,600,000
Cuba and Puerto Rico	4,430	800,000
Colombia: Venezuela,	33,700	785,000
New Granada, and Quito	58,250	2,000,000
Peru	41,420	1,400,000
Chile	14,240	1,100,000
Buenos Aires	126,770	2,300,000
II. PORTUGUESE-AMERICAN POSSESSIONS (BRAZIL)	256,990	4,000,000
III. ANGLO-AMERICAN POSSESSIONS (UNITED STATES)	174,300	10,220,000

When comparing the numerical values of surface area in Mr. Humboldt's work, one should keep in mind that this traveler consistently used *common square leagues* of twenty-five to a degree in his *Political Essay [on the Kingdom of New Spain]*, as all statistical works published in French have done until now, whereas in the *Relation historique [Personal Narrative of the Voyage to the Equinoctial Regions]*, he used *nautical leagues* of twenty to a degree, equivalent to three minutes (in arc), much more suitable to scientific discussions, especially those of astronomical and physical geography. One square nautical league equals 1.5625 common square leagues. E—R.

I.201

CHAPTER II

Configuration of the Coastline—Points Where the Two Seas Are
Closest—General Remarks on the Possibility of Connecting the South Sea
and the Atlantic Ocean—The Peace and the Tacoutché-Tessé Rivers—The
Sources of the Río Bravo and the Río Colorado—The Isthmus
of Tehuantepec—The Lake of Nicaragua—The Isthmus of Panama—
The Bay of Cúpica—The Chocó Canal—The Río Guallaga—
The Gulf of St. George.

The Kingdom of *New Spain*, the northernmost part of all Spanish America, extends from the sixteenth to the thirty-eighth degree of latitude. From south-southeast to north-northwest, the length of this vast region is almost 270 myriameters (or 610 common leagues); its greatest width is at the thirtieth parallel. The distance from the *Red River* in the province of *Texas* (the *Río Colorado*) to the island of *Tiburón*, on the coast of the intendancy of *Sonora,* is 160 myriameters (or 364 leagues) from east to west.

The part of Mexico where the two oceans, the Atlantic and the South Sea, come closest to each other is unfortunately not where the two ports of Acapulco and Veracruz and the capital of Mexico are located. According to my astronomical observations, there is an oblique distance of 2°40′19″ in a large circle from *Acapulco* to *Mexico City* (or 155,885 toises); from Mexico City to Veracruz, 2°57′9″ (or 158,572 toises); and from the port of *Acapulco* to the port of *Veracruz*, in a direct line, 4°10′7″. It is in these distances that the old maps are the most flawed. According to the observations Mr. *Cassini* published in the account of *Chappe*'s voyage, Mexico City and Veracruz would be separated by 5°10′ in longitude, instead of the 2°57′ that one finds in more exact observations. Taking *Chappe*'s longitude for Veracruz, and that of the French Naval Office (published in 1784) for Acapulco, the width of the Mexican isthmus between the two ports would be 175 leagues, seventy-one leagues too long. The small *critical map* in the Mexican atlas illuminates these differences.

The isthmus of *Tehuantepec*, southeast of the port of *Veracruz*, is the point in New Spain where the continent is narrowest. The distance from the

Atlantic Ocean to the South Sea is forty-five leagues. The nearby sources of the *Huasacualco* and *Chimalapa* rivers favor the project of a canal for interior navigation, a project on which the Count of *Revillagigedo*, one of the viceroys most ardent about public welfare, has been working for some time. When we come to describe the intendancy of *Oaxaca*, we shall return to this subject, which is so important to all of civilized Europe. Here, we shall simply discuss the *problem of communication between the two seas* in every general way possible. We shall present nine points in the same tableau, many of which are unknown in Europe, and each of which has a relatively good potential either for canals or for interior communication by river. At a time when the New Continent, profiting from the troubles in Europe and its traditional quarrels, is making rapid advances toward civilization; at a time when trade with China and with the northwest coast of the Americas is becoming more profitable by the year, the subject we are touching on here is of the greatest interest for the trade balance and the political predominance of nations.

I.204

The nine points I have assembled on plate IV of my geographical and political Atlas have, at different times, held the attention of statesmen and enlightened merchants who have spent large amounts of time in the Colonies. They present very different advantages. They are arranged according to their geographic positions, beginning with the northernmost part of the New Continent and descending the coastline all the way to the south of the island of *Chiloé*. Only after examining *all* the projects heretofore undertaken to facilitate communication between the two seas can one decide which is preferable. Before such an examination, for which precise material has not yet been collected, it would be unwise to dig canals in the isthmuses of *Guasacualco*, *Nicaragua*, *Panama*, or *Cúpica*.

I.205

1. Below 54°37′ northern latitude, on the same parallel as *Queen Charlotte* Island, the sources of the River OF PEACE or the *Ounigigah* (*Unijigah*) are within seven leagues of the source of the TACOUTCHÉ-TESSÉ, which is presumably the same as the *Columbia* River. The first of these rivers empties into the Polar Sea, after blending its waters with those of the *Slave Lake* and the *Mackenzie* River. According to *Vancouver*, the celebrated explorer, the latter river, the *Columbia*, flows into the Pacific Ocean near Cape *Disappointment*, south of *Nootka* Bay, at 46°19′ latitude. In some places, Mr. ▼ Fidler found the elevation of the Cordillera of the *Rocky Mountains* (*Stony Mountains*) to be 3,520 English feet,[1] or 550 toises *above the nearby*

---

1. If it is true that this mountain range reaches the limit of perpetual snows (see Mackenzie, [*Voyages*,] vol. III, p. 331), then their *absolute elevation* must be at least 1,000 to 1,100

plains. This mountain range separates the sources of the Peace River and the Columbia River. According to *Mackenzie's* account—he crossed this

I.206   Cordillera in August 1793—it lends itself to *portage*, and the mountains do not appear to be of great height. To avoid the large detour that the Columbia makes, another commercial route could be opened from the sources of the *Tacoutché-Tessé* to the *Salmon* River, whose mouth is east of the Princess Royal Islands, below 52°26′ latitude. Mr. *Mackenzie* rightly observes that the government that opens this communication between the two oceans by setting up regular outposts in the interior of the country and at the extremities of the rivers would, in so doing, dominate the fur trade in North America from 48° latitude to the pole, with the exception of that part of the coast that has long been part of *American Russia*. With the multitude and the course of its rivers, *Canada* offers a level of ease of domestic trade similar to that of *eastern Siberia*. The mouth of the *Columbia* River seems to invite Europeans to establish a thriving colony there. The banks of this river offer fertile land covered with magnificent lumber forests. It must be acknowledged, however, that despite ▼ *Mr. Broughton's* examination, only a very small part is known of the Columbia, which, like the *Severn* and the

I.207   *Thames*, appears to narrow greatly[1] as it leaves the coast. Any geographer who carefully compares *Mackenzie*'s maps with *Vancouver*'s will be amazed that the *Columbia River*, as it descends from these *Stony Mountains* [*Rocky Mountains*?] (which one is tempted to consider as an extension of the *Andes of Mexico*), is able to cross the mountain range that approaches the coast of the Great Ocean, and whose main peaks are *Mount St. Helens* and *Mount Rainier*. But ▼ Mr. *Malte-Brun* has already cast serious doubts on the assertion that the *Tacoutché-Tessé* and the *Columbia* rivers are one and the same. He even assumes that the former flows into the Gulf of California, an assumption that would make the *Tacoutché-Tessé* enormously long. Admittedly, this entire western part of North America is still only barely known.[2]

---

toises. This would suggest that either the nearby plains where Mr. Fidler placed himself to take his measurements are at an elevation of 450 to 550 toises above sea level, or that the summits whose elevation this explorer mentions are not the highest ones in the range that Mackenzie crossed.

    1. Vancouver, *Voyage*, vol. II, p. 49 and vol. III, p. 521.

    2. Since the first edition of this work, we know that the Columbia or Oregon River is entirely different from the Tacoutché-Tessé, or Fraser's River. The Columbia River springs forth in the mountainous terrain that connects the great central range of the *Rocky Mountains* with the maritime Alps of New Albion via a transversal ridge; near its sources it is strangely sinuous. The origin of the Columbia is not, as very recent works of geographers in the United States would have it, at 55° latitude but, rather, at 50°3′. The river first runs NNW to the

At 50° latitude, the NELSON River, the SASKATCHEWAN, and the MIS-   I.208
SOURI, which may be considered one of the main branches of the Mississippi,
also provide opportunities for communication with the Pacific Ocean. All
these rivers originate at the foot of the *Stony Mountains* [Rocky Mountains].
We do not yet have sufficient information about the nature of the terrain
through which *portage* is supposed to be established to judge the usefulness
of these connections. ▾ Captain Lewis's journey on the *Mississippi* and the
Missouri, undertaken at the expense of the Anglo-American government,
may one day shed considerable light on this interesting problem.   I.209

2. At 40° latitude, the sources of the RÍO DEL NORTE, or *Río Bravo*,
which empties into the Gulf of Mexico, are separated from the source of the
Río COLORADO by mountainous terrain that is twelve to thirteen leagues
wide. This terrain is a continuation of the Cordillera of the Grües [Cranes],
which continues toward the *Sierra Verde* and the lake of *Timpanogos* [Lake
Utah] famous in Mexican history. The Río San Rafaël and the Río San
Xavier are the principal sources of the Zaguananas River, which combines
with the Río de Nabajoa to form the Río Colorado and blends its waters
with the Gulf of California. The regions that these rivers traverse are abundant in rock salt; two travelers who were full of enthusiasm and bravery,
Father Escalante and Father Antonio Vélez, both monks of the Order of
St. Francis, explored these regions in 1777. However important the Río Zaguananas and the Río del Norte may one day become for the domestic trade
of this northern part of New Spain, and no matter how easy portage across

---

Arthabasai station (52° lat.), where it is only six to seven leagues away from the main source
of the Tacoutché-Tessé. From there, it turns southward, and is joined by the Flat-Bow and
then by the Flat-Head River (lat. 49°), also called *Clarke's River*, the Saptin or Lewis River
(lat. 46°5′), and the Multnomah (lat. 45°20′). The sources of the Flat-Bow and the Columbia
rivers are four to five thousand toises apart, which means that a vast triangular piece of land,
between 46° and 50° latitude, is almost entirely surrounded by running waters. Before the
colony of ▾ Astoria (at the mouth of the Columbia) was abandoned, this area was much more
frequented. It is possible for 300-ton ships to sail 125 nautical leagues upstream to the confluence of the Multnomah. The Fraser River, or the Tacoutché-Tessé, originates near 52°20′
latitude and, like the Columbia, runs first northward (to 54°30′), then SSW before flowing
into Birch Bay, which is part of the sea inlet that separates the islands of Quadra and Vancouver from the mainland. A distance of over sixty leagues separates the mouth of the Columbia
from that of the Tacoutché-Tessé. A third river, the Caledonia, lies between these two great
rivers that some day may become very important for human civilization. Considering the rivers that rise on the slopes of the Rocky Mountains, one sees that as they leave the mountains,
they follow a course parallel to the axis of the range. This phenomenon, the causes of which
I have examined elsewhere, is characteristic of several Cordilleras in India and China (▾ Rit-
ter, *Erdkunde,* vol. I, p. 248; Humboldt, *Relation historique*, vol. III, p. 518).

the mountains might be, there will never be any connection that would be as beneficial as an interoceanic canal.

3. THE ISTHMUS OF TEHUANTEPEC, at 16° latitude, includes the sources of the Río Huasacualco (or Goazacoalcos), which empties into the Gulf of Mexico, and the sources of the Río Chimalapa. The waters of the latter river empty into the Pacific Ocean near the Barra de San Francisco. I consider the Río del Paso to be the main source of the Huasacualco, although the latter only goes by that name at the Paso de la Fábrica, after one of its branches, which flows from the Los Mixes mountains, joins the Río del Paso. This Isthmus of Tehuantepec is the place that Hernán Cortés, in his letters to emperor Charles V, called the *secret of the strait*, a name that sufficiently proves the importance attached to it since the beginning of the sixteenth century. It drew the attention of mariners once again when the hostilities waged by the castle of San Juan de Ulúa drove the Veracruz trade back toward the Barra de Álvarado and the coast of Tabasco, both of which are near the mouth of the Río Huasacualco. A valley interrupts the line of the summit that forms the watershed between the two Oceans. But I doubt that in the season of the great floods, this valley would be filled (as has recently been suggested) with enough water to allow the indigenous peoples' boats to navigate naturally there. There are similar *temporary connections* between the basins of the Mississippi and the St. Lawrence River, in other words, between Lake Erie and the Wabash River, and between Lake Michigan and the Illinois River. We shall return below to the possibility of digging a canal[1] six to seven leagues long in the forests of *Tarifa*. Since 1798, when an overland road was built that led from the port of *Tehuantepec* to the *Embarcadero de la Cruz* (the road was completed in 1800), the *Río Huasacualco* has been a trade connection between the two Oceans. During the war with the British, the indigo of *Guatemala*, the most precious of all known indigos, came to the port of *Veracruz* by way of this *isthmus* and from there was shipped to Europe.

4. The great LAKE NICARAGUA is connected not only with the Lake *León* but also, on the east, with the Antillean Sea via the *San Juan* River. Communication with the Pacific Ocean would be made possible by digging a canal through the isthmus that separates the lake from the Gulf of Papagayo. The isolated volcanic summits of *Bombacho* (at 11°7′ latitude), Granada,

I.210

I.211

---

1. The opening of this canal was decreed by the Spanish Courts in 1814. The construction of the canal was assigned to the Consulado de Guadalajara, which proposed to call on European capitalists.

and Papagayo (at 10°50′ latitude) are located on this narrow isthmus. Old maps even indicate a water connection across the isthmus. Other, somewhat more recent maps show a river called the Río Partido,[1] one of whose branches flows into the Pacific Ocean and the other into Lake Nicaragua. But this bifurcation is very uncertain and does not appear on the most recent maps that the Spanish and the British have published.

In the archives in Madrid, there are several papers, both French and English, on the possibility of connecting the Lake Nicaragua with the Pacific Ocean. British trade along the Mosquito Coast has greatly contributed to the awareness of this project of connecting the two seas. None of the papers that have come to my attention have shed any light on the main point, namely, the elevation of the terrain on the isthmus.   I.212

Between the kingdom of *New Granada* and the outskirts of the capital of *Mexico*, there is not a single mountain, plateau, or town for which we know the elevation above sea level. Is there a mountain chain that runs uninterrupted through the provinces of Veragua and Nicaragua? Is the *central range* of this Cordillera—which, one assumes, connects the Peruvian Andes with the mountains of Mexico—to the west or to the east of Lake *Nicaragua*? Does the isthmus of Papagayo have a mountainous terrain or just a *sill*, a simple ridge? These are problems whose resolution is of concern to both the statesman and the naturalist geographer! None of the various works that have appeared since the wars of independence in Spanish America ventures beyond the same notions that the first edition of the present work contained, with the exception of some useful information that ▼ Mr. Davis Robinson[2] has given on the sandbar of the Río San Juan of Nicaragua. He confirms that "this sandbar is covered by twelve feet of water and has a narrow passage twenty-five feet deep at only one point." The depth of the Río San Juan is four to six fathoms and that of Lake Nicaragua is three to eight fathoms. According to Mr. Robinson, brigantines and schooners can navigate the Río San Juan.   I.213

There is no other place on the globe dotted more with volcanoes than this part of the Americas, from 11° to 13° latitude. But the trachytic mountains through which the underground fires burst form only isolated groups, and, since they are separated from each other by valleys, they appear to

1. ▼ La Bastide, *Mémoire sur [un nouveau] passage de la mer du Nord à la mer du Sud*, 1791. Marchand, *Voyage*, vol. 1, p. 565. Tomás López [de Vargas Machuca] and Juan de la Cruz [Cano y Olmedilla], *Mapa [maritimo] del Golfo de Mexico*, 1755.

2. [Robinson,] *Memoirs of the Mexican Revolution*, 1821, p. 263. *Edinburgh Review*, 1810, January, p. 47. [Bello,] *Biblioteca Americana*, vol. I, pp. 115–29.

soar up from the plain itself. It is unsurprising that we were ignorant of
such important facts; for we shall soon see that even the height of the mountain range that crosses the Isthmus of Panama is as little known today as it
was before the invention of the barometer and the use of this instrument
in measuring mountains. Transit between Lake Nicaragua and the Pacific
Ocean might also be possible via the lake of León by way of the Tosta River,
which, on the road from León to Realexo, descends from the volcano of
Télica. The terrain there is actually very low in elevation, and the account
of ▼ Mr. *Dampier*'s voyage [*A Collection of Voyages*] leads us to suppose that
there is no *mountain chain* between Lake Nicaragua and the South Sea.
I.214 "The coastline of Nicoya," that great navigator writes, "is low and flooded
at high tide. To reach León from [El] Realejo, one must travel twenty miles
through flat country covered with mangroves." The town of León itself is
situated in a savannah. There is a small river that flows into the sea near
[El] Realejo, which might facilitate transit between this port and the port
of León.[1] From the western bank of Lake Nicaragua, it is only four nautical
leagues to the shore of the Gulf of Papagayo and seven leagues to the Gulf
of Nicoya, which sailors call La Caldera. Dampier expressly states that the
terrain between La Caldera and the lake is not very hilly, mostly level, and
like a savannah.

Because of the position of its interior lake and its connection with the
Antillean Sea via the Río San Juan, the isthmus of Nicaragua shares many
similarities with the gorge in the Scottish Highlands where the River Ness
forms a natural link between the mountain lakes and the gulf of Murray [the
Moray Firth]. In Nicaragua, as in the Scottish Highlands, there is only a
narrow strait to be crossed to the west; perhaps it would suffice to *canalize*
the Río San Juan to the east, without deviating from the riverbed, which
is dammed only in the dry season. Although it is true that the isthmus to
be crossed is dotted with a few hills in its narrowest section, between the
I.215 west bank of Lake Nicaragua and the Gulf of Papagayo, it is also composed
of savannahs and uninterrupted plains that provide an excellent route
for coaches (*camino caretero*) between the city of León and the coastline of
[El] Realejo. This is the main route by which goods are sent from Guatemala to León, disembarking in the Gulf of Fonseca, or Amalapa, at the port
of Conchagua. Lake Nicaragua is at a higher elevation than the South Sea
across the entire thirty-league descent of the Río San Juan: the elevation of
this basin is thus so well-known in the country that it was once considered

1. [Dampier,] *Collection of Voyages*, vol. I, pp. 113, 119, and 218.

an invincible obstacle to the construction of a canal. People feared either a raging spill of water to the west or a decrease of water in the Río San Juan, which, during droughts, presents several rapids above the old Castillo de San Carlos[1] and whose banks are extremely insalubrious in their present fallow state. The construction engineer's technique is advanced enough today to dispel fear of such dangers. Lake Nicaragua could serve as an upper basin, like Lake Oich for the Caledonian Canal. Regulatory locks would allow into the canal only as much water as needed. The slight difference in level that supposedly exists between the Antillean Sea and the Pacific Ocean is probably due only to the uneven height of the tides. One sees a similar difference between the two seas that the great Canal of Scotland connects; even if this were a difference of six toises and as permanent as that of the Mediterranean and the Red Sea, it would be no less favorable for an oceanic connection. The winds that blow over Lake Nicaragua are strong enough that steamboats would not be required to tow the ships that need to pass from one sea to the other; but the power of steam would be most useful for crossings from [El] Realejo or Panama to Guayaquil. During the months of August, September, and October, periods of calm alternate in these areas with winds that blow in the opposite direction of this trajectory.

I.216

Because of terrible storms and rains,[2] the coastline of Nicaragua is rather dangerous during the months of August, September, and October, as well as in January and February, because of the furious northeasterly and east-northeasterly winds called *Papagayos*. This circumstance is most inconvenient for navigation. The port of Tehuantepec on the isthmus of Huasacualco is not more favored by nature; it gives its name to the hurricanes that blow from the northwest, which make all vessels flee from landing at the small ports of *Sabinas* and *Ventosa*. These considerations suggest that there are three possibilities for an *interoceanic canal* in Nicaragua—either

I.217

1. This small fort, taken by the British in 1665, is commonly called *El Castillo del Río San Juan*. According to Mr. Juarros [y Montúfar], it is situated at a distance of ten leagues from the eastern edge of Lake Nicaragua. Built on a crag at the mouth of the river, the fort is referred to as the *Presidio del Río San Juan*. Already in the sixteenth century, the *Desagüadero de las Lagunas* had attracted the attention of the Spanish government that ordered ▼ Diego López Salcedo to found the town of Nueva Jaén near the left bank of the *Desagüadero* or Río San Juan. This town was soon abandoned, as was the town of *Bruselas*. See Humboldt, *Relation historique*, vol. III, p. 138.

2. In the third volume of his *Relation historique* [*Personal Narrative of the Voyage to the Equinoctial Regions*] which has just been published (chap. XXVI, p. 119), Mr. von Humboldt writes that "according to Mr. Davis Robinson, the western coastline of Nicaragua is not as stormy as it was depicted to me during the crossing from Guayaquil to Acapulco." E—R.

from Lake Nicaragua to the gulf of Papagayo or from that same lake to the gulf of Nicoya, or else from the lake of León,[1] or Managua, to the mouth of the Río de Tosta. The distance from the southeast extremity of Lake Nicaragua to the gulf of Nicoya is very different (from twenty-five to forty-eight miles) on Arrowsmith's map of South America and on the fine map of the *Depósito hidrográfico* in Madrid, entitled *Mar de las Antillas*, 1809.

5. The ISTHMUS OF PANAMA was crossed for the first time by ▼ Vasco Núñez de Balboa in the year 1513. Since that memorable period in the history of geographical discoveries, the project of a canal has inspired the imagination of many; even today, three hundred years later, no leveling of the terrain exists, nor does a precise determination of the positions of Panama and Portobello. The longitude of the first of these ports was based on that of Cartagena; the longitude of the second was determined from Guayaquil's. The calculations of Fidalgo and Malaspina are certainly worthy of great confidence; but errors multiply unnoticeably when, by chronometric calculations that encompass the entire coastline of ▼ Terra Firma, from the island of Trinidad to Portobello, and from Lima to Panama, one position becomes dependent on another. It would be important to bring the time of Panama into accordance with that of Portobello and thus to link the calculations made in the South Sea with those that the Spanish government had made in the Atlantic Ocean. ▼ Mr. Fidalgo, Mr. Tiscar, and Mr. Noguera may one day venture onward with their instruments to the southern coast of the isthmus, while Mr. Colmenares, Mr. Isasbirivil, and Mr. Quartara [Quadra?] might advance their work[2] as far as the northern coast. To get a sense of the uncertainty that still remains concerning the shape and width of the isthmus (for example, near Natá), one has only to compare López's maps with Arrowsmith's and with the more recent maps of the *Depósito hidrográfico* in Madrid. Despite its sinuosity and its rapids, the Chagre River, which empties into the Antillean Sea west of Portobello, has great possibilities for trade; it is 120 toises wide at its mouth and twenty toises wide near Cruces, where it becomes navigable. One can sail up the Río Chagre today from its mouth to *Cruces* in four to five days. If the waters are very high, one has to struggle against the current for ten to twelve days. Goods

I.218

I.219

---

1. And not from Lake Léon to the Gulf of Nicoya, as the otherwise very astute editor of the *Biblioteca Americana* [Andrés Bello] states, 1823, August, p. 120.

2. In 1803, these officers of the Spanish navy were ordered to survey the northern and western coastline of South America. *Fidalgo*'s expedition was destined for the coast between the island of *Trinidad* and *Portobello*, *Colmenares*' for the coast of *Chile*, and *Moralada*'s and *Quartara*'s expedition for the part from *Guayalquil* to [*El*] *Realejo*.

are transported on mules from Cruces to Panama over a distance of five short leagues. The barometric elevations given in Ulloa's *Voyage*[1] lead me to believe that, in the *Río Chagre*, there is a difference in level of thirty-five to forty toises from the Antillean Sea to the *Embarcadero*, or the Venta de Cruces. This difference must seem quite small to those who have sailed up the Río Chagre, but they forget that the strength of the current depends as much on a great accumulation of water near the source as on the *general* slope of the river, that is, its slope above *Cruces*. When one compares Ulloa's barometric survey with the one I did on the *Magdalena* River, one notices that, far from being small, the elevation above sea level of *Cruces* is, on the contrary, quite considerable. The slope of the Río de la Magdalena from Honda to the Mahates seawall near Barancas is 160 toises; but this distance is not, as one might assume, four times but eight times greater than I.220 the distance from *Cruces* to *Fort Chagre*.

Proposing to the court of Madrid that a connection be established between the two Oceans via the Río Chagre, the engineers planned to dig a canal from Venta de Cruces to Panama. This canal would have to pass through mountainous terrain whose elevation is entirely unknown. All we do know is that from Cruces, one first climbs rapidly before descending for several hours toward the South Sea coast. It is quite astonishing that in crossing the isthmus neither *La Condamine* nor *Bouguer*, nor *Don Jorge Juan*, or *Ulloa* were curious enough to observe their barometer in order to tell us the elevation of the highest point on the road from the small fortress of Chagre to Panama. These scientists stayed for three months in this region, which is of such great commercial interest; but their long stay contributed little more to the observations that we owe to Dampier and ▼ Wafer. The main Cordillera, or rather a range of hills, that one may consider an extension of the Andes of New Granada, is clearly located between Cruces and Panama, closer to the South Sea than to the Antillean Sea. It is from the top of this Cordillera that people have claimed to see both Oceans at once, an observation that would only presuppose an absolute elevation of 290 meters. Lionel Wafer complains that he was unable to enjoy this spectacle, and he claims that the hills that form the central range are separated from each I.221 other by valleys that allow the rivers to *flow freely* there.[2] If this last asser-

1. [Juan and] Ulloa, *Observationes astronomicas*, p. 97.
2. [Lionel Wafer, *A New Voyage and*] *Description of the Isthmus of America*, 1729, p. 297. Near the town of Panama, a little north of the port, lies the mountain of *Ancon*, which, according to a geometrical measurement, is 101 toises high. Ulloa [and Juan, *Noticias secretas de America?*], vol. I, p. 101.

tion were valid, then it would lend credence to the possibility of a canal that would connect Cruces with Panama, a canal whose navigation would be interrupted by only a very few locks.

The scant data that exists on the temperature of these places and the geography of indigenous plants would lead me to believe that the ridge on the road from Cruces to Panama is lower than five hundred feet; Mr. Robinson assumes it to be four hundred feet at most. Furthermore, whenever one examines nearly any mountainous terrain closely, one finds examples of natural openings through the ridges. The elevation of the hills between the basins of the Saône and the Loire that the Centre canal would have had to cross is eight hundred to nine hundred feet, but there is a gorge or break in the range near the small lake of Long-Pendu which has a sill that is three hundred fify feet lower.

Reports drafted in 1528 reveal that there were many other points where people proposed cutting through the isthmus, for example, by joining the sources of the rivers called Caimito and Río Grande with the Río Trinidad. The eastern part of the isthmus is narrower, but the terrain also seems much higher there. At least this is what one observes on the terrible mail road from Portobello to Panama, a two-day route that passes through the village of *Pequeni* and presents the greatest difficulties.

I.222

In every age and in every clime, people have believed that, of two neighboring seas, one is higher than the other. Traces of this widespread theory are found as far back as the ancients. Strabo[1] relates that the level of the Gulf of Corinth near Lechaeum was believed to be higher than the waters of the Gulf of Chenchrenae. He imagines that it would be very dangerous to cut through the isthmus of the Peloponnesus at the place where the Corinthians, using special machines, had set up a *portage*. In the Americas, on the Isthmus of Panama, it is commonly believed that the South Sea is higher than the Antillean Sea. This theory is based on mere appearance. After struggling against the current of the Río Chagre for several days, one has the impression of having ascended much more than one has descended from the hills near Cruces to Panama. In fact, there is nothing more deceptive than one's judgment of the difference in level on a slope that is protracted and, therefore, quite gradual. In Peru, I could hardly believe my eyes upon discovering, through a barometric measurement, that the city of Lima is ninety-one toises higher than the port of Callao. The rock of the

1. Strabo, [*Strabonis rerum geographicarum*], Book I (ed. Siebenkees), vol. 1, p. 146. Livius, [*Patvini historiarvm*] Book 42, chap. 16.

island of San Lorenzo would have to be completely covered with water during an earthquake for the Ocean to reach the capital of Peru. *Don Jorge Juan* has already challenged the theory of a difference of level between the Antillean Sea and the Great Ocean; he has found that the height of the column of mercury is the same at the mouth of the Chagre as in Panama City. $\quad$ I.223

Doubts may have emerged because of the inaccuracy of the meteorological instruments in use at the time, as well as the absence of any thermometric correction of the calculation of elevations. These doubts seem to have acquired even more weight when the French engineers who were attached to the Egyptian expedition found that the level of the Red Sea was six toises higher than the average height of the Mediterranean. In the absence of any geometrical survey conducted on the Isthmus of Panama itself, one must resort to barometrical measurements. Those that I conducted at the mouth of the Sinu River on the Antillean Sea and on the South Sea coasts of Peru prove that, taking into account all temperature corrections, it is impossible for any difference between the level of the two Oceans to exceed six or seven meters.

Considering the effect of the *rotation current*[1] that carries waters on the northern coast from east to west and gathers them near the coasts of *Costa Rica* and *Veragua*, one is tempted to admit, contrary to received opinion, that the Antillean Sea is slightly higher than the South Sea. Some trivial local causes—the contours of the coastline, currents, and winds (as in the strait of Bab-el-Mandeb)—may disturb the overall balance that must necessarily exist between all parts of the Ocean. The tides at Portobello are one-third of a meter high, and four or five meters high at Panama City, from which follows that the water level [the tides] of the two neighboring oceans must have varied during the periods when these ports were founded. But these slight differences, far from preventing hydraulic construction, will instead favor the operation of locks. $\quad$ I.224

If the Isthmus of Panama were split apart by a great catastrophe like the one that opened the Pillars of Hercules,[2] the *rotation current*, instead of ascending toward the Gulf of Mexico and emptying through the [Old] Bahama Channel, would likely follow the same parallel from the coast of Paria to the Philippine Islands. The effect of this opening or new strait

---

1. What I call the *rotation current* is the general east-west motion of water that one observes in that part of the ocean that lies between the two tropics.

2. ▼ Diodorus Siculus, [*Bibliothecæ historicæ,*] book IV, p. 226; book XVII, p. 553, (ed. Rhodom[ani].).

would extend far beyond the banks of Newfoundland; it would either cause the complete disappearance or diminish the swiftness of the warm current I.225 that is called the *Gulf Stream*[1] and that, first oriented north-northeast from Florida toward the banks of Newfoundland, then at 43° latitude moves eastward toward the coast of Ireland and southeast toward the coast of Africa. A cut several leagues wide formed by earthquakes or volcanic upheaval would produce physical changes on the Isthmus of Panama comparable to those whose memory is preserved in the legend of the *Samothracians*. But dare one compare the puny efforts of men to channels carved by nature herself, like the straits of the *Hellespont* and the *Dardanelles*!

Strabo[2] seems inclined to believe that the sea will one day open the Isthmus of *Suez*. One need not expect a similar catastrophe on the Isthmus of *Panama*, unless enormous natural convulsions, which are unlikely given our planet's current state of repose, cause extraordinary upheavals. A peninsula that extends from east to west in a direction that is almost parallel to the rotation current escapes, as it were, the shock of the waves. The Isthmus of *Panama*, with its south-north orientation, would be under threat if it I.226 were located in the province of Costa Rica between the port of Cartago and the mouth of the *Río San Juan*; in other words, if the narrowest part of the new continent were between 10° and 11° latitude.

Navigation on the *Chagre* River is difficult, because of its many meanders and the swiftness of its current, frequently from one to two meters per second. The meanders, however, offer the advantage of a *counter-current* formed by eddies near its banks, thanks to which small craft called *Bongos* and *Chatas* are able to travel upriver by using either oars or poles or by being towed. This benefit would disappear were these meanders cut, and it would be very difficult to travel from the Antillean Sea to Cruces.

The *minimum* width of the Isthmus of Panama is not—as the early maps of the *Depósito hidrográfico* indicated—fifteen miles, but rather twenty-five and three-quarter miles (at 950 toises each, or sixty miles to a degree), in other words, eight and one-half nautical leagues, or 24,500 toises. This is

1. The *Gulf Stream*, about which ▼ *Franklin* and, later, *Williams* (in his treatise on *Thermometrical Navigation*) have left us such valuable observations, carries tropical waters to the northern latitudes. It must originate in the rotation current that beats up against the coasts of Veragua and Honduras, and that, flowing up toward the Gulf of Mexico between Cabo Catoche and Cabo San Antonio, exits through the [Old] Bahama Channel. The moving waters carry plant products from the Antilles to Norway, Ireland, and the Canary Islands. See *Relation historique*, vol. I, pp. 64–70.

2. Strabo, [*Strabonis rerum geographicarum*] (ed. Siebenkees), vol. I, p. 156.

because the dimensions of the Gulf of San Blas, also called the Ensenada de Mandinga because of the little eponymous stream that empties into it, have caused serious errors. This gulf penetrates seventeen miles fewer inland than was assumed in 1805, when the archipelago of the *Mulatas Islands* was surveyed. Despite the apparent trustworthiness of the latest astronomical calculations (which are the basis of the map of the isthmus published by the *Depósito hidrográfico* in Madrid in 1817), one must nonetheless not forget that these calculations only cover the northern coastline, and that this part of the coast has not yet been related to the southern coastline, either through concatenation or chronometry (by the measurement of time). Thus, the problem of the width of the isthmus does not depend only on the determination of latitudes.[1]

I.227

All of the information that I was able to procure during my stay in Cartagena and Guayaquil leads me to believe that one should abandon the hope of a canal that is seven meters deep and from twenty-two to twenty-eight meters wide and that, like a cut or a strait, would cross the Isthmus of Panama from sea to sea and receive the same ships that sail from Europe to the East Indies. The elevation of the terrain would force the engineer to resort either to underground tunnels or a system of locks. Consequently, goods expected to cross over the Isthmus of Panama would only be transportable on flat-bottomed boats that could not withstand the sea. Warehouses would be needed in Panama City and Portobello. Every nation that wished to trade over this route would be dependent on the nation that governed the isthmus and the canal. This inconvenience would be especially great for

I.228

1. *Relation historique*, vol. III, p. 126. Mr. von Humboldt writes: "Comparing the two maps at the *Depósito hidrográfico* in Madrid that bear the title *Carta esférica del Mar de las Antillas y de las Costas de Tierra Firme desde la isla de la Trinidad hasta el golfo de Honduras* (1806 [1809]) and the *Quarta Hoja que comprehende las costas de la Provincia de Cartagena* (1817) one can see how well-founded were the doubts that I voiced fifteen years ago about the relative orientation of the most important points on the northern and southern coastlines of the isthmus. The southern coast between the mouth of the Río San Juan Díaz and the Río Lúcuma, east of Panama [City?], on the meridian of Punta San Blas extends farther, according to the 1809 map, by 8°54′ latitude, and, according to the map of 1817, by 9°2′. The northern coast, which forms the base of the Mandinga or San Blas Gulf, south of the Mulatas Islands, is located on the first of these maps at 9°9′ on the second at 9°27′. Since Cabo San Blas, in the northwest part of Mandinga Gulf, was not moved north to the same degree as was the base of the gulf near the mouth of the Río Mandinga, the gulf is displaced by twenty-four minutes of arc on the 1807 map and by seven minutes of arc on the 1817 map. Farther west, the average width of the isthmus, between the Castillo de Chagre, Panama [City?], and Portobello, is fourteen nautical leagues. The *minimum* width is two to three times less than the width of the Isthmus of Suez, which ▼ Mr. LePère gives at 59,000 toises.

ships dispatched from Europe. Even in the event that the canal were dug, it is likely that most ships would fear the delays caused by too many locks and would continue their voyage around the Cape of Good Hope. We see that passage through the Sound [Øresund or Öresund, between Denmark and Sweden] is very frequent, despite the presence of the Eyder Canal that connects the Ocean with the Baltic Sea.

This would not be the case for products from western America or goods that Europe sent to the coastline of the Pacific Ocean and to the coasts of Quito and northern Peru. These goods would cross the isthmus at less expense and, especially in wartime, with less danger than by sailing around the southern tip of the New Continent. In the present state of the route, the cost of transporting three quintals by mule back from Panama [City] to Portobello is three to four piasters (fifteen or twenty francs). But the undeveloped state in which the government has left the isthmus is such that the total number of pack animals between Panama City and Cruces is much too small for the Chilean copper, Peruvian chinchina bark, and above all the 70,000 fanegas[1] of cocoa that Guayaquil exports annually to be able to cross this strip of land. The dangerous, slow, and costly navigation around Cape Horn is, therefore, preferable.

In 1802 and 1803, when British corsairs were obstructing Spanish trade in all parts, much of the cocoa from Guayaquil was sent across the Kingdom of New Spain and was shipped out from Veracruz to Cádiz. The crossing from Guayaquil to Acapulco and the overland route of 135 leagues from Acapulco to Veracruz were considered preferable to the dangers of a long journey around Cape Horn and the difficulty of ascending the coastlines of Peru and Chile against the current. This example demonstrates that, should the construction of a canal—across either the Isthmus of Panama or that of Guasacualco—prove too difficult because of the need for numerous locks, the commerce of western America would already profit immensely from fine roads leading from Tehuantepec to the Embarcadero de la Cruz, and from Panama [City?] to Portobello. It is true that the grazing lands on the isthmus have, to date[2], been unfavorable for feeding and breeding livestock, but on such fertile soil, it would be easy to form savannahs by cutting down forests, or to grow *Paspalum purpureum*, *Milium nigricans*, and above all *alfalfa* (*Medicago sativa*), which thrives in the warmest parts of

I.229

I.230

---

1. One *fanega* weighs 110 Castilian pounds.

2. Raynal's claim ([*Histoire . . . des deux Indes*,] vol. IV, p. 150) that the domestic animals transported to Portobello would lose their fertility is absolutely devoid of truth.

Peru. ▼ Introducing camels would be an even more suitable way of reducing hauling expenses. These *land ships*, as the Orientals call camels, are currently found only in the province of Caracas, where the ▼ Marquis del Toro imported them from the Canary Islands.

The advances in population, agriculture, commerce, and civilization on the Isthmus of Panama should not be impeded by any political considerations. The more this strip of land is developed, the more resistance it will offer a foreign enemy. At present, if any enterprising nation wished to control the isthmus, it would likely succeed. Many fine fortifications there have been stripped of men capable of defending them. The insalubrity of the climate (although already decreased in Portobello) makes it difficult to establish a military mission on the isthmus. It is possible to attack Peru not from Panama City but from San Carlos de Chiloé. It takes three or four months to travel upriver against the current from Panama City to Lima, whereas navigation from Chile to Peru is easy and always quick. Despite the disadvantages of the isthmus, its possession never ceases to be of great importance to an enterprising nation. Whale and sperm whale hunting (which, by 1803, had brought sixty British ships to the South Sea); the ease of trade with China; and the fur trade on the Nootka Sound are all seductive lures that will suffice sooner or later to attract the masters of the Ocean to a point on the globe that nature seems to have destined to change the face of [inter]national trade.[1]

6. Southeast of *Panama City,* along the coastline of the Pacific Ocean, from *Cabo San Miguel* to *Cabo Corrientes*, one finds the small port and bay of Cúpica. The name of this bay has become famous in the kingdom of New Granada because of a planned link between the two seas. From Cúpica, one crosses over five or six nautical leagues of terrain that is flat[2] and very

I.231

I.232

1. The Isthmus of Panama, the central point of Spanish America, has recently drawn the attention of the free governments of the New World for a different reason. In the fifth article of the treaty of friendship between the Republic of Colombia and the Confederation of the United Mexican States on October 3, 1823, one finds an expression of hope that the plenipotentiaries of all the Spanish-American states will come together from time to time at a general convention on the Isthmus of Panama, "a convention that might be considered," writes ▼ Mr. Alamán, the Secretary of State in Mexico City (*Informe al Congreso soberano de México del 8 noviembre* 1823, p. 11) "to be based on a *family pact* among peoples of a common origin."

2. "Since you ascended the Río Magdalena on your way to Santa Fé de Bogotá and Quito," an inhabitant of Cartagena de Indias, ▼ Don Ignacio Pombo, author of several very respectable statistical reports, wrote to me in February of 1803, "I have not stopped collecting information on the Isthmus of Cúpica: only five to six leagues separate this port from the embarcadero of the Río Naipi; the entire terrain is flat (*terreño enteramente llano*)."

conducive to building a canal that would end at the Embarcadero of the Río Naipí or Naipipi. The latter is navigable and empties below the village of Zitara into the great Río Atrato, which flows into the Antillean Sea. Mr. Goyeneche, a very intelligent Biscayan pilot, deserves the honor of being the first to draw the government's attention to the Bay of Cúpica; he wished to prove that it could represent for the New Continent what the *Suez* has formerly been for Asia. Mr. Gogueneche has proposed to transport all

I.233   of the cocoa from Guayaquil to Cartagena via the *Río Naipi*. The same route presents the advantage of swift communication between Cádiz and Lima. Instead of sending couriers via Cartagena, Santa Fe [de Bogotá], and Quito, or through Buenos Aires and Mendoza, one should send mail via the mouth of the Atrato to Cúpica and send pack-boats, good small sailboats, from Cúpica to Peru. Had this route been opened, the viceroy of Lima would not have had to wait sometimes five to six months for orders from his court. Furthermore, the outskirts of the *Bay of Cúpica* might contain superb lumber for construction purposes, quite capable of being transported to Lima. The terrain between Cúpica and the mouth of the Atrato is perhaps the only part of the Americas where the chain of the Andes is completely broken. To get a more precise idea of the extraordinary depression formed in the western Cordillera of New Granada, one must recall that, at 2° latitude, in the *knot of mountains* that encloses the sources of the Río Magdalena, the Andes are divided into three ranges. The easternmost range extends, deviating northeast, through Timana, Bogotá, and Pamplona as far as the snow-covered mountains of Mérida: between the lake of Maracaibo and the town of Valencia, it connects with the Cordillera on the Venezuelan coast. The middle range—that of Panama, Guanacas, and Quindiu—separates

I.234   the longitudinal Río Cauca valley from the Río Magdalena valley. It joins the westernmost range of New Granada in the province of Antioquia. In

---

The geographical position of Cúpica is as uncertain as the position of the confluence of the Naipi with the Atrato. It is very important, however, to know whether schooners can sail up from the mouth of the Atrato all the way to this confluence. I was unable to find the port of Cúpica on any Spanish map, but I did find Puerto Quemado or Túpica at 7°15′ latitude. An unpublished sketch of the province of Chocó in my possession confuses Cúpica and Río Sabaleta, lat. 6°30′; however, Río Sabaleta, according to the maps of the *Depósito hidrográfico* in Madrid, is located south and not north of Cabo San Francisco; thus, forty-five seconds in arc south of Puerto Quemado. According to the map of the province of Cartagena published by ▼ Don Vicente Talledo, the confluence of the Naipipi (Naipi) and the Atrato is at 6°40′ latitude; according to ▼ Mr. Restrepo (*Semanario de Bogotá*, vol. II, p. 96) it is at 7°25′. One hopes that these uncertainties will be removed by on-site astronomical observations in these locations.

the Chocó, this range gradually disappears at 7° latitude, slightly west of Zitara, between the left bank of the Atrato and the coastline of the Pacific Ocean. It would be of interest to know the shape of the terrain between Cabo Garachine or the Gulf of San Miguel and Cabo Tiburón, especially toward the sources of the Río Tuyra and the Chucunaque or Chuchunque, so that one might determine precisely where the mountains of the Isthmus of Panama begin to rise, mountains whose *summit line* does not appear to be over one hundred toises in height. The interior of Darfur [westernmost portion of present-day Sudan] is hardly more unknown to geographers than the humid, insalubrious terrain covered with thick forests that extends, northwest of Betoi and the confluence of the Bevara and the Atrato, toward the Isthmus of Panama. At present, the only thing we know concretely is that, between Cúpica and the left bank of the Atrato, there is either a *land strait* or the complete absence of any Cordillera. Because of their direction and geographical position, the mountains of the Isthmus of Panama may be considered a continuation of the mountains of Antioquia and the Chocó, but there is scarcely a sill or a slight ridge to be found on the plains west of the Lower Atrato. There is no group of mountains interposed between the isthmus and the Cordillera of Antioquia that is comparable to the one that connects (between Barquisimeto, Nirgua, and Valencia) the eastern chain of New Granada (the Sierra de la Summa Paz and the Sierra Nevada de Mérida) to the Cordillera on the Venezuelan coast.

I.235

7. In the interior of the province of Chocó, the small *Ravine* (Quebrada) *de la Raspadura* connects the Río de Noanama, commonly called the Río San Juan, with a small river, the Quibdó. The latter, swollen with the waters of the Andagueda and the Río Zitara, forms the Río Atrato, which empties into the Antillean Sea, whereas the Río San Juan flows into the South Sea. A robust monk, the priest of the village of Novita, had his parishioners dig a small canal in the ravine of the Raspadura. By means of this canal, which is navigable when the rains are abundant, canoes loaded with cocoa *traveled from one sea to the other.* Here, then, is an inland link that has existed since 1788 but is unknown in Europe. The small canal of the Raspadura connects the two Ocean coasts between two points that are separated by over ninety-five leagues. This will always be only a *canal for limited navigation*; but it might easily be enlarged, if one joined to it the streams known as the Caño de las Animas, del Caliche, and Aguas Claras. It is easy to set up reservoirs and replenishing channels in a country like the Chocó, where it rains throughout the year and where one hears thunder every day. According to information I acquired in Honda and Vilela, near Cali, from

persons employed in the trade (*rescate*) of gold powder from the Chocó,

I.236 the Río Quibdó, which connects with the Mina de Raspadura canal, joins the Río de Zitara and the Río Andagueda near the village of Quibdó (commonly called Zitara). But according to an unpublished map that I have just received from the Chocó—on which the canal of the Raspadura joins both the Río San Juan and the Río Quibdó a little above the Animas mine (at latitude 5°20′?)—, the village of Quibdó is located at the confluence of the small eponymous river and the Río Atrato, into which the Río Andagueda empties, three leagues higher, near Lloro. From its mouth (latitude 4°6′) south of Punta Charambira, the great Río San Juan, ascending toward the NNE, is joined by the Río Calimal the Río del No (above the village of Noanama); by the Río Tamana, which passes near Novita; by the Río Iro; by the Quebrada de San Pablo; and finally, near the village of Tadó, by the Río de la Platina. The province of Chocó is populated only in these river basins. It has trade connections with Cartagena to the north, via the Atrato, whose banks are completely uninhabited beyond 6°45′ latitude; to the south with Guayaquil and (prior to 1786) with Valparaíso, via the Río San Juan; to the east with the province of Popayán via the Tambo de Calima and the Cali. The ravine of La Raspadura, which serves as a canal and which, I believe, I was the first to make known in Europe, is often confused on maps with the portage of Colima and San Pablo. The *Arastradero* of San Pablo also

I.237 leads to the Río Quibdó, but this is several leagues above the mouth of the canal of La Raspadura. Goods that are sent from Popayán via Cali, Tambo de Calima, and Novita to Chocó del Norte—in other words, to Quibdó—usually take the route of this Arastradero de San Pablo. The connection of two neighboring ports at any point in equinoctial America—either on the Isthmus of Chocó or on those of Panama, Nicaragua, and Huascualco—via a *canal in small sections* (of four to six feet deep) or via a *canalized river* would likely give rise to important commercial activity. A small canal [secondary waterway] would serve the same function as a *railway*, and no matter how small, it would animate and abbreviate communication between the western [Central] American coastline and the coast of the United States and Europe.[1] But no matter how useful such undertakings might be, they

1. The coastline of Verapaz and Honduras also has several ports that are suitable for *canals of limited or secondary navigation*. On the meridian of Sonsonate, the *Golfo Dulce* enters over twenty leagues inland, so that the distance from the village of Zacapa (in the province of Chiquimula, near the southern edge of the *Golfo Dulce*) to the shores of the Pacific Ocean, is only twenty-one leagues. The rivers in the north flow toward the waters that the Izalco and Zacatepec Cordilleras empty into the South Sea. East of the *Golfo Dulce* in the *partido* of

cannot have as powerful an effect on the trade between of these two worlds as a true *interoceanic canal*.

8. At 10° southern latitude, two or three days' journey from Lima, one reaches the edge of the GUALLAGA (or *Huallaga*) RIVER, by which one can reach the coasts of the Grande Pará in Brazil without having to round Cape Horn. The sources of the Río Huanuco,[1] which empties into the *Guallaga*, are four to five leagues away from the source of the Río Huaura, near *Chinche*, which empties into the Pacific Ocean. Even the Río Xauxa, a tributary of the Apurimac or Ucayali, originates near Jauli, close to the sources of the Río Rimac, which traverses the city of Lima. The elevation of the Peruvian Cordillera and the nature of the terrain make the construction of a canal there impossible; but the construction of a convenient road from the capital of Peru to the Río de Huánuco would facilitate the shipping of goods to Europe. The Ucayali and the Guallaga, two great rivers, could carry Peruvian products from the mouth of the *Amazon* to the nearest shores of Europe in five or six weeks, whereas it would require a crossing of four months to bring these same goods to the same location by rounding Cape Horn. The agriculture of the fertile regions on the eastern slope of the Andes and the prosperity and wealth of its inhabitants depend on free navigation on the Amazon River. This liberty, which the court of Portugal refuses the Spanish, could have been attained after the events that preceded the ▼ peace of 1801.

9. Before the coastline of Patagonia was sufficiently explored, it was believed that the GULF OF SAN JORGE, located between 45° and 47° southern latitude, reached far enough inland to connect with the sounds that interrupt the continuity of the western coastline; that is, the coast opposite the Chayamapu archipelago. If this supposition were well-founded, ships en route to the South Sea might cross South America 175 leagues north of the

I.238

I.239

Comayagua, one finds the Río Grande de Motagua or the *Río de las Bodegas de Gualán*, the Río Camalecón, the Ulúa, and the Leán, which are navigable forty to fifty leagues inland, by large pirogues [dugout canoes]. It is very likely that the Cordillera that forms the dividing ridge here is broken up by some transversal valleys. We learn in Mr. Juarros's interesting work, which was published in Guatemala, that the beautiful valley of Chimaltenango sends its waters to both the southern and northern coasts. Steamboats will, I hope, soon reactivate trade on the Motagua and Polochic Rivers. See my *Relation historique*, vol. III, p. 127.

1. See the map that ▼ Father Sobreviela provides in the third volume of an excellent literary journal published in Lima, entitled the *Mercurio Peruano*. ▼ Skinner's work on Peru is an excerpt from this journal, of which a few volumes—unfortunately not the most interesting one—have been procured in London. I placed the complete work in the King's library in Berlin.

Strait of Magellan and thus shorten their route by over 700 leagues. Seafarers would thus avoid the dangers that, despite the perfection of nautical science, still accompany the voyage around Cape Horn and along the western

I.240   coast of Patagonia, from Cabo Pilares to the Chonos Islands archipelago. By 1790, these ideas had attracted the attention of the court of Madrid. The viceroy of Peru, ▼ Mr. Gil Lemos, an honest and zealous administrator, sent a small expedition, under Mr. Moraleda's command,[1] to examine the southern coast of Chile. I have seen that, in the instructions he received in Lima, he was ordered to keep utmost secrecy, were he so fortunate as to discover a *link between the two seas*. In 1793, Mr. Moraleda recognized that the Estero Aisén [Ayssen Estuary], which the Jesuit fathers ▼ José García and Juan Vicuña had visited before him in 1763, is, of all the sea inlets, the one by which the Pacific Ocean stretches the farthest eastward. However, this Estero is only eight leagues long and ends abruptly just beyond *the island of La Cruz*, where, near a hot spring, a narrow river empties into it. The Estero Aisén, located at 45°28' latitude, is therefore eighty-eight leagues away from

I.241   the Gulf of Saint George. This gulf was precisely surveyed by Malaspina's expedition. As early as 1746, Europeans suspected another link between the bay of *San Julián* (latitude 50°53') and the Pacific Ocean.

I have drawn on a single plate the nine points that seem to offer links between the two seas by connecting neighboring rivers, either via canals or via roads that would facilitate transportation to the places where the rivers become navigable. From an astronomical perspective, these sketches are not all equally precise; my intention was merely to spare my reader from having to consult several maps for what could be collected on a single one. It remains for the government that possesses the most beautiful and fertile part of the globe to perfect what I have merely indicated in this discussion. Two Spanish engineers, the ▼ Lemaurs, drew a detailed map of the *Los Güines* canal,[2] which was to cross the entire island of Cuba, from Batabanó to Havana. A similar survey done on the Isthmus of Guasacualco, at Lake

1. Don José de Moraleda y Montero visited the Chiloé and Chonos archipelagos and the western coast of Patagonia between 1787 and 1796. Two interesting manuscripts written by Mr. Moraleda are preserved in the archives of the viceroyalty in Lima. One bears the title "Viage al reconocimiento de las Islas de Chiloé," 1786; the other includes the "Reconocimiento del Archipélago de los Chonos y Costa occidental Patagónica, 1792–1796." It would be interesting to publish excerpts from these journals, which contain curious details about the towns of Los Césares and Arguello, which were supposedly founded in 1554 and which apocryphal accounts place between 42° and 49° southern latitude.

2. See the second note.

Nicaragua, of the area between Cruces and Panama [City], and between Cúpica and the Río Naipi,[1] would guide the statesman in his choice; one would learn if this great project, designed to immortalize a government that I.242 would be devoted to the true interests of humanity, should be carried out in Mexico, Nicaragua, or Darién.

The long circumnavigation of South America would thenceforth be less frequent, and a route would be opened, if not for ships, then at least for goods that need to travel from the Atlantic Ocean to the South Sea. We should like to believe that the time has ended when "Spain, through its unreasonably defiant politics, saw fit to refuse to others a road across its possessions, knowledge of which it has long concealed from the entire world."[2] Enlightened men at the head of government will appreciate the projects for the common good that are proposed to them: a foreign presence will no longer be regarded as a danger to the motherland.

When a canal connects the two Oceans, the products of Nootka Sound and China will be over two thousand leagues closer to Europe and the United States. Only then will great changes take place in the political state of eastern Asia, for this strip of land, against which the great waves of the Atlantic Ocean break, has been the highway of independence for China and I.243 Japan for centuries.

Since Mr. von Humboldt has recently given greater elaboration to the ideas he presented in this chapter on the possibility of an interoceanic canal, we refer the reader to the third volume of the *Relation historique*, pp. 117–47. We limit ourselves here to some numerical information: "it remains to be proven (writes our author) by means of analogy with what men have already built, whether the possibility of a link between the two seas can be realized at the current level of our modern civilization. The more complicated a problem becomes and the more dependent it is on many intrinsically variable elements at once, the more difficult

1. The information that Major Álvarez has recently conveyed to ▼ Captain Cochrane is not favorable to the usefulness of a canal between the Río Naixo, or Naipipi (a tributary of the Atrato) and the Bay of Cúpica or Túpica. This traveler claims that the Naipipi is full of obstructions, and that three ranges of hills cross the isthmus between the river and the coastline of the Pacific Ocean. (*Journal of a residence and travels in Columbia during the years* 1823 *and* 1824 *by Captain Charles Stuart Cochrane*, vol. 2, p. 448.)

2. Mr. de Fleurieu, in his scholarly notes on Marchand's *Voyage [autour du monde]*, vol. I, p. 566.

it becomes to determine the *maximum* effort of intelligence and physical prowess that people are able to exercise. If it were merely a question here of average size canals less than three to six feet deep and used only for inland navigation, I could mention canals built a long time ago that have crossed mountain ridges 300 to 580 feet high. Engineers have long had such little regard for 580 feet as the *maximum* elevation—that is, the height of the connector between the Naurouze [Pass] with the Midi canal—that one can reasonably expect that the famous ▼ Mr. Perronet had considered the Burgundy canal a very feasible project. This canal runs between the Yonne and the Saône and must pass over (near Pouilly) an elevation of 621 feet above the Yonne at low water. Yet such projects, however important for the prosperity of a country's domestic trade, can hardly be called oceanic navigation canals. *We already know three of these canals that were built on a large scale: the Eyder or Holstein canal, which receives 140-to-160-ton ships; the North Holland Canal [Groot Noordhollandsch Kanaal]; and the Caledonian Canal, which, though perhaps not the most useful, is the most magnificent hydraulic work that has been built to date. The North Holland Canal, whose construction does the greatest honor to the government of the Low Countries [Netherlands], is navigable by forty-four-cannon frigates drawing sixteen feet of water. It is fifteen leagues long and 120 strides wide at its narrowest point. The Caledonian Canal was completed in sixteen years: it provides passage to thirty-two-cannon frigates and cargo ships for overseas trade. Its average depth is eighteen feet, eight inches and its width at the base line is 47 feet. The locks, which number twenty-three, are 160 feet long by 37 feet wide. The cost of the Caledonian Canal was almost four million piasters, or 2,700,000 piasters less than the Languedoc Canal, converting one mark of silver to the current exchange rate. During Bonaparte's Egyptian campaign, Mr. LePère projected that expenses for the Suez Canal would rise to five or six million piasters, a third of which would be spent on the secondary Cairo and Alexandria canals. The interoceanic canal planned for South America could be even shallower than the Caledonian Canal. During the last fifteen years, new systems of trade and navigation have led to notable changes in the capacity, or cargo, of the ships most commonly used for trade with Calcutta and Canton. A close examination of the official list of ships from London and Liverpool engaged in the India and China trade over a two-year period (from July 1821 to June 1823) shows that, out of 216 ships, two-thirds were below 600 tons, one-quarter between 900 and 1,400 tons, and one-seventh below 400 tons. In France, the *average tonnage* of ships engaged in the India trade from the ports of Bordeaux, Nantes, and Le Havre is 350 tons. The system of small shipments is especially active in the United States, where one recognizes all the benefits of swift loading and unloading and the quick circulation of capital. The average American ship bound either for India around the

I.244

Cape of Good Hope or for Peru around Cape Horn carries 400 tons. South Sea whalers have a capacity of only two or three hundred tons. In the current state of world trade, these data adequately prove that a canal such as the one planned between the Atlantic Ocean and the South Sea would be sufficiently large, provided that *the size of its segments* and the capacity of its locks can accommodate ships between 300 and 400 tons. These are the *minimum* dimensions that the canal must have. In keeping with what we have indicated earlier, this minimum assumes a capacity almost double that of the Eyder canal but less than that of the Caledonian Canal. It is true that tonnage has only an indirect relationship with a ship's *draft*, because how a ship is built changes both how it handles and what it can carry. Admittedly, however, an average depth of fifteen to seventeen and a half feet (the old French measure) would suffice for a canal linking two oceans.

I.245

"The gigantic works of Europe, like the Caledonian Canal, the North Holland Canal, and the Forth and Clyde, had only small heights to cross, less than 160 feet. Until now, canals that cross ridges of 400 to 600 feet only afford a depth of four to six feet. Difficulties increase with the ridges' elevations, with the depth of the excavations, and with the width of the locks—not with their number. It is not simply a question of digging a canal. It is necessary to ensure that the amount of water taken from higher areas on the summit level will always be sufficient to fill the canal and replace what is lost to locks, evaporation, and filtration. *Neptune's Staircase* in the Caledonian Canal gives us an example of conjoined locks that quickly lift frigates to a height of sixty feet. That hydraulic work costs no more than 257,000 piasters, that is, five times less than the three shafts of the Valenciana mine, near Guanajuato. Ten *Neptune's Staircases* would permit 500-ton ships to pass over a 600-feet ridge, a point higher than the chain of the Corbières between the Mediterranean and the Atlantic Ocean. We are discussing here only the *possibility* of structures that might be built without saying that they must be built.

"When comparing the various routes around the Cape of Good Hope, around Cape Horn, or across one of Central America's isthmuses, one must carefully distinguish among the goods and the different positions of the peoples wishing to take part. The problem posed by trade routes is different for a British trader than for an Anglo-American merchant. Similarly, this important problem is resolved in different ways by those involved in direct trade with Chile, India, and China, and those whose speculations are directed either toward northern Peru and the west coasts of Guatemala and Mexico, or toward China by way of the American northwest, or toward the sperm whale hunting grounds in the Pacific Ocean. The creation of a canal most clearly favors these latter three objects of the maritime trade of both Europe and the United States. By way of the proposed Nicaraguan canal, one travels 2,100 nautical leagues from Boston to Nootka,

I.246

the old center of the fur trade on the northwestern coast of the Americas. The same journey takes 5,200 leagues via Cape Horn, the old—and current—trade route. For a ship departing from London, these respective distances are 3,000 and 5,000 leagues. These data would suggest a shorter route by 3,100 leagues for the Americans of the United States; for the British, by 2,000 leagues, excepting either the possibility of crosswinds or the very different dangers of the two trade routes that we are comparing here. When it comes to direct trade with India or China, the comparison is far less favorable to sailing across Central America, in terms of the relation between time and distance. Ships that normally sail from London to Canton around the Cape of Good Hope, crossing the equator twice, travel 4,400 leagues; from Boston to Canton, it takes them 4,500 leagues. If the Nicaragua canal were built, these respective trade routes would be 4,800 and 4,200 nautical leagues. With the current advances in navigation, the typical duration of a voyage to China from the United States or from England, around the southern tip of Africa, is between 120 and 130 days. Basing one's calculations on voyages from Boston and Liverpool to the Mosquito Coast, and from Acapulco to Manila, one finds that it takes 105 to 115 days to sail from the United States or England to Canton, staying entirely within the northern hemisphere without ever crossing the equator, that is, by taking advantage of a canal through Nicaragua and of the constant trade winds in the most peaceful section of the Great Ocean. The difference in time would be barely one-sixth. One cannot return by the same route, but, the voyage out to China would be safer in every season. I think that a nation with fine settlements on the southern tip of Africa and on the Île de France [Mauritius] would generally prefer the route from west to east, which also avoids the maladies that can befall sailors in the Nicaraguan canal. The canal's principal and true benefits would be faster passage to the western coasts of New Continent, a trade route from Havana and from the United States to Manila, and speedier expeditions from England and Massachusetts to the fur coast (the northwestern coast) or the islands of the Pacific Ocean, from whence one can then visit the market of Canton and Macau.

I.247

"Enlightened members of the new governments of equinoctial America recently consulted me on how to carry out this plan. I think that a stock-issuing company should not be formed until the possibility of building, between 7° and 18° northern latitude, an interoceanic canal large enough to accommodate ships of three to four hundred tons has been put in place and its location determined. It would be dangerous to make a choice before having studied, on a uniform map, the Isthmuses of Tehuantepec, Nicaragua, Panama, Cúpica, and the Chocó. When the maps and profiles of the five areas have been submitted to the public, an open and frank discussion will clarify the advantages and disadvantages of each location, and the construction of this important work will be entrusted to engineers who have competed for similar projects in Europe. The *junction company* will find shareholders among those governments and citizens who, unswayed by the lure of profit and yielding to more noble instincts, will pride themselves at the thought of contributing to a project that is worthy of the civilization of the nineteenth century.

"Of course, it is wise to recall here that the profit motive, the basis of all financial speculations, is inevitably part of this enterprise, which I enthusiastically support. The dividends from companies that obtained concessions to open canals in England prove the profitability of these enterprises for stockholders. In an interoceanic canal, cargo duties can be much higher for ships that wish to benefit from this new passageway to travel either to Guayaquil and Lima, or the sperm whale hunting grounds, or America's northwest coast, or Canton, thereby shortening their route and avoiding the high southern latitudes, which are frequently dangerous during the bad-weather season. Traffic on the canal would grow as traders further familiarized themselves with the new route from one Ocean to the other. Even if dividends were not especially high and the capital invested in this enterprise did not yield returns equal to those offered by many government bonds, from the Mosquito Coast all the way to Europe's final outposts, it would still be in the interest of Spanish America's large states to support this undertaking. Restricting the usefulness of canals and international routes to the duties paid for the transportation of goods, and completely discounting the influence of canals on industry and national ownership, means consigning to oblivion what experience and political economy has taught for centuries."

I.248

CHAPTER III

The Physical Aspect of the Kingdom of New Spain Compared
to Europe and South America—Irregularities of the Terrain—The
Influence of These Irregularities on the Climate, Culture, and Military
Defense of the Country—The Condition of the Coasts.

To this point, we have considered the vast expanse and the borders of the
Kingdom of New Spain. We have examined its relationships with the other
Spanish possessions and the advantages that the shape of its coastline might
present for communication between the Antillean Sea and the Great Ocean.
Let us now draw the physical portrait of the country, fixing our attention on
the irregularities of its terrain and on the influence of these irregularities on
the climate, agriculture, and the military defense of Mexico. We shall limit
our presentation to general results. The meticulous details of descriptive
natural history have no part in statistics, but one could not form a specific
idea of the territorial wealth of a state without knowing the outline of its
mountains, the elevation of its great interior plateaus, and the local temperatures of these regions where successive climates are found as if stacked
one on top of another.

A general view of the entire surface of Mexico shows that only one-third
is located between the tropics, and the other two-thirds lie in the temperate zone. The latter section encompasses 82,000 square leagues; it comprises the *Provincias internas*, some of which are under the direct control
of the viceroy of Mexico (for example, the new Kingdom of León and the
province of Nuevo Santander), while the others are governed by a specific
commander-general. This commander controls the Intendancies of Durango and Sonora and the provinces of Coahuila, Texas, and New Mexico,
sparsely inhabited regions that are collectively designated by the name *Provincias internas de la Comandancia general*, to distinguish them from the
Provincias internas del Vireynato.

On the one hand, small portions of the northern provinces of Sonora and Nuevo Santander extend beyond the tropic of Cancer; on the other, the southern intendancies of Guadalajara, Zacatecas, and San Luis Potosí (especially the area surrounding the famous Catorce mines) extend slightly north of this boundary. We know that the physical climate of a country does not solely depend on its distance from the pole but also on its elevation above sea level, its proximity to the Ocean, the configuration of its terrain, and a great number of other local circumstances. Because of these same factors, over three-fifths of the 36,000 square leagues that are in the Torrid Zone also enjoy a cold or temperate climate. The interior of the viceroyalty of Mexico, the country formerly known as Anahuac and Michoacan, and all of Nueva Vizcaya, form an immense plateau that is 2,000 to 2,500 meters above the level of the nearby seas.

There is hardly a single point on the globe whose mountains have such an extraordinary shape as those of New Spain. In Europe, Switzerland, Savoy, and Tyrol are considered very high countries, but this opinion is only based on the appearance of large groups of summits that are perpetually covered with snow and set out in ranges that are often parallel to the central range. The summits of the Alps are 3,000, even 4,700 meters high, while the neighboring plains in the canton of Bern are only 400 to 600 meters high. This first, quite moderate elevation may be considered that of most of the large-sized plateaus in Swabia, Bavaria, and New Silesia, near the sources of the Wartha and the Piliza. In Spain, the soil of the two Castiles is a little higher than 580 meters (300 toises). In France, the highest plateau is that of the Auvergne, where the Mont-d'Or, Cantal, and Puy-de-Dôme lie; the elevation of this plateau, according to ▼ Mr. de Buch's observations, is 720 meters (370 toises). These examples prove that, in general, the high terrains of Europe that resemble plains are barely over 400 to 800 meters above sea level.[1]

Perhaps in Africa, near the sources of the Nile,[2] and in Asia, at 34° and 37° northern latitude, there are plateaus similar to those of Mexico, but the travelers who have explored these regions have left us completely ignorant of the elevation of Tibet. Himalayan passes are generally at the same height as the summit of Mont Blanc, and ▼ Captain Webb found the Alpine lake,

I.251

I.252

1. According to the most recent measurements (Humboldt, *Relation historique*, vol. III, p. 208), the interior plateau of Spain is 330 to 360 toises high; that of Switzerland, between the Alps and the Jura, 270 toises; that of Bavaria, 260 toises; that of Swabia, 150 toises.

2. According to ▼ Bruce ([*Travels to discover the source of the Nile,*] vol. III, pp. 642, 652, and 712), the sources of the Blue Nile near Gojam are 3,200 meters above the level of the Mediterranean.

Rawun Rudd, where the Sutledge River originates (very near the famous Lake Manasarovar), at over 4,600 meters above sea level. People have heretofore measured the mountain summits and *cols* or *passes* of Tibet, rather than its high plains, such as those around Lhasa and Ladakh; but I have no doubt that the average height of the plateau between the ranges of the Himalayas and the Zangling [Qinling], or Kunlun, does not exceed 3,500 meters. Farther north, the great Gobi Desert northwest of China only reaches a height of 1,400 meters, according to ▼ Father Duhalde's work. ▼ Colonel Gordon had claimed to Mr. Labillardière that, from the Cape of Good Hope to 21° southern latitude, the soil of Africa rises imperceptibly to a height of 2,000 meters (1,000 toises),[1] and more recent measurements have proven the correctness of this claim. The entire African plateau north of the thirty-first parallel, inhabited by Bechuana, Korana [Griqua], and Batswana [Bushmen], is 880 toises above sea level.[2]

The mountain range that forms the vast plateau of Mexico is the same one that, under the name the Andes, runs across all of South America. The makeup or structure of this range, however, is very different south and north of the equator. In the southern hemisphere, crevices that resemble open seams and are not filled with diverse substances, tear and interrupt the Cordillera everywhere. If there are plains there at a height of 2,700 to 3,000 meters (1,400 to 1,500 toises)—as in the Kingdom of Quito and, farther north, in the Province of Los Pastos—they are not of a size comparable to those in New Spain. They are more like longitudinal valleys bordered by two branches of the great Andes Cordillera. In Mexico, on the contrary, it is the ridge of the mountains itself that forms the plateau; the orientation of the plateau indicates, so to speak, that of the whole range. In Peru, the highest summits form the crest of the Andes; in Mexico, these same summits are less gigantic, it is true, but still 4,900 to 5,400 meters high (2,500 to 2,770 toises) and dispersed over the plateau or set out in lines that have no parallel relation to the main axis of the Cordillera. Peru and the Kingdom of New Granada have transversal valleys whose perpendicular depth sometimes reaches 1,400 meters (700 toises). The existence of these valleys prevents the inhabitants from traveling, except on horseback, by foot, or carried on the backs of Indians called *cargadores*. In the Kingdom of New Spain, on the contrary, carriages run from the capital of Mexico to Santa Fe in the

I.253

I.254

---

1. [Houtou de] Labillardière, [*Relation du voyage,*] vol. I, p. 89.
2. ▼ Barrow, [*An Account of]* Travels in[to] the interior of South[ern] Africa*, vol. I, p. 10. [283]. ▼ Lichtenstein, *Reisen im südlichen Afrika*, vol. II, p. 544.

province of New Mexico, over a distance of more than 2,200 kilometers or 500 common leagues. On this entire road, there were no serious obstacles that required technical skill to overcome.

In general, the Mexican plateau is so uninterrupted by valleys, its slope so uniform and gentle that, as far as the town of Durango in Nueva Vizcaya, at a distance of 140 leagues from Mexico City, the terrain is at a constant elevation of between 1,700 and 2,700 meters (850 to 1,350 toises) above the nearby sea level. This is the same height as the Mount Cenis, Saint Gotthard, and Great Saint Bernard mountain passes. To give as clear and complete a presentation as possible of such a curious and novel geological phenomenon, I conducted five barometrical surveys. The first crosses the Kingdom of New Spain from the South Sea coasts to those of the Gulf of Mexico, from Acapulco to Mexico City, and from the capital to Veracruz. The second survey runs from Mexico City to Guanajuato via Tula, Querétaro, and Salamanca; the third includes the intendancy of Valladolid, from Guanajuato to Patzcuaro and the volcano of Jorullo; the fourth leads from Valladolid to Toluca, and from there to Mexico City; the fifth includes the area surrounding Morán and Actopán. I determined the elevation of 208 points, either using a barometer or trigonometrically. All these points are distributed across a terrain between 16°50′ and 21°0′ northern latitude and 102°8′ and 98°28′ longitude (west of Paris). Beyond these boundaries, I know of only one location whose elevation has been precisely determined. This is the town of Durango, whose elevation above sea level, deduced from the average height of the barometer, is 2,000 meters (1,027 toises). The Mexican plateau thus maintains its extraordinary elevation even as it extends northward, well beyond the tropic of Cancer.

These barometrical measurements, together with the astronomical observations that I conducted on the same expanse of terrain, have enabled me to construct the physical maps that accompany this work. They contain a series of vertical cross sections or profiles. I have tried to represent entire countries using a method that until now has only been employed for mines or small portions of land through which canals are intended to pass. In the statistics of the Kingdom of New Spain, it was necessary to limit drawings to those that attract interest from the viewpoint of political economy. The physiognomy of a country, the grouping of mountains, the total surface area of plateaus, the elevations that determine its temperatures and aridity— whatever constitutes the physical aspect of the globe—is fundamentally related to the advances of the population and the well-being of its inhabitants. The modifications of the surface of the earth affect the state of agriculture,

I.255

I.256

which varies depending on differences in climate and the direction of isothermal lines, the ease of interior commerce, communication that is more or
less favored by the nature of the terrain, and finally military defense, upon
which the external security of the country depends. In these respects alone
do broad geological views interest the statesman, as he seeks to evaluate the
strength and territorial wealth of nations.

In South America, the Cordillera of the Andes exhibits—and this at immense heights—completely level terrains. The plateau where the town of
I.257   Santa Fe of Bogotá is located is 2,658 meters (1,365 toises) high, planted
with European wheat, potatoes, and *Chenopodium Quinoa*, falls in this category, as does also the Cajamarca plateau in Peru, the former residence of
the unfortunate Atahualpa, which I found to be at a height of 2,860 meters (1,464 toises). The great plains of Antisana, in the midst of which rises
the part of the volcano that reaches the limit of perpetual snows, are 4,100
meters (2,100 toises) above sea level. These plains are 389 meters (200
toises) higher than the Peak of Tenerife; they are so level that those who
dwell in these high regions under the local sun are unaware of the extraordinary situation in which nature has placed them. But all these plateaus of
New Granada, Quito, and Peru occupy no more than forty square leagues.
Because they are difficult to reach and separated from each other by deep
valleys, they do not lend themselves well to the hauling of goods and domestic trade. Crowning these isolated summits, they form, so to speak, islands in the middle of an airy Ocean. Thus, those who live on these cool
plateaus remain concentrated there; they are reluctant to descend to neighboring lands, where suffocating heat prevails that is harmful to the primitive
inhabitants of the high Andes.

The terrain of Mexico is completely different. The plains are more ex-
I.258   pansive than in Peru, with a surface no less uniform; they are so close to one
another that on the extended ridge of the Anahuac Cordillera, they form
but a single plateau. Such is the plateau that is found between 18° and 40°
northern latitude. Its length is equal to the distance that one would travel
from Lyon to the tropic of Cancer across the great African desert. The Mexican plateau drops imperceptibly to the north. As we have noted above, no
measurement has been taken in New Spain beyond the town of Durango,
but travelers observe that the terrain descends visibly toward New Mexico and the sources of the Río Colorado. The profiles that accompany this
work show three cross sections, one of which is longitudinal and oriented
from south to north: it represents the ridge of the mountains as they extend toward the Río Bravo. The two other drawings show transversal cross

sections from the coastline of the Pacific Ocean to that of the Gulf of Mexico. These three vertical sections demonstrate at a glance the obstacles that the extraordinary configuration of the terrain poses to the transportation of products, specifically to trade between the interior provinces and the commercial towns on the coast.

As one travels from the capital of Mexico to the great mines of Guanajuato, one remains in the valley of Tenochtitlan for ten leagues, at an elevation of 2,277 meters (1,168 toises) above the neighboring Ocean. The floor of this beautiful valley is so uniform that the village of Gueguetoque, situated at the foot of the Sincoque Mountain, is only twenty meters higher than Mexico City. The hill of Barrientos is merely a promontory that stretches into the valley. From Gueguetoque, one ascends near Batas to Puerto de los Reyes, and from there one descends into the Tula valley, which is 222 meters lower than the Tenochtitlan valley, and through which the large drainage canal of Lakes San Cristóbal and Sumpango carries its waters to the Río de Moctezuma and the Gulf of Mexico. To reach the great plateau of Querétaro from the base of the Tula valley, one must pass over the Calpulalpan Mountain, which is only 2,687 meters (1,379 toises) above sea level and which, as a result, is lower than the town of Quito, although it appears to be the highest point on the whole road from Mexico City to Chihuahua. The broad plains of San Juan del Río, Querétaro, and Zelaya, fertile plains dotted with villages and good-sized towns, begin north of this mountainous country. They are called the *bajío* (lowlands), yet their average elevation is equal to that of the summit of Puy-de-Dôme in the Auvergne; they are almost thirty leagues long and extend as far as the foot of the metalliferous mountains of Guanajuato. Those who have traveled to New Mexico claim that the rest of the road resembles the section that I have just described, which I have represented in a special profile. There is a succession of immense plains, which resemble dried-up basins of former lakes and are separated only by hills that rise barely 200 to 250 meters above the bottoms of these basins. I shall present in another work (in the Atlas appended to the *Personal Narrative* of my journey) the profile of the four plateaus that surround the capital of Mexico. The first, which includes the Toluca valley, is 2,600 meters high (1,340 toises); the second, the Tenochtitlan valley, is 2,274 meters (1,168 toises) high; the third, or the Actopán valley, is 1,966 meters (1,009 toises) high; and the fourth, the Istla valley, is 981 meters (504 toises) high. These four basins differ as much in climate as in their elevations above sea level. Each one has a different agriculture: the fourth, the lowest, is suited for growing sugarcane; the third, for cotton; the second, for

I.259

I.260

European wheat; and the first, the Toluca basin, for Agave plantations, and it may be considered the vineyards of the Aztec Indians.

The barometric survey that I conducted between Mexico City and Guanajuato demonstrates how favorable the configuration of the terrain—that is, in the interior of New Spain—is, longitudinally, to the transportation of commodities, to navigation, and even to the construction of canals. This is not the case for the transversal cross sections between the South Sea and the Atlantic Ocean. These cross sections expose the difficulties that nature imposes on communication between the interior of the kingdom and its coastlines; they show an enormous difference in level and temperature everywhere, whereas between Mexico City and Nueva Vizcaya, the height of the plateau remains the same and therefore has a cold rather than a temperate climate. The descent from the capital of Mexico to Veracruz is shorter and faster than from the same point to Acapulco. One might venture to say that the country has a better natural defense militarily against European peoples than against attacks by an Asian enemy. But the constancy of trade winds and the great rotation current, which never ceases in the tropics, practically negate any political influence that, over the centuries, China, Japan, or Asiatic Russia might wish to exert over the New Continent.

Traveling eastward from the capital of Mexico on the Veracruz road, one must advance sixty nautical leagues before reaching a valley whose floor is under 1,000 meters above sea level, where, as a result, oaks no longer grow. On the Acapulco road, descending from Mexico City toward the South Sea, one reaches these same temperate regions in fewer than seventeen leagues. The eastern slope of the Cordillera is so steep that, once the descent from the great central plateau has begun, one continues to descend until one arrives at the eastern coast, the coastline of Alvarado and Veracruz.

I.262   The western slope is furrowed by four longitudinal valleys that are very striking and so evenly distributed that the ones nearest the Ocean are also deeper than the farthest ones. If one looks closely at the profile that I have drawn up, using precise measurements, one sees that from the Tenochtitlan plateau, the traveler first descends into the Istla valley, then into the Mescala valley, then the Papagallo valley, and finally into the Peregrino valley. The bottoms of these four basins are 981 meters, 514 meters, 170, or 158 meters (504, 265, 98, or 82 toises), respectively, above sea level. The deepest basins are also the narrowest. A curve drawn through the mountains that separate these valleys—through the Cerro La Cruz del Marqués (Cortés's former camp), and the summits of Tasco, Chilpanzingo, and the Posquelitos—would have a very regular shape. One might even be tempted to think that this regularity conforms to the general model that nature followed in constructing the mountains there, but the appearance of the Andes of South America is enough to dispel such systematic fantasies. We know from a large number of geological considerations that when these mountains were formed, a few, seemingly minor causes determined that matter would accumulate into colossal peaks, at some points near the center of the Cordilleras and at others near their edges.

The Asia route is quite different from the one that leads out from the coastline facing Europe. Over the space of seventy-eight and a half leagues in a straight line from Mexico City to Acapulco, one is always ascending and descending and constantly passing from a cold climate to excessively hot regions. The Acapulco route, however, can be made suitable for haulage. Of the eighty-four and a half leagues that separate the capital from the port of Veracruz, fifty-six are taken up by the great Anahuac plateau. The rest of the route is but a difficult and continuous descent, especially from the small fortress of Perote to the town of Jalapa, and from that point, one of the most beautiful and picturesque in the known world, to Rinconada. It is the difficulty of this descent that makes the cost of hauling flour from Mexico City to Veracruz more expensive and has heretofore prevented this flour from competing in Europe with flour from Philadelphia. A superb road[1] is now being built along the eastern declivity of the Cordillera. This structure, which is entirely due to the great and commendable activity of the Veracruz merchants, will have the profoundest effect on the well-being of the inhabitants of the entire Kingdom of New Spain. Thousands of mules will be replaced by carts hauling merchandise from one Ocean to the other; they will thus bring the Asian trade of Acapulco closer to the European trade of Veracruz.

As we have stated above, 23,000 square leagues of the Mexican provinces in the Torrid Zone enjoy a climate that is rather cold than temperate. Moreover, this entire vast expanse of country is traversed by the Mexican Cordillera, a range of colossal mountains that may be considered an extension of the Peruvian Andes. Despite the low height of the mountains west of the Río Atrato, in the Chocó, and in the province of Darién, the Andes that run through the Isthmus of Panama appear to be connected to the western chain of New Granada: they reach large heights again in the province of Veragua and in the Kingdom of Guatemala. At some points, their crest approaches the Pacific Ocean and at other points occupies the center of the country; sometimes it even bears toward the coastline of the Gulf of

I.263

I.264

---

1. Since the first edition of this *Political Essay*, that road has been completed.

Mexico. In the Kingdom of Guatemala, for example, this crest, spiked with

I.265 volcanic cones,[1] skirts the western coastline from Lake Nicaragua to the Bay of Tehuantepec. But in the province of Oaxaca, between the sources of the Chimalapa and the Guasacualco Rivers, it occupies the center of the Mexican isthmus. From 18 and one-half to 21° latitude, in the Intendancies of Puebla and Mexico City, from Mixteca to the Zimapán mines, the Cordillera runs from south to north and approaches the eastern coastline.

In this part of the great Anahuac plateau, between the capital of Mexico and the small towns of Córdoba and Jalapa, rises a group of mountains that rival the highest summits of the New Continent. It is enough to name four[2] of these colossi, the heights of which were unknown prior to my journey: Popocatepetl (5,400 meters or 2,771 toises), Iztaccihuatl (or the White

I.266 Woman, 4,786 meters or 2,455 toises), Citlaltepetl (or the Peak of Orizaba, 5,295 meters, or 2,717 toises), and Naucampatepetl (or the Cofre de Perote, 4,089 meters or 2,089 toises). This group of volcanic mountains exhibits strong similarities to the group in the Kingdom of Quito that is located one and a half degrees south and one-quarter degree north of the equator. If the height attributed to Mount Saint Elias[3] today is exact, then it is admissible

---

1. Humboldt, *Relation historique*, vol. III, p. 206. From the Gulf of Nicoya to the parallel of Soconusco (from 9° and one half to 16° latitude), there are twenty-one volcanoes, some extinct and some active. The *Volcán de agua* between the Pacaya volcano and the *Volcán de Fuego* (also called the *Volcán de Guatemala*) is covered with snow for most of the year and appears to be over 1,750 toises high. On the torrents of "water and rocks" that the *Volcán de agua* spewed on September 11, 1541, destroying Almolonga or the Ciudad Vieja (which should not be confused with Antigua Guatemala), see Remesal, *Historia de la provincia de San Vicente*, Book IV, chap. 5; and [Juan Domingo] Juarros [y Montúfar], *Compendio de la Historia de Guatemala*, vol. I, pp. 72–85 and vol. II, p. 551.

2. With the exception of the Cofre de Perote, these four measurements are all geometrical, but because the bases are at elevations ranging from 1,100 to 1,200 toises above sea level, the first part of the total height has been calculated using Mr. Laplace's barometrical formula. The word Popocatepetl derives from *popocani,* smoke, and *tepetl*, mountain; *Iztaccihuatl* from *iztac*, white, and *ciuatl*, woman. Citlaltepetl means "a mountain that is as bright as a star," from *citlaline,* star, and *tepetl*, mountain, because when it spews fire, the Peak of Orizaba appears from a distance like a fiery star. Naucampatepetl derives from *nauhcampa*, a square object. It is an allusion to the shape of a small trachytic rock that is found at the summit of the Perote Mountain, which the Spanish compared to a chest. (See ▼ Father Alonzo de Molina, *Vocabulary of the Aztec Language*, published in Mexico in 1571, p. 63.)

3. This colossal mountain is not technically on the extension of the central Mexican range: it belongs to the *Maritime Alps* of California and the entire northwest coast, a range that is connected to the Rocky Mountains by ridges and foothills (around 48° latitude). In 1791, Spanish seafarers using exact methods found the elevation of Mount Saint Elias to be 2,793 toises above sea level, whereas in the account of Lapérouse's Voyage, the elevation is indicated as being only 1,980 toises. We would like to point out here that north of the *nexus*

that only at 19° and 60° latitude in the northern hemisphere do the mountains reach the enormous height of 5,400 meters above sea level. I.267

Farther north beyond the nineteenth parallel, near the famous Zimapán and Doctor mines located in the Intendancy of Mexico City, the Cordillera takes the name of *Sierra Madre*. Once again it veers away from the eastern part of the kingdom and bears northwest toward the towns of San Miguel el Grande and Guanajuato. North of Guanajuato, which may be considered the Potosí of Mexico, the *Sierra Madre* expands to an extraordinary width. Shortly thereafter it divides into three branches, the easternmost of which stretches toward Charcas and the Real de Catorce then thins out in the New Kingdom of León. The western branch covers a portion of the intendancy of Guadalajara. It drops sharply after Bolaños and extends through Culiacan and Arispe in the Intendancy of Sonora all the way to the banks of the Río Gila. At 30° latitude, it soars again to great heights in the Tarahumara, near the Gulf of California, where it forms the mountains of the Pimería Alta, famous for their large gold washes. The third branch of the Sierra Madre, which must be considered the central range of the *Mexican Andes,* occupies the entire Intendancy of Zacatecas. One can follow it through Durango and the Parral (in Nueva Vizcaya) to the *Sierra de los Mimbres* (west of the *Río Grande del Norte*). From there, this branch traverses New Mexico I.268 before connecting with the Crane mountains [las montañas de la grulla] and the Sierra Verde. This mountainous area, located at 40° latitude, was explored in 1777 by Fathers Escalante and Font. The Río Gila originates there, near the source of the Río del Norte. The crest of this central branch of the Sierra Madre divides the waters between the Pacific Ocean and the Antillean Sea. Fidler, Mackenzie, Pike, Long, and James explored its continuation between 37° and 68° latitude. In these northern regions, the Anahuac Andes bear the names *Rocky Mountains* (*Stony* or *Chippaweya Mountains*). They are spiked with granitic peaks that American travelers

---

of the Loja mountains (3°–5° southern latitude), the Andes rise above the majestic height of 2,600 toises only three times, to wit: in the Quito group, from 0° to 2° southern latitude (Chimborazo, Antisana, Cayambe, Cotopaxi, Collanes, Iliniza, Sangay); in the Cundinamarca group, 4° and three-quarters northern latitude (the Tolima Peak north of the Quindiu pass); and in the Anahuac or central Mexican group, 18°59′ to 19°12′ latitude (Popocatépetl and the Peak of Orizaba). In the immense expanse of the Cordilleras, from 8° southern latitude to the Strait of Magellan, there is not a single mountain covered with perpetual snow whose elevation above sea level has been determined, either by simple geometrical measurement or by combined barometrical and geometrical methods. The highest Andean peak north of the equator is Tolima Peak (latitude 4°46′), whose name is practically unknown in Europe and which I found to have an elevation of 2,865 toises.

have called *Spanish Peak*, *James Peak*, and *Bighorn* (from latitude 37°20′ to 40°13′) at elevations ranging between 1,600 and 1,870 toises.[1] Farther on, north of the source of the Platte River, the *Rocky Mountains* appear to drop off around 46° and three-quarters and 47°; they rise again between 48° and 49°. Their crests reach from 1,200 to 1,300 toises, and their passes nearly 950 toises, between the source of the Missouri and the Lewis River (one of the tributaries of the Oregon or the Columbia River). The Cordilleras widen prodigiously and form an elbow that recalls that of *Knot of Cuzco* (lat. 14° and one-half south). With an orientation of N 24° W they extend toward the mouth of the Mackenzie River (69° northern latitude), after reaching— from Tierra del Fuego or, more precisely, from the Diego Ramírez Reef [or Islands] (56°33′ southern latitude)—a length of 3,700 leagues at twenty-five to the degree. The length of the Andes is equivalent to the distance from Cabo Finisterre in Galicia to the Northeast Cape (Tschuktschoi-Noss) of Asia. Of all the mountain ranges of the Globe, that of the Andes is the most continuous, the longest, and the most constant in its direction from south to north and to the northwest.[2]

I.269

1. See *Introduction*, p. 40 [in this edition].
2. It is the longest, but not the highest. Below are the ratios that I find between the Himalayas, the Andes, the Alps, and the Pyrenees:

NAMES OF MOUNTAIN RANGES	HIGHEST PEAKS	MEDIUM ELEVATION OF THE PEAKS
Himalayas (between 30°18′ and 31°53′ northern latitude, and between 75°23′ and 77°38′ longitude)	4,026 t.	2,450 t.
Cordilleras of the Andes (between 5° northern latitude and 2° southern latitude)	3,350 t.	1,850 t.
Swiss Alps	2,450 t.	1,150 t.
Pyrenees	1,787 t.	1,150 t.

The Himalayan passes that lead from Chinese Tartary to Hindustan are from 2,400 to 2,700 toises high. The pass of Guanacas, through which passes the road from Santa Fé de Bogotá to Popayán, is 2,300 toises high. As for the highest summit in the Himalayas, I have chosen the Iawahir (latitude 30°22′19″, longitude 77°35′7″) which Mr. ▼ Hodgson and Mr. Herbert found to have an elevation of 4,026 toises. Only by using angles taken from great distances can one give a height of 4,390 toises to the Dhawalagiri Peak, which is south of Mustung, near the sources of the Gunduck (see my "Mémoire sur la hauteur des montagnes de l'Inde" in the *Annales de chimie et de physique*, 1816, vol. III, p. 313; my *Relation historique*, vol. III, p. 191, and [Hodgson and Herbert, "VI. An Account of Trigonometrical and

We have just sketched the tableau of the Andes Mountains and their relationships with the mountains of New Spain. We have shown that, with few exceptions, only the coastline of this vast land enjoys a warm climate suitable for growing the products traded by the Antilles. The Intendancy of   I.270
Veracruz (with the exception of the plateau that extends from Perote to the Peak of Orizaba), the Yucatán Peninsula, the coastline of Oaxaca, the maritime provinces of Nuevo Santander and Texas, the entire New Kingdom of León, the province of Coahuila, the uncultivated country called the *Bolsón de Mapimi*, the coastline of California, the western part of Sonora, Sinaloa, and Nueva Galicia, the southern regions of the Intendancies of Valladolid, Mexico City, and Puebla—all these are on low terrain that is only punctuated with small hills. The average temperature on these plains is similar to the temperature found in the tropics between 17° and 23° latitude wherever the elevation is below three to four hundred meters above sea level: it ranges between twenty-five and twenty-six degrees Centigrade[1]; in other words, it is eight to nine degrees higher than the average temperature of Naples.

These fertile regions, which the indigenous peoples call *tierras calien-*   I.271
tes, produce sugar, indigo, cotton, and bananas in abundance. When unacclimated Europeans stay there for long periods of time, whenever they gather there in densely populated towns, these regions become a breeding ground for yellow fever, which is known in Mexico as "black vomit," or *vómito prieto*. The port of Acapulco and the Papagayo and Peregrino valleys are among those places on earth where the air is consistently both the hottest and the most insalubrious. On the eastern coastline of New Spain, the great heat breaks during the winter months. The north winds carry layers of cold air from Hudson Bay down to the parallel of Havana and Veracruz. These impetuous winds blow from October to March; they presage their arrival by disturbing the regularity of the small atmospheric tides[2] and the hourly variations of the barometer. They often chill the air to such an extent that the Centigrade thermometer drops[3] to four degrees near Havana and even as low as sixteen degrees in Veracruz.

Astronomical Operations for determining the Heights and Positions of the principal Peaks of the Himalaya Mountains,"] *Asiat[ick] Researches*, vol. 14, pp. 187–373 [371].)

1. In this work, I have consistently used the centesimal scale of the mercury thermometer, and *centesimal degree* must be understood unless the opposite is expressly stated.

2. I have written more about this phenomenon in my *Essai sur la Géographie des plantes* and in my *Tableau physique des régions équinoxiales*, 1807, pp. 92–94.

3. In the three years of 1810, 1811, and 1812, Mr. Ferrer did not see the Centigrade thermometer in Havana rise above thirty degrees or drop below sixteen degrees; but in

On the slope of the Cordillera, at an elevation between 1,200 and 1,500 meters, there is always a mild spring-like temperature that only varies by four to five degrees. Neither extreme heat nor excessive cold are known there. The indigenous peoples call this region *tierra templada*, where the average yearly temperature is between eighteen and twenty degrees. This is the delightful climate of Jalapa, Tasco, and Chilpanzingo, three towns famous for their healthy climates and the abundance of fruit trees cultivated in their vicinity. Unfortunately, the average elevation of 1,300 meters is also nearly the same elevation at which clouds hover above the plains near the sea. As a result, these temperate regions halfway up the slope (for example, near the town of Jalapa) are often shrouded in thick fog.

We have yet to discuss the third zone, known as the *tierras frías*. It includes the plateaus that are higher than 2,200 meters above sea level and whose average temperature is below seventeen degrees. In the capital of Mexico, the Centigrade thermometer has occasionally been seen to drop to a few degrees below freezing, but this is a rare occurrence. Most often the winters are as mild as in Naples. In the coldest season, the average daytime temperature is between thirteen and fourteen degrees; in summer, the thermometer does not rise above twenty-six degrees in the shade. In general, the average temperature across the great Mexican plateau is seventeen degrees, slightly below that of Naples and Sicily. But according to the indigenous peoples' classification system, this plateau belongs (as we have mentioned above) to the *tierras frías*, because cold and hot have no absolute value. Under a blazing sky in Guayaquil, people of color complain of excessive cold when the Centigrade thermometer suddenly drops to twenty-four degrees, even though it stays at thirty degrees for the rest of the day.

The plateaus that are higher than the Valley of Mexico, for example those whose absolute elevation is greater than 2,500 meters, have a harsh, disagreeable climate for the tropics, even to an inhabitant of the north. Such are the plains of Toluca and the heights of Guchilaque, where the air is no warmer than six or eight degrees for most of the day. Olive trees bear no fruit here, whereas they are successfully cultivated a few hundred meters below, in the Valley of Mexico.

I.273

January 1811, at sunrise in Río Blanco, south of Havana, on a plain only a few toises above sea level, I found the thermometer at 7.5° Centigrade. The astronomer Don Antonio Robredo even saw ice formed in a *batia* (a water-filled vase) in the interior of the island of Cuba (latitude 22°56′) at an absolute elevation of forty toises. This ice was probably the effect of light shining on the surface of the fluid, and the temperature of the atmosphere had probably not dropped to below plus-three degrees during the night.

All these so-called cold regions enjoy an average temperature of eleven to thirteen degrees, equal to that of France and Lombardy. Their vegetation, however, is much less vigorous and European plants do not grow as rapidly there as in their native soils. At an elevation of 2,500 meters, winters are not extremely harsh; but even in the summer, the sun does not heat up the rarified air on these plateaus enough to accelerate the blooming of flowers and to bring fruits to a perfect ripeness. The evenness of the temperature and the absence of temporary bouts of great heat have a peculiar influence on the climate of the high equinoctial regions. Certain vegetables are more difficult to grow on the ridge of the Mexican Cordilleras between the tropics [of Cancer and Capricorn] than on the plains located at much more northerly latitudes. The average annual temperature on those plains may be lower than on the plains between the nineteenth and the twenty-first parallels; but the ripeness of fruits and the development of more or less hearty vegetation do not depend as much on the average annual temperature as on the distribution of heat across the various seasons.

These general remarks on the physical division of New Spain are of great political interest. In France, as in most of Europe, both land use and agricultural divisions depend almost entirely on geographical latitude; in the equinoctial regions of Peru, as well as in those of New Granada and Mexico, it is elevation above sea level that solely determines the climate, the type of products, and the appearance—dare I say the physiognomy—of the country. The effect of the elevation detracts from the significance of geographical position. Agricultural lines, like those ▼ Arthur Young and Mr. Decandolle drew on horizontal projections of France, can only be shown on *profiles* [cross sections] of New Spain. At 19° and 21° latitude, sugar, cotton, and especially cocoa and indigo grow abundantly only up to elevations of 600 to 800 meters.[1] European wheat is cultivated in an area that generally begins on the mountain slopes at 1,400 meters and ends at 3,000 meters. The banana tree (*Musa paradisiaca*), a valuable plant that constitutes the primary food of many of the inhabitants of the tropics, yields almost no fruit above 1,550 meters; Mexican oaks grow only between 800 and 3,100 meters; pines descend toward the coastline of Veracruz only

I.275

1. These remarks only concern the general distribution of plant cultivation. I shall later list those places where sugar and cotton, favored by a special exposure, are grown up to an elevation of 1,700 meters above sea level.

down to 1,850 meters; but they also grow only up to an elevation of 4,000 meters,[1] near the perpetual snow line.

Like the rest of North America, the provinces that are called *internas* and are located in the temperate zone (especially those between 30° and 38° latitude) enjoy a climate that is essentially different from the one found at the same parallels in the old continent. These regions experience striking differences in seasonal temperatures. There, Neapolitan and Sicilian summers are followed by German winters. I discussed these phenomena in my *Mémoire sur les inflexions des lignes isothermes.* It is unnecessary to cite any other causes than the great width of the continent and its extension toward the North Pole. Many enlightened travelers have discussed this subject with all the care it deserves; Mr. de Volney has done so again in his recent work on the soil and climate of the United States. I limit myself here to adding that the difference in temperature observed on the same latitude in Europe and the Americas is much less striking in those parts of the New Continent that are near the Pacific Ocean than it is in the eastern parts. By studying agricultural conditions and the natural distribution of plants, ▼ Mr. Barton has attempted to prove that the Atlantic Provinces are much colder than the wide plains located west of the Allegheny Mountains.

One very notable advantage for the advances of national industry arises from the elevation at which nature has deposited a great wealth of precious metals in New Spain. In Peru, the largest silver deposits—those of Potosí, Pasco, and Chota—are at huge elevations very near the eternal snow line. To mine these deposits, workers, supplies, and animals must be brought from afar. Towns located on treeless plateaus where water freezes all year long do not make for pleasant stays. Only the hope of great wealth can persuade a free man to abandon the delightful climate of the valleys to live in isolation on the ridge of the Peruvian Andes. In Mexico, on the other hand, the wealthiest silver veins—those of Guanajuato, Zacatecas, Tasco, and Real del Monte—are found at average heights of 1,700 to 2,000 meters. The mines there are surrounded by cultivated fields, towns, and villages; forests crown the nearby peaks, and everything there makes it easy to tap the underground riches.

Even amidst the many advantages that nature has bestowed on the Kingdom of New Spain, this country, like Old Spain, suffers from a general

I.276

I.277

---

1. On this subject, one may consult the profile of the road from Mexico City to Veracruz (plate XII of my Atlas) and the agricultural scale in my *Essai sur la géographie des plantes,* p. 139.

scarcity of water and navigable rivers. The great northern river (Río Bravo del Norte) and the Río Colorado are the only rivers capable of attracting the traveler's attention, because of their length and the great volume of water that they pour into the ocean. The Río del Norte runs a course of 512 leagues, from the mountains of the Sierra Verde (east of Lake Timpanogos [Lake Utah]) to its mouth in the province of Nuevo Santander. The course of the Río Colorado is 250 leagues. But these two rivers, located in the most uncultivated part of the kingdom, will never hold any interest for trade until great changes in the social order and other favorable events compel colonists to settle these fertile and temperate regions. These changes are perhaps not far off. In 1797, the banks of the Ohio[1] were still so sparsely populated that an area of 130 leagues was occupied by barely thirty families, whereas today, the settlements there are separated by only one or two leagues.

I.278

Throughout the equinoctial part of Mexico, one finds only small rivers with rather wide mouths. The aridity of the plateau and the narrow shape of the continent there make it impossible for a large volume of water to gather. The steep slope of the Cordillera gives rise to torrents rather than rivers. In this respect, Mexico is similar to Peru, where the Andes are also very close to the coastline, and where this great proximity has the same effect on the aridity of the neighboring plains. Among the few rivers that exist in the southern part of New Spain, the only ones that may in time become interesting for domestic trade are: (1) the Río Guasacualco and the Alvarado, both southeast of Veracruz and suitable for facilitating communication with the kingdom of Guatemala; (2) the Río de Moctezuma, which carries the waters of the lakes and the valley of Tenochtitlan to the Río de Panuco, and by which (putting aside the fact that Mexico City is 2,277 meters above sea level) navigation is planned from the capital to the eastern coast; (3) the Río Zacatula; and (4) the great Santiago River, which originates from the confluence of the Lerma and the Lajas Rivers and which could carry flour from Salamanca, Zelaya, and perhaps from the entire intendancy of Guadalajara to the port of San Blas on the Pacific Ocean coastline.

I.279

Mexico's abundant lakes, most of which appear to be shrinking from year to year, are but the remains of the immense water basins that seem to have existed long ago on the high plains of the Cordillera. In this physical tableau, I shall merely mention the great lake of Chapala in Nueva Galicia, which covers nearly 160 square leagues, twice the size of Lake Constance; the lakes of the Valley of Mexico, which today occupy only one-tenth of the

1. ▼ Michaux, *Voyage à l'ouest des monts d'Alléghanys*, p. 115.

surface of that valley; Lake Patzcuaro in the intendancy of Valladolid, one of the most picturesque sites that I know on either continent; Lake Mextitlan; and Lake Parras in Nueva Vizcaya.

The interior of New Spain, especially a large part of the high Anahuac plateau, is devoid of vegetation: in some places, its aridity recalls the plains of the two Castiles. Several causes combine to produce this extraordinary effect. The Mexican Cordillera is too high for this elevation not to increase noticeably the evaporation that takes place on all large plateaus. On the other hand, the country is not high enough for a large number of summits to reach above the perpetual snow line. Below the equator, this line is at 4,800 meters (2,460 toises), below 45° latitude at 2,700 meters (1,400 toises) above sea level. In Mexico, at 19° and 20° latitude, the eternal snows begin at 4,600 meters (2,350 toises). Thus, of the six colossal mountains that nature has arranged on the same line between the nineteenth and the nineteenth and one-quarter parallels, only four—the Peak of Orizaba, Popocatepetl, Iztaccihuatl, and the Nevado de Toluca—are perpetually covered with snow, while the other two—the Cofre de Perote and the Colima volcano—are without it for most of the year. To the north and south of this parallel of great elevations, beyond the unique zone where the new volcano of Jorullo is also found, there are no mountains with perpetual snows.

At the parallel of Mexico City, there is no snow below 4,500 meters during the minimum snow period, in the month of September. But in the month of January, the snow line is at 3,700 meters: this is the maximum snow period. The fluctuation of the perpetual snow line at 19° latitude is therefore 800 meters from one season to the next, whereas on the equator, it is only between sixty and seventy meters. One should not confuse eternal snow with the snow that sporadically falls during the winter in much lower-lying regions. The latter phenomenon, like everything in nature, is subject to immutable laws worthy of scientific research. On the equator, in the province of Quito, one sees ephemeral snow only at elevations of 3,800 to 3,900 meters. In Mexico, by contrast, snow can be seen up to an elevation of 3,000 meters below 18° and 22° latitude. It has even snowed on the streets of the capital of Mexico, at 2,277 meters, and even 400 meters lower, in the town of Valladolid.

In general, the soil, climate, and the physiognomy of plants in the equinoctial regions of New Spain bear the character of temperate zones. The elevation of the plateaus, the strong radiation of heat toward an extremely clear sky, the proximity of Canada, the great width of the New Continent above 28° latitude, and the snow mass that covers it all cause drops in the

I.280

I.281

temperature of the Mexican atmosphere that one would hardly expect in regions so close to the equator.

Although the plateau of New Spain is unusually cold in winter, its summer temperature is much higher than what Bouguer and La Condamine found in their thermometric observations in the Peruvian Andes. Both the great mass of the Mexican Cordillera and the huge expanse of its plains reflect sun rays to a degree that one does not find at the same elevation in mountainous countries with a more uneven surface. This heat and other local causes have an influence on the aridity that plagues these beautiful regions.

I.282

North of 20°, especially from 22° to 30° latitude, the rains (which occur only during the month of June, July, August, and September) are very infrequent in the interior of the country. We have already shown earlier that the high elevation of this plateau and the low barometric pressure of the thin air accelerate evaporation. The ascending air current, or the column of hot air that rises from the plains, prevents the clouds from precipitating as rain that would water a dry, saline land devoid of brush and shrubs. Streams are rare in these mountains composed mainly of porous amygdaloid and finely cracked trachyte. Instead of collecting in small underground basins, the water trickles into the crevices that date back to ancient volcanic activity; it emerges only at the foot of the Cordillera near the coastline; there, it creates many rivers, which are somewhat short due to the configuration of the land.

The aridity of the central plateau and the lack of trees, to which the water's prolonged stay in the great valleys has perhaps contributed, are unfavorable to mining. These problems have grown since the Europeans' arrival in Mexico. Not only did the colonists destroy without planting, but by artificially draining broad expanses of terrain, they caused other more important damages: muriate of soda and lime, nitrate of potash, and other saline substances have taken over the arid surface of the soil and spread at a rate that chemists are at a loss to explain. Because of the abundance of salts and efflorescents, which are detrimental to farming, the Mexican plateau in some places resembles the plateau of Tibet or the salt steppes of Central Asia. Since the time of the Spanish conquest, both its sterility and its lack of robust vegetation have worsened, especially in the valley of Tenochtitlan, for that valley was graced with a verdant beauty when the lakes covered more terrain and the clayey soil was more frequently flooded.

I.283

Fortunately, it is only on the highest plains that one encounters such an arid soil, the physical causes of which we have just detailed. A large portion of the vast Kingdom of New Spain is among the most fertile lands on earth. The slope of the Cordillera is exposed to humid winds and frequent mists:

I.284 its vegetation, constantly nourished by these aqueous vapors, has an imposing beauty and strength. The humidity on the coastline, while favoring the decay of a large mass of organic substances, causes illnesses to which only Europeans and other non-acclimated persons are susceptible; the insalubrity of the air beneath the blazing tropical sun almost always signals an extraordinarily fertile soil. The amount of rain that falls in Veracruz over the course of a year is 1.62 meters, whereas in France, it is scarcely .80 meters. But excluding a few seaports and a few deep valleys where the indigenous peoples suffer from intermittent fevers, New Spain must be considered a remarkably healthy country.

The inhabitants of Mexico are less worried about earthquakes and volcanic eruptions than are those of the kingdom of Quito and the provinces of Guatemala and Cumaná. In all of New Spain, there are but five active volcanoes—Orizaba, Popocatepetl, and the mountains of Tuxtla [San Martín Tuxtla], Jorullo, and Colima. Earthquakes, which are rather frequent on the coastline of the Pacific Ocean and on the outskirts of the capital, do not cause as much damage there as they do in Lima, Riobamba, Guatemala, and Cumaná. On September 14, 1759, a horrible catastrophe caused the Jorullo volcano to emerge from the ground, surrounded by a multitude of small, smoking cones. Subterranean rumblings—almost more terrifying

I.285 since they were not followed by anything else—were heard in Guanajuato in January 1784. All these phenomena seem to prove that the area between the eighteenth and twenty-second parallels conceals an active fire that occasionally pierces the earth's crust, even at great distances from the Ocean's coasts.

The physical situation of Mexico City offers incalculable advantages, if one considers it from the viewpoint of its connections with the rest of the civilized world. Situated between Europe and Asia on an isthmus that is washed by the South Sea and the Atlantic Ocean, Mexico City seems destined one day to exert a great influence on the political events that stir the two continents. A king of Spain established in the Mexican capital could transmit his orders to the European Peninsula in five weeks and to the Philippine Islands in six weeks. If carefully cultivated, the vast Kingdom of New Spain could alone produce what trade now gathers from the rest of the globe—sugar, cochineal, cocoa, cotton, coffee, wheat, hemp, flax, silk, oils, and wine. It could furnish all metals, even mercury. Excellent lumber for construction and an abundance of iron and copper would favor the progress of Mexican navigation, but the state of the coastline and the lack of ports between the mouths of the Río Alvarado and the Río Bravo del Norte are obstacles difficult to surmount.

These obstacles, it is true, do not exist on the Pacific Ocean side. San Francisco in New California, San Blas in the intendancy of Guadalajara, near the mouth of the Santiago River, and, especially, Acapulco are magnificent ports. The latter, most likely created by an earthquake, is one of the most splendid basins that a navigator could find in the entire world. In the South Sea, only Coquimbo on the coast of Chile may be preferable to Acapulco; in winter, however, during the hurricane season, the sea is very rough at the latter port. Farther south, one finds the port of [El] Realejo in the kingdom of Guatemala, formed, like the port of Guayaquil, by a large, beautiful river. Sonsonate, which is much frequented in the summer months, has an open roadstead like the port of Tehuantepec; consequently, it is very dangerous in winter.

I.286

If we turn our attention to the eastern coast of New Spain, we see that it lacks the same advantages as the western coast. As we have seen above, there is, strictly speaking, no port there, because the port of Veracruz, where trade amounting to fifty to sixty million piasters takes place annually, is merely an inadequate mooring among the shallows of Caleta, Gallega, and Lavandera. The physical cause of this disadvantage is easily explained. The coast of Mexico along the eponymous gulf may be considered a harbor wall against which trade winds and the perpetual movement of water from east to west hurl the sands that are suspended in the rough Ocean. This rotation current runs along South America from Cumaná to Darién; it ascends toward Cabo Catoche and, after swirling at length in the Gulf of Mexico, exits through the Florida channel and runs toward the Newfoundland bank. The sands amassed by the swirling waters from the Yucatán peninsula to the mouths of the Río del Norte and the Mississippi imperceptibly narrow the basin of the Mexican gulf. Some very remarkable geological facts prove the enlargement of the continent's landmass; we see the Ocean withdrawing everywhere. Near Soto la Marina, east of the small town of Nuevo Santander, Mr. Ferrer found quicksands filled with deep-sea shells ten leagues into the interior of the country. I made the same observation on the outskirts of Antigua and Nueva Veracruz. The Rivers that flow down from the Sierra Madre to empty into the Antillean Sea contribute greatly to this increase in shoals. It is curious to observe that the eastern coastline of both Old and New Spain present the same drawbacks to seafarers. The coastline of New Spain, from 18° to 26° latitude, abounds in *sandbars*; ships that draw on more than thirty-two decimeters (ten feet) of water cannot pass over any of these bars without running the risk of grounding. These hindrances, so unfavorable to trade, would nonetheless

I.287

facilitate the defense of the country against the ambitious plans of a European conqueror.

I.288   Discontent with the port of Veracruz (if one may call the most dangerous of all anchorages by that name), the inhabitants of Mexico console themselves with the hope of opening more secure trade routes with the metropole. I shall only name the mouths of the Alvarado and the Guasacualco Rivers south of Veracruz; the Río Tampico and, above all, the village of Soto la Marina north of Veracruz, near the Santander sandbar. For a long time, these four points have held the attention of the government, but even in these otherwise very promising areas, the shallows prevent the entry of large ships. The ports would have to be *dredged*, if the localities inspire confidence that this solution would have a lasting effect. At this point, too little is known about the coastline of Nuevo Santander and Texas, especially the part that extends northward from Lake San Bernardo or La Carbonera, to ascertain whether nature presents the same obstacles and sandbars in this entire area. Mr. de Cevallos and Mr. Herrera, two Spanish officers distinguished by their zeal and astronomical knowledge, have conducted research of interest both to trade and to navigation. At present, Mexico is militarily dependent on Havana, the only nearby port capable of receiving squadrons; this is the most important point for the defense of the eastern coastline of New Spain. Since the last capture of Havana by the British, the government has gone to enormous expense to increase its fortifications. Cognizant of its true interests, the court of Madrid has declared in principle that to retain possession of New Spain, it must remain in control of the island of Cuba.

One very serious inconvenience is common to both the eastern coast and the coast washed by the Great Ocean, erroneously called the Pacific Ocean. Violent storms make both unapproachable for several months and prevent almost any navigation in the area. The north winds (*los Nortes*), which are actually northwest winds, blow in the Gulf of Mexico between the fall and spring equinoxes. These winds are usually weak in September and October; they are strongest in March and sometimes last until April. Navigators who have long frequented the port of Veracruz recognize the signs that presage the storm, more or less as a doctor knows the symptoms of an acute illness. According to Mr. Orta's observations, a dramatic movement of the barometer and a sudden interruption of the regular pattern of hourly variations are the most certain signs of a storm. The following phenomena accompany these changes: first, a slight land-wind (*terral*) blows from the west-northwest; a breeze from the northeast and then the south follows the *terral*. During this time, a suffocating heat prevails; water dissolved in the

air precipitates on brick walls, the pavement, and iron or wooden balus-   I.290
trades. The summit of the Peak of Orizaba and the Cofre de Perote, the
Villa Rica mountains, and especially the Sierra de San Martín, which runs
from Tuxtla to the mouth of the Río Guasacualco, appear to have no cloud
cover at the same time that their feet are hidden beneath a veil of translucent
vapor. The summits of the Cordilleras stand out against a beautiful azure
background. In these atmospheric conditions, a storm breaks loose, and
sometimes with such force that for the first quarter of an hour, it would be
dangerous to remain on the pier in the port of Veracruz. All communication
between the town and the castle of San Juan de Ulúa is interrupted. These
north wind storms usually last three or four days, sometimes ten to twelve.
If the north wind changes to a southern breeze, the latter is very irregular.
It is likely then that the storm will resume, but if the north wind veers to the
east by the northeast, then the breeze or good weather will last. In winter,
one can count on the breeze's continuing for three or four days, an interval
that is more than necessary for a ship leaving Veracruz to reach the high
seas and be free from the danger of the shallows near the coast. Sometimes,
even in the months of May, June, July, and August, strong gales are felt in
the Gulf of Mexico: they are called *Nortes de hueso Colorado*, but fortunately
they are not very common. Moreover, the times when the black vomit [yel-   I.291
low fever] and the northern storms prevail in Veracruz do not coincide.
Europeans who arrive in Mexico and Mexicans obliged by their affairs to
descend to the coasts from the high plateau of New Spain must choose between the dangers of navigation and those of a fatal illness.

In the months of July and August, navigation is quite dangerous on the
western coastline of Mexico, which faces the Great Ocean; terrible hurricanes blow there from the southwest. During that period, and throughout
September and October, moorings at San Blas, Acapulco, and all the ports
of the kingdom of Guatemala are among the most difficult; but even from
October to May, during the fair season (*verano de la Mar del Sur*), the calm
of the Pacific Ocean is occasionally disturbed by sudden winds from the
northeast and north-northeast, called *Papagallo* and *Tehuantepec*.

Having experienced one of these tempests firsthand, I shall examine
elsewhere whether, as some navigators believe, the *Papagallos* as purely local winds are caused by the proximity of volcanoes or if they result from
the narrowness of the Mexican isthmus on the parallel of Lake Nicaragua.
One could imagine that if the atmospheric balance were disturbed on the
coastline of the Antillean Sea in the months of January and February, the   I.292
agitated air would flow back toward the Great Ocean as if in a fit of rage.

According to this supposition, the Tehuantepec would only be the effect, or rather the continuation, of the *Nortes* from the Gulf of Mexico and the *brisotes* from Santa Marta. It makes the coast of Salinas and Ventosa almost as inaccessible as the coastline of Nicaragua and Guatemala, where violent southwest winds known as *Tapayaguas* prevail in August and September. Thunder and rains accompany these southwest winds, whereas the *Tehuantepecs* and *Papagallos*[1] wield their fury beneath a clear blue sky. This is why, at certain times, virtually the entire coastline of New Spain is dangerous for navigators.

At the end of this chapter, we shall recall the numerical results related to the Mexican climate that Mr. von Humboldt recorded in his *Mémoire sur les lignes isothermes*, pp. 120–23, and in his work *De distributione geographica plantarum secundum coeli temperiem et altitudinem montium*. All indications of temperature are in degrees of the centesimal thermometer.

On the eastern coastline of New Spain, the average annual temperature is 25°4'. Veracruz latitude 19°11', generally during the day in the hottest season 27° to 30°; at night 25.7° to 28°; in the cold season, 19° to 24°; at night, 18° to 22°. Highest heat of the year 36°, lowest 16°. The mean temperature of the month of December differs from that of the month of August by 5.6°.

I.293 On the western coast of New Spain, the average temperature is 26.8°. *Acapulco*, latitude 16°50', day 28° to 31°; night 23° to 25°, around sunrise often 18°.

(By way of comparison: *Cumaná*, latitude 10°27', average temperature 27.7°, in the hottest month 29.2°; in the coldest month 26.2°. By day generally 26° to 30°; by night 22° to 23.5°. The maximum Mr. von Humboldt observed in the course of a year was 32.7°, the minimum 21.2°. *Havana*, latitude 23°8'. Average temperature 25.6°; hottest month 28.5°; coldest 21.1°. When the north wind blows below eight minutes of arc, the Centigrade thermometer drops. *Cairo*, latitude 30°2', average temperature 22.4°, hottest month 29.9°; coldest month 13.4°. *Funchal*, latitude 32°37', average temperature 18.8°. *Rome*, latitude 41°53', average temperature 15.8°.)

The temperate region of New Spain: *Jalapa*, latitude 19°30'; elevation 677 toises, average temperature 18.2°. In winter, the thermometer drops to 14°.

---

1. The Papagallos are concentrated between the Cabo Blanco of Nicoya (latitude 9°30') and the Ensenada of Santa Catalina (latitude 10°45').

Chilpanzingo, latitude 18°11; elevation 708 toises; average temperature probably 20.6° because of the heat on the plateau where the town is located.

Valladolid latitude 19°42′; elevation 1,000 toises. The average temperature is thought to be 20°, but the thermometer has been known to drop to 3.4° below zero.

The *central plateau*, highly fertile, between Querétaro, San Juan del Río, Zelaya, and Guanajuato, elevation 940 to 1,070 toises, average temperature 19.3°.

The cold region: *Mexico City*, latitude 19°25′; elevation 1,168 toises, average temperature 17°; in the hottest months, day temperature 16° to 21°; night 13° to 15°; in the coldest months, by day 11° to 15°; by night 0° to 7°. It is sometimes several degrees below zero. Snow falls every thirty to forty years. Maximum heat is approximately 26°. The summer heat in Mexico City is like the late June heat in Paris; the winter warmth in Mexico City is like the warmth in Paris at the end of April. The average temperature in the hottest month in Mexico City differs by six to seven degrees from the average temperature in the coldest month.

Toluca, latitude 19°16′, elevation 1,380 toises, average temperature appears to be 15°.

(By way of comparison: *Caracas*, latitude 10°31′, elevation 450 toises, average temperature 20.8°. *Guaduas*, latitude 5°3′, elevation 590 toises, average temperature 19.7°; *Popayán*, latitude 2°26′, elevation 911 toises, average temperature 18.7°. *Santa Fe de Bogotá*, latitude 4°35′, elevation 1,365 toises, average temperature 14.3°. *Quito*, southern latitude 0°14′, elevation 1,492 toises, average temperature 14.4°. *Micuipampe*, southern latitude 6°43′, elevation 1,856 toises, average temperature in town probably 8°.

Marseille, latitude 43°17′, average temperature 15.0°. *Philadelphia*, latitude 39° 56′, average temperature 11.9°. *Paris*, latitude 48°50′, average temperature 10.6°.)

I.294

BOOK II

General Population of New Spain—Division of Its Inhabitants by Caste

CHAPTER IV

The General Census of 1793—Population Growth in the Ensuing
Ten Years—Birth-to-Death Ratio.

The physical tableau that we have just sketched demonstrates that in Mexico, like elsewhere, nature has unevenly distributed her blessings. Humanity, ignoring the wisdom of this distribution, has profited but little from the wealth that is provided for it. Gathered together on a small expanse of terrain in the center of the kingdom, on the plateau of the Cordillera, men and women have left uninhabited both the most fertile regions and those nearest the coastline.

In the United States, the population is concentrated on the Atlantic coast, in other words, in the long narrow area that runs from the ocean to the Allegheny Mountains. In the Capitanía general of Caracas, the only inhabited and well-cultivated terrains, so to speak, are those in the maritime regions. In Mexico, on the contrary, agriculture and civilization are consigned to the interior of the country. The Spanish conquerors only followed in the footsteps of the conquered people. Coming from a country north of the Río Gila, the Aztecs, who originally perhaps even migrated from the northernmost part of Asia, continued their migration southward along the ridge of the Cordillera, preferring the cold, alpine regions to the excessive heat of the coastline.

At the time of Hernán Cortés's arrival, the portion of Anahuac that comprised the kingdom of Montezuma II did not equal one-eighth of the surface of the present New Spain. The Acolhua kings from Tlacopan and Michoacan were independent princes. The great Aztec cities and the best cultivated lands were on the outskirts of the Mexican capital, especially in the beautiful Tenochtitlan valley. This reason alone would have sufficed for

I.296

the Spanish to set up the center of their new empire there; but they were also content to live on the plateaus whose climate, similar to that of their homeland, was conducive to growing wheat and European fruit trees. Indigo, cotton, sugar, and coffee—the four great trade commodities of the Antilles and all the hot regions in the Tropics—were of little interest to the sixteenth-century conquerors; they avidly sought only precious metals, and the search for these metals kept them on the ridge of the central mountains of New Spain.

It is as difficult to assess with any certainty the number of inhabitants in Montezuma's kingdom as it would be to pronounce on the ancient populations of Egypt, Persia, Greece, or Latium. The extensive ruins of the towns and villages found below the eighteenth and twentieth parallels in the interior of Mexico most likely prove that the population of this part of the kingdom was far greater than the present one. Cortés's letters addressed to Emperor Charles V, ▼ Bernal Díaz's memoirs, and a great number of other historical monuments confirm this interesting fact.[1] But when one reflects on how difficult it is to arrive at a precise idea of the statistics of a country, one should not be surprised by the ignorance in which sixteenth-century authors have left us regarding the ancient population of the Antilles, Peru, and Mexico. History presents us with conquerors eager to take advantage of the fruit of their exploits; on the other side, we have the archbishop of Chiapas and a small number of well-intentioned men who employed the arms of eloquence with noble ardor against the cruelty of the first colonists. All parties were equally interested in exaggerating the flourishing state of the recently discovered countries: the Franciscan fathers alone boasted of having baptized more than six million Indians from 1524 to 1540, and (what is more) only Indians who lived in the areas closest to the capital!

I.298

One striking example proves how careful one must be in placing faith in the numbers found in the old descriptions of America. It has often been printed that the census of the inhabitants of Peru, carried out on the orders of the archbishop of Lima, ▼ Father Gerónimo de Loaysa in the year 1551, found 8,285,000 Indians. This result must pain those who know that in 1793, in the very precise census ordered by the viceroy Gil Lemos, the Indians of the present Peru (after the separation of Chile and Buenos Aires) did not exceed 600,000 individuals. So, one might have believed that

---

1. See the Abbot Clavijero's judicious observations on the ancient population of Mexico, which were directed against Robertson and the Abbot [de] Pauw. *Storia antica di Messico*, vol. IV, p. 282.

7,600,000 Indians had disappeared from the face of the earth. Fortunately, the Peruvian author's assertion has been found devoid of truth. With the help of ▼ Father Cisneros's research in the archives in Lima, it has been discovered that the existence of the eight million Indians in 1551 is not based on any historical document. ▼ Even Dr. Feijoo, the author of the statistics of Trujillo, has since admitted that his risky claim was based only on a fictitious calculation, a census of so many towns ruined since the time of the conquest. These ruins seemed to suggest to him the existence of an immense population in the remotest times. The examination of an erroneous opinion often leads to an important truth. Searching among the archives of the sixteenth century, Father Cisneros discovered that the viceroy Toledo, who was rightly considered the foremost Spanish legislator in Peru, counted only 1,500,000 Indians upon his personal visit to the kingdom from Tumbez to Chuquisaca (approximately the present extent of Peru) in 1575.

I.299

Generally, nothing is vaguer than approximating the population of a recently discovered country. Cook estimated the inhabitants of the island of Tahiti to be 100,000; Protestant missionaries from Great Britain only suppose the population there to be 49,000 souls; ▼ Captain Wilson sets it to 16,000: Mr. Turnbull believes he can prove that the number of inhabitants does not exceed 5,000. I doubt that these differences are the result of a progressive reduction. This depopulation is probably the result of the deadly maladies with which the civilized people of Europe have infected these once-happy lands, but it cannot have been so rapid as to have killed off nineteen-twentieths of the population in forty years.

I.300

We have already mentioned that the outskirts of the capital of Mexico, and perhaps the entire country under Montezuma's domination,[1] were once more populated than they are today, but that this great population was concentrated in a very small area. One knows—and knowledge of this fact is a consolation to humanity—that the number of indigenous peoples had not only not ceased to increase in the past century, but also that all the vast region that we are calling by the general name of New Spain is more inhabited today than it was when the Europeans arrived. The first assertion is proven by the state of the *capitation tax*, and the second is based on a very simple consideration. At the beginning of the sixteenth century, the Otomi and other barbarian people occupied the lands north of the Panuco and the Santiago rivers. Since land cultivation and civilization have advanced toward

1. Clavijero, *Storia antica di Messico*, vol. I, p. 36.

Nueva Vizcaya and the *provincias internas*, the population has increased in these northern provinces with a rapidity that is visible wherever a nomadic people is replaced by farming colonists.

Even in Spain, studies of political economy based on exact numbers were infrequent prior to work by the illustrious ▼ Campomanes and the

I.301 ministry of Count Floridablanca. One should not be surprised, then, that the archives of the viceroyalty in Mexico do not contain any censuses before that of 1794, which was ordered by the Count of Revillagigedo, one of the most active and wisest administrators of the eighteenth century. In the work on the population of Mexico at the time of viceroy ▼ Pedro Cebrián, Count of Fuenclara, in 1742, it was considered sufficient to count the number of families; but what ▼ Villaseñor has preserved of this census is vague and incomplete. Those who know the difficulties of taking a census in the most civilized parts of Europe (recalling that *economists* attributed only eighteen million inhabitants to all of France and that even recently there were disputes as to whether the population of Paris numbered 500,000 or 800,000 inhabitants) can easily imagine the powerful obstacles to be overcome in a country where bureaucrats have no skill whatsoever in statistical research. Despite his zeal and the extent of his activity, the viceroy Count de Revillagigedo was unable to complete his work. The census that he undertook was left unfinished in the intendancies of Guadalajara and Veracruz and in the province of Coahuila.

The following table gives the status of the population[1] of New Spain, ac-

I.302 cording to the information that the intendants and the provincial governors had forwarded to the viceroyalty by May 12, 1794:

NAMES OF INTENDANCIES AND GOVERNMENTAL DISTRICTS IN WHICH THE CENSUS WAS CARRIED OUT IN 1795	POPULATION OF	
	INTENDANCIES AND CAPITALS	GOVERNMENT DISTRICTS
MEXICO CITY	1,162,856	112,926
PUEBLA	566,443	52,717
TLAXCALA	59,177	3,357

1. I am publishing this statistic according to the copy preserved in the viceroy's archives, but I note that other copies in circulation in the country give dubious numbers, for example, 638,771 souls for the intendancy of Puebla, including the former republic of Tlaxco [Tlaxcala].

NAMES OF INTENDANCIES AND GOVERNMENTAL DISTRICTS IN WHICH THE CENSUS WAS CARRIED OUT IN 1795	POPULATION OF	
	INTENDANCIES AND CAPITALS	GOVERNMENT DISTRICTS
OAXACA	411,366	19,069
VALLADOLID	289,314	17,093
GUANAJUATO	397,924	32,098
SAN LUIS POTOSÍ	242,280	8,571
ZACATECAS	118,027	25,495
DURANGO	122,866	11,027
SONORA	93,396	
NUEVO MEXICO	30,953	
THE TWO CALIFORNIAS	12,666	
YUCATÁN	358,261	28,392
Total population of New Spain, as derived from the 1793 census	3,865,559	
In a report to the king, Count de Revillagigedo estimated the population of the intendancy of Guadalajara to be	485,000 inhabitants	
that of Veracruz to be	120,000	618,000
Coahuila province to be	13,000	
Approximate result of the 1793 census: 4,483,559 inhabitants		

These results show the minimum population that was admissible at the time. The central government and, above all, the provincial administrations soon recognized how far they had fallen short of their desired goal. On the New Continent as well as the Old, the people assume that any census is a sinister augury of some financial operation. Living in constant fear of a tax increase, all heads of households try to decrease the number of persons in their household. It is easy to demonstrate the validity of this claim. Prior to Count [de] Revillagigedo's census, it was believed, for example, that there were 200,000 inhabitants in the Mexican capital. This estimate might have been exaggerated, but tables of consumption, the number of baptisms and burials, and the comparison of these numbers with those of large European cities, tend to prove that the population of Mexico City was at least 135,000. However, the table

I.303

that the viceroy published in 1790 shows only 112,926. In smaller towns that are more easily surveyed, the error was larger and even more obvious. Those who closely followed the counting of the registers drawn up in 1793 decided then that the number of inhabitants who had escaped counting in the general census could not be compensated for by those who roamed about without a permanent address and had been counted several times. It was generally admitted that to obtain a satisfactory result, one-sixth or one-seventh would have to be added to the complete total which meant a population of 5,200,000 souls for all of New Spain. For the year 1804, I use[1] 5,837,100.

I.304

It is unfortunate that the viceroys who succeeded Count [de] Revillagigedo as administrators of the country did not renew the complete census. Since that time, the government has given no further thought to statistical research. Several intendants' reports on the present situation in the country entrusted to their care contain exactly the same numbers as the 1793 table, as though the population could have remained the same over a ten-year period. The population is surely increasing at a phenomenal rate. The increase in tithes and in the Indian capitation, as well as in all duties on consumption, advances in agriculture and civilization, and the vision of a countryside covered with newly built houses all herald considerable growth in almost all parts of the kingdom. How, then, can we imagine that social institutions could be so imperfect in a fertile land and temperate climate, that a government could disrupt the natural order to prevent the progressive multiplication of the human species in periods of calm, with no wars in its territory? The indigenous peoples no longer groan under the yoke of the ▼ *Encomiendas*, and a peace of three centuries has almost erased the memory of the atrocities that the first conquerors committed in their fanaticism and insatiable greed!

In order to draw up a population table for 1803 and present numbers that are as accurate as possible, it was necessary to augment the results of the last census (1) by the portion of the population omitted from the lists, and (2) by the portion resulting from the excess of births over deaths. *I preferred to settle on a number that was below the present population* rather than hazard assumptions that might seem too optimistic. I therefore decreased the estimated number of inhabitants omitted from the general census: instead of one-sixth, I gauged it at only *one-tenth*.

---

1. This number also seems the most likely one to the statesmen whom the sovereign Congress of Mexico chose to form a plan for a *federative constitution*. "Our responsibility," they wrote in their report of November 20, 1820, "is to propose institutions that will improve the lot of six million free men who live in the Mexican provinces, all of whom speak the same language and share the same religion."

As for the progressive increase of the population between 1793 and the time of my journey, I was able to determine it using fairly precise information. The special kindness shown me by a respectable prelate, ▼ the present archbishop of Mexico City,[1] enabled me to conduct detailed studies   I.306 on the birth-to-death ratio in relation to the difference in climate between the central plateau and the regions near the coast. Several parish priests interested in resolving the question of the increase or decrease of our species undertook a laborious task and conveyed to me the number of baptisms and burials, year by year, from 1752 to 1802. Taken together, these meticulous registers, which I have preserved, prove that the birth-to-death ratio is nearly 170 to 100. I shall give but a few examples here to justify this claim; they are all the more interesting in so far as we lack statistical information on the of death-to-birth ratio in the Torrid Zone.

From 1750 to 1801, in the Indian village of Singuilucan, which is located eleven leagues north of the capital, there were 1,950 deaths and 4,560 births: the surplus of births was therefore 2,610.

From 1767 to 1797, in the Indian village of Axapusco, located thirteen leagues north of Mexico City, there were in all 3,511 deaths and 5,528 births from the time when this village separated from the parish of Otumba. Births thus exceeded deaths by 2,017.

From 1752 to 1802, in the Indian village of Malacatepec, located twenty-eight leagues west of the Tenochtitlan valley, there were 13,734 births and   I.307 10,529 deaths, hence a surplus of 3,205 births.

From 1756 to 1801, in the village of Dolores, there were 24,123 deaths and 61,258 births, which produced the extraordinary surplus of 37,135 births.

From 1797 to 1802, in the city of Guanajuato, there were 12,666 births and 6,294 deaths over the course of five years, hence a surplus of 6,372 births.

In the same period of time, there were 3,702 births and 1,904 deaths in the village of Marfil near Guanajuato, hence a surplus of 1,798 births.

In five years, there were 3,629 births and 1,857 deaths in the village of Santa Ana near Guanajuato, hence a surplus of 1,772 births.

In ten years, there were 3,373 births and 2,395 deaths in Yguala, a village located in a very hot valley near Chilpanzingo, hence a surplus of 978 births.

In ten years, there were 5,475 births and 2,602 deaths in the Indian village of Calimaya, located on a cold plateau, hence a surplus of 2,673 births.

1. Don Francisco Javier de Lizana. I also owe very useful information to ▼ Don Pedro de Fonte, the headmaster of the archbishopric (and since then successor to his uncle, Monsignor de Lizana). *See* note B at the end of this work.

In 1793, there were 5,064 births and 2,678 deaths in the jurisdiction of the town of Querétaro, hence a surplus of 2,386 births.

These examples demonstrate that the ratio of deaths to births differs greatly depending on the local climate and the salubrity of the air. This ratio is:

at Dolores	=	100:253
at Singuilucan	=	100:234
at Calimaya	=	100:202
at Guanajuato	=	100:201
at Santa Ana	=	100:195
at Marfil	=	100:194
at Querétaro	=	100:188
at Axapuzco	=	100:157
at Iguala	=	100:140
at Malacatepec	=	100:134
at Panuco	=	100:123

The average of these eleven places yields a ratio of 100 to 183; but the ratio one may realistically consider in regard to the population as a whole is only 100 to 170.

It seems that on the high plateau of the Cordillera, the surplus of births is larger than it is closer to the coastline or in very hot regions. What a difference between the villages of Calimaya and Yguala! In Panuco, where the climate is as scorching as in Veracruz, although the deadly illness of *black vomit* has not yet been identified there, the number of births from 1793 to 1802 was 1,224, and the number of deaths 988, which yields the unfavorable ratio of 100 to 123. Both Hindustan and South America (especially the province of Cumaná, the coastline of Coro, and the plains, or *llanos*, of Caracas) prove that heat alone is not the cause of such high mortality. In countries that are at once very hot but also dry, the human species perhaps enjoys greater longevity than what we observe both in the temperate zones and wherever the temperature and climate are extremely variable. Slightly older Europeans who move to the equinoctial part of the Spanish colonies usually reach a ripe and happy old age there. In Veracruz, in the midst of the epidemics of

I.309

black vomit, the indigenous peoples and the already acclimated foreigners enjoy perfect health.

Both the coastline and the arid plains of equatorial America must generally be regarded as healthy, despite the excessive heat of the sun, whose perpendicular rays are reflected by a soil nearly devoid of vegetation. Mature adults, mainly those who are approaching old age, have little to fear in these regions, which are at once scorching and dry. Insalubrity is wrongly attributed to these regions. In places where very high temperatures are accompanied by excessive humidity, only infants and young people face high mortality rates. Intermittent fevers are prevalent all along the coast of the Gulf of Mexico from the mouth of the Alvarado to Tamiagua, Tampico, and the plains of Nuevo Santander. The western slope of the Mexican Cor-   I.310
dillera and the coastline of the South Sea from Acapulco to the ports of Colima and San Blas are all equally unhealthful. This humid, fertile, but insalubrious terrain is comparable to the coastal part of the province of Caracas that runs from Nueva Barcelona to Portocabello. Tertian fevers are the scourge of these regions, which nature has nonetheless graced with vigorous vegetation rich in useful products. These illnesses are all the more devastating because the indigenous peoples leave their sick in the most appalling abandon; the Indian children are especially victimized by this shameful neglect. Mortality is so high in the hot and humid regions that significant population growth is virtually impossible, while in the cold and temperate regions of New Spain (and these regions make up the largest part of the kingdom), the ratio of births to deaths is approximately 183 to 100, or even 200 to 100.

The ratio of the entire population to either births or deaths is more difficult to gauge than the ratio of births to deaths. In a country where the laws only tolerate the practice of one religion and where the parish priest receives a portion of his income from baptisms and burials, one can be fairly certain of having an exact figure for the surplus of births over deaths. But the number that represents the ratio of deaths to the overall population is affected by some of the uncertainty that envelops the population itself. The city of Querétaro and its territory have a population of 70,600. Dividing   I.311
this number by 5,064 births and 2,678 deaths, one finds that one person is born for every fourteen and that one person dies for every twenty-six. In Guanajuato, which includes the nearby mines of Santa Ana de Marfíl, there are 3,998 births and 2,011 deaths among a population of 60,100 in a regular year (on a five-year average). Thus, one person is born for every fifteen,

and one dies for every twenty-nine. Europe has a birth-to-death ratio relative to the entire population which is much less favorable to the propagation of the species: in France, for instance, in 1800 there was one birth for every twenty-eight and three-tenth persons, and one death for every thirty and nine-tenth persons. ▼ Mr. *Peuchet* deduced this precise result from the tables drawn up in ninety-eight departments. In 1823, there was one birth for every thirty-one and two-third persons, versus one death for every thirty-nine and two-third persons.

In England, births are to deaths as twenty-five and three-tenths persons to nineteen; in the Prussian Monarchy, this ratio is approximately twenty-eight to nineteen. In Sweden, a country less favored by nature, one person is born for every thirty, and one dies for every thirty-nine, according to ▼ Mr. *Nicander*'s tables, the most precise and extensive that have ever been compiled.

I.312   If it were admissible that, in the Kingdom of New Spain, the birth-to-death ratio were one to seventeen, and the ratio of deaths to the population as a whole were one to thirty, then one would arrive at approximately 350,000 for the number of births and 200,000 for the number of deaths. The surplus of births under favorable conditions—in other words, in years without famine, smallpox epidemics, and *matlazahuatl*, the deadliest of the Indian illnesses—would thus be 150,000. Everywhere on the globe, one sees that under the most varied forms of government (whenever absolute power does not degenerate into tyranny), the population grows rapidly in countries that are still underinhabited, have eminently fertile soil, a mild climate and an even temperature, and, above all, a race of robust men and women whom nature calls to marriage at a young age.

Those parts of Europe where civilization began to develop only very late present striking examples of birth surpluses. In West Prussia, in 1784 there were 27,134 births and 15,669 deaths among a population of 560,000. These figures give a birth-to-death ratio of 36 to 20, or 180 to 100, a ratio that is almost as favorable as that of the Indian villages located on the central Mexican plateau. In the Russian Empire, there were 361,134 births and 818,433 deaths in 1806. The same causes produce the same effects every-

I.313   where. The younger the culture of a country, the easier it is to subsist on newly cleared ground, and the growth of the population is more rapid. To confirm this axiom, one has only to look at the birth-death ratios in the following table:

In France, in 1823	=	125:100
In England[1]	=	137:100
In Sweden	=	130:100
In Finland	=	160:100
In the Russian Empire	=	166:100
In western Prussia	=	180:100
In the government district of Tobolsk, according to Mr. *Hermann*	=	210:100
In various parts of the Mexican plateau	=	230:100
In the United States, in the state of New Jersey	=	300:100

1. ▼ Malthus, *An Essay on the Principle of Population*, one of the most profound works of political economy ever published.

The information that we have collected on birth-to-death ratios relative to the entire population proves that if natural order were not disturbed from time to time by some extraordinary and disruptive cause, the population of New Spain would double[1] every nineteen years. It should increase by 44 percent over a ten-year period. In the United States, the population has doubled every twenty to twenty-three years since 1784. ▼ Mr. Samuel Blodget's fascinating tables published in his *Statistical Manual for the United States of America* prove that in some states, this felicitous cycle occurs every thirteen to fourteen years. In France, provided that no war or contagious illness reduced the annual surplus of births over deaths, one would see the population double over a period of 109 years. Such is the difference between countries that are already very populated and those whose industry has only just begun!

The only true sign of real and permanent population growth is an increase in the means of subsistence. This increase or growth in agricultural products is evident in Mexico; it seems to indicate an even much more rapid population

I.314

1. If p represents the present population of a country, n the ratio of the population to births, d the ratio of deaths to births, and k the number of years at the end of which one wishes to measure the population, then the population level in period k will be expressed as $p\,(1 + n\,(1 - d)^k$; it follows that if one wishes to know in how many years the population will double, this number of years will be expressed by

$$k = \frac{log.2}{log.(1 + n\,1 - d)}$$

growth than was supposed by deducing the population in 1803 from the inaccurate 1793 census. In a Catholic country, ecclesiastical tithes are the thermometer, so to speak, by which one may measure agricultural development; I.315 as we shall see below, these tithes double at least every twenty-four years.

All these considerations suffice to prove that, by allowing for 5,800,000 inhabitants in the kingdom of Mexico at the end of the year 1803, I have used a number that, far from being exaggerated, is probably even lower than the existing population. No public calamity has beset the country since the census of 1793. If one added (1) one-tenth for the persons not included in the census, and (2) two-tenths for the growth in the population over a ten-year period, one would assume a surplus of births half as small as the one in parish registers. According to this assumption, the number of inhabitants should double only every thirty-six to forty years. But well-informed persons—those who have closely observed the development of agriculture, the expansion of villages and several towns, the increase in all the revenue of the crown that depends on the consumption of commodities—are tempted to believe that the population of Mexico has made much more rapid strides. Far be it for me to pronounce on such a sensitive topic; it is enough to have presented the details of the material that has heretofore been collected and that may lead to accurate results. I consider it highly likely that in 1808, Mexico's population was over 6,500,000. In the Russian Empire, whose political and moral state bears many striking similarities with the country I.316 we are studying here, the growth of the population due to the surplus of births is much greater than in Mexico. According to Mr. Hermann's statistical work, the census of 1763 showed 14,726,000 souls. The result of the census of 1783 was nearly 25,677,000, and in 1805, the total population of Russia was already estimated at 40,000,000. But what obstacles does nature itself throw in the way of population growth in the northernmost parts of Europe and Asia! What a contrast between the fertility of the Mexican soil, enriched by the most precious plant growth from the Torrid Zone, and the sterile plains that remain buried in snow and ice for over half the year!

---

*Since the publication of the first edition of this *Political Essay*, the population of New Spain and that of the countries with which its population was compared have increased notably. In France, for instance, the birth-to-death ratio for the entire population has changed markedly. For a long time, twenty-eight to one had been used to express the ratio of population to births and thirty to one to express the ratio of population to deaths. According to ▼ Mr. Villermé's excellent work of political economy, based on the average results from 1817–1821, in the whole of France there was one birth for every thirty-one and thirty-five hundredths and one death for every thirty-nine and twenty-nine-hundredths. These results conform entirely to those obtained by the author of a scholarly paper on the population of France inserted in volume XXV of the *Revue encyclopédique* (March 1825). The ratio of the entire population of France to annual growth is approximately 157 to 1, or 193,000 persons. If one supposes that this ratio is sustained, one finds that the annual growth rate is 0.0063, and that the population will double in 109 years; but this growth in population also affects the causes by which it is shaped, then gradually weakened, and, finally, destroyed. Population growth in France, which was long and mistakenly thought to be 0.003, has now almost doubled. In the Americas, where the population takes almost twenty-four years to double, it is nearly 3 percent, or, more precisely, 0.2915, according to what we know about population growth there from 1810 to 1820. The total population of the United States was 7,239,903 in 1810; 9,649,999 in 1820; while the slave population for these two periods was 1,191,364 and 1,623,124, respectively. The growth rate for slaves is 0.02611.

I.317

If one had precise knowledge of the birth-to-death ratio in the vast expanses of land in the hot, temperate, and cold climates of Mexico, it would be easy to determine the increase in the number of inhabitants that must have taken place since the inaccurate surveys attempted in 1793. But the data that I owe to Monsignor the Archbishop of Mexico City are insufficient to yield average results that might apply to the entire country. It is only through induction that one can come close to the truth. I believe to have proven, using positive data, that in 1804 the population of the former viceroyalty of New Spain, including the *Provincias internas* and the Yucatán (but not the *capitanía general* of Guatemala) had at least 5,840,000 inhabitants, of which two and a half million were copper-skinned indigenous peoples, one million Spanish Mexicans, and 75,000 Europeans. I even stated that in 1808, the population must be close to six and a half million, of which two- to three-fifths, or 3,250,000, were Indians. The political movements that have led to unrest in the intendancies of Mexico City, Veracruz, Valladolid, and Guanajuato have likely slowed the annual growth of the Mexican population, which was probably around 150,000 during the time of my stay in the country. Studies conducted in the country itself have recently proven the validity of the estimates that I made twelve years ago. Don Fernando Navarro y Noriega published in Mexico City the results of a wide-ranging study on the number of *curatos y missiones* in Mexico; he estimated the population in 1810 to be 6,128,000 (*Catálogo de los curatos que tiene la Nueva España*, 1813, p. 38,

I.318   and *Rispuesta de un Mexicano al n° 200 del Universal*, p. 7). The same author, whose position in the ministry of finances (*Contador de los ramos de arbitrios*) enables him to analyze statistical information on site, thinks (*Memoria sobre la población de Nueva España, México* 1814; and *Semanario político y literario de la Nueva España*, number 20, p. 94) that in 1810, the population of New Spain, not including the provinces of Guatemala, was composed of the following:

1,097,928	Europeans and Spanish Americans.
3,676,281	Indians.
1,338,706	Mixed castes or races.
4,229	Secular clergymen.
3,112	Regular clergymen.
2,098	Nuns.
6,122,354	

I am inclined to believe that New Spain today has a population of nearly seven million inhabitants. This is also the opinion of a reputable prelate, the archbishop of Mexico City, Don José de Fonte, who has visited a large part of his diocese and whom I recently had the honor of meeting again in Paris. One of the most important results of Mr. Navarro's work is his estimate of the number of pure-blooded Indians [Indiens de race pure] (3,676,000). It seems to be all the more reliable due to the author's long employment in the administrative branch, which enabled him to examine the registers of the Indian *tributaries*. In the whole of New Spain (*Catálogo de los curatos y missiones* 1813, p. 39), there are nearly 1,500 baptismal fonts (*pilas bautismales*), that is, 1,037 for *curatos*, 157 at *missions*, and 270 in *paroquias auxiliaries* or *vicarias*, such that there are over 4,000 persons for each baptistry; in Spain, by contrast, where parishes are smaller, each one has fewer than 600 persons of all ages and sexes. At the meeting of the first Mexican congress, elections were set according to the statistical table that I am providing below; the figures have been taken from the table that

I.319   I published in the eighth chapter of my *Political Essay*:

STATES OF THE MEXICAN CONFEDERATION[1]	POPULATION	SQUARE LEAGUES
Mexico City	1,300,000	5,926
Jalisco	650,000	9,612
Zacatecas	250,000	2,225

STATES OF THE MEXICAN CONFEDERATION[1]	POPULATION	SQUARE LEAGUES
San Luis Potosí	180,000	2,357
Veracruz	174,000	4,141
Puebla	750,000	2,696
Oaxaca	600,000	4,447
Guanajuato	500,000	911
Michoacán	400,000	3,446
Querétaro	180,000	"
Tamaulipas	70,000	5,193
Interno del Norte	240,000	19,143
Yucatán	500,000	5,997
Interno de Occidente	170,000	20,271
Interno de Oriente	110,000	"
Tlascala	70,000	"
Tabasco	60,000	"
	6,204,000	

1. [This and the three entries below were corrected by Humboldt; see Errata in his *Essai politique*, IV.311.]

Although it is not based on a new census, this official document includes 6,204,000 inhabitants, not counting the two territories of Colima and the Californias. The surface areas have been copied from my *statistical table*; the population figures are taken from the same table, slightly modified to bring them up-to-date. Allowing for a population of 6,800,000 at the end of 1823, one may estimate the number of Indians at 3,700,000; mixed-race [race mixtes] individuals at 1,860,000; whites [blancs] at 1,230,000; blacks [nègres] at 10,000 maximum. These are, I repeat, only approximate results; numbers are limited to *minima*. Doubts will only be dispelled only when the government has found a way to take an accurate census and, most importantly, to report the ratio of the population to births and deaths in the various regions (*tierra caliente, templada, y fría*) of Mexico. What follows at the end of this chapter are five tables, in which I consider the population of the Americas in terms of caste, race, religion, and language.

DISTRIBUTION OF THE RACES IN CONTINENTAL AND INSULAR AMERICA

1) Whites

Spanish America	3,276,000
Antilles, excluding Cuba, Puerto Rico and Margarita Island	140,100
Brazil	920,000
United States	8,575,000
Canada	550,000
British, Dutch, and French Guiana	10,000
	13,471,000

2) Indians

Spanish America	7,530,000
Brazil: landlocked Indians of the Rio Negro, the Rio Branco, and the Amazon	260,000
Free Indians east and west of the Rocky Mountains, at the borders to New Spain, the Mosquitos, etc., etc.	400,000
Free Indians in South America	420,000
	8,610,000

3) Blacks

Antilles including Cuba and Puerto Rico	1,960,000
Continental Spanish America	387,000
Brazil	1,960,000
British, Dutch, and French Guianas	206,000
United States	1,920,000
	6,433,000

4) Mixed Races

Spanish America	5,328,000
Antilles, excluding Cuba, Puerto Rico, and Margarita Island	190,000
Brazil and United States	890,000
British, Dutch, and French Guianas	20,000
	6,428,000

DISTRIBUTION OF THE RACES IN CONTINENTAL
AND INSULAR AMERICA

I.321

Summary

Whites	13,471,000 or	38 p[er] c[ent]
Indians	8,610,000	25
Blacks [Nègres]	6,433,000	19
Mixed Races [Races mixtes]	6,428,000	18
	34,942,000	

DISTRIBUTION OF THE RACES IN SPANISH AMERICA

1) Natives (Indians, Red Men, American Copper-Colored or Primitive Race, with No Mixture of White and/or Black [Negre])

Mexico	3,700,000
Guatemala	880,000
Colombia	720,000
Peru and Chile	1,030,000
Buenos Aires, including the Sierra provinces	1,200,000
	7,530,000

2) Whites (Europeans and Descendants of Europeans, with No Mixture of Black [Negre] and/or Indian, the So-Called Caucasian Race)

Mexico	1,230,000
Guatemala	280,000
Cuba and Puerto Rico	339,000
Colombia	642,000
Peru and Chile	465,000
Buenos Aires	320,000
	3,276,000

3) Blacks (African Race, with No Mixture of White and/or Indian; Free and Enslaved Blacks [Noirs])

Cuba and Puerto Rico	389,000
[American] Continent	387,000
	776,000

(*continued*)

DISTRIBUTION OF THE RACES IN SPANISH AMERICA

4) Mixed Races of Blacks [Noirs], Whites, and Indians (Mixed-Race [Mulâtres],
Mestizos, Zambos, and Mixture of Mixtures)

Mexico	1,860,000
Guatemala	420,000
Colombia	1,256,000
Peru and Chile	853,000
Buenos Aires	742,000
Cuba and Puerto Rico	197,000
	5,328,000

Summary According to the Predominant Races

Indians	7,530,000	or	45 p[er] c[ent]
Mixed races	5,328,000		32
Whites	3,276,000		19
Blacks [Noirs], African race	776,000		4
	16,910,000		

I.322

BLACK POPULATION OF CONTINENTAL AND INSULAR AMERICA

1) Enslaved Blacks [Negres]

Antilles, insular America	1,090,000
United States	1,650,000
Brazil	1,800,000
Spanish Continental colonies	307,000
British, Dutch, and French Guianas	200,000
	5,047,000

2) Free Blacks [Negres]

Haiti and the rest of the Antilles	870,000
United States	270,000
Brazil, roughly	160,000
Spanish Continental colonies	80,000
British, Dutch, and French Guianas	6,000
	1,386,000

BLACK POPULATION OF CONTINENTAL AND INSULAR AMERICA			I.323

Summary

Blacks [Noirs] without any mixture, that is, excluding mulattos

5,047,000	enslaved	79	P[er] c[ent]
1,386,000	free	21	
6,433,000			

DISTRIBUTION OF THE TOTAL POPULATION OF THE AMERICAS IN TERMS OF RELIGIOUS DIVERSITY

I. Roman Catholics			22,486,000
a) Continental Spanish America		15,985,000	
Whites	2,937,000		
Indians	7,530,000		
Mixed races and Blacks [Nègres]			
	5,518,000		
	15,985,000		
b) Portuguese America		4,000,000	
c) United States, Lower Canada, and French Guiana		537,000	
d) Haiti, Cuba, Puerto Rico, French Antilles		1,964,000	
		22,486,000	
II. Protestants			11,636,000
a) United States		10,295,000	
b) British Canada, Nova Scotia, Labrador		260,000	
c) British and Dutch Guiana		220,000	
d) British Antilles		777,000	
e) Dutch, Danish, etc. Antilles		84,000	
		11,636,000	
III. Free Indians, Non-Christian			820,000
			34,942,000

PREDOMINANT LANGUAGES ON THE NEW CONTINENT

1) English

United States	10,525,000
Upper Canada, Nova Scotia, New Brunswick	260,000
Antilles and British Guiana	862,000
	11,647,000

2) Spanish

Spanish America, that is:

Whites	3,276,000
Indians	1,000,000
Mixed and black [nègres] races	6,104,000
Haiti's Spanish part	124,000
	10,504,000

3) Indian Languages

Spanish and Portuguese America, including sovereign tribes	7,593,000

4) Portuguese

Brazil	3,740,000

5) French

Haiti	696,000
Antilles dependent on France:	
Louisiana and French Guiana	290,000
	1,242,000

6) Dutch, Danish, Swedish, and Russian

Antilles	84,000
Guiana	117,000
Russians on the NW coast	15,000
	216,000

Summary

English	11,647,000
Spanish	10,504,000
Indian	7,593,000

(continued)

Portuguese	3,740,000	
French	1,242,000	
Dutch, Danish, Swedish	216,000	
	34,942,000	
Languages of Latin Europe	15,486,000 ⎫	European languages
Languages of the Germanic races	11,863,000 ⎭	27,349,000
Indian languages	7,593,000	

I did not make separate mention of German, Gallic (Irish), and Basque, **I.325** because the rather large number of individuals who retain these languages as their mother tongues also know either English or Castilian. The number of individuals who regularly speak indigenous languages are currently in a ratio of 1 to 3.4 speakers of European languages. Due to the more rapid growth of United States, Germanic languages will imperceptibly gain on Romance languages in the total numerical proportion. But because of the growing Spanish and Portuguese presence in Indian villages, where barely one twentieth of the population understands even a few words of Castilian or Portuguese, Romance languages will also spread. I believe that there are over seven and a half million indigenous people throughout the Americas who still speak their own languages and have virtually no knowledge of European languages. This is also the opinion of the archbishop of Mexico City and of several equally respectable clerics who have lived in Upper Peru for a long time and with whom I was able to confer on this topic. The small number of Indians (perhaps one million) who have completely forgotten the indigenous languages dwell in the large cities and densely populated villages that surround these towns. Among those who speak French on the New Continent there are over 700,000 blacks of the African race [nègres **I.326,** de race Africaine]. Despite the highly commendable efforts of the Haitian Government to educate the people, this circumstance does not contribute to preserving the purity of the language. In general, of the 6,223,000 blacks [noirs] in continental and insular America, over one-third (at least 2,360,000) speak English, over one-fourth speak Portuguese, and one-eighth speak French.

These tables of the population of the Americas considered in terms of the different races, languages, and religions, contain extremely variable elements that represent the current state of American society in numerical approximations. A work of this type deals only with large numbers of people; only over time can partial estimates gain more rigorous precision. The language of numbers, the only hieroglyphics preserved amidst the signs that represent thought, does not require interpretation. There is something gravely prophetic in these inventories of the human species: the New World's entire future seems inscribed within them.

CHAPTER V

Diseases That Periodically Halt Population Growth—Natural
and Inoculated Smallpox—The Vaccine—Matlazahuatl—
Food Shortage—Miners' Health

It remains for us to examine the physical causes that periodically halt the
growth of the Mexican population. These causes are smallpox, the cruel
disease that the indigenous peoples call *matlazahuatl*, and especially food
shortages, the effects of which can be felt for a long time.

Introduced around 1520, smallpox seems to inflict its ravages only every
seventeen to eighteen years. Like the *black vomit* and many other diseases
in the equinoctial regions, it has fixed periods in which it strikes regularly.
One might say that in these regions, the indigenous peoples only regain
susceptibility to certain miasmas in widely separated periods, because
although the ships arriving from Europe often reintroduce the smallpox
virus, it only becomes epidemic after long intervals of time. This unusual
circumstance makes the disease all the more dangerous for adults. Smallpox ravaged Mexico in 1763 and especially in 1779: in that year, it struck
down more than nine thousand persons in the capital city alone. Tumbrels
passed through the streets every evening to collect corpses, as in Philadelphia during the yellow fever epidemic. A large part of Mexico's youth was
razed in that fatal year.

The epidemic of 1797 was less deadly because of the zeal with which
inoculation was performed on the outskirts of Mexico City and in the bishopric of Michoacan. The capital of the bishopric, Valladolid, lost only 170
out of 6,800 inoculated persons, or two and a half persons for every one
hundred. It should also be noted that many of those who perished were inoculated at a point after they had probably already been infected by the disease through the effect of natural contagion. Death took fourteen of every

hundred persons who had not been inoculated, regardless of age. Several private individuals, especially among the clergy, displayed remarkable patriotism by halting the progress of the epidemic through inoculation. I shall name only two equally enlightened men, ▼ Mr. de Riaño, the intendant of Guanajuato, and Don Manuel Abad, the canon penitentiary of the cathedral of Valladolid, whose generous and selfless plans were constantly directed toward the common good. At that time, over fifty to sixty thousand individuals in the kingdom were inoculated.

I.329

But since January 1804, the vaccine itself has been introduced in Mexico, thanks to the efforts of a respectable citizen, ▼ Mr. Tomás Murphy, who had several shipments of the virus sent from North America. Its introduction met with few obstacles; the vaccine only appeared to cause a slight illness, and smallpox inoculation had already accustomed the Indians to the idea that it could be useful to submit to a passing illness to protect oneself from a more serious one. If either the preventive vaccine or even the regular smallpox inoculation had been known in the New World from the sixteenth century, several million Indians would not have perished from smallpox and, in particular, from the senseless treatment that made this disease so dangerous. It was smallpox that was responsible for decreasing, at such a terrifying rate, the number of indigenous peoples in California.

Royal navy ships charged with carrying the vaccine to the American and Asian colonies arrived in Veracruz shortly after my departure. ▼ Don [Francisco Javier de] Balmis, the chief doctor of the expedition, visited Puerto Rico, the island of Cuba, Mexico, and the Philippine Islands; his stay in Mexico, where there was knowledge of the vaccine even prior to his arrival, singularly facilitated the dissemination of this most beneficial prevention. In the major cities of the kingdom, vaccination committees (*juntas centrales*) were formed, composed of the most enlightened individuals, who, by vaccinating people month after month, see to it that the miasma of the vaccine is not lost. It is even less likely to disappear, because it exists in the country itself; Mr. Balmis discovered it on the outskirts of Valladolid and in the village of Atlixco, near Puebla, in the udders of Mexican cows. Now that the commission has carried out the beneficent plans of the king of Spain, there is hope that, through the influence of the clergy and above all that of the missionary monks, the vaccine can be gradually introduced into the interior of the country. Thus will Mr. Balmis's journey be forever remembered in the annals of history. For the very first time, the Indies saw the ships that had previously born only the instruments of carnage and destruction instead bring the seeds of relief and consolation to human suffering!

I.330

In many coastal sites, the arrival of the armed frigates that carried Mr. Balmis across the Atlantic Ocean and the South Sea prompted one of the simplest and therefore most touching religious ceremonies. The bishops, military governors, and most distinguished persons of rank went down to the shore, carrying in their arms the children who were to take the vaccine to the indigenous peoples of the Americas and to the Malay race of the Philippine Islands. Amid public applause, they laid the precious tributes of beneficial preventives at the foot of the altar, giving thanks to the Supreme Being for having witnessed such a joyful event. In fact, one must know firsthand the ravages that smallpox wreaks in the Torrid Zone—especially among a people whose physical constitution seems adverse to skin rashes— to realize how much more important ▼ Mr. Jenner's discovery is for the equinoctial part of the New Continent than it has been for the temperate part of the Old Continent.

I.331

It will be useful to record here a fact important to those who study the history of vaccination. Until November 1802, the vaccine was unknown in Lima. At that time, smallpox was rampant on the South Sea coasts. The merchant ship Santo Domingo de la Calzada stopped over at Lima on its way from Spain to Manila. A resident of Cádiz had had the good sense to send the vaccine to the Philippine Islands via this ship. Taking advantage of this favorable opportunity, ▼ Mr. Unanué, professor of anatomy in Lima and author of an excellent physiological treatise on the Peruvian climate,[1] vaccinated several persons with the virus that the merchant ship had brought. No pustules appeared: the virus appeared to have been altered or weakened. Mr. Unanué did, however, observe that the persons who had been vaccinated all had a particularly benign form of smallpox, so he used this pox venom in regular inoculations to make the epidemic less deadly. Thus, he indirectly recognized the effects of a vaccination that was assumed to have failed.

I.332

During the course of this same epidemic in 1802, it was accidentally discovered that the beneficial effect of the vaccine had long been known to the country folk of the Peruvian Andes. A black slave [nègre esclave] in the household of the ▼ Marquis of Valleumbroso had been inoculated for smallpox but showed no symptoms of the disease. When he was about to be re-inoculated, the young man declared that he was sure that he would

---

1. This work shows an intimate familiarity with the French and British literature; its title is *Observaciones sobre el clima de Lima y sus influencias en los seres organizados, en especial el hombre*, by Dr. Hipólito Unanué, Lima, 1806.

never contract smallpox, because while milking cows in the Cordillera of the Andes, he had developed a skin rash that, according to some old Indian shepherds, had been caused by contact with some tubercles on the cows' udders. The black man [le nègre] said that whoever had already had this rash would never contract smallpox. Africans, and especially Indians, display a great sagacity for observing the character, habits, and diseases of the animals among which they are accustomed to living. One should thus not be surprised that following the introduction of horned cattle into the Americas, the lower classes noticed that the pustules found on cows' udders gave shepherds a type of benign smallpox; those who had it escaped the general contagion at the time of the great epidemics.

I.333

Matlazahuatl, a disease specific to the Indian race, rarely appears from one century to the next; it was especially rife in 1545, 1576, and 1736; Spanish authors called it a plague. Since the most recent epidemic occurred at a time when, even in the capital, medicine was not considered a science, we lack precise information about the *matlazahuatl*. It is probably similar to yellow fever or black vomit, but it does not attack white people,[1] either Europeans or the descendants of the indigenous peoples. Members of the European race do not seem subject to this fatal typhus, whereas yellow fever, or the black vomit, only rarely assails the Mexican Indians. The main base of the *vómito prieto* is the maritime region, where the climate is excessively hot and humid. *Matlazahuatl*, on the contrary, carries its deadly terror into the very interior of the country, onto the central plateau, and to the coldest, most arid regions of the kingdom.

I.334

The Franciscan ▼ Father Torribio, better known by his Mexican name Motolinia, claims that the smallpox introduced in 1520 by a black man [nègre], a slave of [Pánfilo de] Narváez's, carried off half of the inhabitants of Mexico. ▼ Torquemada's somewhat baseless opinion is that of the two epidemics of *matlazahuatl*, in 1545 and 1576, 800,000 Indians died in the first and 2,000,000 in the second. When one reflects on the difficulty, even today, of estimating the number of plague victims in Eastern Europe, one has good reason to doubt that in the sixteenth century, the two viceroys Mendoza and ▼ Almanza, who were governing a recently conquered land, were able to ascertain just how many Indians had fallen victim to the

1. When the first Puritans landed in New England in 1614 via Santander, a European colony, a plague that caused the dying to turn yellow carried off nineteen-twentieths of the native population of Massachusetts; *foreigners* (▼ Richard Vines and his companions) were not affected by this North American *matlazahuatl*. This fact seems most noteworthy. (▼ Morse and Parish, *Hist[ory] of New England*, 1820, p. 59.)

matlazahuatl. I do not accuse the two monk historians of untruthfulness; but it is rather unlikely that their count is based on precise data.

An interesting problem remains to be solved. Was the plague that is said to have ravaged the Atlantic regions of the United States before the Europeans arrived—and which the famous ▼ Rush and his followers assumed to be yellow fever—identical to the *matlazahuatl* of the Mexica Indians? One can only hope that if the latter disease reappears in New Spain, doctors there will carefully monitor it.

A third obstacle to the population growth in New Spain—perhaps the cruelest of them all—is famine. Like the inhabitants of Hindustan, the American Indians are accustomed to being satisfied with the least amount of food necessary to sustain life; their numbers grow without a necessary increase in the means of subsistence proportional to the growth of the population. Naturally indolent, especially because of their situation in a pleasant climate with a generally fertile soil, the Indians grow only the corn, potatoes, and wheat they need for their own nourishment, or at most what is needed for consumption by nearby towns and mines. It is true that agriculture has made rapid strides in the past twenty years; but there has also been an extraordinary rise in consumption due to population growth; an unbridled lavishness previously unknown among the mixed-blood castes; and the mining of so many new veins, which require men, horses, and mules. Of course, factories in New Spain employ only a few men, but many workers are taken away from agriculture by the need to transport merchandise—mining products, iron, powder, and mercury—on the backs of mules from the coast to the capital, and from there to the mines along the ridge of the Cordilleras.

Thousands of men and animals spend their lives on the main roads between Veracruz and Mexico City, between Mexico City and Acapulco, and between Oaxaca and Durango, as well as on the back roads used to carry provisions to the plants set up in the arid, uncultivated regions. This class of inhabitants, which the economists, in their system, designate as sterile and unproductive, is consequently larger in America than one would expect in a country where the manufacturing industry is still so underdeveloped. The disproportion between population growth and the growth in the amount of food produced by agriculture rekindles the horrifying spectacle of famine whenever a great drought or some other local cause spoils the corn harvest. In all parts of the globe, food shortage has always been accompanied by epidemics that have devastated the population. In 1784, the lack of food caused asthenic illnesses among the poorest class of people.

I.335

I.336

These combined calamities cut down a large number of adults, and especially children; it is estimated that over eight thousand individuals perished in the town and in the mines of Guanajuato. A striking meteorological phenomenon also contributed to the famine: after surviving an extraordinary drought, the corn froze during the night of August 28, due to radiation cooling under very clear skies. An estimated 300,000 inhabitants perished from the surface of the kingdom due to this fatal combination of famine and disease. This number will seem less exaggerated if one recalls that even in Europe, famine can sometimes reduce the population more in a single year than the surplus of births to deaths enlarges it over a four-year period. In Saxony, for instance, nearly 66,000 people died in 1772, while from 1764 to 1784, births exceeded deaths by more than 17,000 on average.

I.337

The effects of famine are the same in almost all the equinoctial regions. In South America, in the province of Nueva Andalucía, I saw villages whose inhabitants were forced by famine to scatter from time to time throughout uncultivated regions and to look for food there among the wild plants. Missionaries use their authority to prevent such dispersals, but to no avail. In the province of Los Pastos, the Indians who lack potatoes, the basis of their diet, sometimes take refuge on the highest ridge of the Cordillera, where they eat the center of the *achupalla* [Aymara: pineapple], a plant similar to the genus Pitcairnia. For several months of the year, the Otomi of Uruana, on the banks of the Orinoco, swallow clay soil to absorb the gastric and pancreatic juices whose profuse secretion this soil activates; in this way, they assuage the hunger that torments them.[1] On South Sea islands, on fertile soil and in the midst of a vigorous and beautiful natural realm, famine drives the inhabitants to the cruelest form of anthropophagism. In the Torrid Zone, where a kind hand seems to have spread the seed of abundance, humankind, careless and phlegmatic, experiences a periodic lack of sustenance, which the industry of civilized peoples banishes from even the most barren regions of the north.

I.338

Mining work has long been considered one of the major causes of the depopulation of the Americas. It would be difficult to deny that, in the early period of the conquest and even into the seventeenth century, many Indians succumbed to the excessive work they were forced to do in the mines. They perished without progeny, like the thousands of African slaves who die every year on the plantations of the Antilles, weakened by lack of food

---

1. See my *Tableaux de la Nature*, vol. I, pp. 62, 191, and 209, and *Relation historique*, chap. XXIV, pp. 609–20.

and sleep. In Peru, at least in the southernmost part, mining has depopulated the countryside, because of ▼ *La Mita*, which exists there even now (in 1804), a barbaric law that forces the Indian to leave his home and to travel to distant provinces where manpower is needed to extract the subterranean riches. But it is not so much the work as the sudden change of climate that makes the *Mita* so pernicious for the Indians' survival. Their race lacks the organizational flexibility characteristic of Europeans. The health of the copper-skinned man suffers greatly when he is transported from a   I.339 hot to a cold climate, especially when he is forced to descend from the high Cordillera into the narrow, humid valleys, where all the miasmas from the nearby regions seem to collect.

In the Kingdom of New Spain, at least in the past thirty to forty years, work in the mines is free work. There is no trace of the *Mita*, although a deservedly famous author, Robertson,[1] has suggested the opposite. Nowhere does the lower class enjoy the fruits of its labor more than in the mines of Mexico; no law forces the Indian to choose this type of work or to prefer one mining operation to another. If he is unhappy with the proprietor of a mine, the Indian will leave it to offer his services to someone else who pays more regularly or in ready money. These valid and consoling facts are not widely known in Europe. In the entire Kingdom of New Spain, the number of persons employed in underground works—which is divided into several classes (*Barenadores, Faeneros, Tenateros, Bareteros*)—is less than 30,000. Only one two-hundredth of the populations is therefore currently employed in the mining of metallic riches.

In Mexico, the mortality rate among miners is generally not much higher than what is observed among the other classes of people. One is   I.340 easily convinced of this by examining the lists of the deceased from the various parishes of Guanajuato and Zacatecas. This phenomenon is all the more striking since, in many of these mines, the miner is exposed to a temperature six degrees higher than the average temperature of Jamaica or Pondicherry. At the bottom of the Valenciana mine (en los planes), at the great perpendicular depth of 513 meters, I found thirty-four degrees on the Centigrade thermometer, whereas near the well, in the fresh air, the same thermometer drops to four or five degrees above zero in winter. This shows that the Mexican miner can tolerate a temperature difference of over thirty degrees. The extreme heat of the Valenciana mine is not, however, due to

1. Robertson, *Hist[ory] of America*, vol. II, p. 373.

such a large number of men and lights occupying a small space, but, rather, to geological causes that we shall examine later on.

It is interesting to see how the Mestizos and Indians who are employed to carry the ore on their backs, and who go by the name of *Tenateros*, are weighed down for six hours at a time with 225 to 350 pounds, while being exposed to a very high temperature; they climb stairways of 1,800 steps eight to ten times in a row without resting. The sight of these hardworking, robust men might have changed the opinion of Raynal and [de] Pauw, and many other authors (however esteemed) who were content to criticize
I.341   the degeneration of our species in the Torrid Zone. In the Mexican mines, children ages ten to twelve already carry rock loads weighing one hundred pounds. The occupation of the *Tenateros* is considered unhealthy, if they go into the mine more than three times a week. But the labor that rapidly ruins even the most robust constitutions is that of the *Barenadores*, who blow up rock with gunpowder; they rarely live beyond the age of thirty-five but are so eager to earn money that they perform their strenuous work all week long. They generally spend no more than five or six years in this occupation, after which time they prefer work that is less dangerous to their health.

The miner's craft is gradually improving; the students of the mining school are slowly expanding their knowledge of air circulation in shafts and galleries; machinery is being introduced to replace the previous method of carrying ore and water on men's backs over steep stairways. As the mines of New Spain begin to look more like the mines of Freiberg, Clausthal, and Schemnitz, the miners' health will be less affected by the fumaroles and by excessively prolonged muscular exertion.[1]

I.342   Nearly five to six thousand persons are employed either in the amalgamation of ore or in the process that precedes it. Many of these individuals spend their lives walking barefoot over heaps of crushed metal that has been dampened and mixed with muriate of soda, iron sulfate, and mercuric oxide through its contact with the atmosphere and the rays of the sun. It is remarkable to see these men enjoying such perfect health. Doctors who practice in the mining areas unanimously confirm that afflictions of the nervous system attributable to absorbing mercuric oxide are quite rare. People in Guanajuato sometimes drink the water in which the amalgam has been washed (*agua de los lavaderos*) without any effect on their health. This fact

---

1. It would be unnecessary here to expound on the extent to which the associations recently formed in Europe for mining work in free Spanish America would contribute to the miners' health by introducing machinery and by *boring* aerated tunnels.

has often perplexed Europeans unfamiliar with basic chemistry. The wash water is bluish-gray at first; suspended in it are black mercuric oxide, small globules of pure mercury, and an amalgam of silver. This metallic mixture gradually precipitates, and the water becomes clear: it can dissolve neither the mercury oxide nor the mercury muriate, which is one of the most insoluble salts known to us. The donkeys like to drink this water because it contains a slight amount of dissolved muriate of soda [common salt].

In discussing the population growth of Mexico and the reasons for which it has slowed, I have mentioned neither the arrival of new European colonists nor the mortality affected by the black vomit. We shall discuss these two topics later on in this work. Suffice it to say here that the *vómito prieto* is a scourge that is only felt on the coastline and that does not kill more than two to three thousand persons in the whole kingdom annually. Europe sends only eight hundred persons to Mexico. Political writers have always exaggerated what they call the depopulation of the Old Continent for the New one. ▼ Mr. Page,[1] for instance, in his work on trade in Saint-Domingue, claims that European emigration provides over 100,000 persons annually to the United States. This estimate is ten times too high, for in 1784 and 1792, their number[2] did not exceed 5,000. ▼ Mr. Gallatin[3] confirms that the annual average number of immigrants from Europe to the United States in the past few years has been 10,000. The number vacillates between 4,000 and 22,000. The population increases in Mexico and North America are the simple result of the growth in domestic prosperity.

I.343

1. [Page, Pierre François, and Auguste Dubois, *Traité d'économie politique et de commerce des colonies*,] vol. II, p. 427.

2. Blodget, *Economica*, 1806, p. 58.

3. See my *Relation historique*, vol. III, p. 179.

CHAPTER VI

Difference between Castes—Indians or Indigenous Americans—
Their Number and Migrations—Diversity of Languages—
Degree of Civilization of the Indians

The Mexican population is composed of the same elements as the other Spanish colonies. There are seven different races: (1) those born in Europe, commonly called *Gachupines*; (2) the Spanish Creoles or American-born whites of European stock; (3) those of mixed blood (*Mestizos* [Métis]), descendants of whites and Indians; (4) the Mulattos [Mulâtres], descendants of whites and blacks [nègres]; (5) the *Zambos*, descendants of blacks [nègres] and Indians; (6) the Indians themselves, or the copper-skinned indigenous race; and (7) the African Blacks [Nègres]. Setting aside subdivisions results in four castes: the whites included under the general denomination of Spaniards; the Blacks [Nègres]; the Indians; and people of mixed race—mixtures of Europeans, Africans, American Indians, and Malays. Because of the frequent contacts between Acapulco and the Philippine Islands, many people of Asian origin, either Chinese or Malay, have settled in New Spain.

There is a widespread belief in Europe that only a very small number of copper-skinned indigenous peoples, descendants of the ancient Mexicans, survive to this day. The Europeans' cruelties have led to the complete extinction of the former inhabitants of the Antilles. Fortunately, this horrible result has not been reached on the American continent. In New Spain, the number of Indians is two and a half to three million, counting only those who are of pure race, with no admixture of European or African blood.[1] What is even more comforting, we repeat, is that far from becoming extinct,

---

1. We have shown earlier (p. 318) that, according to Mr. Navarro [y Noriega], the Indian population of New Spain likely exceeds 3,600,000.

the indigenous population has grown considerably in the past fifty years, as the capitation, or tribute, registers prove.

Overall, the Indians seem to make up over two-fifths of the entire population of Mexico. In the four intendancies of Guanajuato, Valladolid, Oaxaca, and Puebla, this population even rises to three-fifths. The 1793 census produced the following table:

I.346

NAME OF INTENDANCY	TOTAL POPULATION	NUMBER OF INDIANS
Guanajuato	398,000	175,000
Valladolid	290,000	119,000
Puebla	638,000	416,000
Oaxaca	411,000	363,000

It follows from this chart that, in the intendancy of Oaxaca, eight-eight of every one hundred persons are Indians. The high number of indigenous peoples proves beyond a doubt how old the culture of this country is. Near Oaxaca, and especially in the southwest, in Chiapas, one thus finds ruins of architectural monuments that attest to a highly advanced civilization in Mexico.

Indians, or copper-skinned persons, are very rare in the northern part of New Spain; they are rarely found in the provinces called *internas*. History attributes this phenomenon to several causes. At the time of the Spanish conquest of Mexico, only a few inhabitants were found in the regions above the twentieth parallel. These provinces were home to the Chichimec and the Otomi, two nomadic peoples whose sizable hordes occupied vast territories. Agriculture and civilization, as we have seen, were concentrated on the plateaus that extend southward from the Santiago River, especially between the Valley of Mexico and the province of Oaxaca.

In general, from the seventh to the thirteenth century, the population seems to have flowed continuously toward Guatemala. The warlike peoples who, one after the other, invaded the land of Anahuac came from the regions north of the Río Gila. We do not know whether this was their place of origin, or whether they had originally come from Asia or the northwest coast of America and had then crossed the savannahs of the Nabajoa [Navajo] and the Moqui [Hopi] before arriving at the Río Gila. The hieroglyphic scripts of the Aztecs have bequeathed to us the memory of the major periods of the great migrations of the American peoples. This migration is

I.347

somewhat similar to the one that, in the fifth century, plunged Europe into a state of barbarism, the dire repercussions of which are still felt in some of our social institutions. On the contrary, the people who traversed Mexico left behind traces of culture and civilization. The Toltecs first appeared there in the year 648, the Chichimecs in the year 1170, the Nahualteca in the year 1178, and the Acolhua and the Aztecs in 1196. The Toltecs introduced corn and cotton farming, built towns, roads, and, above all, the great pyramids that we still admire today, and whose faces are very accurately oriented. They knew hieroglyphic script, how to smelt metals and cut the hardest stones; they had a solar calendar year that was more accurate than that of the Greeks and Romans. Their form of government shows that they were descended from a people that had itself already experienced great vi-

I.348   cissitudes in their social structure. But what was the origin of this culture? And from what land did the Toltecs and the Mexica come?

Legend and historical hieroglyphics cite Huehuetlapallan, Tollan [Tula], and Aztlan as the first home of these wandering nations. Nothing today points to an ancient human civilization north of the Río Gila or in the northern regions explored by Hearne, Fidler, and Mackenzie. But on the northwest coast, between Nootka and the Cook River, especially below 57° north latitude in Norfolk Bay and the Cox Canal, the indigenous peoples show a definite proclivity for hieroglyphic pictures.[1] A distinguished scholar, Mr. de Fleurieu, suspects that these peoples could be the descendants of a Mexican colony, who took refuge in the northern regions during the conquest. This clever theory seems less valid when one considers the great distance that these colonists would have had to cross, and if one recalls that Mexican culture did not extend northward of 20° latitude. I am more inclined to believe that, during the Toltecs' and the Aztecs' migrations southward, some tribes settled on the coastline of New Norfolk and New Cornwall, while others continued their southward march. One can imagine how peoples that traveled en masse—for example, the Ostrogoths or the Alani—were able to reach Spain from the Black Sea; but is it also believable that a portion of these same people were able to return from west to east at a time when other hordes had already taken over their original homes on the banks of the Don and the Borysthenes [Dnieper]?

1. [Marchand,] *Voyage [autour du monde],* vol. I, pp. 258, 261, 375; ▼ Dixon, [*A Voyage Round the World,*] p. 332. On the major questions of ancient culture and the migration of the American nations, see Humboldt, *Vues des Cordillères et Monuments des peuples indigènes [de lÁmerique],* vol. I, p. 85; vol. II, p. 214. *Relation historique,* vol. III, pp. 155–63.

We cannot discuss here the large issue of the Asian origin of the Toltecs or the Aztecs. The general question of the earliest origin of the inhabitants of a continent is beyond the prescribed limits of history; it may not even be a philosophical question. Other peoples probably already existed in Mexico when the Toltecs appeared there on their migration. Exploring the possibility of the Toltecs as an Asian race is therefore not the same as asking if all Americans descended from the high plateaus of Tibet or eastern Siberia. Using the Chinese annals, ▼ De Guignes believed to have proven that these people began coming to the Americas in 458. ▼ Horn, in his ingenious work, *De originibus Americanus* (published in 1699), Mr. Schérer, in his *Recherches historiques [et géographiques] sur le Nouveau-Monde,* and other more recent writers, have made ancient relations between Asia and America highly probable.

I suggested elsewhere[1] that the Toltec or the Aztec might be a branch of the Hiong-Nu [Xiongnu], who, according to Chinese history, emigrated, following their chief Punon, and were lost in northern Siberia. This people of warlike shepherds has changed the political face of eastern Asia more than once: mixed with Huns and other peoples of the Finnish or Uralian race, they devastated the most beautiful parts of civilized Europe. All these conjectures will acquire greater probability with the discovery of salient similarities between the languages of Tartary and those of the New Continent, similarities that, according to the research of Mr. Smith Barton, ▼ Mr. Vater, and Mr. Wilhelm von Humboldt, extend to only a very small group of words. The lack of wheat, oats, barley, and rye—the nutritious grains generally called cereals—seems to prove that, if Asiatic tribes migrated to the Americas, they must have been descended from a nomadic or pastoral people. On the Old Continent, the cultivation of cereals and the use of milk were introduced at the earliest moment of recorded history. The only grain that the inhabitants of the New Continent grew was corn (▼ *Zea*). Their diet included no milk products, even though the llamas, alpacas, and the two types of native oxen in northern Mexico and Canada would have provided abundant milk. These are some of the striking contrasts between the peoples of the Mongol and the American race.

Without losing ourselves in suppositions about the homeland of the Toltecs and the Aztecs, and without determining the geographical position of the ancient kingdoms of the Huehuetlapallan and Aztlan, we shall limit ourselves to what the Spanish historians have taught us. In the

1.350

1.351

1. *Tableau de la Nature,* vol. 1, p. 53.

sixteenth century, the northern provinces—Nueva Vizcaya, Sonora, and New Mexico—were only sparsely inhabited. The indigenous peoples were nomadic hunting peoples; they retreated as the European conquerors advanced northward. Agriculture alone ties men to the land and fosters a love of country. We see that, in the northern part of Anahuac, in the cultivated regions near Tenochtitlan, the Aztec colonists, patiently enduring the cruel humiliations that the conquerors imposed on them, suffered everything rather than give up the land that their fathers had worked with their own hands. In the northern provinces, on the contrary, the indigenous peoples ceded to the conquerors the uncultivated savannahs that served as pasture land for buffalo. The Indians took refuge beyond the Gila, in the direction of the Río Zaguanas and the Las Grullas mountains. The Indian tribes that formerly occupied the territory from the United States to Canada followed the same course of action: to avoid being forced to live among the Europe-

I.352 ans, they preferred to retreat, first to the west of the Allegheny Mountains, then to the west of the Ohio, and finally to the west of the Missouri. For the same reason, one finds the race of copper-skinned indigenous peoples neither in the *provincias internas* of New Spain nor in the cultivated regions of the United States.

Since the migrations of the American peoples have always been from north to south, at least from the sixth to the twelfth century, it is clear that the Indian population of New Spain must been very heterogeneous. As the population migrated southward, some tribes stopped on the way and mixed with the peoples who followed closely behind them. The great many languages that are still spoken today in the kingdom of Mexico attest to a great variety of races and origins.

These languages number more than twenty, fourteen of which already have rather complete grammars and dictionaries. These are their names: Mexican or Aztec [Nahuatl]; Otomi; Tarascan [Purepecha]; Zapotecan; Mixtecan; Mayan, or the language of the Yucatán; Totonacan; Popolocan; Matlazincan; Huastecan; Mixean; Caquiquel [Cakchiquel]; Tarahumara; Tepehuan; and Coran. It seems that many of these languages—far from being dialect of a single language (as some authors have wrongly suggested)—

I.353 are more different from one another than Persian is from German, or French from the Slavic languages. This is at least the case with the seven languages in which I have some competency. ▼ The sheer variety of the languages spoken among the peoples of the New Continent, and which, with little exaggeration, one might put at several hundred, is a very remarkable phenomenon, especially if one compares it to the few languages spoken in Asia and Europe.

The Mexica language—that of the Aztecs—is the most widely spoken. Its range extends from the thirty-seventh parallel to Lake Nicaragua, a distance of four hundred leagues. The Abbot Clavijero has proven[1] that the Toltecs, the Chichimecs (whose descendants are the inhabitants of Tlaxcala), the Acolhua, and the Nahuatlaca all spoke the same language as the Mexica. This language is less sonorous[2] than the Inca language but is nearly as widespread and as rich. After the Mexican or Aztec language, for which there are eleven published grammars, the most pervasive language of New Spain is that of the Otomi.

I am quite sure that a detailed description of the customs, character, and physical and intellectual state of these indigenous peoples of Mexico, to whom Spanish law refers as Indians, would be of interest to readers. At the root of the general fondness that Europeans have for the descendants of the early populations of the New Continent is a generous sentiment that honors humanity. The history of the conquest of the Americas is portrayed as the unequal struggle between those with developed skills and others who were taking only their first steps toward civilization. The unfortunate races of the Aztec and Otomi Indians who had escaped the carnage seemed destined for extinction after being oppressed for several centuries. It is difficult to imagine that over three million aboriginals could have survived these prolonged calamities. The inhabitants of Mexico and Peru, the Indians of the Philippines, and the Africans dragged into slavery in the Antilles draw attention to themselves for reasons completely different from those that make journeys to China or Japan so fascinating. The interest that the misfortune of a conquered people sparks in us is such that it often prejudices us against the descendants of the conquerors.

To paint the portrait of the indigenous inhabitants of New Spain, it is not enough to represent their present state of degradation and misery; one must go back to that distant time when their nation, governed by its own laws, could wield its own force. One must consult the hieroglyphic paintings, the buildings of cut stone, and the works of sculpture that have been preserved to this day; these works bear witness to the infancy of the art, but they are strikingly similar to many monuments of the most civilized

I.354

I.355

---

1. [Clavijero,] *Storia [antica] di Messico*, vol. I, p. 153.

2. "Notlazomahuizteopixcatatzin" means "venerable priest whom I cherish as I do my father." The Mexicans use this twenty-seven letter word, or rather title (since the philosophy of grammar is at odds with calling it a *word*), when speaking to their curates.

nations. These studies are reserved for another work,[1] the nature of which, however, does not allow us to go into details that are of equal importance to the history and the psychological study of the human species. I shall limit myself here to describing the most prominent features of the vast tableau of the indigenous peoples of the Americas.

The Indians of New Spain generally resemble those who live in Canada and Florida, Peru, and Brazil: they have the same swarthy, coppery skin, smooth and straight hair, and little facial hair; stocky bodies and slanted eyes, the corners of which turn up toward the temples; prominent cheekbones and wide lips; their mouths have a gentle expression that contrasts with their solemn, severe gaze. After the hyperborean race, the American race is the least numerous, but it occupies the most space on the globe. Across 1,700,000 square leagues (twenty-five to a degree), from the islands of Tierra del Fuego to the Saint Lawrence River and the Bering Strait, one is, at first glance, struck by the resemblance in the features of the inhabitants. One seems to recognize that they are all descended from the same stock, despite the linguistic differences that divide them. Yet, when one reflects on this family resemblance, after an extended stay among the indigenous peoples of the Americas, one realizes that those famous travelers who were only able to observe a few individuals on the coasts have singularly exaggerated the physical similarities among peoples of the American race.

Intellectual culture is what contributes the most to diversifying features. Among barbarians, there is a physiognomy of the tribe and the horde rather than an individualized physiognomy. The same seems to apply when one compares domesticated animals to those who live in our forests. But the European who decides that all darker-skinned races resemble each other also falls prey to a particular illusion: he is struck by their complexion, so different from ours, and, for a long time, this uniform coloring distracts his eyes from the difference between individual features. The recently arrived colonist has trouble distinguishing among the local people because his eyes are less fixed on the facial expression—either gentle, melancholy, or fierce— than on that coppery-red color and that shiny, black, coarse hair, so sleek that it always seems to be wet.

In the faithful table that Mr. de Volney, an excellent observer, has drawn of the Canadian Indians, one probably recognizes the tribes scattered

I.356

---

1. This work, which was translated into several languages, appeared as *Views of the Cordilleras and Monuments of the Indigenous Peoples of the Americas*, in two folio volumes with sixty-nine plates.

across the prairies of the Río Apure and the Caroni. The same type exists in both Americas; but Europeans who, like me, have navigated the great rivers of South America and have had the opportunity to witness different tribes gathered under the monastic hierarchy of the Missions, will also have seen that the American race includes many peoples whose features are as fundamentally different from each other as the numerous racial varieties among the Caucasians, Circassians, Moors, and Persians. The slender form of the Patagonians, who live at the southern tip of the New Continent, is also found, so to speak, among the Caribs, who inhabit the plains from the Orinoco Delta to the source of the Río Blanco. But what a difference there is between the shape, physiognomy, and physical constitution of those Caribs,[1] who must be counted among the most robust people on earth (and who should not be confused with the degenerate *Zambos* who were formerly called Caribs on the island of Saint Vincent) and the squat body of the Chaymas Indians in the province of Cumaná! What a difference between the shape of the Tlaxcala Indians and the Lipan [Apaches] and Chichimecs in the northern part of Mexico!

I.357

The indigenous peoples of New Spain have a darker complexion than the inhabitants of the hottest countries of South America. This characteristic is even more remarkable since, among the Caucasian race that one may call the Arabo-European race, the skin of southern people is not as fair as that of northerners. While many individuals from the Asian nations who flooded Europe in the sixth century had very dark coloring, it seems that the shades of complexion that one observes among the peoples of the white race is due less to their *origin* and their mixture than to the influence of the local climate. Climate, however, seems to have almost no effect on either Americans or Blacks [Nègres]. Those races with abundant hydrogen carbide in what ▼ Malpighi calls the reticular or mucous body are particularly resistant to the effects of the ambient air. The Blacks [Nègres] of the mountains of Upper Guinea are no less black than those who live near the coast. Among the indigenous peoples of the New Continent, there are likely very light-skinned tribes whose complexion is closer to that of the Arabs or the Moors. We found that the peoples of the Río Negro are darker than those of

I.358

1. In the sixteenth century, the great nation of the Caribs, or Caraïbes, who, after exterminating the Cabres [persons of mixed descent], conquered a large part of South America, from the equator to the Virgin Islands (see my *Relation historique*, vol. III, ch. 25, pp. 5, 22, 161, and 163). The few families who still existed on the eastern Antillean Islands in our time, and who were deported by the British to Ratan Island, were a mixture of true Caribs and Blacks [Nègres].

the Lower Orinoco, although the climate on the banks of the former river is cooler than it is in the more northerly regions. In the forests of Guiana, especially near the source of the Orinoco, there are many whitish-skinned [blanchâtres] tribes—the Guiaca, Guaharibs [Guaharibos], Guainares, and Maquiritares—among whom there are a few robust individuals who, showing no signs of the asthenic illness that characterizes *Albinos*, have the color
I.359   of true Mestizos [Métis].[1] Yet, these tribes have never mixed with the Europeans and are surrounded by other small groups of blackish-brown [brunnoirâtre] peoples. The Indians who, in the Torrid Zone, live on the highest plateaus of the Cordillera of the Andes, and those who, below 45° southern latitude in the Chonos Archipelago, live by fishing, have a complexion as coppery as the Indians who, under a scorching sky, grow bananas in the narrowest and deepest valleys of the equinoctial region. It must be added that the mountain-dwelling Indians wear clothes and have done so since long before the conquest, while the indigenous peoples who roam the plains are completely naked; consequently, they are always exposed to the vertical rays of the sun. I have never observed that, on a single individual, the covered parts of the body were less brown than those exposed to the warm, humid air. One notices everywhere that the Americans' color depends but little on the locales in which we actually find them. As we have seen above, Mexicans are less swarthy than the Indians of Quito and New Granada, who live in a very similar climate; one even notices that the people who are scattered north of the Río Gila are browner than those who live near the kingdom of Guatemala. This dark color is prevalent up to the coasts nearest to Asia. But at Cloak Bay, below 54°10′ northern latitude, amidst copper-
I.360   skinned Indians with small, almond-shaped eyes, there is a tribe with large eyes, European features, and skin that is less brown than that of the peasants in our own countryside. They may be descended from the Indo-Germanic peoples, the Ousun and the Tingling, who were introduced to us by ▼ Mr. Klaproth,[2] and who dwelled in the central and northern part of Asia almost two hundred years before the Christian era. All these facts tend to prove that, despite the variety of climes and elevations where the various human races dwell, nature does not deviate from the type that has dominated for thousands of years.

My observations on the innate or natural color of the indigenous peoples are in part contrary to the claims made by Michikinikwa, the famous chief

1. *Relat[ion] hist[orique]*, vol. I, pp. 498, 503; vol. II, pp. 572, 574.
2. [Klaproth,] *Tableaux historiques de l'Asie*, 1825, pp. 162–74.

of the Miamis, whom the Anglo-Americans called *Little Turtle*, and who gave so much valuable information to Mr. de Volney. He claimed that "the children of the Canadian Indians are white at birth," that "the adults turn brown because of the sun and the oils and plant juices with which they rub their skin," that "the part of a woman's waist that is always covered by clothing is white."[1] I have never seen the Canadian peoples that the chief of the Miamis describes, but I can verify that in Peru, in Quito, on the coastline of Caracas, on the banks of the Orinoco, and in Mexico, children are never born white, and that all parts of the bodies (except the insides of their hands and the soles of their feet) of the Indian caciques, who have an easier life and remain clothed inside their homes, are the same brownish-red or coppery color.[2]

I.361

The Mexica, especially those of the Aztec and Otomi race, have more facial hair than I have seen among the other indigenous peoples of South America. Almost all the Indians on the outskirts of the capital sport small mustaches; this is also a characteristic mark of tributary caste. These mustaches, which modern travelers have also found among the inhabitants of the northwest coast of the Americas, are even more unusual because renowned naturalists have left undecided the question of whether Americans are naturally beardless and without hair on the rest of their bodies, or if they carefully remove their hair. Without going into physiological details, I can confirm that the Indians who live in the Torrid Zone of South America generally have some facial hair. This beard thickens when they shave, as we have seen in several examples in the missions of the Capuchin monks of Caripe, where the Indian sextons strive to resemble the monks who are their absolute masters. Still, many are born entirely without any facial or body hair.

I.362

1. Volney, *Tableau du climat et du sol des États-Unis*, vol. II, p. 435.
2. This claim by *Little Turtle*, whose tomb near Fort Wayne was recently visited by Major Long (*Narration of an Expedition to the Lake of Winnepeek*, 1824, vol. I, p. 85), was successfully refuted by travelers who had the opportunity to observe all the Indian nations from the Ohio to the Rocky Mountains (Long, *[Account of an] Expedition to the Rocky Mountains*, vol. I, p. 285). Earlier, ▼ Vespucci had already ventured the opinion that some of the American indigenous peoples would be as white as the Europeans if only they wore clothes (*Grynæus Orbis Novus*, p. 224). ▼ Father Dobrizhoffer notes that the Puelche Indians [Mapudungun: *pwelche*, people of the east] and the Patagonians, who live in cold climates, are much less dark-skinned than the Abipons, Mocobi, and Toba [Guarani: toba, big forehead], and all the hordes of the Chaco who roam the scorching plains between 9° and 10° southern latitude (*Historia de Abiponibus*, vol. II, p. 17).

In his *Relacion del ultimo viage al estrecho de Magallanes*,[1] Mr. Galiano tells us that, among the Patagonians, there are old men with beards, albeit short and not very full. If one compares this statement with the evidence that Marchand, ▼ Meares, and especially Mr. de Volney collected in the northern temperate zone, one might be tempted to suggest that the Indians have more facial hair in proportion to their distance from the equator. Moreover, the apparent lack of facial hair is a characteristic that is not specific to the American race: many of the hordes of eastern Asia, and especially some of the Black African peoples [Nègres africaines], have such little facial hair that one might be tempted to deny that they have any at all. The Blacks [Nègres] of the Congo and the Caribs—two races of exceptionally robust men—prove well enough that it is a physiological fantasy to consider a beardless chin a sure sign of the degeneration and physical weakness of the human species. One forgets all too easily that everything that has been observed about the Caucasian race applies neither to the Mongol nor to the American race, nor to that of the Blacks of Africa [Nègres de l'Afrique].

I.363

The indigenous peoples of New Spain, at least those subject to European domination, generally live to a rather advanced age. Peaceful farmers who have lived in villages for six hundred years, they are not exposed to the hazards of the roving life of the hunters and warriors of the Mississippi and the savannahs of the Río Gila. Following a regular and almost completely plant-based diet consisting mainly of corn and cereals, the Indians would attain considerable longevity if drunkenness did not weaken their constitution. Their inebriating drinks are spirits distilled from sugarcane, corn, and the roots of the Jatropha, and especially the local wine called *Pulque*, made from the juice of the century plant, or Agave americana. This drink, which we shall describe in the following section [livre], can even be nourishing because of its non-decomposed sugar. Many indigenous peoples who are addicted to *pulque* feel no need for solid food for a long time. Taken in moderation, the drink is very beneficial: it fortifies the stomach and supports the gastric functions.

I.364

The vice of drunkenness, however, is less common among the Indians than is generally believed. Europeans who have traveled east of the Allegheny Mountains, between the Ohio and the Missouri Rivers, will find it difficult to believe that we saw indigenous peoples in the forests of Guiana, on the banks of the Orinoco, who were repulsed by the spirits that we had them taste. There are some very sober Indian tribes, whose fermented drinks are

---

1. [Vargas Ponce, *Relación del último] Viaje al estrecho de Magallanes*, p. 331.

too mild to intoxicate. In New Spain, drunkenness is especially common among the indigenous peoples who dwell in the Valley of Mexico, the outskirts of Puebla and Tlaxcala, and wherever maguey, or agave, is widely grown. In the Mexican capital, the police send around tumbrels to collect the drunks found lying in the streets. These Indians, who are treated like corpses, are taken to the main guardhouse; the next day, an iron ring is placed around one of their feet, and they must work for three days cleaning the streets. By releasing them on the fourth day, one is certain to round up several more in the course of the same week. Excessive consumption of liquor is also very detrimental to the health of the lower classes in the hot countries and near the coastline where sugarcane is grown. One must hope that this vice will decrease as civilization advances among a caste of people whose manners are extremely crude.

Travelers who judge the Indians only by their physiognomy will be tempted to believe that it is rare to see any elderly among them. In fact, without consulting parish records, which, in the hot regions, are devoured by termites every twenty to thirty years, it is very difficult to form an idea of the age of the local people; they themselves (I am speaking of the poor Indian peasant) are completely oblivious to it. Their hair only rarely turns gray. It is infinitely rarer to encounter a white-haired Indian than a white-haired Black [Nègre], and the lack of a beard keeps always the former looking young. The Indians' skin is also much less likely to wrinkle. In the temperate zone of Mexico located halfway up the Cordillera, it is not uncommon to find indigenous peoples, especially women, who reach the age of one hundred. Their old age is usually happy, because both the Mexican and Peruvian Indians retain their muscular strength until death. During my stay in Lima, in the village of Chiguata, four leagues from the town of Arequipa, the Indian Hilario Pari died at the age of one hundred and forty-three; for ninety years, he was married to the Indian woman Andrea Alea Zar, who had reached the age of one hundred and seventeen. This aged Peruvian man walked three or four leagues every day until the age of one hundred and thirty; he became blind thirteen years prior to his death, and of his twelve children, only a seventy-three-year-old daughter survived him. I.365

The copper-skinned indigenous peoples have a physical advantage that most likely has to do with the extreme simplicity of their ancestors' daily life over thousands of years; they are prone to almost no physical deformity. I have never seen a hunchbacked Indian; it is very rare to see any who squint, limp, or are one-armed. In the countries where the inhabitants suffer from goiters, this disease of the thyroid gland is not found among the I.366

Indians, and seldom among the Mestizos. ▼ The famous Mexican giant, Martín Salmerón, who is wrongly considered an Indian, belongs to the latter caste. His waist measures 2.224 meters, or six feet, ten inches, two and two-thirds lines of *pied de Paris*. He is the son of a Mestizo who married an Indian woman from the village of Chilapa el Grande near Chilpanzingo.[1]

If one considers only the savages who are hunters or warriors, one might think that there are only strong men among them, because those who have natural deformities either die from fatigue or are abandoned by their parents. But the Mexican and Peruvian Indians in Quito and New Spain, among whom I lived for a long time, are farmers who can only be compared to our European peasant class. The absence of natural deformities among them is probably the result of their way of life and the natural constitution of their race. All men with very swarthy skin, those of Mongol and American origin, especially Blacks [Nègres], share the same advantage. One is tempted to think that the Arab-European race has a greater structural flexibility, and that this structure, which is easily modified by a great number of external factors—differences in food, climate, and habits—tends to vary more often from its original type.

I.367

What we have just stated about the external appearance of the indigenous peoples of the Americas confirms many travelers' claims about the similarities between Americans and the Mongol race. One finds these similarities especially in the color of the skin and hair, the lack of facial hair, prominent cheekbones, and the shape of the eyes. One cannot deny that the human species has no closer races than those of the Americans, the Mongols, the Manchus, and the Malays. But the resemblance of a few features does not constitute the same racial identity. Simply because the hieroglyphic script and traditions of the inhabitants of Anahuac collected by the early conquerors seem to indicate that a group of nomadic tribes spread from the northwest toward the south, one must not conclude that all indigenous peoples on the new continent are of Asian origin. In fact, osteology teaches us that the American skull is noticeably different from that of the

I.368

---

1. This is the actual size of this giant who has the best proportions of any I have seen. He is one inch taller than the giant from Torneo who was exhibited in Paris in 1735. The American paper gives Salmerón seven feet, one inch according to the Paris measure. *Gazeta de Guatemala*, [vol. IV,] 1800. [Cavanilles, "Noticia de un gigante, inserta en dicha gazeta [de Guatemala] de 25 de Agosto de 1800."] *Anales de Madrid* [*Anales de Ciencias Naturales*] vol. IV, no.12.

The human species appears to vary in height from two feet, four inches to seven feet, eight inches, or from 0.757 to 2.490 meters (▼ Schreber. *Die Säugthiere*, vol. I, p. 27).

Mongol: the former has a more slanted facial line, although it is straighter than that of the Black [Nègre]; there is no race on the globe whose frontal lobe is more receded or that has a less prominent forehead.[1] The American's cheekbones are almost as prominent as the Mongol's, but their contours are more rounded and the angles less sharp. The lower jawbone is wider than the Black's [Nègre], but there is less space between its points than there is among the Mongols. The occipital bone is less curved, and the protuberances that correspond to the cerebellum, to which ▼ Mr. Gall's system gives great importance, are scarcely noticeable. This race of copper-skinned people whom we generally call American Indians is perhaps a mixture of Asiatic tribes and the aboriginal populations of this vast continent. Is it possible that the faces with enormous aquiline noses that one sees in the Mexican hieroglyphic paintings preserved in Vienna, Velletri, and Rome, as well as in the historical fragments that I have presented, demonstrate the physiognomy of some extinct races? The Canadian savages persist in calling themselves Metoktheniakes—born of the earth—despite efforts by the *black robes*[2] (as they call the missionaries) to persuade them otherwise.

As for the moral faculties of the indigenous peoples of Mexico, it is difficult to do them justice, if one considers this caste, which has long suffered under tyranny, only in its present state of degradation. At the beginning of the Spanish conquest, a large part of the most well-off Indians, those among whom one might expect a certain level of intellectual culture, perished as victims of the Europeans' brutality. Christian fanaticism especially attacked the Aztec priesthood, exterminating the Teopixqui, or ministers of the divinity, all those who lived in the Teocalli,[3] or god-houses, and who were the keepers of local historical, mythological, and astronomical knowledge—for it was the priests who observed the meridian shadow cast by the gnomons

I.370

---

1. This extraordinary flatness is found among people who have never been able to create artificial deformities, as proven by the skulls of the Mexican Indians, Peruvians, and the others that Mr. Bonpland and I have examined, many of which have been consigned to the Museum of Natural History in Paris. I am inclined to believe that the barbarous custom of some of these savage hordes of flattening a child's head between two boards arises from the notion that beauty consists in a shape of the frontal lobe that is racially characteristic in an exaggerated manner. Blacks [Nègres] show a preference for large, very prominent lips, the Kalmyk for upturned noses. The Greeks, in their heroic statues, unduly raised the facial angle from 85 to 100° (▼ Cuvier, *Anat[omie] comparé*, vol. 2, p. 6). The Aztecs, who never disfigured the heads of their infants, represented their major divinities, as the hieroglyphics scripts prove, with a head much flatter than what I have seen among any of the Caribbean peoples.

2. Volney, [*Tableau du climat et du sol des États-Unis d'Amérique,*] vol. 2, p. 438.

3. From "Teotl," god, θεοῦ [theos].

and regulated the calendar. The monks ordered the burning of the hieroglyphic paintings, by which knowledge of all sorts was passed on from generation to generation. Deprived of these tools of learning, the people fell into a state of ignorance that was so deep that the missionaries, who were unskilled in the Mexican languages, replaced old ideas with few new ones. The Indian women who had preserved some wealth preferred to ally themselves with the conquerors rather than share the contempt in which the latter held the Indians. The Spanish soldiers were very intent on these alliances, since very few European women had accompanied the army. All that remained of the indigenous people then were the most indigent—poor farmers, artisans (among whom there were a large number of weavers), porters who were used as pack animals, and above all the dregs of humanity, the crowd of beggars who, bearing witness to the imperfections of social institutions and the feudal yoke, have thronged in the streets of all large cities of the Mexican empire since the time of Cortés. How, then, are we to judge from this miserable residue of a mighty people both the degree of culture they had attained from the twelfth to the sixteenth century and the intellectual development to which they are prone? If, one day, all that remained of the French or German peoples were a few poor peasants, could one read in their features that they once belonged to nations who had produced a Descartes, a ▼ Clairaut, a Kepler, or a Leibniz?

I.371

We have seen that even in Europe, the lower classes have only made very slow progress toward civilization for centuries. The Breton or Norman peasant or the inhabitant of northern Scotland is only slightly different from what he was at the time of Henry IV and James I. When we consider what Cortés's letters and Bernal Díaz's memoirs, written with such an admirable naïveté, along with other contemporary historians, tell us about the state in which the inhabitants of Mexico, Texcoco, Chollollan, and Tlaxcala were found at the time of King Moctezuma II, we think we are seeing the portrait of the Indians of our own time: the same nakedness in the hot regions; the same type of clothing on the central plateau; and the same customs in their domestic life. How might any significant changes possibly be wrought on the indigenous peoples when they are kept isolated in villages where whites dare not settle; when the language difference erects an almost insurmountable barrier between them and the Europeans; when they are oppressed by magistrates who, for political reasons, make appointments from within their own circle; and, finally, when the only person from whom they might expect any moral improvement is a man who speaks to them of mysteries, dogma, and ceremonies, whose purpose they do not understand?

There is no point in discussing here what the Mexicans were before the   I.372
Spanish conquest; we touched on this interesting subject at the beginning
of this chapter. Considering that the indigenous inhabitants had an almost
perfect knowledge of the length of the calendar year, which they intercalated at the end of their great cycle of 104 years with more precision than
the Greeks,[1] the Romans, and the Egyptians, one is tempted to believe that
this advancement was not the result of the intellectual development of the
Americans themselves, but, rather, that they owed it to their contact with
some highly civilized people of Central Asia. The Toltecs appeared in New
Spain in the seventh century, the Aztecs in the twelfth; they were already
drawing geographical maps of the area they had traversed, and they built
cities, roads, dikes, canals, and immense pyramids that were very precisely
positioned, with a base of up to 438 meters long. Their feudal system and
their civil and military hierarchy were so complex from then on that one
must assume a long succession of political events in order for the peculiar
sequence of their noble and clerical authorities to establish itself and for a   I.373
small portion of the people, themselves enslaved to the Mexican sultan, to
subjugate the vast majority of the peoples. We find distinct forms of theo-
cratic government in South America: those of the Zaque [Muisca][2] in Bogotá (the ancient Cundinamarca) and of the Inca in Peru, two vast empires
where despotism concealed itself behind the appearance of a mild and pa-
triarchal regime. In Mexico, on the contrary, small tribes who had tired of
tyranny created republican constitutions for themselves. It is only after long
popular struggles that such free constitutions can be formed. The existence
of republics is not a sign of a very recent civilization. How could one doubt
that a part of the Mexican nation had reached a certain level of civilization,
considering the care with which the books of hieroglyphics[3] were written

1. Mr. Laplace recognized that, in Mexican intercalation (on which I gave him the material that Gama had collected), the length of the Mexicans' tropical year is almost identical to
the length found by the astronomers of ▼ Al-ma'mūn. For this observation, which is important to the history of the origin of the Aztecs, see his *Exposition du système du monde*, third
edition, p. 554; *Vues des Cordillères et monumens des peoples indigènes de l'Amérique* (octavo),
vol. I, pp. 332–92; vol. II, pp. 1–99, and the historical note at the end of the sixth chapter.

2. The empire of the Zaque, which included the kingdom of New Granada, was founded
by ▼ Idacanzas or Bochica, a mysterious person who, according to the traditions of the
Mosca, lived in the temple of the sun of Sogamoso for two thousand years.

3. Aztec manuscripts are written either on agave paper or on deer skins; they are often
twenty to twenty-two meters (or sixty to seventy feet) long, and the surface of each page is
100 to 150 square inches. These manuscripts are folded here and there into a diamond-like
shape; very thin strips of wood attached to either end form their binding and make them
resemble our quarto volumes. No other known nation on the old continent has made such

I.374   and recalling that, amidst the clash of armies, ▼ a citizen of Tlaxcala took advantage of the facility our Roman alphabet afforded him to write in his own language five hefty volumes of the history of a nation whose subjection he deplored?

This is not the place to resolve the problem, so important for history, of whether the Mexicans of the fifteenth century were more civilized than the Peruvians, and if, left respectively to their own devices, neither would have made more rapid advances in their intellectual culture than it did under the domination of the Spanish clergy. Nor shall we examine whether, despite the despotic nature of the Aztec princes, individual development encountered fewer obstacles in Mexico than in the Empire of the Incas. Among the Incas, the legislator only wished to force his will on the people en masse. Keeping them in monastic obedience and treating them as if they were animated machines, he forced them to undertake projects, the order and magnitude of which astound us, especially the perseverance of those who directed them. If we analyze the mechanism of Peruvian theocracy, which is

I.375   generally too highly regarded in Europe, we shall see that wherever people are divided into castes—where each one is allowed to perform only certain kinds of work, and wherever the inhabitants do not own any private property but work only for the benefit of their community—there one will find canals, roads, aqueducts, pyramids, and immense structures. But we shall also see that these same people, who preserved the same outward appearance of comfort for thousands of years, have hardly developed their moral culture, which is the product of individual liberty.

In the portrait we are drawing here of the different races that comprise the population of New Spain, we are limiting ourselves to a consideration of the Mexican Indian in his present condition. We find in him neither the vivacity of feelings, gestures, and features, nor the lively mind that so positively characterizes many peoples in the equinoctial regions of Africa. There is no greater contrast than that between the impetuous liveliness of the Blacks [Nègres] in the Congo and the seeming apathy of the coppery-skinned Indian. It is, above all, their awareness of this contrast that makes Indian women prefer Blacks [Nègres] not only to men of their own race but

_____

extended use of hieroglyphic writing, and no other has left us real bound books like the ones we have just described. One should not confuse these books with other Aztec paintings composed with the same signs but in the form of tapestries that measure more than sixty square feet. I have seen some of them in the archives of the viceroyalty in Mexico City, and I myself own some fragments of them, which I have had engraved for the *Picturesque Atlas* that accompanies the *Personal Narrative* of my journey.

even to Europeans. As long as intoxicating liquor has not affected him, the Mexican native is solemn, melancholy, and taciturn. This somberness is especially noticeable among Indian children, who, at the age of four or five, exhibit much more intelligence and maturity than white children. Mexicans like to envelop their most trivial acts in an air of mystery; violent passions do not register in their features—it is rather alarming when they suddenly shift from absolute repose to violent, frantic activity. The native of Peru is more mild-mannered, while the Mexican's energy devolves into hardness. These divergences may well arise from the different religions and former governments of the two countries. This type of energy is especially noticeable among the Tlaxcala. Amid their present-day squalor, the descendants of these republicans stand out because of a certain dignity and pride that the memory of their former greatness inspired in them.

I.376

Like the inhabitants of Hindustan and all peoples who have long suffered under civil and religious tyranny, Americans are most stubbornly attached to their customs, manners, and beliefs. I say beliefs, because the introduction of Christianity has had almost no other effect on the indigenous peoples of Mexico than the substitution of new ceremonies, the symbols of a mild and humane religion, for the ceremonies of a bloody religion. The shift from an ancient to a new rite was the result of duress rather than persuasion. Political events brought about this change. On the new continent, as on the old, semi-barbarian nations were accustomed to receiving new laws and new divinities from their conquerors; the vanquished local gods seemed to yield to their foreign counterparts. In a mythology as complex as that of the Mexicans, it was not difficult to find a genealogical relationship between the divinities of Aztlan and those of the Orient. Cortés even took clever advantage of a popular legend, according to which the Spanish were but the descendants of king Quetzalcohuatl, who had left Mexico to bring culture and laws to lands in the east. The ritual books that the Indians wrote in hieroglyphic script at the beginning of the conquest, a few fragments of which are in my possession, prove that at that time, Christianity was blended with Mexican mythology: the Holy Spirit was associated with the Aztecs' sacred eagle. To a certain extent, the missionaries not only tolerated but even encouraged this confusion of ideas so as to spread the Christian religion more easily among the indigenous peoples. They persuaded them that even in ancient times, the gospel had already been preached in the Americas; they sought traces of it in Aztec ritual, just as in our day, scholars studying Sanskrit discuss similarities between Greek mythology and myths from the shores of the Ganges and the Buramputer.

I.377

These circumstances, which will be detailed in another work, explain
I.378 how the Mexican Indians, despite the stubbornness with which they clung
to everything they received from their forefathers, easily forgot their ancient
rites. It was not a dogma that ceded to a dogma; it was merely one ceremony
that gave way to the other. What the indigenous peoples know of religion
is only its external trappings. Since they are fond of anything related to a
prescribed ceremonial order, they take particular pleasure in Christian
religious practice. For the Indians of the lower class, church festivals, the
fireworks that accompany them, and the processions that include dancing
and baroque costumes are a rich source of entertainment. In these festivals,
the national character displays all of its individuality. Everywhere, Christian ritual has taken on the nuances of the country where it has been transplanted. In the Philippine Islands and the Mariana Islands, the Malaysians
have mixed it with their own ceremonies. In the province of Pastos, on the
ridge of the Andean Cordillera, I saw masked Indians decorated with small
bells performing wild dances around the altar, as a Franciscan monk raised
the host.

Long accustomed to slavery, either under the domination of their own
kings or the yoke of the early conquerors, the Mexican indigenous peoples
patiently endure the hardships to which the whites still frequently expose
them. Their only resistance is a ruse veiled by the most deceptive appearance of apathy and stupidity. Since he can only rarely take revenge on the
I.379 Spanish, the Indian joins forces with them to oppress his own countrymen. Harassed for centuries and forced into blind obedience, he seeks his
own turn at tyranny. The Indian villages are governed by magistrates from
the copper-colored race, and an Indian *alcalde* holds sway with a severity
that is so relentless that he can be sure of the support of the priest or the
Spanish *subdelegate*. Oppression has the same effects and corrupts morals
everywhere.

Since almost all the indigenous peoples belong to the peasantry and
the lowest social class, it not easy to assess their aptitude for the fine arts.
I know of no race of humans that appears more devoid of imagination. When
an Indian attains a certain degree of culture, he displays a high aptitude for
learning, a judicious mind, natural logic, and a special penchant for subtlety
and discerning differences when comparing objects. He reasons in a cool
and orderly manner, but he does not show the highly active imagination, the
emotional color, or the creative art that characterize the people of southern
Europe and several tribes of African Blacks [Nègres]. I voice this opinion,
however, with reserve; one must be infinitely circumspect in criticizing

what one might call the moral and intellectual disposition of peoples from whom we are separated by the multiple barriers that arise from differences of language, customs, and manners. A philosophical observer finds inaccuracy in what has been published on the national character of Spaniards, French, Italians, and Germans in the middle of civilized Europe. How could a traveler stranded on an island, or who had lived in a distant country for some time, presume to judge the different faculties of the soul and the prevalence of reason, wit, and imagination of other peoples?

I.380

The indigenous peoples' music and dances betray a distinctive lack of joy. Mr. Bonpland and I observed the same thing throughout South America. The singing is mournful and melancholy. The Indian women are livelier than the men, but they share the same misfortune of servitude to which their sex is condemned among all peoples who are still only partially civilized. Women usually do not take part in dancing; they attend in order to offer the dancers fermented drinks that they have prepared.

Mexicans have a special liking for painting and the arts of sculpting in stone and wood. It is amazing to see what they can do with a bad knife and on the hardest woods. They are particularly skilled at painting images and carving statues of the saints. For three hundred years, they have slavishly imitated the models that the Europeans brought with them at the beginning of the conquest. This imitation even obeys a religious principle that dates back quite far. In Mexico, as in Hindustan, the faithful were not allowed to change even the slightest detail in the appearance of their idols. Everything that had to do with the rites of the Aztecs and the Hindus was subject to immutable laws. For this same reason, one is likely to misjudge the state of the arts and the national taste of these peoples if one only considers the monstrous figures through which they represent their deities. In Mexico, Christian images have maintained some of the stiff, hard features typical of the hieroglyphic paintings of Montezuma's century. Some Indian children, those educated in colleges in the capital or taught at the Academy of Painting founded by the king, have clearly distinguished themselves; but this is due less to their genius than to their industriousness. Never straying from the beaten path, they show considerable aptitude for the arts of imitation; but they demonstrate much more skill in the purely mechanical arts. One day, when manufacturing enjoys a boom in this country, where everything remains for a reformative government to create, this skill will become very valuable.

I.381

The Mexican Indians still have the same fondness for flowers that Cortés found among them in his day. A bouquet of flowers was the most precious gift that one could give the ambassadors who visited Montezuma's

I.382

court. This monarch and his predecessors had collected a great number of rare plants in the gardens of Iztapalapan. The famous *tree of hands*, the Cheirostemon[1]—which ▼ Mr. Cervantes describes, and of which, for a long time, only one very old specimen was known—appears to indicate that the Toluca kings also cultivated foreign trees in that part of Mexico. In his letters to Emperor Charles V, Cortés praises the Mexicans' industry as gardeners and complains that no one has sent him the seeds of the ornamental flowers and alimentary plants that he requested of his friends in Seville and Madrid. This love of flowers must indicate a sense of the beautiful. It is astonishing to find this sentiment in a people among whom a bloody religion and frequent sacrifices appear to have extinguished any sensitivity of the soul and tenderness of feeling. At the main market of Mexico City, the indigenous peoples will not sell peaches, pineapples, vegetables, or pulque (the fermented juice of the agave) unless their stall is decorated with flowers that are changed every day. The Indian vendor appears to sit in a veritable trough of greenery. A meter-high hedge of fresh turf, especially grasses with

I.383

delicate leaves, surrounds the fruits that are stacked up like a semi-circular wall. The floor is all green and divided by garlands of flowers that sit parallel to one another. Small bouquets, arranged symmetrically between the festoon, give the stall the appearance of a carpet strewn with flowers. The European who enjoys studying the customs of the lower classes will also be struck by the care and elegance with which the indigenous peoples arrange the fruits that they sell in small boxes made of a very light wood. The sapodilla (*achras*), mammea, pears, and grapes are placed in the back, while the upper part is decorated with scented flowers. This art of intermingling flowers and fruits may well date from that happy time when, long before the introduction of an inhumane ritual, the inhabitants of Anahuac, like the Peruvians, offered the first fruits of their harvest to the Great Spirit Teotl.

The traits I have collected here as being characteristic of the indigenous peoples of Mexico belong to the Indian farmer, whose civilization, as mentioned above, is similar to that of the Chinese and Japanese. I could give only an imperfect sketch of the customs of the nomadic Indians to whom the Spanish refer as *Indios bravos*, and of whom I have only seen a few

1. Mr. Bonpland reproduces it among our equinoctial plants, [Humboldt and Bonpland, *Plantae Aequinoctiales*,] vol. 1, p. 75, plate XXIV. Most recently, we have found shoots of the *árbol de las manitas* in the gardens of Montpellier and Paris. The Cheirostemon is also remarkable because of the shape of its corolla and its fruit, the Mexican *Gyrocarpus* which we introduced into the gardens of Europe and whose flower the illustrious ▼ Jacquin was unable to find.

individuals brought to the capital as prisoners of war. The Mecos (of the Chichimec tribe), Apaches, and Lipan are hordes of hunting peoples who, in the course of their often-nocturnal forays, overrun the borders of Nueva Vizcaya, Sonora, and New Mexico. Like the savages of South America, they have a livelier spirit and a stronger character than the farming Indians. Some tribes even speak languages whose grammar is proof of an ancient civilization. They have great difficulty in learning our European languages, although they express themselves with great ease in their own. These same Indian chiefs, whose grim reticence strikes the observer, will hold forth for several hours when keen interest arouses them to break their habitual silence. We have observed the same volubility of speech in the missions of Spanish Guiana and among the Caribs of the Lower Orinoco, whose language is singularly rich and sonorous.[1]

I.384

After examining the physical constitution and intellectual faculties of the Indians, we must quickly survey their social status. The history of the lowest classes of a people is the account of the dire events that slowly placed one part of the nation under the control and domination of the other, creating at once great inequality of fortune, pleasure, and individual happiness. We search almost in vain for this account in the annals of history: they preserve the memory of great political revolutions, wars, conquests, and other scourges that have crippled humanity, but they teach us little about the more or less wretched lot of the poorest and largest class of society. Only in a very small part of Europe does the peasant freely enjoy the fruits of his labor; and we are forced to admit that this civil liberty is not so much the result of a progressive civilization as the effect of violent crises during which one class or state has benefited from dissent among the others. The true perfection of social institutions likely depends on enlightenment and intellectual development, but the chain of springs that moves a state is such that in one part of the nation, this development may make remarkable progress without any improvement of the situation of the lower classes. Nearly all of northern Europe confirms this sad experience: there, we find countries where, despite the lauded civilization of the upper classes of society, the peasant still lives in almost the same degradation under which he groaned three or four centuries ago. We might find the Indians' lot a happier one if we compared it to that of the peasants in Courland, Russia, and a large part of northern Germany.

I.385

1. I have described the Carib tribes in my *Relation historique* (quarto edition), vol. I, pp. 461 and 496; vol. II, p. 97, 260, 395; and especially vol. III, pp. 16 and 161.

I.386 The indigenous peoples whom we see today spread throughout the towns and especially throughout the Mexican countryside, and whose number (excluding those of mixed blood) has reached three and a half million, are either the descendants of earlier farmers or the remnants of a few great Indian families. Disdaining an alliance with the Spanish conquerors, these families preferred to cultivate with their own hands the fields where formerly their own vassals had labored for them. This difference has a noticeable effect on the political status of the indigenous peoples and divides them into either tributaries or nobles (Caciques). According to Spanish law, the latter should enjoy the privileges of the Castilian nobility, but in their present situation, this advantage is merely an illusion. It is difficult to distinguish the Caciques externally from the other indigenous peoples who, already in the time of ▼ Montezuma II, constituted the lowest caste of the Mexican nation. The noble, by the simplicity of his clothing and food and the miserable countenance he prefers to display, is easily confused with the tributary Indian. The latter shows the former a respect that is indicative of the social distance prescribed by the ancient constitutions of the Aztec hierarchy. Far from protecting the caste of indigenous tributaries, the families who enjoy the hereditary rights of the *Cacicasgos* abuse their influence more often than not. Since they occupy the magistracy in the Indian villages, they levy the capitation tax. Not only do they take pleasure in becoming instruments of humiliation for the whites, but they also use their power

I.387 and authority to extort small sums for their own gain. Wise intendants, who have long studied the inner workings of the Indian regime, confirm that the caciques bear down heavily on the indigenous tributaries. Similarly, in some parts of Europe, where Jews are still deprived of the rights of citizens, rabbis bear down on the members of their own community. The Aztec nobility displays the same crude manners, the same lack of civilization as the lowest class of Indians; the nobles dwell, as it were, in the same isolation, and examples of indigenous Mexicans who, benefiting from the *Cacicasgo*, follow a career in the law or the military are very rare. More Indians embrace the ecclesiastical way of life, especially as parish priests: the solitude of the convents seems only to attract young Indian girls.

When the Spanish conquered Mexico, they found the people already in the state of submission and poverty that accompanies despotism and feudalism everywhere. The emperor, princes, nobility, and the clergy (the *Teopixqui*) were the ones who owned the most fertile land; the provincial governors made the most egregious demands with impunity; the farmer was downtrodden. As we have seen earlier, the big thoroughfares were crawling with beggars, and the scarcity of large domestic quadrupeds forced thousands of Indians to serve as pack animals and to transport corn, cotton, hides, and other commodities sent as tribute from the farthest provinces to the capital. The conquest made the situation of the lowest class even more deplorable; farmers were uprooted from the land and dragged to the mountains, where they began working the mines. Many Indians were forced to follow the armies and, without food and rest, haul burdens that exceeded their strength over mountainous roads. All Indian property, both movable property and land, was regarded as belonging to the conqueror. A law assigning to the indigenous peoples a small plot of land surrounding the newly built churches sanctioned this outrage. I.388

Seeing that the new continent was rapidly becoming depopulated, the Spanish court took measures that were beneficent in appearance, but which the conquerors (*Conquistadores*), with their greed and cunning, were able to turn against those whose misfortunes the court hoped to alleviate. The system of *Encomiendas* was introduced. The indigenous peoples, whose freedom Queen Isabella had proclaimed in vain, had until then been the slaves of the whites who seized them indiscriminately. With the institution of the *Encomiendas*, slavery became more structured. To put an end to the quarrels among the Conquistadores, what remained of the conquered people was parceled out: the Indians were divided into tribes of several hundred families and given to masters who were appointed in Spain. These were soldiers who had distinguished themselves in the conquest and men of the law[1] whom the court sent to govern the provinces and serve as a counterweight to the usurpatory power of the generals. Many of the largest *Encomiendas* were given to monks. Religion, which by its principles should have favored freedom, was degraded when it profited from the people's servitude. This distribution of the Indians attached them to the glebe, and their work thus belonged to the *Encomenderos*. The serf often took the master's name. Many Indian families still have Spanish names, even though their blood has never mingled with European blood. The court in Madrid thought it was providing protectors for the Indians, but it only made the evil worse and systematized their oppression. I.389

Such was the condition of Mexican farmers in the sixteenth and seventeenth centuries. Since the eighteenth century, their lot has gradually improved. The families of the *Conquistadores* have partially died out. The

---

1. These powerful men were often simply called *Licenciados*, according to the degree they had taken in their faculties.

Encomiendas, considered fiefs, have not been redistributed. The viceroys, especially the *Audiencias*, looked out for the Indians' interests; their freedom, and in some provinces even their comfort, has increased bit by bit. Above all, it was ▼ King Charles III [of Spain] who, taking sensible and resolute measures, became the indigenous peoples' benefactor: he abolished the *Encomiendas* and prohibited the *Repartimientos* by which the *Corregidors* arbitrarily made themselves the creditors of the indigenous peoples, and thus the masters of their labor, by selling them horses, mules, and clothing (*ropa*) at exaggerated prices. The establishment of the intendancies during the administration of the ▼ Count of Gálvez was an especially momentous period for the Indians' well-being. Under the close scrutiny of the intendants, the petty humiliations to which low-ranking Spanish and Indian magistrates constantly exposed the farmers radically decreased; the Indians began to enjoy the advantages that the generally mild and humane laws granted them but of which they had been deprived during centuries of barbarity and oppression. The first selection of those upon whom the court conferred the important positions of intendant, or provincial governor, was most felicitous. Not a single one of the twelve intendants who administered the country in 1804 was accused by the public of corruption or lack of integrity.

I.390

Mexico is the land of inequality. Nowhere is there a more alarming imbalance in the distribution of wealth, civilization, agriculture, and population. The interior of the country has four towns that are only one or two days' journey apart from each other, with 35,000; 67,000; 70,000; and 135,000 inhabitants. From Puebla to Mexico City and from thence to Salamanca and Zelaya, the central plateau is dotted with villages and hamlets, like the most cultivated parts of Lombardy. East and west of this strip of land, there are extensive uncultivated tracts with fewer than ten or twelve people per square league. The capital and several other cities have scientific establishments that are comparable to those of Europe. The architecture of both the public and the private buildings, the elegance of furnishings and retinues, the luxury of women's dress, and the tone of society all announce a refinement that contrasts with the nakedness, ignorance, and vulgarity of the lowest class. This immense imbalance of wealth is not found only among the white caste (Europeans or Creoles); one also finds it among the indigenous peoples.

I.391

Considered *en masse*, the Mexican Indians offer a picture of great misery. Banished to the most barren lands, indolent by nature and even more so as the result of their political situation, the indigenous peoples only live from day to day. One would look among them almost in vain for examples of individuals who enjoy even a meager fortune. Instead of a happy prosperity, we find a few families whose fortune seems even more colossal for being less expected among the lowest class of the people. In the intendancies of Oaxaca and Valladolid, the Toluca valley, and especially in the great city of La Puebla de los Ángeles, there are a few Indians whose wretched appearance conceals great wealth. When I visited the small town of Cholula, an old Indian woman was being buried who had left to her children *maguey* (agave) plantations worth over 360,000 francs. These plantations are the vineyards and the entire wealth of the country. There are, however, no caciques in Cholula; the Indians there are all tributaries characterized by great sobriety, as well as mild and peaceful manners. These attitudes contrast dramatically with those of their neighbors in Tlaxcala, where many insist that they are the descendants of the most titled nobility; the Tlaxcala exacerbate their misery through their fondness for trials and a restless, quarrelsome spirit. The richest Indian families are the Axcotla, the Sarmientos, and the Romeros in Cholula; the Sochipiltecatl in Guajocingo; and, above all, the Tecuanouegues in the Village of Los Reyes. Each of these families has assets in the amount of 800,000 to 1,000,000 *livres tournois*. As we have seen above, they are highly regarded by the tributary Indians, although they usually go barefoot, covered only with a Mexican tunic of coarse, blackish-brown cloth; they dress like the lowest of the indigenous race.

The Indians are exempt from any indirect taxes and do not pay the *alcabala* [excise tax]: the law grants them complete freedom to sell their products. The supreme council of finances of Mexico City, called the *Junta superior de Real Hacienda*, has occasionally (especially in the past five to six years) attempted to force the indigenous peoples to pay the *alcabala*. One must hope that the court of Madrid, which has always protected this unfortunate class, will preserve their immunity as long as they continue to be subjected to the direct tax of the tributes (*tributos*). This tax is a true capitation tax that all Indian males, from the age of ten to fifty, must pay. The tribute is not the same in all provinces of New Spain: it has been decreased over the past two hundred years. In 1601, the Indian paid thirty-two silver reales as tribute and four reales as *servicio real*, a total of about twenty-three francs. This sum was gradually reduced in some intendancies to fifteen and even five francs.[1] In the bishopric of Michoacan, as in most

I.392

I.393

---

1. ▼ Don Joacquín Maniau, *Compendio de la Historia de la Real Hacienda de Nueva España*, an unpublished work presented in 1793 to the secretary of state minister Don Diego de Gardoqui, a copy of which is in the archives of the viceroyalty.

of Mexico, the capitation tax today is eleven francs. Furthermore, the Indians pay as rights of parish (*derechos parroquiales*) ten francs for a baptism, twenty francs for a marriage certificate, and thirty-two francs for a burial. To this must be added sixty-two francs that the church exacts as a tax on each Indian, twenty-five to thirty francs for offerings that are called voluntary and are designated by the names *Cargos de cofradias, Responsos,* and *Misas para sacar ánimas.*[1]

I.394

While the laws of Queen Isabella and Emperor Charles V seem to favor the indigenous peoples in terms of taxes, they also deprive them of the most important rights enjoyed by other citizens. In a century when it was formally debated if the Indians were rational beings, it was supposed to be for their benefit to treat them like minors, putting them under the perpetual sponsorship of whites and declaring null and void any act signed by a copper-colored indigenous person and any debt that this native contracted for a value over twenty-five francs. These laws are resolutely enforced and raise insurmountable barriers between Indians and the other castes, with whom they are also forbidden to mix. Thousands of inhabitants cannot enter into a binding contract (*no pueden tratar y contractar*); condemned to the status of minors forever, they become a charge to themselves and the state where they live. I can think of no better way to round out the political table of the Indians of New Spain than to lay before the reader an excerpt from a report filled with the wisest opinions and most liberal ideas, which the bishop and head of Michoacan[2] presented to the king in 1799.

I.395

The highly regarded bishop[3]—whom I had the advantage of knowing personally and whose helpful and laborious life ended at the age of eighty— pointed out to the monarch that, under present circumstances, it is impossible to perfect the Indian morally unless the obstacles to the progress of

---

1. Confraternity fees, responses, and masses that the Indians are told will redeem souls from purgatory.

2. *Informe del Obispo y Cabildo eclesiástico de Valladolid de Mechoacán al Rey sobre Jurisdicción del Clero Americano.* The report, of which I have a manuscript that is over ten sheets long, was drawn up on the occasion of the famous royal warrant of October 25, 1795, which allowed secular judges to try the *delitos enormes* [great crimes] of the clergy. Sure of its rights, the *Sala del crimen* in Mexico City inveighed against the priests and threw them into the same prisons with the lowest classes of the people. In this struggle, the Audiencia was on the clergy's side at the hearing. Juridical disputes are common in faraway countries. They are pursued even more fiercely because, ever since the first discovery of the New World, European policies have considered the disintegration of castes, families, and constitutive authorities as a way of maintaining the colonies' dependence on the metropole.

3. ▼ *Fray Antonio de San Miguel,* monk of Saint Jerome of Corvan and native of the *Montañas de Santander.*

national industry are removed. He uses several passages from Montesquieu and ▼ Bernardin de Saint-Pierre to prove the principles he asserts. These citations must come as a surprise from the pen of a prelate who is a member of the regular clergy, who has spent part of his life in monasteries, and who holds an Episcopal seat on the shores of the South Sea. "The population of New Spain," declared the bishop near the end of his narrative, "is composed of three classes —white or Spanish, Indians, and *Castes*. I estimate that the Spanish represent one-tenth of the whole population. Almost all the property and wealth in the kingdom is in their hands. The Indians and *Castes* work the soil and are in the service of the wealthy; they live only by manual labor. As a result, there is a conflict of interest between the Indians and the whites, a mutual hatred that easily arises between those who have everything and those who have nothing, between masters and those who live in servitude. On the one hand, we see the effect of envy and discord, deceit, theft, the inclination to damage the interests of the rich; and on the other hand, we have arrogance, severity, and the desire to take advantage of the Indian's vulnerability at every turn. I am aware that these evils arise wherever there is a great inequality of social conditions. This is even more alarming in the Americas, where there is no middle ground. One is either rich or poor, noble or degraded by laws and the strength of opinion (*infame de derecho y hecho*)."

I.396

"In fact, the Indians and races of mixed blood (*Castas*) live in a state of profound humiliation. The indigenous peoples' own color, their ignorance, and especially their wretchedness put them at an almost infinite distance from the whites, who occupy the first rank of the population of New Spain. The privileges that laws appear to grant to the Indians are of little advantage to them; one ought to admit that they are rather detrimental. Restricted to a narrow space in a radius of 600 varas (500 meters), which an ancient law assigns to Indian villages, the indigenous peoples have no private property to speak of; they are required to farm communal property (*bienes de comunidad*). This farming has become an even more unbearable burden, since in the past few years, the indigenous peoples have had almost no hope of profiting from the fruits of their labor. The new regulation for the intendancies states that indigenous peoples may not receive assistance from the community coffers without the specific permission of the Mexican College of Finances (*Junta superior de la Real Hacienda*)." (The intendants have farmed out communal property; revenue from the indigenous peoples' work is deposited in the royal treasury, where the *Officiales reales* keep an account under a special heading of what they called the property of each

I.397

village. I say "what they call" because for over twenty years, this property has been virtually fictitious. The intendant himself cannot even dispose of it to the benefit of the indigenous peoples, who tire of asking for help from their community coffers. The *Junta de Real Hacienda* requests *informes* [reports] from the tax office and the viceroy's *Assessor*. Years pass before all information is gathered, but the Indians receive no answer. Thus, people have become so used to thinking of the funds in the *Cajas de Comunidades*

I.398   as a sum that has no set purpose, that in 1798 the intendant of Valladolid sent to Madrid nearly one million francs that had accumulated over a twelve-year period. It was presented to the king as an unbinding and patriotic gift from the Indians of Michoacan to the sovereign to assist him in furthering the war with England!)

"The law forbids the mixing of castes; it prohibits whites from living in Indian villages and prevents the indigenous peoples from settling among the Spanish. This state of isolation presents obstacles to civilization. The Indians govern themselves, and all lower magistrates are of the copper-skinned race. In every village, there are eight or ten Indians who live in complete sloth at the expense of others; their authority is based either on a so-called birthright or on a political shrewdness handed down from father to son. It behooves the chiefs, who are usually the only village-dwellers who speak Spanish, to keep their fellow citizens in the deepest ignorance; they contribute the most to perpetuating the prejudices, ignorance, and the former barbarity of manners.

"Since, by the laws of the Indies, the indigenous peoples are unable to sign a contract before a notary or to incur debts of more than five piasters, they cannot improve their lot and enjoy a comfortable existence as workers

I.399   or artisans. ▼ Solorzano, Fraso, and other Spanish authors have searched in vain for the hidden reason why the privileges granted to the Indians always have such detrimental effects to their caste. I am astonished that these celebrated jurists do not realize that what they call a hidden reason is based on the nature of the privileges themselves. They are weapons that have never served to protect those whom they are intended to defend, and which the citizens of the other castes skillfully use against the indigenous race. The combination of such deplorable circumstances has generated in them a laziness of mind and a state of apathy and indifference in which a person is affected neither by hope nor by fear.

"The *Castes*, who are the descendants of Black slaves [Nègres esclaves], are branded with infamy by the law and subject to the payment of the *tribute*. This direct tax brands them with an indelible stain; they consider it as a mark of slavery that is transmitted to all future generations. Among the mixed-blood race [race de sang-mêlé], the mestizos [métis] and mulattos [mulâtres] there are many families who, by their coloring, physiognomy, and culture, could be mistaken for Spaniards; but the law keeps them in degradation and contempt. Endowed with an energetic, ardent character, these men of color [hommes de couleur] live in a constant state of irritation against whites: it is even surprising that their resentment does not lead them to take revenge more often.

"The Indians and *Castes* are in the hands of the district magistrates (*Justicias territoriales*), whose immorality has greatly contributed to their misery. As long as the *Alcaldías mayores* have existed in Mexico, the alcades have considered themselves merchants who have had the exclusive privilege of buying and selling in their province; taking advantage of this privilege, they earn between 30,000 and 200,000 piasters (150,000 to 1,000,000 francs) and this, moreover, in the short span of five years. These usurious magistrates force the Indians to purchase a certain number of animals from them at arbitrary prices. In so doing, the indigenous peoples become indebted to them. On the pretext of recovering capital and interest, the *Alcalde mayor* uses the Indians as de facto serfs during the entire year. Individual happiness has certainly not increased among these unfortunates, who have sacrificed their freedom to have a horse or a mule with which to work for their master's profit. But in the midst of this state of things brought on by abuse, agriculture and industry have progressed. I.400

"When the intendancies were created, the government wished to put an end to the vexations that resulted from the *repartimientos*. Instead of *Alcaldes mayores*, *subdelegados* were appointed as lower-rank magistrates who were strictly prohibited from every sort of trade. Because they were not given salaries or any sort of fixed remuneration, the abuses have almost I.401 worsened. The *Alcaldes mayores* administered justice with impartiality only whenever their own interests were not at stake. Since the subdelegates to the intendants had no other income except from contingencies, they felt entitled to use illicit methods to procure more prosperity for themselves: hence the continuous oppressions, the abuse of authority regarding the poor, and the indulgence toward the rich in a shameful trafficking of justice. The intendants encountered great difficulties in choosing the *subdelegados*, from whom the Indians, in the current state of things, can rarely expect protection and support. They seek them instead from the clergy. As a result, the clergy and the subdelegates live in constant conflict. But the indigenous peoples have more confidence in the clergy and higher-rank magistrates, the intendants and *Oídores* (members of the *Audiencia*). Now, Sire, what allegiance to the government can the scorned, vilified Indian have, since he has almost no property or hope of improving his lot? He is attached to the social body by a link that offers him no benefits. Let no one tell Your Majesty that the fear alone of punishment is sufficient to preserve the peace in these countries; there must be other, more powerful incentives. If the new legislation that Spain impatiently awaits does not concern itself with the fate of the Indians and people of color, the influence of the clergy over the hearts

I.402   of these unfortunate people, however great it may be, will not suffice to contain them in the submission and respect that they owe to their sovereign.

"Let the odious personal tax of the *tribute* be abolished, and the stigmatization (*infamía de derecho*) with which unfair laws have marked people of color cease; let people of color be declared capable of holding any civil position that does not require a special title of nobility; let the indigenous peoples' communal and collective possessions be shared; let a portion of sovereign lands (*tierras realengas*), which are usually uncultivated, be given to the Indians and *Castes*; let an agrarian law similar to the laws of Asturias and Galicia be granted to Mexico, a law according to which the poor farmer is allowed to work the land that large proprietors have left fallow for centuries, to the detriment of national industry; let full liberty be granted to Indians, *Castes*, and whites to settle in the villages that now belong to one of these classes; let regular salaries be assigned to all district judges and magistrates: these, Sire, are the six principal points upon which depends the happiness of the Mexican people.

"It will probably come as a surprise to hear that, at a time when state finances are in a deplorable situation, someone presumes to propose the

I.403   abolition of the tribute to Your Majesty. A simple calculation, however, will prove that, by taking the suggested measures and granting each Indian all rights of citizenship, state revenue (*Real Hacienda*), far from decreasing, will increase considerably." The bishop supposes that in all of New Spain, there are 810,000 Indian and persons of color. Many of these families, especially among those of mixed blood, are clothed and enjoy some comfort; they live more or less like the lower class on the Iberian Peninsula; they amount to one-third of the whole group. The annual consumption of this third amounts to 300 piasters per family. Counting only sixty piasters[1] for

---

1. One figures that in the hot regions of Mexico, a day laborer requires, for himself and for his family, seventy-two piasters per year for food and clothing. Around twenty piasters less is considered luxury in the country's cold regions.

the other two-thirds, and supposing that the Indians pay the *alcabala* of 14 percent, as do the whites, this represents an annual revenue of five million piasters, which is more than four times the present value of the *tributes*. We cannot guarantee the exactness of the number on which this calculation is based. But a simple glance is enough to prove that by establishing equal rights and taxes for the different classes of people, not only would the abolition of the capitation tax not create any deficit in the revenue of the crown, but this revenue would perforce increase with the increase in material comfort and well-being among the indigenous peoples."

I.404

One had hoped that the administration of three enlightened viceroys—the Marquis of Croix, the Count of Revillagigedo, and the Chevalier of Asanza—animated by the noblest zeal for the common good, would have brought about the beneficial changes to the political situation of the Indians; but these hopes were dashed. The viceroys' powers have been singularly diminished of late; their every move is blocked, not only by the financial Junta (*de Real Hacienda*) and the high court of justice (*Audiencia*) but especially by the metropolitan government's obsession with governing in minute detail provinces that are two thousand leagues away, and whose moral and physical circumstances are unknown in Spain. Philanthropists claim that it is for the Indians' own good that they are being neglected in Europe, because sad experience has proven that most measures adopted to improve their lives have produced the opposite effect. Judges and jurists, who detest innovations, and Creole proprietors, who often take advantage of keeping the farmer in misery and degradation, insist that the indigenous peoples must not be disturbed, because if they are granted more freedom, the whites would have no choice but to face the Indians' arrogance and vindictive spirit. The language is the same wherever it is suggested that the peasant be allowed to enjoy the rights of a free man and a citizen. In Mexico, Peru, and in the kingdom of New Granada, I have heard the same arguments repeated that are used in parts of Germany, Poland, Livonia, and Russia against the abolition of slavery among the peasants.

I.405

Recent examples teach us how dangerous it is to let the Indians form a *status in statu*, of perpetuating their isolation, the barbarity of their customs, their wretchedness, and consequently the reasons for their hatred of the other castes. The same witless and indolent Indians who allow themselves patiently to be flayed at the door of a church appear clever, energetic, impetuous, and cruel whenever they act *en masse* during popular disturbances. It will be useful to give proof of this assertion. In the great revolt of 1781, the king of Spain lost nearly all the mountainous parts of Peru, at the same time that Great Britain was losing nearly all its colonies on the American continent. ▼ José Gabriel Condorcanqui, known by the Inca name Tupac-Amaru, led an Indian army up to the walls of Cuzco. He was the son of the cacique of Tongasuca, a village in the province of Tinta; or, rather, he was the son of the cacique's wife, because it seems certain that the supposed Inca was a Mestizo and his real father was a monk. The Condorcanqui family traces its origin to the Inca Sayri-Tupac, who disappeared in the thick forests east of Willkapampa and to the Inca Tupac-Amaru who, contrary to the orders of ▼ Phillip II, was beheaded in 1578, under the viceroy Don Francisco de Toledo.

I.406

José Gabriel received a thorough education in Lima; he returned to the mountains after having in vain attempted to claim from the Spanish court the title of Marquis de Oropesa, which belongs to the family of the Inca Sayri-Tupac. His implacable spirit led him to incite the mountain Indians who were angry with the *corregidor* ▼ Arriaga. The people recognized him as the descendant of their true sovereigns, whose father was the Sun. The young man took advantage of the popular enthusiasm he had excited by displaying the symbols of the former grandeur of the empire of Cuzco; he often wore the imperial ribbon of the Incas as a headband and skillfully mixed Christian ideas with memories of the worship of the Sun.

At the start of his campaigns, he protected both the clergy and Americans of all colors. Only inveighing against the Europeans, he acquired a following, even among the Mestizos and Creoles, but the Indians mistrusted the sincerity of their new allies and soon began a war of extermination against anyone who was not of their race. José Gabriel Tupac-Amaru, of whom I have in my possession the letters in which he proclaims himself the Inca of Peru, was not as cruel as his brother Diego and especially his nephew, Andrés Condorcanqui, who, at the age of seventeen, displayed great talents but a bloodthirsty disposition. ▼ The uprising, which seems little known in Europe, and about which I shall give more detailed information in the personal narrative of my journey, lasted for nearly two years. Tupac-Amaru had already conquered the provinces of Quispicanchi, Tinta, Lampa, Azangara, Caravaja, and Chumbivilcas, when the Spanish took him and his family prisoners. All were drawn and quartered in the town of Cuzco.

I.407

The respect that the so-called Inca inspired in the indigenous peoples was so great that despite their fear of the Spanish, and although they were surrounded by the soldiers of the victorious army, they prostrated themselves at the sight of the last child of the Sun when he passed through the streets on the way to his execution. José Gabriel Condorcanqui's brother, known as Diego Cristóbal Tupac-Amaru, was not executed until long after the end of the Peruvian Indians' revolutionary movement. When the leader fell into the hands of the Spanish, Diego surrendered, taking advantage of the pardon that he was promised in the king's name. A formal agreement was signed between him and the Spanish general on January 26, 1782, in the Indian village of Siquani in the province of Tinta. He lived peacefully with his family until, as the result of insidious, mistrustful policies, he was arrested on the pretext of a new conspiracy.

Some of the horrors that the Peruvian indigenous peoples inflicted on the whites in 1781 and 1782 in the Cordillera of the Andes were repeated during the minor uprisings that took place twenty years later on the Rio-   I.408
bamba plateau. It is of the greatest importance, even for the tranquility of the European families who have lived on the continent of the New World for centuries, that an interest be taken in the Indians, and that they be rescued from their present barbarous, abject, and miserable condition.

---

*A CHRONOLOGICAL TABLEAU OF THE HISTORY OF MEXICO
(Excerpted from Mr. von Humboldt's work on the *Monuments of the Indigenous Peoples of America* [sic], vol. II, pp. 118, 136, and 385; in 1811 ed: pp. 202–3; 210–11, 318–20).

The most astonishing among all the similarities that one can observe in the monuments, the customs, and the traditions of the peoples of Asia and of the Americas, is what Mexican mythology offers with its cosmogonic myth of the periodic destructions and regenerations of the Universe. This mythic fiction, which links the return of the great cycles to the notion of the renewal of matter presumed to be indestructible, and which transfers to space characteristics that seem to belong only to time,[1] dates back to the earliest time of antiquity. The sacred books of the Hindus, especially the *Bhagavata Purana*, already speak of the four ages and the *pralayas*, or cataclysms, which, in various epochs, have led to the destruction of the human species.[2] A tradition of *five ages*, similar to that of the Mexicans, can

---

1. Hermann, *Mythology der Griechen*, vol. II, p. 332.
2. ▼ Hamilton and Langlès, *Catalogue des Manuscrits sanskrits de la Bibl[iotèque] impér[iale]*, p. 13. [William Jones, *Researches Asiatick*,] vol. II, p. 17 [171]. Moor, *Hindu Pantheon*, pp. 27 and 101.

also be found on the high plateau of Tibet.[1] If it is true that this astrological fiction,
I.409 which became the foundation of a particular cosmogonic system, was conceived
in Hindustan, then it is also likely that it spread from there, through Iran and
Chaldea, to the peoples in the west. One cannot disregard a certain resemblance
between the Indian tradition of the *yuga* and the *kalpa*, the cycles of the ancient
inhabitants of Etruria, and the series of annihilated generations that ▼ Hesiod
characterized through the emblem of the four metals.

"The peoples of Culhua or Mexico," said ▼ Gómara,[2] who wrote in the mid-
sixteenth century, "believe, following their hieroglyphic paintings, that before the
sun that now shines upon them, there had already been four suns that had become
extinguished one after the other. These five suns constitute the ages during which
humankind has been wiped out by floods, by earthquakes, by an all-consuming
blaze, and by the effect of fierce storms. After the destruction of the fourth sun,
the world was plunged into shadows for a period of twenty-five years. It was in the
midst of this deep night, ten years before the fifth sun reappeared, that humanity
was regenerated. At that time, the gods, for the fifth time, created one man and
one woman. The day on which the last sun appeared bore the sign *tochtli* (rab-
bit), and the Mexicans count eight hundred and fifty years from that time to 1552.
Their annals date back all the way to the fifth sun. They used historical paintings (*escritura pintada*) even in the four preceding ages; but these paintings, they
claim, were destroyed, because with each new age everything must be renewed."
According to Torquemada,[3] this fable about the revolution of time and the regen-
eration of nature is of Toltec origin; it is a national tradition that belongs to the
group of peoples we know as the Toltecs, Chimicheca, Acolhua, Nahua, Tlaxcalteca, and Aztecs and who, speaking the same language, had migrated from the
north to the south since the middle of the sixth century of our era.

While examining in Rome the *Cod[ex] Vaticanus*, number 3,738, copied in
I.410 1566 by the ▼ Dominican friar Pedro de los Ríos, I discovered that he gives the
duration of each sun in Aztec hieroglyphic numerals. The four periods lasted:

5,206	years	=	13	×	400	+	6
4,804	years	=	12	×	400	+	4
4,010	years	=	10	×	400	+	10
4,008	years	=	10	×	400	+	8
18,028	years						

1. Georgi, *Alphab[etum] Tibetanum*, p. 220.
2. Gómara, *Conquista [de México]*, folio CXIX.
3. Torquemada, [*Monarchía yndiana*,] vol. I, p. 40; vol. II, p. 83.

In the Mexica system, the four great revolutions of nature are caused by the four elements: the first catastrophe is the devastation of the productive force of the earth, while the other three are due to the effect of fire, air, and water. The human species regenerates after each destruction, and anyone from the former race who did not perish is transformed into birds, monkeys, or fish. These transformations also recall the traditions of the East, yet in the Hindus' system, the ages, or *yugas*, all end in floods, while in that of the Egyptians,[1] the cataclysms alternate with conflagrations, and men find refuge either on the mountains or, at other times, in the valleys. We would digress from our topic were we to explain here the small revolutions that occurred on several occasions in the mountainous part of Greece[2] and discuss the famous passage from the second book of Herodotus, which has so challenged the sagacity of commentators. It seems quite clear that, in this passage, it is not a matter of *apocatastases* but, rather, of four (visible) changes that occurred at the places of sunset and sunrise[3] and were brought about by the precession of the equinoxes.[4]

Since one might be surprised to find five ages, or *suns*, among the peoples of Mexico, while the Hindus and the Greeks recognize only four, it is useful to observe here that the Mexicans' cosmogony aligns closely with that of the Tibetans, which also regards the present age as the fifth one. By examining closely the lovely section from Hesiod[5] in which he explains the Eastern system of the renewal of nature, one sees that this poet actually counts five generations within four epochs. He divides the Bronze Age into two parts that cover the fourth and fifth generations,[6] and one may be surprised that such a clear passage has been occasionally misinterpreted.[7] We do not know the number of ages mentioned in the books of the Sibyl,[8] but we believe that the similarities we just pointed out are not accidental, and that it is not unimportant for the philosophical history of humanity to see that the same fictional stories are widespread, from Etruria to Latium to Tibet and from there all the way to the ridge of the Cordilleras of Mexico.

I.411

1. [Plato,] *Timaeus*, chap. 5 (*Oper[a] omni]*, 1578, ed. Serran[us], vol. III, p. 22). *De legib[us]*, book III (*Op[era] omn[i]*, bol. II, pp. 678–79). ▼ Origenes, *Contra Celsum*, book I, chap. 20; book IV, chap. 20 (ed. Delarue, 1733, pp. 332 and 514).

2. Arist[otle], *Meteor[ologia]*, Book I, chap. 14 (*Op[era] omn[i]*, ed. Duval, 1639, p. 770).

3. Herodotus, [*Histoire d'Hérodote*,] Book II, chap. 142 (ed. Larcher, 1802, vol. II, p. 482).

4. ▼ Dupuis, *Mémoire explicatif du zodiaque*, pp. 37 and 59.

5. Hesiod, *Oper[um] et die[rum]*, verse 174 (*Op[era] omn[ia]*, ed. Cleric[us], 1701, p. 224.

6. Hesiod, [ibid.,] verses 143 and 155.

7. ▼ Fabrici[us], *Bibl[iotheca] graeca, Hamb[urg]*, 1790, vol. I, p. 246.

8. Virgil, *Bucol[ica]*, IV, verse 4 (ed. Heyne, Lond[on] 1793, vol. I, pp. 74 and 81).

Like the Caucasus, the mountainous region of Mexico had been inhabited by a great many peoples of different races since the remotest times. Some of these peoples can perhaps be considered the remainders of the numerous tribes that passed through the land of Anahuac during their migrations from north to south. There, some families separated from the rest of the group, held in Anahuac by the love of the land, which they cultivated, and they maintained their language, their customs, and the original form of their government.

The most ancient peoples of Mexico, those who regarded themselves as autochthonous, are the Olmecs, or Hulmecs, who went as far in their wanderings as the Gulf of Nicoya and León in Nicaragua; the Xicalanca, the Cora, the Tepaneca, the Tarasca, the Mixteca, the Zapoteca, and the Otomi. The Olmecs and the Xicalanca, who lived on the Tlaxcala plateau, prided themselves for having conquered or destroyed, on their arrival, the giants, or *quinametin*; this belief was probably founded on the findings of fossilized Elephant bones in the high regions of the mountains of Anahuac. (Tor[quemada], [*Monarchia yndiana*,] vol. I, pp. 37 and 364.) Boturini states that the Olmecs were chased away by the Tlaxcalteca and proceeded to people the Antilles and South America.

I.412

The Toltecs, who left their motherland Huehuetlapallan, or Tlapallan, in the year 544 of our era, arrived in Tollantzinco in the land of Anahuac in 648 and in Tula in 670. During the reign of the Toltec king Ixtlicuechahuac, the astrologer Huematzin composed, in 708, his famous divine book, the Teoamoxtli, which contained the history, mythology, calendar, and laws of the nation. The Toltecs are also the ones who appear to have built the pyramid of Cholula, after the model of the pyramids of Teotihuacán. The latter pyramids are the oldest of them all, and Sigüenza believes them to be the work of the Olmecs. (Clav[ijero, *Storia antica del Messico*], vol. I, pp. 126 and 129; vol. IV, p. 46.)

The Mexican Buddha appears during the time of the Toltec monarchy, or even during earlier centuries: Quetzalcohuatl, a white, bearded man accompanied by other strangers who were wearing black garb that looked like cassocks. The people used these Quetzalcohuatl clothes until the sixteenth century to dress up in disguise during festivals.

In the Yucatán, this holy man was called Cuculca; in Tlaxcala, he was known as Camaxtli (Torq[uemada, *Monarchia yndiana*,] vol. II, pp. 55 and 307). His robe was covered with red crosses. As High Priest of Tula, he founded religious congregations. "He ordered the sacrifice of flowers and fruits and covered his ears when there was talk of war." His fellow traveler, Huemac, was in possession of worldly power, while he himself enjoyed spiritual powers. This form of government was similar to the ones in Japan and Cundinamarca (Torq[uemada, *Monarchia yndiana*,] vol. II, p. 237). But the first monks, Spanish missionaries, seriously debated the question of whether Quetzalcohuatl was from Carthage or from Ireland. From Cholula, he sent colonies to Mixteca, Huaxayacac, Tabasco, and Campeche. People assume that the palace of Mitla had been built on the orders of this unknown person. At the time when the Spanish arrived, some green stones were kept in Cholula as if they were precious relics; they were said to have belonged to Quetzalcohuatl. And Father Toribio de Motolinea even witnessed a sacrifice in honor of the saint on the summit of the mountain of Matlalcuye, near Tlaxcala. In Cholula, Father Toribio also participated in penitential practices ordered by Quetzalcohuatl, during which the penitents had to pierce their tongues, ears, and lips. The High Priest of Tula made his first appearance I.413 in Panuco; he left Mexico, intending to return to Tlalpallan; on this voyage, he disappeared but not, as one must have assumed, in the north but in the east, on the banks of the Río Huasacualco. (Torq[uemada, *Monarchia yndiana,*] vol. II, pp. 307–11.) The people hoped for his return for many centuries. "When, upon my way to Tenochtitlan, I passed through Xochimilco," recounts the monk ▼ Bernardino de Sahagún, "all the world wanted to know if I was coming from Tlalpallan. At the time, I did not understand the meaning of this question, but I found out later that the Indians believed us to be descendants of Quetzalcohuatl" (Torq[uemada, *Monarchia yndiana,*] vol. II, p. 53.) It would be fascinating to trace the life of this mysterious figure, which belonged to heroic times probably before the Toltecs, down to its most minute details.

Pestilence and destruction of the Toltecs in the year 1051. They move farther south during their migrations. Two children of the last king and some Toltec families remain in Anahuac.

After they left their homeland Amaquemecan, the Chichimecs arrived in Mexico in 1170. Migration of the Nahuatlaca (Anahuatlaca) in 1178. This nation included the seven tribes of the Xochimilca, the Chalca, the Tepaneca, the Acolhua, the Tlahuica, the Tlaxcalteca, or Teochichimecs, and the Aztecs, or Mexicans, all of whom, like the Chichimecs, used the Toltec language. (Clav[ijero, *Storia antica del Messico*], vol. I, p. 151; vol. IV, p. 48.) These tribes called their homeland Aztlan or Teo-Acolhuacan, and they said that it bordered on Amaquemecan. (García, *Origen de los Indios*, pp. 182 and 502.) According to Gama, the Aztecs had left Aztlan in 1064; according to Clavijero in 1160. The Mexicans properly speaking separated from the Tlaxcaltecas and the Chalcas in the mountains of Zacatecas. (Clav[ijero, *Storia antica del Messico*], vol. I, p. 156; Torq[uemada, *Monarquia yndiana*], vol. I, p. 87; Gama, *Descripción de [las] dos piedras*, p. 21.)

Arrival of the Aztecs in Tlalixco or Acahualtzinco in 1087; reform of the calendar and first festival of the new fire since the departure from Aztlan.

The Aztecs arrive in Tula in 1196, in Tzompanco in 1216, and in Chapultepec in 1245.

I.414 "During the reign of Nopaltzin, the king of the Chichimecs, a Toltec named Xiuhtlato, master of Quaultepec, instructed the people in the cultivation of maize [corn] and cotton and taught them how to bake bread from cornmeal. The few Toltec families who lived on the banks of the lake of Tenochtitlan had completely neglected the cultivation of this cereal, and American wheat would have been lost forever had Xiuhtlato not preserved some grains of it from his youth" (Torq[uemada, Monarquia yndiana], vol. I, p. 74).

The alliance of the Chichimecs, the Acolhua, and the Toltecs. Nopaltzin, son of king Xolotl, weds Azcaxochitl, daughter of a Toltec prince, Pochotl, and the three sisters of Nopaltzin enter liaisons with the heads of the Acolhua. There are few nations whose annals show such a multiplicity of family and place names as the hieroglyphic annals of Anahuac.

In 1314, the Mexicans are enslaved by the Acolhua, but, due to their valor, they soon succeed in freeing themselves from slavery.

Founding of Tenochtitlan in 1325.

Mexican Kings: I. Acamapitzin, 1352 to 1389; II. Huitzilihuitl, 1389 to 1410; III. Chimalpopoca, 1410 to 1422; IV. Itzcoatl, 1423–1436; V. Moctezuma Ilhiucamina or Moctezuma the First, 1436–1464; VI. Axayacatl, 1464–1477; VII. Tizoc, 1477–1480; VIII. Ahuitzotl, 1480–1502; IX. Moctezuma-Xocoyotzin or Moctezuma the Second, 1502–1520; X. Cuitlahuatzin, who ruled only for three months; XI. Quauhtemoztin, who ruled for nine months of the year 1521 (Clav[ijero, *Storia antica del Messico*], Vol. IV, pp. 55–61.)

During the reign of Axayacatl, Netzahualcoyotl, king of Culhuacan or Texcoco died; he is as memorable for the culture of his mind as he is for the wisdom of his laws. This king of Texcoco wrote sixty hymns in honor of the Supreme Being in the Aztec language, along with an elegy about the destruction of the town of Azcapotzalco and another about the fickleness of human power, as was proven by the example of the tyrant Tezozomoc. ▼ The great nephew of Netzahualcoyotl, baptized as Fernando Alva Ixtlilxochitl, translated some of these verses into Spanish, and the Chevalier Boturini owned the originals of two of his hymns; they had been composed fifty years before the conquest and, in Cortés's day, were written down in Roman letters on *metl* paper. I have searched in vain for these hymns amidst the remains of Boturini's collection in the palace

I.415 of the viceroy in Mexico City. It is also quite noteworthy that the famous botanist ▼ Hernández [de Toledo] made use of many of the drawings of plants and animals, with which king Netzahualcoyotl had decorated his house in Texcoco and which had been created by Aztec painters.

Cortés lands on the beach of Chalchicuecan in 1519.

Conquest of the city of Tenochtitlan in 1521.

The Counts of Moctezuma and of Tula, who lived in Spain, are descended from Ihuitemotzin, the grandson of King Moctezuma-Xocoyotzin, who had married Doña Francisca de la Cueva. The origin of the illustrious houses of Cano Moctezuma, of Andrade Moctezuma, and of the Count of Miravalle (in Mexico City) goes back to Tecuichpotzin, the daughter of King Moctezuma-Xocoyotzin. This princess, baptized Elisabeth, survived five spouses, among them the last two kings of Mexico, Cuitlahuitzin and Quauhtemotzin, as well as three Spanish officers.

CHAPTER VII

Whites, Creoles, and Europeans—Their Civilization—
Wealth Inequality among Them—Blacks—Mixing of the Castes—
The Relationship between the Sexes—Longevity and Racial
Differences—Sociability

Among the racially pure inhabitants, the whites would be in second place, if considered only in terms of their number. They are divided into whites born in Europe and descendants of Europeans, born in the Spanish colonies of the Americas or in the Asiatic islands. The first group is called *Chapetones*, or *Gachupines*, and the second Criollos. The indigenous peoples of the Canary Islands, who are usually called *Isleños* (Islanders), and who are the *overseers* of the plantations, consider themselves Europeans. Spanish laws give the same rights to all whites; but those who are called upon to carry out these laws seek to destroy an equality that offends European I.417 pride. The government, which mistrusts the Creoles, gives important positions exclusively to indigenous peoples of old Spain. For some years now, even in Madrid, they have held some of the most trivial positions in the customs administration or the control of tobacco. During a period when state monitoring was fairly relaxed, the system of venality made alarming progress. More often than not, it was pecuniary interest, rather than a policy of suspicion and distrust, that put all employment in European hands. This resulted in grounds for constant jealousy and hatred between the Chapetones and the Creoles. The most wretched European, with no education or intellectual culture, thinks himself superior to the whites who were born in the new world. Protected by his compatriots and favored by opportunities that are frequent in countries where fortunes are acquired as rapidly as they are ruined, he knows that he may one day reach heights whose access is almost forbidden to indigenous peoples, even to those who distinguish themselves by their talents, knowledge, and moral standards. These indigenous

peoples prefer to be called *Americans* rather than Creoles. Since the peace of Versailles, and especially since 1789, one often hears a voice proudly proclaim: "I am not *Spanish*, I am *American*," words that conceal the effect of a longstanding resentment. In the eyes of the law, any white Creole is Spanish; but the abuse of the law, false measures taken by the colonial government, the example of the confederate states of North America, and the influence of the *Zeitgeist* have relaxed the ties that had formerly bound   I.418
Spanish Creoles and European Spanish more closely together. A prudent administration may be able to restore harmony, calm passions and resentments, and perhaps preserve for yet a long time the union of the members of a single large family scattered across Europe and the Americas, from the shores of Patagonia to northern California.

The number of individuals who make up the white race (*casta de los blancos* or *de los Españoles*) is probably 1,200,000 in all of New Spain, of which a quarter live in the *Provincias internas*. In Nueva Vizcaya, or the intendancy of Durango, there is not a single individual who is subject to the *tribute*. Nearly all of the inhabitants of these northernmost regions insist that they are pure Europeans.

In 1793, there were, among the total population of the intendancy of Guanajuato,

	SOULS	SPANIARDS
total population of	398,000	103,000
in Valladolid	290,000	80,000
in Puebla	638,000	63,000
of Oaxaca	411,000	26,000

This is unaltered census data, with none of the changes necessitated by that inaccurate work, which we have discussed in chapter five. In the four intendancies nearest the capital, one thus finds 272,000 whites, be   I.419
they Europeans or descendants of Europeans, out of a total population of 1,737,000 souls. For every one hundred inhabitants, there were:

In the intendancy

of Valladolid	27	Whites
of Guanajuato	25	"
of Puebla	9	"
of Oaxaca	6	"

Such considerable differences underscore the degree of civilization that the ancient Mexica had attained south of the capital. The southernmost regions had always been the most populous. In the north, as we have seen several times throughout the course of this work, the Indian population was sparser; agriculture has not made noticeable progress there since the time of the conquest.

It is interesting to compare the number of whites in the Antillean Islands with those in Mexico. Even in its heyday, in 1788, the French part of Saint-Domingue had a population that was less than that of the Intendancy of Puebla on an area of 1,700 square leagues (at twenty-five to the degree). Page estimates the former at 420,000 inhabitants, among whom there were 40,000 whites, 28,000 emancipated slaves, and 452,000 slaves. For every one hundred souls in Saint-Domingue, there were eight whites, six free persons of color [hommes de couleur libres], and eighty-six African slaves [esclaves africains]. For every one hundred inhabitants in Jamaica, there were ten whites, four persons of color [hommes de couleur], and eighty-six slaves; however, that English colony had a population one-third less than that of the intendancy of Oaxaca. Hence, the disproportion between Europeans or their descendants and the castes of Indian or African blood is still greater in the southern parts of New Spain than in the French and British Antillean Islands. The island of Cuba, by contrast, shows to this day a large and reassuring difference in its racial distribution. According to careful statistical research conducted in 1821, the total population of the island was 630,980, of which 290,021 were whites; 115,691 free persons of color [hommes de couleur libres]; and 225,268 slaves.

I.420

It is likely that, in 1823, there were 317,000 whites, 127,000 free persons of color [hommes de couleur libres], and 256,000 slaves, for a total of 700,000. I believe that the population of the Spanish Antilles (Cuba and Puerto Rico) is generally composed of black slaves [esclaves noirs] with some mulattos [mulâtres] (281,400); free persons of color [libres de couleur], mulatto [mulâtre] and black [noirs] (319,500); and whites (342,100), for a total of 943,000. The ratio of free persons to slaves is dramatically different from that of the entire archipelago of the Antillean Islands, where, out of a total population of 2,843,000, there are 1,147,500 (or 40 percent) black and mulatto slaves [esclaves noirs et mulâtres]; 1,212,900 (or 43 percent) free persons of color [libres de couleur] (mulattos and blacks [mulâtres et noirs]); and 482,600 (or 17 percent) whites. As a result, the number of whites is much larger on the island of Cuba than in Mexico, even in the regions where there are the fewest Indians.

The following table shows the average proportion of the other castes to whites in the various parts of the new continent. For every one hundred inhabitants, there are: I.421

in the United States of North America	83	Whites
on the island of Cuba	45	"
in the Kingdom of New Spain (excluding the *Provincias internas*)	16	"
in the kingdom of Peru	12	"
on the island of Jamaica	10	"

According to the Count of Revillagigedo's census, for every one hundred inhabitants in the Mexican capital, there are forty-nine Spanish Creoles, two Spaniards born in Europe, twenty-four Aztec and Otomi Indians, and twenty-five persons of mixed blood. Precise knowledge of these ratios is of great political interest to those in charge of safeguarding the peace of the colonies.

It would be very difficult to estimate exactly how many Europeans there are among the 1,200,000 whites who live in New Spain. As in the Mexican capital itself, where the government gathers more Spaniards than elsewhere, out of a population of over 135,000 souls, there are not even 2,500 persons who were born in Europe; it is more than likely that the entire kingdom does not have more than 70,000 to 80,000 of them. They therefore make up only one-seventieth of the entire population, and the ratio of Europeans to white Creoles is one to fourteen.

Spanish law prohibits any European not born on the Peninsula from entering Spain's American possessions. In Mexico and Peru, the words "European" and "Spanish" have become synonymous. The inhabitants of the remote provinces can thus hardly believe that there are Europeans who do not speak their language; they consider this ignorance a mark of lowly extraction, for in their midst only the lowest class does not speak Spanish. Since sixteenth-century history is more familiar to them than our own times, they imagine that Spain continues to exert a marked influence on the rest of Europe. In their eyes, the Peninsula is the center of European civilization. But this is not the case for the Americans who live in the capital. Those who have read works of French or English literature easily fall prey to the opposite extreme and have a less favorable idea of the metropole than the French had at a time when contact between Spain and the rest of Europe was less frequent. They prefer foreigners from other countries to the Spanish and like to believe that intellectual culture has made more rapid advances in the colonies than on the Peninsula. I.422

These advancements are quite remarkable in Mexico City, Havana, Lima, Santa Fe, Quito, Popayán, and Caracas. Of all these cities, Havana resembles European cities the most in terms of its customs, refinement of luxury, and the taste of its society. The Habaneros are the best informed about politics and its influence on trade. But despite the efforts of the *Patriotic Society of the Island of Cuba*, which supports the sciences with the most generous enthusiasm, science develops slowly in a country where agriculture and the cost of colonial products consume the inhabitants' entire attention. The study of mathematics, chemistry, mineralogy, and botany is most widespread in Mexico City, Santa Fe, and Lima. There is a strong intellectual movement everywhere now, and young people have a rare gift for grasping scientific principles. This facility is supposedly even more pronounced among the inhabitants of Quito and Lima than it is in Mexico City and Santa Fe. In the first two places people appear to be more nimble-minded, while the Mexicans and the indigenous peoples of Santa Fe have a reputation for greater perseverance in their studies.

I.423

No city on the new continent, including those of the United States, has such exceptional and solid scientific establishments as does the capital of Mexico. I limit myself here to mentioning the School of Mines, whose director is the learned Delhuyar (we shall return to it when we discuss the mining of metals), the botanical garden, and the Academy of Painting and Sculpture. This academy is called the *Academía de los Nobles Artes de México*. It owes its existence to the patriotism of several Mexican individuals and the protection of ▼ Minister Gálvez. The government granted it a spacious building where one finds a finer and more complete collection of plaster casts than in any part of Germany. One is amazed to see that the Apollo Belvédère [the Pythian Apollo], the Laocoön group, and even more colossal statues were able to travel across mountain roads that are at least as narrow as those of the Saint Gotthard. It is surprising to find these masterpieces of antiquity assembled here in the Torrid Zone, where the plateau is at a higher elevation than the Great Saint Bernard monastery. It cost the king almost two hundred thousand francs to transport the collection to Mexico City. The other remnants of Mexica sculpture, colossal basalt, and porphyry statues that are covered with Aztec hieroglyphics and often show similarities with the Egyptian and Hindu style, should be collected in one of the courtyards of the Academy building, if not in the building itself. It would be strange to place these monuments of the early civilization of our species, the work of a semi-barbarous people living in the Mexican Andes, beside the handsome shapes created beneath the skies of Greece and Italy.

I.424

The revenue of the Mexican Academy of Fine Arts is 125,000 francs, of which the government gives 60,000, the Mexican mining corps nearly 25,000, and the capital's *Consulado*, or merchant guild, over 15,000. One I.425 should not underestimate the impact that this establishment has had on the nation's taste. This influence is especially visible in the arrangement of the buildings, the perfection of the stone cutting, the decoration of the columns' capitals, and the stucco relief work. What lovely buildings are already to be found in Mexico, even in the provincial towns, such as Guanajuato and Querétaro! These monuments, which often cost from one to one and a half million francs, would not be out of place on the most beautiful streets of Paris, Berlin, or St. Petersburg. ▼ Mr. Tolsá, a professor of sculpture in Mexico City, has even been able to cast an equestrian statue there of King Charles IV, which, with the exception of that of Marcus Aurelius in Rome, surpasses in its beauty and stylistic purity anything in this genre that remains in Europe. Instruction at the Academy of Fine Arts is *free of charge*. It is not limited only to landscape and figure drawing; someone was mindful enough to use other ways in which to animate national industry. The Academy labors successfully among artisans to spread a taste for elegance and refined forms. Every evening, a few hundred young people gather in large rooms well-lighted by Argand lamps, where some draw from relief or from a live model, while others copy drawings of furniture, chandeliers, or other bronze decorations. In this gathering (and this is quite amazing in a country I.426 where the nobility is so inveterately prejudiced against the castes), rank, color, and race do not matter; one sees Indians or Mestizos [Métis] next to whites, and the son of a poor artisan competes with the children of the great lords of the realm. It is comforting to see that in all parts of the world, the cultivation of the arts and sciences creates a certain equality among men, making them forget, if only for a short time, the petty passions whose effects are a hindrance to social well-being.

Since the end of the reign of Charles III and, now, under that of Charles IV, the study of the natural sciences has made tremendous progress, not only in Mexico but throughout the Spanish colonies. No other European government has sacrificed more considerable sums to advance knowledge of plants than the Spanish government. Three *botanical expeditions*—in Peru, Granada, and New Spain—supervised by ▼ Mr. Ruiz and Mr. Pavón, by Don José Celestino Mutis, and by Mr. Sessé and Mr. Moziño, cost the state nearly two million francs. Botanical gardens have also been planted in Manila and the Canary Islands. The commission appointed to draw up the plans for the Güines Canal was also directed to study plant cultivation on

the island of Cuba. All this research, conducted for twenty years in the most
I.427 fertile regions of the new continent, has not only enriched the scientific field with more than four thousand new plant species, but has also contributed greatly to spreading the taste for natural history among the country's inhabitants. Mexico City has a very interesting botanical garden within the compound of the viceroy's palace.[1] Professor Cervantes teaches courses there that are very popular. Besides his herbariums, this learned man has an excellent collection of Mexican mineral specimens. Mr. Moziño, whom we have just mentioned along with Mr. Sessé, and who has extended his difficult excursions from the kingdom of Guatemala as far as the northwest coast, to Vancouver Island and Quadra; and ▼ Mr. Echeverría, a painter of plants and animals whose work is equal in perfection to whatever Europe has produced in this genre, are both indigenous peoples of New Spain. They had both risen to a distinguished rank among scientists and artists before leaving their home country.[2]

I.428 The principles of new chemistry, which are known in the Spanish colonies by the somewhat equivocal term "the new philosophy" (*la nueva filosofía*) are more widespread in Mexico than in many parts of the Peninsula. A European traveler would surely be surprised to find young Mexicans in the interior of the country, near the border with California, studying the decomposition of water in the process of its amalgamation with open air. The School of Mines has a chemistry laboratory, a geological collection classified according to ▼ Werner's system, a physics case in which one finds not only Ramsden's, Adams's, Lenoir's, and Louis Berthoud's valuable instruments but also models made in the capital itself, with the greatest precision and the finest woods in the country. The best mineralogical work in the Spanish language, the Manual of Oryctognosy, edited by Mr. del Río in accordance with the principles of the Freiberg School [of Mines] (where

1. There has recently (1823) been a project for a National Museum and a School of Medicine in the *Naturales* hospital, and another to replace the small botanical garden in the courtyard of the viceroys' palace with two others, in the cemetery of the *Naturales* hospital and in the Ejido de Velasco. The Academy of Fine Arts has been closed for lack of funds since the political revolutions that followed in close succession; on the other hand, the notes from Boturini's and ▼ Dupaix's Mexica collections have been assembled and organized.

2. At present, the public only enjoys access to the discoveries made by the botanical expedition to Peru and Chile. Mr. Sessé's large herbariums, and the immense collection of drawings of Mexican plants made under his supervision, have been in Madrid since the year 1803. We impatiently await the publication of the *Flora Mexicana* [by Mizoño and Sessé] and the *Flora of Santa Fe de Bogotá*. The latter is the fruit of forty years of research and observations by one of the greatest botanists of the century, the famous Mutis.

its author was educated), was published in Mexico City. The first Spanish translation of ▼ Lavoisier's *Élémens de chimie* was also published in Mexico City. I refer to these isolated facts because they reflect the zeal with which the exact sciences are beginning to be embraced in the capital of New Spain. This passion is much greater than any enthusiasm there for the study of ancient languages and literature.

I.429

Instruction in mathematics is less thorough at the University of Mexico than at the School of Mines. Students at the latter school do more in-depth analysis and are taught integral and differential calculus. With the return of peace and open communications with Europe, the use of astronomical instruments (chronometers, sextants, and Borda's repeating circles) will become more common; in the most remote parts of the kingdom, young people will be able to make observations and calculate them, using the most current methods. I have already indicated in the Analysis of the Atlas the benefit that the government would draw from this extraordinary aptitude, namely in commissioning a map of the country. In any case, the interest in astronomy has a long history in Mexico. Three distinguished men, Velázquez, Gama, and Alzate, honored their country toward the end of the last century. All three made several astronomical observations, especially of the eclipses of Jupiter's satellites. Alzate, the least scholarly of them, was a corresponding member of the Academy of Sciences in Paris. An inaccurate observer whose work was often bull-headed, he tried to do too many things at once. In the geographical introduction to this work, we discussed the merit of his astronomical work. His other real merit is to have incited his countrymen to study the physical sciences. The *Gazeta de Literatura*, which he published for a long time in Mexico City, was great factor in the encouragement and motivation of Mexican youth.

I.430

The most remarkable geometrician that New Spain has seen since Sigüenza's day is Don Joaquín Velázquez Cárdenas y León. All of that tireless scholar's astronomical and geodesic work is of the greatest precision. Born July 21, 1732, in the interior of the country (on Santiago Acebedocla's farm near the Indian village of Tizicapan), he was, as it were, completely self-educated. At the age of four, he gave the smallpox to his father, who died of it. An uncle, who was the parish priest of Xaltocan, saw to his education and had him taught by an Indian called Manuel Asentzio, a man of great natural strength of mind, who was well versed in Mexican history and mythology. In Xaltocan, Velázquez learned several Indian languages as well as the use of the Aztec hieroglyphic script. It is a shame that he published nothing on this interesting branch of antiquity studies. Enrolled in the

Tridentin College in Mexico City, he found virtually no professors, books, or instruments there. With a modicum of assistance, he enriched himself

I.431 with the study of mathematics and ancient languages. A lucky chance led him to discover Newton's and Bacon's work. From the former, he acquired a taste for astronomy, and from the latter a knowledge of true philosophical methods. Since he was poor and could not find any instruments in Mexico City, he began to build telescopes and quarter circles with his friend, Mr. Guadalajara (who today is professor of mathematics at the academy of painting). At the same time, he worked as a lawyer, an occupation that in Mexico, like everywhere else, is more lucrative than observing the stars. He used what he earned professionally to purchase instruments in England. Appointed university professor, he accompanied the *Visitador* Don José de Gálvez[1] on his journey to the Sonora. Sent to California on a commission, he took advantage of the beauty of the sky over that Peninsula to make a great number of astronomical observations. He was the first to observe that, for

I.432 centuries, all maps had drawn this part of the new continent several degrees farther west than it actually is, through an enormous error in longitude. When the Abbot Chappe, who was more famous for his courage and his devotion to science than for the accuracy of his work, arrived in California, he found the Mexican astronomer already established there. Velázquez had an observatory built for himself from planks of mimosa at Santa Ana. When he had determined the position of this Indian village, he informed the Abbot Chappe that the lunar eclipse of June 18 would be visible in California. The French geometer doubted this claim until the announced eclipse occurred. Acting alone, Velázquez made a very good observation of Venus's transit across the solar disk on June 3, 1769. The very next day, he conveyed the result of the transit to the Abbot Chappe and to two Spanish astronomers, Don Vicente Doz and ▼ Don Salvador de Medina. The French traveler was surprised by the accordance between Velázquez's observation and his own. He was likely amazed to meet a Mexican in California who, without belonging to an academy or ever having left New Spain, could observe as well as any academician. In 1773, Velázquez conducted an important geodesic

---

1. Before becoming minister of the Indies, Count de Gálvez crossed the northern part of New Spain with the title of *Visitador*. This name is given to those employed by the court to collect information on the state of the colonies. Their journey (*visita*) generally has no other effect than to counterbalance for a time the viceroys' and *Audiencias*' power to receive an infinite quantity of papers, petitions, and projects, and to mark their stay with the introduction of new taxes. The people await the arrival of the *Visitadores* with the same impatience as they look forward to their departure.

study, some of the results of which we have provided in the analytical part of the Mexican Atlas; we shall return to it when we discuss the outflow conduit from the lakes in the Valley of Mexico. The most vital service this tireless man rendered his country was to establish the *Tribunal* and the School of Mines, the proposals for which he presented at court. He ended his productive career on March 6, 1786, as the first general director of the *Tribunal de Minería* with the title of *Alcalde del Corte honorario*. I.433

Having cited the work of Alzate and Velázquez, it would be unfair not to record Gama's name here, since he was both a friend and collaborator of the latter. Without any fortune, he was forced to provide for a large family through difficult, almost mechanical labor. Unknown and neglected throughout his life by his fellow citizens[1] who lavished praise on him after his death, Gama trained himself to become a skilled astronomer. He published several papers on lunar eclipses, Jupiter's satellites, the almanacs and chronology of the ancient Mexica, and the climate of New Spain; these papers all convey accurate ideas and precise observations. If I have digressed in detail on the literary merit of three learned Mexicans, it is only to prove by their example that the ignorance that European pride complacently pins on the Creoles is neither the effect of the climate nor due to a lack of moral energy. This ignorance, where it is still evident, is solely the effect of isolation and defects in the social institutions of the colonies. I.434

If, under present circumstances, the caste of whites is the one where intellectual development is almost exclusively found, it is also practically the only caste that enjoys great wealth. Unfortunately, this wealth is distributed even more unequally in Mexico than in the *Capitanía general* of Caracas, in Havana, and, especially, in Peru. In Caracas, the heads of the richest families have an income of 200,000 livres tournois; on the island of Cuba, one finds those who have revenues of more than 600,000 to 700,000 francs. In these two industrious colonies, agriculture has created greater wealth than working the mines has in Peru. In Lima, an annual income of 80,000 francs is quite rare. I know of no Peruvian family that currently has a stable and secure income of 130,000 francs. In New Spain, however, there are individuals who own no mines but whose annual income is as high as one million francs. The family of the Count *de la Valenciana* alone, for instance, owns property on the ridge of the Cordillera that is worth more than twenty-five million francs,

---

1. During his stay in Mexico, the famous navigator Alessandro Malaspina made observations together with Gama. He recommended him with great warmth at court, as is proven by Malaspina's official letters, preserved in the viceroy's archives.

not counting the Valenciana mine near Guanajuato, which in a normal year produces a net profit of one and a half million livres tournois. This family, whose present head, the young Count of [la] Valenciana, is generous and devoted to education, is divided into only three branches. Together, even in years when the mine is not very lucrative, they have an income of more than 2,200,000 livres. The ▼ Count of Regla, whose youngest son, the Marquis de San Cristóbal,[1] distinguished himself in Paris for his knowledge of physics and physiology, had, at his own expense, two very large sailing ships built from mahogany and cedar (*cedrella*) in Havana, which he presented to his sovereign. The Vizcaína vein, near Pachuca, was the foundation of the fortune of the house of Regla. The *Fagoaga* family, known for its benevolence, knowledge, and passion for the common good, offers the example of the maximum wealth that a mine could provide for its owners. In five or six months, after all expenses were deducted, a single vein that the Fagoaga family owns in the Sombrerete district produced a net profit of twenty million francs.

On the basis of these data, one might assume that Mexican families have infinitely more capital than meets the eye. The deceased Count de la Valenciana, the first to hold this title, occasionally received up to six millions livres of net revenue from his mine in a single year. For the last twenty-five years of his life, this annual revenue was never less than two or three million livres tournois. Yet, when he died, this extraordinary man, who came to the Americas with no fortune and continued to live with the greatest simplicity, left only ten million in property and capital besides his mine, which is the richest in the world. There is nothing surprising about this very accurate fact to those who have seen how the great Mexican houses are run. Money rapidly earned is spent with the same facility. Mining becomes a game that one plays with boundless passion. The rich mine owners squander enormous sums on charlatans who convince them to invest in new ventures in the remotest provinces. In a country where work is done on such a large scale, where it often costs two million francs to sink a mine shaft, a risky project that is poorly executed may absorb in only a few years whatever was earned from working one of the richest mines. It must be added that because of the internal disorder that prevails in most of the great houses of Old and New Spain, a head of family is often financially at odds, although he has an income of half a million livres and appears to enjoy no other luxury than several teams of mules.

I.435

1. Mr. Terreros (the name by which that modest scholar is known in France) long preferred the education he received during his stay in Paris to a great fortune which he could only enjoy if he lived in Mexico City itself.

The mines have been the principal source of the great fortunes of Mexico. Many miners have made good use of their wealth by purchasing land   I.437
and devoting themselves enthusiastically to agriculture. But a considerable
number of very powerful families have never had lucrative mines to work.
The rich descendants of Cortés and the Marquis del Valle belong to these
families. ▼ The Duke of Monteleone, a Neapolitan lord who is now the head
of the Cortés household, owns superb estates in the province of Oaxaca,
near Toluca, and in Cuernavaca. His net income is now only 550,000 francs,
since the king deprived the duke of the privilege of collecting the *Alcabalas*
and duties on tobacco. Regular administrative expenses exceed 125,000
francs. Moreover, several governors of the *Marquesado* have become exceptionally rich. If the descendants of the great *Conquistador* wished to live
in Mexico, their income would soon increase to more than one and a half
million.

To finish the inventory of the immense wealth that is now in the hands of
a few individuals in New Spain—rivaling fortunes in Great Britain and the
European possessions in Hindustan—I shall add a few specific statements
concerning the income of the Mexican clergy and the pecuniary sacrifices
that the corps of miners (*Cuerpo de Minería*) makes each year to perfect
the exploitations of metals there. In the three years from 1784 to 1787, this
body, which is composed of a group of mine owners and is represented by   I.438
deputies who preside over the *Tribunal de Minería*, has advanced the sum
of four million francs to individuals who lacked the necessary funds to undertake significant projects. The locals believe that this money has not been
put to very good use (*para habilitar*), but its distribution proves the generosity and magnanimity of those who are capable of such largesse. A European reader will be even more surprised if I mention here the extraordinary
fact that, a few years ago, the respectable Fagoaga family loaned more than
three and a half million francs, without interest, to a friend whose fortune
they thought they were solidifying: this enormous sum has been squandered on an unsuccessful new mining venture. The architectural projects
that are undertaken in the Mexican capital to adorn the city are so costly
that despite the low cost of labor, the splendid building being constructed
for the *Tribunal de Minería* to house the School of Mining will cost over
three million francs, almost two-thirds of which had already been allocated
when the foundation was laid. In the year 1803 alone, the Mexican corps of
miners allotted the sum of fifty thousand francs to accelerate construction,   I.439
and, in particular, to enable students to use a laboratory designed for metal-
lurgical experiments on the amalgamation of large mineral masses (*beneficio*

de patio). Such is the facility with which vast projects are undertaken in a country where wealth belongs to a small group of individuals.

This disparity of fortune is even more striking among the clergy, some of whom suffer in utmost wretchedness, while others have income that surpasses the revenue of many sovereign German princes. The Mexican clergy is composed of only ten thousand persons, almost half of whom are regulars who wear a habit, and this number is smaller than one believes in Europe. If one includes lay brothers and sisters, or servants (*Legos, Donados y Criados de los Conventos*) and all those who are not in members of holy orders, one may estimate the size of the clergy at thirteen to fourteen thousand individuals.[1] Yet, the total annual income of the eight Mexican bishops, which we present in the following table, is 2,721,950 francs.

I.440

The bishop of Sonora, the least wealthy of all, has no income from tithes. Like the bishop of Panama, he is directly paid from the royal coffers (*de Caxas reales*). His revenue represents only the twentieth part of that of the bishop of Valladolid de Michoacan. What is truly disturbing is that in an archbishop's diocese where annual revenue is up to 650,000 francs, there are priests in Indian villages who touch less than five to six hundred francs per year! The bishop and canons of Valladolid have sent a series of gifts to the king, especially during the last war with France, totaling 810,000 francs. The value of the lands of the Mexican clergy (*bienes raices*) does not exceed twelve to fifteen million francs, but the same clergy has enormous wealth in mortgage capital on the property of individuals. The total amount of this capital (*capitals de Capellanías y obras pias, fondos dotales*

I.441

---

1. The number of monks of Saint Francis in Spain is 15,600. This is larger than the number of all the ecclesiastics in the kingdom of Mexico. On the Peninsula, the clergy consists of 177,000 individuals. For every thousand inhabitants, there are sixteen ecclesiastics, whereas in New Spain, there are but two. See the detailed breakdown of the clergy in some Intendancies, according to the 1793 census:

IN THE INTENDANCY OF			
Puebla	667 irregular clergymen, or *Clerigos*, and	881	regular clergymen
Valladolid	293	298	
Guanajuato	225	197	
Oaxaca	306	342	
In Mexico City	550	1,646	

If we include the *Donados*, or servant brothers, in this count, then the convents in the capital house over 2,500 persons.

Income of the			
Archbishop of	Mexico City	130,000	piastras fuertes
Bishops of	Puebla	110,000	
	Valladolid	100,000	
	Guadalajara	90,000	
	Durango	35,000	
	Monterey	30,000	
	Yucatán	20,000	
	Oaxaca	18,000	
	Sonora	6,000	

de Comunidades religiosas), which we shall give in detail later, is forty-four and a half million piastres fortes, or 233,625,000 francs.[1] From the earliest days of his conquest, Cortés feared the clergy's extravagance in a country where it is difficult to maintain ecclesiastic discipline. In a letter to Emperor Charles V, he naively writes that he "begs his majesty to send *men of the cloth* and not *canons* to the West Indies, because the latter display boundless luxury, bequeathe enormous wealth to their illegitimate children, and expose the recently converted Indians to their scandals." Dictated by the frankness of an old soldier, this advice was not taken in Madrid. We have transcribed this curious passage from a work that was published a few years ago by a cardinal[2]: it is not our place to accuse the conqueror of New Spain of either preferring ordinary clerics or showing animosity toward canons!   I.442

The rumors of the extravagance of Mexican wealth that have spread in Europe have given rise to exaggerated ideas about the abundance of gold and silver used in New Spain for dishes, furniture, cooking pots, and harnesses. A traveler whose imagination has been piqued by these tales of solid silver keys, locks, and hinges will be surprised, upon arriving in Mexico

1. I used the information given in the *Representación [a nombre de los labradores y comerciantes] de Valladolid* (dated October 24, 1805), an invaluable unpublished work [by Manuel Abad y Queipo]. Throughout this work, I calculate the piastra fuerta as five livres, five sous. Its intrinsic value is five livres, eight and one-third sous tournois. We should not confuse the *peso*, which is also called the *peso sencillo* or *trade piastre*, and is a fictitious coin, with the American piaster forte, or *duro*, or *peso duro*. One piastra fuerta equals twenty reales of vellón, or 170 *quartos*, or 680 *maravedis*, whereas the *peso sencillo*, which is worth three livres, fifteen sous, contains only fifteen reales of vellón, or five *maravedís*.

2. The Archbishop of Lorenzana.

City, not to find any more precious metals used in everyday life there than there are in Spain, Portugal, and other parts of southern Europe. At most, he will be amazed to see in Mexico, Peru, or Santa Fe people of the lowest class wearing large silver spurs on bare feet, or to find silver goblets and dishes somewhat more common here than in they are in France and England. The traveler's surprise will cease, however, when he recalls that porcelain is quite rare in these newly civilized regions; that the nature of the mountain roads makes its transport extremely difficult; and that, in a country where trade is not very active, it scarcely makes a difference whether one has a few hundred piasters in coin or in silver furniture. Moreover, despite the enormous difference in wealth in Peru and Mexico, considering only the fortunes of the great landowners, I would be tempted to believe that there is more true comfort in Lima than in Mexico City. The

I.443   disparity of wealth is much less pronounced in the first of these two capitals. If it is quite rare, as we have seen above, to find individuals who enjoy fifty to sixty thousand francs in income, one does, however, find there many mulatto and freed black craftsmen [artisans mulâtres et de nègres affranchis] whose industriousness earns them well more than the bare necessities of life. Among the members of this class, capital holdings between ten and fifteen thousand piasters are quite common, while the streets of Mexico City teem with twenty to thirty thousand unfortunates (*Saragates, Guachinangos*), most of whom spend the night outdoors and lie in the sun during the day, their naked bodies covered with a flannel sheet. These dregs of society, Indians and Mestizos, are very similar to the Lazaroni of Naples. Lazy, careless, and plain like them, the Guachinangos are not at all violent, and they never ask for any alms; if they work one or two days a week, they earn what they need to buy pulque or some of the ducks that cover the Mexican lagoons, which they roast in the duck's own fat. The Saragates' fortune rarely exceeds two or three reales, whereas the people of Lima, who are more devoted to luxury and pleasure and are perhaps also more industrious, often spend two or three piasters in a single day. People say that the mixture of Europeans and Blacks [Nègres] produces a race of more active and more hardworking people than does the mixture of whites with Mexican Indians!

I.444   Of all the European colonies in the Torrid Zone, the Kingdom of New Spain has the fewest Blacks [Nègres]. One might almost think that there were no slaves, and that one might cross all of Mexico City without seeing a single black face [visage noir]. No house has a staff composed of slaves. Especially from this point of view, Mexico presents a great contrast with

Havana, Lima, and Caracas. According to precise information from persons who worked on the 1793 census, it appears that in all of New Spain, there are not even six thousand Blacks [Nègres], and nine to ten thousand slaves at the most, of whom the majority live in the ports of Acapulco and Veracruz, the hot region near the coastline (*tierras calientes*). Slaves are four times more numerous in the *Capitanía general* of Caracas, which does not have even one-sixth the population of Mexico. The ratio of Blacks [Nègres] in Jamaica to those in New Spain is two hundred fifty to one. In the Antillean islands, Peru, and even Caracas, the general advancement of agriculture and trade now depends on the increase in the number of Blacks [Nègres]. From 1792 to 1803, nearly 55,000 slaves[1] were introduced to the island of Cuba where, in twelve years, annual sugar exports rose from 400,000 to 1,000,000 quintals. In Mexico, on the contrary, the growth of colonial prosperity is not a function of an increase in the slave trade. As recent as twenty years ago, Mexican sugar was virtually unknown in Europe; today, Veracruz alone exports more than 120,000 quintals of sugar; but, luckily, the progress that sugarcane cultivation has made in New Spain since the ▼ revolution in Saint-Domingue has not noticeably increased the number of slaves there. Among the 74,000 Blacks whom Africa[2] provides annually for the equinoctial regions of America and Asia—the equivalent in the colonies themselves of 111,000,000 francs—there are not even one hundred who land on the shores of Mexico.

I.445

According to the law, there are no Indian slaves in the Spanish colonies. But through a striking abuse, two types of wars, very different in appearance, have created conditions for the Indians very much like those of the African slaves. From time to time, the missionary monks in South America make forays into country occupied by peaceful tribes of Indians, who are called savages (*Indios bravos*) because they, unlike the equally unclothed Indians in the Missions (*Indios reducidos*), have not yet learned how to make the sign of the cross. During these nightly raids, which are prompted by the most deplorable fanaticism, the missionaries seize whomever they can surprise, especially children, women, and the elderly. They mercilessly separate children from their mothers, to make sure that they do not conspire to run away together. The monk who leads the expedition then distributes

I.446

1. According to the figures of the Havana customs house, of which I have a copy at hand, 34,500 slaves were imported from 1799 to 1803, of which 7 percent die annually.

2. According to ▼ Mr. Norris and the information given to the British Parliament in 1787 by the merchants of Liverpool.

the children to those Indians at his Mission who have contributed the most to the success of the *Entradas*. On the Orinoco and the shores of the Portuguese portions of the Río Negro, these prisoners are called *Poitos*, and they are treated as slaves until they are of a marriageable age. The desire to have *Poitos* and put them to work for eight or ten years leads the Indians of the Missions themselves to incite the monks to make such raids; the bishops have typically had the good sense to accuse them of vilifying both religion and its ministers. In Mexico, the prisoners who are taken in the petty skirmishes that are almost ongoing in the borderlands of the *Provincias internas* share a much worse fate than the *Poitos*. These prisoners, who are usually from the Indian nation of the Mecos or Apaches, are dragged to Mexico City, where they languish in the dungeons of a prison (*La Cordada*). Solitude and despair make them more violent. When they are sent off to Veracruz and to the island of Cuba, they die there, much like any savage Indian would if he were taken from the high central plateau to the lowest and therefore hottest regions. There have been recent cases of some Mecos prisoners who escaped from the dungeons and committed ghastly acts of cruelty in the nearby countryside. It is high time that the government took care of these unfortunates, who are few in number, and whose lot it would therefore be easy to improve.

I.447

It seems that at the beginning of the conquest, there were many of these prisoners of war, who were treated like slaves. I found a remarkable passage on this subject in Hernán Cortés's will,[1] a historical monument that should be rescued from oblivion. The great captain, who, in the course of his victories, and especially in his conduct toward the unfortunate king Montezuma II, had not shown much sensitivity and conscience,[2] developed

1. *Testamento que otorgó el Excellentissimo Señor Don Hernán Cortés, Conquistador de la Nueva España hecho en Sevilla el 11 del mes de octubre*, 1547. The original of this curious document, which I had copied, is in the archives of the residence *del Estado* (of the Marquis del Valle), located in the great square of Mexico City. It was never published. In the same archives, I also found a report written by Cortés shortly after the siege of Tenochtitlan, with instructions for building roads, setting up inns along the major roads, and other topics of general policy.

2. In his letters dated from the Rica Villa at Veracruz, Cortés depicts the town of Tenochtitlan to Emperor Charles V as if he were describing the wonders of the capital of El Dorado. After relating to him all that he was able to learn about the wealth "of this powerful lord Montezuma," he assures his sovereign that the Mexican king must fall into his hands, dead or alive. "*Certifique a Vuestra Alteza que lo habría preso ó muerto ó subdito á la Real Corona de Vuestra Magestad*" (Lorenzana, [*Historia de Nueva-España,*] p. 39). It is important to note that this plan was conceived when the Spanish general was still on the coast and had not had any communication with Montezuma's ambassadors.

scruples toward the end of his career about the legitimacy of the titles by   I.448
which he held enormous properties in Mexico. He ordered his son to make
the most careful inquiries into the tributes he had received from the great
Mexican lords who had been the proprietors of his majorat prior to the arrival of the Spanish in Veracruz. Cortés even insisted that the value of the
tributes that had been exacted in his name, beyond the taxes formerly paid,
be returned to the indigenous peoples. Speaking of the slaves in the thirty-
ninth and fortieth articles of his will, Cortés appended these memorable
words: "Since it remains doubtful that a Christian can in good conscience
use as slaves indigenous peoples who are prisoners of war, and since until
now this important point has not been clarified, I order my son, Don Martín, and those of my descendants who will possess my majorat after him, to
collect all possible information on the duties that one may legitimately levy
on prisoners. After paying me tributes, the indigenous peoples who have
been forced into personal service must be compensated, if the decision en-
sues that such duties may not be asked of them." From whom were decisions   I.449
on such problematic questions expected, if not from the pope or a council?
Admittedly, and despite the enlightenment of a more advanced civilization
three centuries later, rich landowners in the Americas have a less apprehensive conscience, even on their deathbed. In our day, it is philosophers and
not the devout who debate the question of whether or not slavery is permis-
sible! But the small influence that the empire of philosophy has always had
leads us to believe that it would have been more useful to suffering humanity if this brand of skepticism had been preserved among believers.

Furthermore, the slaves who are fortunate to be few in number in Mexico
are, as in all Spanish possessions, somewhat more protected by laws than the
Blacks [Nègres] who live in the colonies of other European nations. These
laws are always interpreted in favor of liberty. The government wishes to
see the number of freed slaves increase. A slave who has saved some money
through his hard work may force his master to free him by paying him the
modest sum of 1,500 to 2,000 livres. A Black [Nègre] cannot be refused freedom on the pretext that it cost thrice the amount to purchase him, or that he
has a particular skill in a lucrative trade. According to the law, a slave who
has been cruelly mistreated may even acquire his freedom on that account,
if the judge sides with the injured party. One can imagine that this well-
intentioned law must be frequently eluded. But I have seen, even in Mexico   I.450
City, in July 1803, the example of two Black women [Negresses], whom the
magistrate performing the duties of the *Alcalde de Corte* freed, because their
mistress, a woman born in the islands, had covered their bodies with wounds

inflicted by scissors, pins, and penknives. In the course of this horrible trial, the woman was accused of having broken the teeth of her slaves with a key when they complained of a fluxion [?] on their gums, which kept them from working. Roman matrons were no more refined in their acts of vengeance. Barbarity is the same across the centuries, when people give free reign to their passions and governments tolerate an order of things that is contrary to natural law, and consequently to the welfare of society.

We have listed the different races of men that constitute the present population of New Spain. Turning to the physical tableaux in the Mexican Atlas, one sees that the majority of the nation's six million inhabitants may be considered mountain people. On the Anahuac plateau, whose elevation is more than twice the height of the large clouds that are suspended over our heads in summer, one finds copper-skinned peoples who have come from the northwest part of North America, some Europeans, and some Blacks

I.451 [Nègres] from the coastline of Bonny, Calabar, and Melimbo. When we consider that those whom we now call Spanish are a mixture of Alani and other Tartar hordes with Visigoths and the ancient inhabitants of Iberia; when we recall the striking similarity of most European languages, Sanskrit, and Persian; and finally, when we ponder the Asian origin of the nomadic tribes who have entered Mexico since the seventh century, we are tempted to believe that a part of those nations that wandered for so long set out from a single center, but on roads that were diametrically opposed, and after having taken the great tour, so to speak, around the globe, met once again on the ridge of the Mexican Cordillera.

To complete the table of the elements that comprise the Mexican population, it remains for us to point out the differences among the *castes* that arise from the mixture of pure races with each other. These castes represent a social group that is almost as considerable as the Mexican indigenous peoples. We may estimate the total number of persons of mixed blood at nearly 2,400,000. Refining their conceit, the inhabitants of the colonies have enriched their language, categorizing the most intricate shades of skin color produced by the degradation of the original color. It would be

I.452 very useful to explain this nomenclature,[1] since many travelers have misunderstood it, and this causes an awkward confusion when reading Spanish works about the American possessions.

The son of a white man (Creole or European) and a copper-skinned indigenous woman is called *Mestizo* [Métis]. His color is almost perfectly

---

1. Doctor Unanué, *Sobre el Clima de Lima*, p. 58. This work was printed in Peru in 1806.

white, and his skin has a special transparency. It is the lack of facial hair, small hands and feet, and a certain slant of the eyes, rather than the type of hair, that are more indicative of a mixture of Indian blood. If a Mestiza [Métisse] marries a white man, the second generation is scarcely distinguishable from the European race. Since so few Blacks [Nègres] have been introduced into New Spain, Mestizos [Métis] probably account for seven-eighths of all castes. They are generally reputed to have a gentler character than the Mullatos [Mulâtres], the offspring of White men [Blancs] and Black women [Negresses], who are distinguished by the violence of their passions and an unusual volubility of language. In Mexico City, in Lima, and even in Havana, the descendants of Black men [Nègres] and Indian women are called by the strange name of ▼ *Chinos* (Chinese). On the coast-line of Caracas and, as the laws have revealed, even in New Spain, they are also called *Zambos*. This name is now mainly reserved for the descendants of a Black man [Nègre] and a Mulata [Mulâtresse], or a Black man [Nègre] and a China. Among these ordinary Zambos, they distinguish the *Zambos prietos*, the offspring of a Black man [Nègre] and a Zamba. The caste of the *Quarterones* issues from the mixture of a White man [Blanc] and a Mulata [Mulâtresse]. When a *Quarterona* marries a European or a Creole, her son is called a *Quinterón*. A new alliance with the white race causes the loss of so much of the remainder of color that the child of a White man [Blanc] and a Quintcrona [Quinteronne] is also white. The Indian- or African-blooded caste have the odor that is particular to the cutaneous perspiration of the two original races. The Peruvian Indians, who use their refined sense of smell to distinguish among different races at night, have coined three words for the odor of the European, the indigenous American, and the Black [Nègre]: the first is called *pezuña*, the second *posco*,[1] and the third *grajo*. Moreover, the mixtures in which the children's color is darker than their mothers' are called *salta-atras*, "throwbacks."

I.453

In a country governed by whites, the families reputed to have the least amount of black or mulatto blood [sang nègre ou mulâtre] are naturally the most honored. In Spain, it is considered almost a title of nobility to be descended neither from Jews nor from Moors. In the Americas, the degree of whiteness of the skin determines a person's rank in society. A White man [Blanc] who rides barefoot on horseback takes himself for local nobility. Color even establishes a certain equality between men who, as is the case wherever civilization is little advanced or backward, complacently refine

---

1. An old qquichua [Quechua] word.

the prerogatives of race and origin. Whenever a local man has an argument with one of the titled lords of the country, he is frequently heard to say, "Could you possibly think that you were whiter [plus blanc] than I am?" This word characterizes the present state and origin of the aristocracy. Consequently, it is of great interest for public pride and respect to estimate accurately the fractions of European blood that one assigns to the different castes. The following proportions are recognized by custom:

CASTES	BLOOD MIXTURE			
Quarteroon	1/4	Black	3/4	White
Quinteroon	1/8	Black	7/8	White
Zambo	3/4	Black	1/4	White
Zambo prieto	7/8	Black	1/8	White

Families suspected of being of mixed blood will often request that the high court of justice (*Audiencia*) declare that they belong to the Whites [Blancs]. These declarations do not always correspond to the judgment of the senses. One sees very dark-skinned Mulattos [Mulâtres très basanés] who have skillfully *whitened* themselves [se faire *blanchir*] (this is the vulgar expression). When a person's skin color is too contrary to the solicited judgment, the petitioner is satisfied with a somewhat problematic expression. The judgment then reads that "such and such individuals may consider themselves white (*que se tengan por Blancos*)."

I.455   It would be very interesting to be able to discuss thoroughly the influence of the diversity of castes on the relations between the sexes. From the census taken in 1793, I saw that among the Indians in the towns of La Puebla and Valladolid, there were more men than women, while among the Spanish, or the white race, there were more women than men. In the intendancies of Guanajuato and Oaxaca, one finds the same surplus of men in all castes. I was unable to collect sufficient material to resolve the problem of the diversity of the sexes in relation to racial difference, the heat of the climate, or the elevations of the regions where these people live; we shall thus have to content ourselves with general results.

According to a partial census taken in France with the utmost care, the ratio of living women to men is nine to eight, among a population of 991,829 souls. Mr. Peuchet[1] appears to have decided on the ratio of thirty-four to

1. [Peuchet,] *Statistique élémentaire de la France*, p. 242.

thirty-three. There are certainly more women than men in France; it is most remarkable that more boys are born in the countryside and in the south than in the towns and departments located between 47° and 52° latitude.

In New Spain, on the contrary, the calculations of political arithmetic give a completely different result. Men are typically more numerous than women, as the following table that I have drawn up proves; it includes the eight provinces, or a population of 1,352,000.

I.456

INTENDANCY OR GOVERNMENT DISTRICTS	RACIAL DIVERSITY	MEN	WOMEN	RATIO OF MEN TO WOMEN
GUANAJUATO	Spaniards or Whites	53,983	49,316	100:91
	Indians or natives	89,753	85,429	100:95
	Mixed castes	59,659	59,604	100:99
VALLADOLID DE MICHOACÁN	Spaniards	40,399	39,081	100:97
	Indians	61,352	58,016	100:94
	Mixed castes	44,704	43,704	100:98
OAXACA	Spaniards	12,923	12,882	100:99
	Indians	182,342	180,738	100:95
	Mixed castes	11,163	10,566	100:98
DURANGO		60,727	59,586	100:98
SONORA[1]	in these five provinces	20,473	17,832	100:87
SINALOA	no distinction was made	27,772	27,290	100:98
NUEVO MEXICO	on the basis of race	15,915	14,910	100:94
CALIFORNIA		6,770	5,946	100:87
	Total	687,935	664,900	average
		1,352,835		100:95

1. One may suppose that the surplus of males in the north of Mexico can be partly attributed to the existence of military posts called *Presidios*, in which no women live. But we will see in the following that all of the *Presidios* combined amount to no more than three million men.

It follows from my calculations, when compared with those made by the Ministry of the Interior in Paris, that the ratio of men to women is one hundred to ninety-five among the general population of New Spain, and

100 to 103 in France. These numbers appear to represent the actual state

I.457 of things, since it is difficult to imagine why it would be more advantageous for Mexican women than for men to reduce their numbers in the census ordered by the Count of Revillagigedo. This is even more improbable, when we consider that the same census shows a ratio between the sexes that is completely different in large cities from what is found in rural areas. We shall soon see that census-taking in the United States also indicates a higher number of living men than women.

This feature of large cities has likely given rise to the false but generally accepted idea in the colonies that more females than males are born in warm climates, and thus in all the lower regions in the Torrid Zone, The few parish registers I have been able to examine show exactly the opposite. From 1797 to 1802 in the Mexican capital, there have been:

IN THE PARISH	MALE BIRTHS	FEMALE BIRTHS
of Sagrario	3,705	3,603
of Santa Cruz	1,275	1,167

In Panuco and Yguala,[1] two places located in a hot and very unhealthful climate, not even in one out of nine consecutive years was there a surplus of male births. In general, the ratio of male to female births appears to me

I.458 to be one hundred to ninety-five, which indicates a surplus of males slightly larger than in France, where ninety-six girls are born for every one hundred boys.

As for the relationship between deaths and the difference between the sexes, it was not possible for me to recognize the law that nature has established. In Panuco, 479 men died in ten years, compared to 509 women. In the single parish of Sagrario in Mexico City, there were 2,393 female deaths, compared to 1,951 male deaths, in five years. According to this meager information, it is true that the surplus of living men should be even larger than what we have found. But it appears that in other regions, male deaths are more frequent than female deaths. In Yguala and Calimaya, the ratio of the former to the latter was 1,204 to 1,191 and 1,330 to 1,292, respectively, over a ten-year period. ▼ Mr. de Pommelles has already observed that, in France itself, the difference between the sexes is more noticeable in

1. In Panuco, the parish registers from 1793 to 1802 show 674 male births and 550 female births. In Yguala, there were 1,738 boys and 1,635 girls.

births than in deaths; one-seventeenth more males are born than females, and the serenity of peasant life shows only one nineteenth more male than female deaths. The result of all this information is that, in Europe as well as in the equinoctial regions which have enjoyed a long period of tranquility, one would find a surplus of men, if the ocean, wars, and the dangerous work in which our gender engages did not tend to diminish their number.

The population of large cities is not fixed and does not by itself maintain an equilibrium with respect to the different sexes. Women from rural areas come to the city to work in houses that have no slaves. Many men leave the countryside to travel the land as muleteers (*arrieros*) or settle in places where there is significant exploitation of mines. Whatever the cause of the disproportion of the sexes in cities, it certainly does exist. The following table, which includes only three cities, presents a striking contrast to the table of the general population of eight Mexican provinces:

I.459

CITY	RACIAL DIVERSITY	MEN	WOMEN	RATIO OF MEN TO WOMEN
MEXICO CITY	Europeans	2,118	217	100:10
	Spaniards or white Creoles	21,338	29,033	100:136
	Indians or natives	11, 232	14,371	100:128
	Mixed race [Mulâtres]	2,958	4,136	100:128
	Other castes or mixed-bloods	7,832	11,525	100:147
QUERÉTARO	Spaniards	2,207	2,929	100:133
	Indians	5,394	6,190	100:115
	Mixed castes	4,639	5,490	100:118
VALLADOLID	Spaniards	2,207	2,929	100:133
	Mulattos	1,445	1,924	100:133
	Indians	2,419	2,276	100:93
	Total	63,789	81,020	average
		144,809		100:127

In the United States of North America, as in Mexico, the censuses of the entire population show a surplus of living men. This surplus is

I.460

very unstable in a country where the emigration of whites, the introduction of many male slaves, and sea trade tend incessantly to disrupt
the natural order of things. In the states of Vermont,[1] Kentucky, and
South Carolina, there are almost one-tenth more males than females,
while in Pennsylvania and New York, this disproportion does not exceed one-eighteenth.

When the Kingdom of New Spain has an administration that supports
knowledge, political arithmetic may provide infinitely more significant information there, both for statistics in general and for the physical history of
humankind in particular. There are so many unsolved problems in a mountainous country that exhibit, on the same latitude, the greatest variety of
climates, a population composed of three or four original races, and the
intermingling of these races in all imaginable combinations! So much research remains to be done on the age of puberty, the fertility of the species,
the difference between the sexes, and longevity, which is greater or lesser
depending on the elevation and temperature of localities; on the variety of
I.461 races; on the time when colonists arrive in the various regions; and finally,
on the dietary differences in the provinces where banana trees, jatropha,
rice, corn, wheat, and potatoes grow in a small space!

A traveler cannot devote himself to this research, which demands much
time, the involvement of the highest authorities, and the participation of
many persons who all share a common goal. Let it suffice to indicate here
what will remain to be done at a time when the government decides to benefit from the fortuitous position in which nature has placed this extraordinary country.

The 1793 population studies on the capital shows results that deserve to
be included at the end of this chapter. This part of the census distinguishes
between individuals over and under fifty years of age along the lines of caste
differences. It was found that this age was exceeded by:

4,128 Creole Whites among a total population of 50,371 individuals of the same race	
539 Mulattos	7,094
1,789 Indians	25,603
1,278 Mixed-bloods	19,357

1. See Samuel Blodget, [*Economica* (1806 ed.),] p. 75.

so that over the age of fifty, one finds:

Per 100	Creole Whites (Spaniards)	8
	Indians	6 4/5
	Mulattos	7
	Individuals of other mixed castes	6

These numbers, which confirm the admirable conformity that characterizes all natural laws, seem to indicate that longevity is somewhat greater among the better nourished races, and among those where puberty sets in later. Of the 2,335 Europeans who lived in Mexico in 1793, no fewer than 442 reached the age of fifty; this, however, scarcely proves that Americans are three times less likely to reach old age than Europeans, since the latter rarely move to the West Indies well into adulthood.

After examining the physical and moral state of the different castes that comprise the Mexican population, the reader will likely want to approach the question of the impact that the intermingling of races has on the general well-being of the society. What degree of enjoyment and individual happiness can a civilized person reach in the present state of this country, amidst conflicting interests, prejudices, and feelings?

We shall not discuss here the advantages that the Spanish colonies offer through the wealth of their natural productions, the fertility of their soil, the ease with which a man is able to choose—as he wishes and with thermometer in hand, within a space of a few square leagues—the temperature or climate he finds most suited to his age and physical constitution, or to the type of crops he wishes to raise. Nor shall we retrace the tableau of these delightful areas situated halfway up in the region of oaks and firs, between 1,000 and 1,400 meters in elevation, where a perpetual spring reigns, where the most delectable fruits of the West Indies are grown next to European fruit trees, and where these pleasures are troubled neither by the multitude of insects nor by the fear of yellow fever (*vómito*) or the frequency of earthquakes. We shall not discuss here whether, outside the tropics, there exists a region where a man can work less but provide better for the needs of a large family. The colonist's physical prosperity alone does not transform his intellectual and moral existence.

When a European who has enjoyed all the attractions of social life in the most civilized countries is transported to the distant regions of the new continent, he feels at each step the influence that the colonial government

I.463

has, for centuries, exerted over the population's morale. The educated man interested only in the intellectual development of his species may suffer less there than the man endowed with great sensitivity. The former remains connected to the metropole; sea-going links furnish him with books and instruments; he is delighted to see the progress that the study of exact sciences have made in the large cities of Spanish America; the contemplation of a vast, marvelously diverse nature compensates his mind for the depriva-

I.464   tions to which his location condemns him. The latter individual finds life in the colonies agreeable only when he withdraws into himself, for that is where his isolation and solitude seem most desirable, if he wishes to benefit peaceably from the advantages offered by the beauty of these climes, their perpetually green appearance, and the political calm of the New World. In stating these ideas candidly, I am not criticizing the moral character of the Mexicans or the Peruvians, nor am I implying that the people of Lima are worse than the inhabitants of Cádiz. I would be more inclined to believe what many authors before me have observed, namely, that Americans are naturally endowed with a graciousness and gentleness of manners that tends toward indolence, just as the energy of some European nations easily degenerates into harshness. The lack of sociability that prevails in the Spanish possessions, the hatreds that divide the closest castes, and whose effects spread bitterness throughout the colonists' lives are due solely to the political principles that have governed these regions since the sixteenth century. A wise government that took the best interests of humanity to heart would propagate enlightenment and education; it would succeed in increasing the colonists' physical well-being by gradually making this monstrous inequality of rights and fortunes disappear; but it would find enormous difficulties to surmount when trying to make the inhabitants sociable and teach them to consider each other as fellow citizens.

I.465   Let us not forget that in the United States, society was formed in a completely different way than in Mexico and in other continental regions of the Spanish colonies. Exploring deep into the Allegheny Mountains, the Europeans found vast forests with some wandering tribes of hunters whom nothing could attach to the uncultivated soil. At the arrival of the new colonists, the indigenous peoples gradually withdrew into the western savannahs that border the Mississippi and the Missouri Rivers. In this manner, free men of the same race and origin became the first components of a new people. "In North America," writes a famous statesman, "a traveler who leaves a major city, where the social contract has attained perfection, will successively cross every stage of civilization and industry, which become increasingly weaker until in a few days' time, he reaches a rough and shapeless cabin built of recently cut logs. Such a journey is a sort of practical analysis of the origin of nations and states. One departs from the most complex group only to arrive at the simplest elements, traveling backwards through the history of the advancement of the human spirit. One encounters in space what is only the result of the passage of time."[1]

Nowhere in New Spain and Peru, with the exception of the Missions, have the colonists returned to a natural state. Settling among agrarian peoples who themselves lived under despotic and complicated governments, the Europeans benefited from the advantages of their more developed civilization, their astuteness, and the authority that the conquest gave them. This particular situation, and the intermingling of races whose interests are diametrically opposed, became an inexhaustible source of hatred and discord. As the Europeans' descendants became more numerous than those whom the metropole sent there directly, the white race was divided into two groups whose mutual resentments even the ties of blood could not assuage. The colonial administration thought that it could take advantage of this dissension through misleading politics. As the colonies grew, the administration's stance became more mistrustful. According to ideas that have unfortunately been accepted for centuries, these distant regions are considered tributaries of Europe. They are not governed in the manner that public interest requires, but, rather, as the fear of seeing the prosperity of their population increase too rapidly dictates. Seeking security in civil dissension, in the balance of powers, and in obstructing all the wheels of the great political machinery, the metropole strives constantly to foment a factional spirit and increase mutual odium among the castes and constituent authorities. From this state of affairs arise uneasiness and a rancor that disturb the enjoyment of social life.

In the first two volumes of this work, I have examined the area of New Spain, the physical appearance of the country and the different races of its inhabitants. In the third volume, I shall assemble what reliable information I have been able to collect about the provinces and intendancies that comprise the vast Mexican territory.

I.467

END OF VOLUME ONE

1. ▼ Prince of Talleyrand [-Périgord], *Essai sur les [avantages à retirer de] colonies nouvelles*, I.

VOLUME 2

BOOK III

Specific Statistics of the Intendancies
That Comprise the Kingdom of
New Spain—Their Territorial Extent
and Population

CHAPTER VIII

On the Political Division of the Mexican Territory and the Relationship between the Population of the Intendancies and Their Territorial Extent—Principal Cities

Before presenting the table that contains the specific statistics of the intendancies of New Spain, we shall discuss the principles on which the new
territorial divisions are based. These divisions are entirely unknown even
to the most up-to-date geographers, and we shall repeat here what we indi- II.2
cated in the Introduction to this work: that our general Map printed in the
Mexican Atlas is the only one that shows the borders of the intendancies
since 1776.

In the second edition of his Modern Geography,[1] ▼ Mr. Pinkerton has
attempted to give a detailed description of the Spanish possessions in North
America. He has combined many precise notions taken from the *Viagero
universal* [by Estala] with vague information from ▼ Mr. Alcedo's Dictionary. This author, who considers himself exceptionally well-informed about

---

1. It has just been announced ([Caritat,] *Bibliothèque americaine*, 1808, number 9)
that Mr. Pinkerton used my manuscripts for his work on Mexico. With the frankness that
is natural to my character, I sent some notes in manuscript to Mr. Bourgoing, Mr. Laborde,
and some other learned men who are equally respectable. I have never corresponded with
Mr. Pinkerton; and the manner in which he had treated me in his *[Modern] Geographye*, before my return to Europe, does certainly not compel me to engage with him. A compiler who
is as imprecise as he is bold, Mr. Pinkerton, in the style that is his own, finds "ridiculous,
disgusting, and absurd" all that run counter to the ideas he developed in his own chamber.
Ignorant of the fact that La Cruz's map is drawn on top of Caulin's, he does not allow any
different course for the rivers than what he found marked out in the former. It is necessary
to be very skeptical of his claims that ▼ Mr. De Pons, the author of *Voyage à la Terre Ferme*,
did not even know the name of the country in which he had traveled for four years! The notes
that accompany the new edition of Mr. Pinkerton's *Geography* contribute to the spread of the
most fatuous ideas about physics and descriptive natural history.

the real territorial divisions[1] of New Spain, considers the provinces of So-
II.3 nora, Sinaloa, and La Pimería as part of Nueva Vizcaya. He divides what
he calls the *Mexican Domain* into the districts of New Galicia, Panuco,
Zacatula, etc. According to the same principle, the important divisions of
Europe would be Spain, Languedoc, Catalonia, and the districts of Cádiz
and Bordeaux.

Before Count José de Gálvez, the minister of the West Indies, introduced the new administration, New Spain comprised (1) the Reino of
Mexico; (2) the Reino of Nueva Galicia; (3) the Nuevo Reino of León; (4)
the Colonía of Nuevo Santander; (5) the Provincia of Texas; (6) the Provincia of Coahuila; (7) the Provincia of Nueva Vizcaya; (8) the Provincia
of Sonora; (9) the Provincia of Nuevo Mexico; (10) Ambas Californias, or
the Provincias de la Vieja y Nueva California. These former divisions are
still frequently used in the country. The same border that separates Nueva
Galicia from the Reino of Mexico, which includes a section of the ancient
kingdom of Michoacan, is also the line of demarcation between the jurisdiction of the two Audiencias of Mexico and Guadalajara. This line, which
I was unable to draw on my general Map, does not, however, follow the outline of the new intendancies exactly. It begins on the coastline of the Gulf
of Mexico, ten leagues north of the Panuco River and the town of Altamira,
near Bara Ciega, and crosses the Intendancy of San Luis Potosí all the way
to the Potosí and de Bernalejo mines; from there, running along the south-
II.4 ern extremity of the Intendancy of Zacatecas and the western border of the
intendancy of Guanajuato, it continues through the intendancy of Guadalajara, between Zapotlán and Sayula, between Ayotitan and the town of La
Purificación, on to Gualán, one of the ports on the Pacific Ocean. Whatever

1. According to the Federal Constitution of the United States of Mexico proclaimed on
October 4, 1824, the present divisions are: the States of Chiapas and Chihuahua; those of
Coahuila and Texas, Durango, Guanajuato, Mexico City, Michoacán, Nuevo León, Oaxaca,
Puebla de los Angeles, Querétaro, San Luis Potosí, Sonora and Sinaloa, Tabasco, Tamauli-
pas (formerly Nuevo Santander), Veracruz, Jalisco, Yucatán, and Zacatecas; the Territory of
Upper California and Lower California, and those of Colima and Santa Fé in New Mexico.
Like the Territories of the Confederation, the two Californias and the Partido de Colima (not
including the village of Tonalá, which is still part of Jalisco) are immediately subject to the
supreme authorities that govern it. The countries that comprise the province of the Isthmus
of Guazacualcos are still included in the Territory of the State of Veracruz, but Lake Términos belongs to the State of Yucatán. In the list of these political divisions, one will recognize
that the borders of the former intendancies have been preserved. It will be easy to remove at a
later date the drawbacks that arise from uneven population growth.

lies north of this line belongs to the Audiencia of Guadalajara; whatever lies south of it, to the Audiencia of Mexico.

In its present state, New Mexico is divided into twelve intendancies, to which must be added three other districts that are quite far from the capital and that have preserved the simple name of provinces. These fifteen divisions are:

I. IN THE TEMPERATE ZONE: 82,000 square leagues, with 677,000 souls, or eight inhabitants per square league.
   A. *The Northern Region*, an interior region.
      1. PROVINCIA OF NUEVO MEXICO, along the Río de Norte, north of the thirty-first parallel.
      2. INTENDENCIA OF NUEVA VIZCAYA, southwest of the Río del Norte, on the central plateau that descends rapidly from Durango toward Chihuahua.
   B. *The Northwest Region*, near the Great Ocean.
      3. PROVINCIA OF NUEVA CALIFORNIA, or the northwest coast of North America, which the Spanish occupy.
      4. PROVINCIA OF ANTIGUA CALIFORNIA. Its southern extremity is already in the Torrid Zone.
      5. INTENDENCIA OF LA SONORA. The southernmost part   II.5
         of Sinaloa, where the famous mines of Copala and Rosario are located, also crosses over the tropic of Cancer.
   C. *The Northeast Region*, near the Gulf of Mexico.
      6. INTENDENCIA OF SAN LUIS POTOSÍ. It includes the provinces of Texas, the colony of Nuevo Santander and Coahuila, the Nuevo Reino of León, the districts of Charcas, Altamira, Catorce, and Ramos. The latter districts comprise the intendancy of San Luis, properly speaking. The southern part, which extends south of the Barra de Santander and the Real de Catorce, is in the Torrid Zone.
II. IN THE TORRID ZONE. 36,500 square leagues, with 5,160,000 souls, or 141 inhabitants per square league.
   D. *The Central Region*.
      7. INTENDENCIA OF ZACATECAS, except the part that extends north of the Fresnillo mines.
      8. INTENDENCIA OF GUADALAJARA.
      9. INTENDENCIA OF GUANAJUATO.
      10. INTENDENCIA OF VALLADOLID.

11. INTENDENCIA OF MEXICO CITY.
12. INTENDENCIA OF PUEBLA.
13. INTENDENCIA OF VERACRUZ.
E. *The Southwest Region.*
14. INTENDENCIA OF OAXACA.
15. INTENDENCIA OF MÉRIDA.

II.6   The divisions shown in this table are based on the physical state of the country. It is clear that almost seven-eighths of the population live in the Torrid Zone. The population becomes even sparser as one moves toward Durango and Chihuahua. In this respect, there are striking similarities between New Spain and Hindustan, which also borders in the north on nearly uncultivated and uninhabited regions. Among the five million who dwell in the equinoctial part of Mexico, four-fifths live on the ridge of the Cordillera or on the plateaus, where the elevation above sea level is as high as the Mont Cenis pass.

If we consider the provinces of New Spain according to their trade relations or the location of their direct coastlines, the country is divided into three regions.

I.   INTERIOR PROVINCES, which do not extend to the shores of the Ocean:
    1. NUEVO MEXICO.
    2. NUEVA VIZCAYA.
    3. ZACATECAS.
    4. GUANAJUATO.

II. MARITIME PROVINCES *on the eastern coastline*, facing Europe:
    5. SAN LUIS POTOSÍ.
    6. VERACRUZ.
    7. MÉRIDA or YUCATÁN.

II.7   III. MARITIME PROVINCES *on the western coastline*, facing Asia:
    8. NUEVA CALIFORNIA.
    9. VIEJA CALIFORNIA.
    10. SONORA.
    11. GUADALAJARA.
    12. VALLADOLID.
    13. MEXICO CITY.
    14. PUEBLA.
    15. OAXACA.

These divisions will be of great political interest one day, when Mexican agriculture is less concentrated on the central plateau or on the ridge of the Cordillera, and when the coastline becomes more populated. The western coastal provinces will send their ships to Nootka, China, and the East Indies. The Sandwich Islands, which are inhabited by a savage, industrious, but ambitious people, seem destined for Mexican rather than European colonists. They offer an important scale to nations that conduct trade between free ports on the Great Ocean. Until now, the inhabitants of New Spain and Peru have not been able profit from the advantages of their location on a coast that faces Asia and New Holland. The do not even know the products of the islands in the Pacific Ocean. Some day, Tahitian breadfruit and sugarcane, the precious reed whose cultivation had such a fortunate effect on the trade in the Antilles (rather than in the nearby islands), will come to them from Jamaica, Havana, and Caracas! In the last few years, the confederated states of North America have made a concerted effort to open a road to their western coastline, the same coastline where Mexicans have their most beautiful ports; but these ports are lifeless and without trade! | II.8

According to the old division of the country, the *Reino of Nueva Galicia* was more than fourteen thousand square leagues wide and had almost one million inhabitants; it included the intendancies of Zacatecas and Guadalajara,[1] as well as a small part of San Luis Potosí. The regions that are now grouped under the heading of the seven intendancies of Guanajuato, Valladolid or Michoacan, Mexico City, Puebla, Veracruz, Oaxaca, and Mérida, with a small portion of the Intendancy of San Luis Potosí,[2] formed the *Reino of Mexico*, strictly speaking. As a result, this kingdom covered over 27,000 square leagues and had a population of almost four and a half million.

Another division of New Spain that is equally old but less vague is the one that distinguishes *New Spain proper* from the *Provincias internas*. With the exception of the two Californias, everything that is north and northwest of the kingdom of New Galicia belongs to the *provincias*: (1) the | II.9 small kingdom of León; (2) the colony of Nuevo Santander; (3) Texas; (4) Nueva Vizcaya; (5) Sonora; (6) Coahuila; and (7) New Mexico. The *Provincias internas del Virreinato*, which cover 7,814 square leagues, are different from the *Provincias internas de la Comandancia* (de Chihuahua),

1. With the exception of the southernmost strip of land, where the Colima volcano and the village of Ayatitlán are found.
2. The southernmost part that the Panuco River crosses.

which were established as a *Capitanía general* in 1779; the latter occupy
59,375 square leagues. Three of the twelve new intendancies are located
in the internal provinces, those of Durango, Sonora, and San Luis Potosí.
One should, however, remember that the intendant of San Luis is subject
directly to the viceroy for León, Santander, and the districts that are near
his residence—Charcas, Catorce, and Altamira. The governments of Coahuila and Texas are also part of the Intendancy of San Luis Potosí, but they
belong directly to the *Comandancia general de Chihuahua*. The following
tableaus will clarify these rather complicated territorial divisions. All of
New Spain is thus divided into:

   A. *Provincias sujetas al Virrey de Nueva España*; 59,103 square leagues,
      with 547,790 souls:
         the ten intendancies of Mexico City, Puebla, Veracruz, Oaxaca,
            Mérida, Valladolid, Guadalajara, Zacatecas, Guanajuato, and
            San Luis Potosí (not including Coahuila and Texas);

II.10   the two Californias;
   B. *Provincias sujetas al comandante general de provincias internas*,
      59,375 square leagues, with 359,200 inhabitants.
      The two intendancies of Durango and Sonora.
      The province of New Mexico.
      Coahuila and Texas.

All of New Spain: 118,478 square leagues, with 5,837,100 inhabitants.

These tables present the surface area of the provinces calculated in
square leagues, at twenty-five to a degree, according to the general map
in my Mexican Atlas. Mr. Oteiza and I did the early calculations in 1803 in
Mexico City itself. Because my geographical works have since been perfected, Mr. Oltmanns agreed to recalculate all the territorial surfaces. He
undertook this work with the precision that characterizes all his undertakings and created squares whose sides have only three minutes in arc.

The population indicated in my tables is what one can assume it was in
1803. I developed above, in the fourth chapter (pp. 57 and 65 [pp. 199ff in
this edition]), the principles on which the changes to the numbers from
the 1793 census are based. I am aware that modern geographers only allow a population of two to three million in Mexico. The population of
Asia has always been inflated, just as that of the Spanish possessions in the
II.11 Americas has been underestimated. One forgets that a population makes
rapid strides in a fair climate and on fertile soil, even in countries that are

the most poorly administrated. One forgets that men who are spread across an immense terrain suffer from the imperfections of the social state less than when the population is very concentrated.

The northern and eastern borders of New Spain are difficult to gauge. It is not enough for a missionary monk to have crossed the country, or for a royal navy ship to have sighted a single coastline, to determine that they belong to the Spanish colonies in the Americas. In 1770, Cardinal Lorenzana announced in print in Mexico City that the diocese of Durango in New Spain possibly bordered on Tartary and Greenland![1] People are too well-versed in geography today to engage in such fanciful assumptions. One Mexican viceroy sent a team from San Blas to visit the Russians' American colonies on the Alaska Peninsula. The Mexican government's attention has long been focused on the northwest coast, especially the settlement at Nootka, which the court in Madrid was compelled to abandon to avoid war with Great Britain. The inhabitants of the United States are pushing their civilization toward the Missouri. They tend to advance toward the coastline of the Great Ocean, where they are lured by the fur trade. The time is drawing closer when the borders of New Spain will touch those of the Russian Empire and the great confederation of American republics. At present, the Mexican government only extends as far as the Mission of San Francisco on the west coast, south of Cape Mendocino, and as far as the village of Taos in New Mexico. To the east, toward the state of Louisiana, the borders of the intendancy of San Luis Potosí are rather vague; the Congress in Washington tends to fix them on the right bank of the Río Bravo del Norte, whereas the Spanish include the savannahs that extend from the Río Mexicano (or Mermentas), east of the Río Sabina, which they call part of the province of Texas.

II.12

The following table presents the surface area and population of the largest political groupings in Europe and Asia. It provides fascinating comparisons with the present state of Mexico.

We can see from this table, which may give rise to some interesting observations on the disproportionate nature of European culture, that New Spain is almost four times as large as France, with a population that even today is almost seven times smaller. A comparison of the United States and

1. [Cortés, *Historia de Nueva-España*,] Lorenzana ed., p. 38: "The empire of all of New Spain, from the Isthmus of Panama to the most remote corner of the Diocese of Durango, encompasses one thousand five hundred leagues of longitude, and yet one overlooks that it borders on Tartary and Greenland: on Tartary because of the Californias, and an Greenland because of New Mexico."

II.13	MAJOR POLITICAL UNITS IN 1804	SQUARE LEAGUES AT TWENTY PER DEGREE	TOTAL POPULATION	INHABITANTS PER SQUARE LEAGUE
	Russian Empire	616,000	54,000,000	87
	1) European part	150,000	52,000,000	345
	2) Asian part	465,600	2,000,000	4
	All of Europe up to the Urals	304,700	195,000,000	639
	The United States of North America	174,300	10,220,000	58
	British Empire in India	90,100	73,000,000	810
	The Austrian Monarchy	21,900	29,000,000	1,324
	France	17,100	30,616,000	1,790
	Spain	15,000	11,446,000	763
	Spanish Colonies in the Americas	371,400	16,785	45
	New Spain	75,830	6,800,000	90

Mexico presents especially striking ratios, if one considers Louisiana and the western territory as the *provincias internas* of the great confederation of American republics.

In this chapter, I have presented the state of the *provincias internas* as it was during my stay in Mexico. Since then, there has been a change in the military government of these vast provinces, whose surface area is almost twice that of France. In the year 1807, two *comandantes generales*, ▼ brigadiers Don Nemesio Salcedo and Don Pedro Grimarest, governed these northern regions.

This is the present division of the *Gobierno militar*, which is no longer exclusively controlled by the governor of Chihuahua:

PROVINCIAS INTERNAS DEL REINO DE NUEVA ESPAÑA:
A. *Provincias internas occidentales*:
    1. SONORA.
    2. DURANGO O NUEVA VIZCAYA.
    3. NUEVO MÉXICO.
    4. CALIFORNIAS.

B. *Provincias internas orientales*:
1. COAHUILA.
2. TEXAS.
3. COLONIA DEL NUEVO SANTANDER.
4. NUEVO REINO DE LEÓN.

The new *comandantes generales* of the internal provinces are considered the directors of the financial administration of the two intendancies of Sonora and Durango in the province of New Mexico and in the part of the Intendancy of San Luis Potosí that includes Texas and Coahuila. As for the small kingdom of León and Nuevo Santander, they are only under the commanders' control with respect to military defense.

Statistical Analysis of the Kingdom of New Spain   II.15

TERRITORIAL DIVISIONS	SURFACE AREA IN SQUARE LEAGUES AT 25 PER DEGREE	POPULATION IN 1803	NUMBER OF INHABITANTS PER SQUARE LEAGUE
NEW SPAIN, (surface area of the entire viceroyalty, excluding the kingdom of Guatemala)	118,478	5,837,100	49
A) PROVINCIAS INTERNAS	67,189	423,200	6
a) *Under the direct control of the viceroy* (Provincias internas del Virreinato)	7,814	64,000	8
1) NUEVO REINO DE LEÓN	2,621	29,000	10
2) NUEVO SANTANDER	5,193	38,000	7
b) *Under the control of governor of Chihuahua*, (provincias internas de la comandancia general)	59,375	359,200	6
1) INTENDENCIA DE LA NUEVA VIZCAYA O DURANGO	16,873	159,700	10
2) INTENDENCIA DE LA SONORA	19,143	121,400	6
3) COAHUILA	6,702	16,900	2
4) TEXAS	10,948	21,000	2
5) NUEVO MÉXICO	5,709	40,200	7

(continued)

TERRITORIAL DIVISIONS	SURFACE AREA IN SQUARE LEAGUES AT 25 PER DEGREE	POPULATION IN 1803	NUMBER OF INHABITANTS PER SQUARE LEAGUE
B) New Spain proper, directly under the viceroy, including los Reinos de México, Michoacán y Nueva Galicia and the two Californias	51,289	5,413,900	105
1) Intendancy of Mexico City	5,927	1,511,900	255
2) Intendancy of Puebla	2,696	813,300	301
3) Intendancy of Veracruz	4,141	156,000	38
4) Intendancy of Oaxaca	4,141	156,000	120
5) Intendancy of Mérida or Yucatán	5,977	465,800	81
6) Intendancy of Valladolid	3,446	476,400	273
7) Intendancy of Guadalajara	9,612	630,500	66
8) Intendancy of Zacatecas	2,355	153,300	65
9) Intendancy of Guanajuato	911	517,300	568
10) Intendancy de San Luis Potosí (excluding New Santander, Texas, Coahuila, and the kingdom of León)	2,357	230,000	98
11) Old California (Antigua California)	7,295	9,000	1
12) New California (Nueva California)	2,125	15,600	7

II.16   The statistical table that we have just presented shows how imperfect the territorial division is. It appears that the purpose of entrusting the administration of the police and finances to intendants was to divide Mexican land according to principles similar to those the French government followed in dividing the land into departments. In New Spain, each intendancy comprises several *subdelegations*. Similarly, departments in France were gov-

II.17   erned by *subdelegates* who performed their duties under the intendant's orders. But in the creation of the Mexican intendancies, there was little concern for the size of the territory or the more or less concentrated state of the population. This new division also took place at a time when the minister of the colonies, the Council of the Indies, and the viceroys completely lacked the necessary information for such important work. And how could one govern in detail a country whose map has not been drawn, where even the simplest calculations of political arithmetic have not yet been attempted!

When comparing the size of the surface area of the Mexican intendancies, one finds that many of them are ten, twenty, even thirty times larger than others. The intendancy of San Luis Potosí, for instance, is larger than all of European Spain, while the intendancy of Guanajuato does not exceed the size of two or three departments in France. Here is an exact inventory of the extraordinary disproportion in territorial area among the Mexican intendancies. We have arranged them in the order of their size:

Intendancy of San Luis Potosí: 27,821 square leagues
Intendancy of Sonora: 19,143 square leagues
Intendancy of Durango: 16,873 square leagues
Intendancy of Guadalajara: 9,612 square leagues
Intendancy of Mérida: 5,977 square leagues
Intendancy of Mexico City: 5,927 square leagues
Intendancy of Oaxaca: 4,447 square leagues II.18
Intendancy of Veracruz: 4,141 square leagues
Intendancy of Valladolid: 3,447 square leagues
Intendancy of Puebla: 2,696 square leagues
Intendancy of Zacatecas: 2,355 square leagues
Intendancy of Guanajuato: 911 square leagues

With the exception of the three intendancies of San Luis Potosí, Sonora, and Durango, each of which occupies more land than the entire empire of Great Britain, the other intendancies have an average surface area of three or four thousand square leagues. They are comparable in size to the kingdom of Naples or Bohemia. It is understandable that the less populated a country is, the fewer subdivisions its administration requires. No department of France is larger than 555 square leagues; the average size of a department is 300 square leagues. In European Russia and in Mexico, on the contrary, governments and intendancies are nearly ten times larger in size.

In France, the heads of the departments, or prefects, see to the needs of a population that rarely exceeds 450,000 souls, and whose average number is closer to 300,000. The governments into which the Russian empire is divided, as well as the Mexican intendancies, have a larger numbcr of inhabitants despite the difference in the state of their respective civilizations. The following table illustrates the disproportion among the populations of the different territorial divisions of New Spain; it begins with the most populous intendancy and ends with the most sparsely populated one.

II.19

> *Intendancy of Mexico City*: 1,511,800 inhabitants.
> *Intendancy of Puebla*: 813,000.
> *Intendancy of Guadalajara*: 630,500.
> *Intendancy of Oaxaca*: 534,800.
> *Intendancy of Guanajuato*: 517,300.
> *Intendancy of Mérida*: 465,700.
> *Intendancy of Valladolid*: 376,400.
> *Intendancy of San Luis Potosí*: 334,000.
> *Intendancy of Durango*: 159,700.
> *Intendancy of Veracruz*: 156,000.
> *Intendancy Zacatecas*: 153,300.
> *Intendancy of Sonora*: 121,400.

When comparing the table of the population of the twelve intendancies with that of their surface areas, one is struck most of all by the uneven distribution of the Mexican population, even in the most civilized part of the kingdom. The intendancy of Puebla, which is near the top of the second table, is almost at the end of the first one. No principle, however, should guide those who assign boundaries to territorial divisions more than the ratio of population to surface area, expressed in square leagues or in myriameters. Only in states (such as France) with the inestimable good fortune of having populations that are almost uniformly distributed across their surface area can such divisions be nearly equal. The third table presents the state of the population which one might call *relative*. To obtain the numerical results that indicate the relationship between the number of inhabitants and the size of the populated land, the *absolute* population must be divided by the size of the intendancy's territory. Here are the results of this work:

II.20

> *Intendancy of Guanajuato*: 568 inhabitants per square league.
> *Intendancy of Puebla*: 301.
> *Intendancy of Mexico City*: 255.

Intendancy of Oaxaca: 120.
Intendancy of Valladolid: 109.
Intendancy of Mérida: 81.
Intendancy of Guadalajara: 66.
Intendancy of Zacatecas: 65.
Intendancy of Veracruz: 38.
Intendancy of San Luis Potosí: 12.
Intendancy of Durango: 10.
Intendancy of Sonora: 6.

This table shows that in the intendancies where agriculture has made the least progress, the *relative population* is fifty to ninety times smaller than in the formerly civilized regions that border on the capital. This extraordinary difference in population distribution is also found in northern and northeastern Europe. In Lapland, there is scarcely one inhabitant per square league, while in other parts of Sweden, for instance in Götland, there are more than 248. In the states governed by the king of Denmark, the island of Zealand has 944, and Iceland, eleven inhabitants per square league. In European Russia, the governments of Archangel, Olonez, Kaluga, and Moscow differ so sharply in the ratio of population to territory size that the first two of these governments have six and twenty-six persons per square league, respectively, while the latter two have 842 and 974 persons per square league. These are the enormous differences that indicate that one province is 160 times larger than another.

II.21

In France, where the total population is 1,094 inhabitants per square league, the most populous departments are those of Escaut, the Nord, and the Lys, with relative populations of 3,869; 2,786; and 2,274. The least populated department, the Hautes-Alpes, a part of the former Dauphiné, has only 471 inhabitants per square league. As a result, the extremes of population in France are in a ratio of eight to one, and Guanajuato, the Mexican intendancy with the highest population density, is scarcely more inhabited than the least populated department of continental France![1]

1. These comparisons do not take into account either the departement of Liamone, which consists of the southern portion of Corsica and has only 277 inhabitants per square league, or the Seine departement. The latter appears to have a relative population of 26,165 inhabitants; it would be useless to reveal the causes that produced such an order of things so unnatural in a departement whose administrative center is the seat of a vast empire.

II.22   I hope that the three tables I have compiled of the surface areas as well as the absolute and relative populations of the intendancies of New Spain give a sufficient demonstration of how imperfect the present division of territory is. A country in which the population is scattered across a vast area requires that the provincial administration be limited to areas smaller than those that form the Mexican intendancies. Wherever the population was found to be less than one hundred inhabitants per square league, the administration of an intendancy or department should not encompass more than 100,000 inhabitants. A number that is two or three times greater could be assigned to regions with greater population density.

The level of industry, the resulting trade activity, and the number of businesses are likely all related to this density, to which the departmental government must be attentive. In this respect, the small intendancy of Guanajuato keeps an administrator busier than the provinces of Texas, Coahuila, and New Mexico, which are six to ten times larger. On the other hand, how could the intendant of San Luis Potosí ever hope to grasp the needs of a province of nearly 28,000 square leagues? Even if he devoted himself to the duties of his position with great patriotic zeal, how could he supervise the *subdelegates* and protect the Indians against the vexations that arise in those communities?

II.23   This aspect of administrative organization cannot be dealt with too carefully. Above all else, a reformist government must make it a priority to change the present borders of the intendancies. Such a political change must be based on precise knowledge of the physical and agricultural state of the provinces that constitute New Spain. In this respect, France offers an example of perfection that is worthy of imitation in the New World. From the beginning of their work, the enlightened men who created the Constituent Assembly demonstrated what great importance they attached to the proper division of territory. This division is correct when grounded in principles that can be considered wise insofar as they are simple and natural.

II.25   ## A Statistical Analysis of the Kingdom of New Spain

Territory: 118,478 square leagues (2,339,400 Myriare).
Population: 5,837,100 inhabitants, or 449 inhabitants per square league (or two and a half per Myriare).

NEW SPAIN includes: II.26

A. *Mexico proper (el Reino de Mexico)*
      Surface area: 52,280 square leagues.
            Population: 5,414,900 inhabitants, or 105 inhabitants per square league.

B. *Las provincias internas orientales y occidentales.*
      Surface area: 67,189 square leagues.
            Population: 423,200 inhabitants, or six inhabitants per square league.

New Spain

Statistical Analysis

I. The Intendancy of Mexico City

POPULATION IN 1803: 1,511,800
SURFACE AREA IN SQUARE LEAGUES: 5,927
INHABITANTS PER SQUARE LEAGUE: 255

This entire intendancy is located in the Torrid Zone. It extends from 16°34' to 21°57' northern latitude. To the north, it borders on the intendancy of San Luis Potosí, to the west on the intendancies of Guanajuato and Valladolid, and to the east on the intendancies of Veracruz and La Puebla. To the south, the waters of the South Sea or the Great Ocean wash the intendancy of Mexico City along a coastline that is eighty-two leagues long, from Acapulco to Zacatula.

Its greatest length, from the port of Zacatula to the Doctor mines,[1] is 136 leagues; its greatest width, from Zacatula to the mountains east of Chilpanzingo, is ninety-two leagues. In its northern part, toward the famous Zimapán and Doctor mines, a narrow ribbon of land separates the intendancy

---

1. The edges are specifically located southeast of Acapulco, near the mouth of the Río Nespa, and north of the Real del Doctor, near the town of Valles, which is already in the intendancy of San Luis Potosí. Since noteworthy sites are rarely located on the borders themselves, we have preferred to mention those closest to them. A glance at my general map of New Spain will suffice to justify this method of indicating the borders of the intendancies.

of Mexico City from the Gulf of Mexico; near Mazatlán, this strip is only nine leagues wide.

Over two-thirds of the intendancy of Mexico City is mountainous country with immense plateaus over 2,000 to 2,300 meters above sea level. From Chalco to Querétaro, there are almost uninterrupted plains that are fifty leagues long and eight to ten leagues wide; in the part closest to the west coast, the climate is hot and very unhealthy. Only one peak, Nevado de Toluca, located on a fertile plain 2,700 meters in elevation, reaches the lower edge of the eternal snow line. But even the porphyritic summit of this very old volcano, whose shape greatly resembles that of Pichincha near Quito, and which seems to have been extremely high long ago, also loses its snow in the rainy months of September and October. The elevation of the Pico del Fraile, the highest peak of Nevado de Toluca, is 4,620 meters (2,370 toises). No mountain in this intendancy is as high as Mont Blanc. II.29

The Valley of Mexico City, that is, Tenochtitlán, of which I have published a very detailed map, is located in the middle of the Anahuac Cordillera, on the ridge of the porphyritic and basaltic amygdaloid mountains that extend from south-southeast to north-northwest. This valley is oval-shaped. According to my observations and those of a distinguished mineralogist, Mr. Don Luis Martín, it is eighteen and one-third leagues long—from the mouth of the Río Tanango at Lake Chalco to the foot of the Cerro de Sincoq, near the Desagüe Real de Huehuetoca—and twelve and a half leagues wide, from San Gabriel near the small town of Texcoco to the source of the Río de Escapusalco, near Guisquiluca.[1] The territorial area of the valley is 244 and a half square leagues, of which the lakes occupy only twenty-two square leagues, which is not quite one-tenth of the entire surface area.

The circumference of the valley, measuring from the ridge of the mountains that surround it like a circular wall, is sixty-seven leagues. This ridge is highest to the south, especially to the southeast, where the two great volcanoes of La Puebla—Popocatepetl and Iztaccihuatl—border the valley. One of the roads that connects the valley of Tenochtitlán with the valley of Cholula and Puebla runs between the same two volcanoes via Tlalmanalco, Ameca, La Cumbre, and La Cruz del Correo. Cortés's small army traveled this road during its first invasion. II.30

1. The maps of the valley of Mexico City that have heretofore been published are so inaccurate that on Mr. Mascaró's map, which is printed every year in the Mexico City almanac, the distances I give above are twenty-five and seventeen leagues, instead of eighteen and twelve. The archbishop Lorenzana is probably referring to this map when he cites a circumference of over ninety leagues for the entire valley, when it is actually one-third less.

Six important roads cross the Cordillera that encloses the valley, with a mean elevation is 3,000 meters above sea level: (1) the road from Acapulco to Guchilaque and Cuervaracca over the high peak called La Cruz del Marqués[1]; (2) the road from Toluca via Tianguillo and Lerma, a magnificent highway that I could not admire enough, beautifully built, partly across natural bridges; (3) the Querétaro, Guanajuato, and Durango road, *el camino de tierra adentro*, which runs through Gauatitlán, Huehuetoca, and Puerto de Reyes, near Bata, over hills that are scarcely eighty meters higher than the pavement of the great square in Mexico City; (4) the Pachuca road. It runs to the famous mines of Real del Monte, over the Cerro Ventoso which is covered with oaks, cypresses, and rose bushes that are almost always in bloom; (5) the old La Puebla road via San Bonaventura and the Llanos de Apán; and finally (6) the new road from La Puebla via Río Frio and Tesmelucos southeast of the Cerro del Telapón, whose distance from the Sierra Nevada, much like the distance from the Sierra Nevada (Iztaccihuatl) to the great volcano (Popocatepetl), formed the basis of Mr. Velázquez and Mr. Costanzó's trigonometric calculations.

Those long accustomed to hearing the capital of Mexico described as a city built in the middle of a lake and connected to land only through causeways will probably be surprised when they glance at my Mexican Atlas and see that the center of the present city is 4,500 meters away from Lake Texcoco, and over 9,000 meters away from Lake Chalco. They will be led to doubt the accuracy of the descriptions in the history of discoveries in the New World or else will think that the capital of Mexico is not built on the same soil as Montezuma's former residence.[2] But the city has certainly not changed its place; the cathedral of Mexico City stands in exactly the same place where the temple of Huitzilopochtli stood; the present street of Tacuba is the old street of Tlacopan on which Cortés made his famous retreat on July 1, 1520, during the *sad night* that is referred to as the *Noche triste*. The difference in location between what the old maps show and

II.31

II.32

1. At the beginning of the conquest, this was a military outpost. When the residents of New Spain say "el Marqués" without adding a family name, they are referring to Hernán Cortés, Marqués de la Valle de Oaxaca. Similarly, in Spanish America, the expression *el Almirante* designates Christopher Columbus. This plain way of speaking proves the respect and admiration that they preserve in memory of these great men.

2. The real Mexican name of this king is *Motecuzuma*. There are two different kings with this name in the genealogy of the Aztec rulers, the first of whom is called Huehue *Motecuzoma*, and the second, who died as Cortés's prisoner, *Motecuzoma Xocojotzin*. The adjectives placed before and after the proper name signify "elder" and "younger."

what I have printed is explained solely by the drop in the water level of Lake Texcoco.

It will be useful here to recall the passage from one of Cortés's letters to Emperor Charles V, dated October 30 of the year 1520, in which he gives a description of the Valley of Mexico. The passage is written with a great simplicity of style and at the same time gives us a good idea of the order that reigned in old Tenochtitlán. "The province where the great lord *Moctezuma's* residence is located," writes Cortés, "is surrounded by a circle of high mountains and crosscut by precipices. The plain is nearly seventy leagues in circumference, and the two lakes on this plain nearly fill the entire valley, for within a radius of over fifty leagues the locals travel in canoes." (It should be noted that the general speaks of only two lakes, because he was not very familiar with Lakes Sumpango and Xaltocan, between which he passed on his hasty retreat from Mexico City to Tlaxcala before the battle of Otumba.) "Of the two large lakes in the Valley of Mexico, one is freshwater and the other salt water. They are separated from each other by a low mountain range [the cone-shaped, solitary hills near Iztapalapa]. These mountains rise in the middle of the plain, and the waters of the lake blend together in a strait between the hills and the high Cordillera [most likely the eastern slope of the Cerros de Santa Fe]. The towns and numerous villages built on both lakes conduct all their trade by canoe, without touching dry land. The large city of Temixtitlan[1] [Tenochtitlán] is built in the middle of the salty lake, which has tides, like the ocean. Only two leagues separate the city from dry land, from whichever direction one may choose to enter it. Four manmade causeways also lead to the city; they are two lances wide. The city is as large as Seville or Córdoba. The streets (I speak only of the main ones) are either very narrow or very wide, some are half-dry and half-filled with navigable canals, and fitted out with well-built wooden bridges that are wide enough for ten men on horseback to cross at the same time. The market is twice the size of the one in Seville and surrounded by a great portico where all sorts of wares are displayed: comestibles; ornaments made of gold, silver, lead, tin, gemstones, bone, shells, and feathers; crockery; leather; and spun cotton. One also finds hewn stone, tiles, and timber for construction. There are separate corridors for wild game, and others for vegetables and gardening tools. There are houses where barbers

II.33

II.34

1. Temistitan, Temixtitan, Tenoxtitlan, Temihtitlan are the corruptions of the real name Tenochtitlán. The Aztecs, or Mexica, also call themselves *Tenochques*, from which the name *Tenochtitlán* is derived.

(with razors made from obsidian) shave people's heads, and other houses that resemble our apothecaries, where prepared medicines, ointments, and plasters are sold. There are other houses where one can purchase food and drink. The market offers such an abundance of things that I could never list them all for Your Highness. To avoid confusion, each kind of merchandise is sold on a different corridor; everything is sold by the ell [yardstick], but until now, no one has been seen weighing anything in the marketplace. In the middle of the great square sits a house that I shall call the *audiencia*, where ten or twelve persons are always seated who settle disputes arising from the sale of wares. There are others who constantly mingle with the crowd to ensure that prices are fair. They have been seen destroying the rigged measures that they confiscated from merchants."

II.35

This was the state of affairs in Tenochtitlán in 1520, according to Cortés himself. I have searched in vain in his family archives, preserved in Mexico City in the Casa del Estado, for the map that the great captain had drawn of the environs of the capital, and which he sent to the emperor, as he describes in his third letter, published by Cardinal Lorenzana. The Abbot Clavijero has ventured to produce a map of Lake Texcoco as he supposes it to have been in the sixteenth century. This sketch is not very precise, although it is far preferable to the one provided by Robertson and other European authors, who are equally unversed in Mexican geography. I have drawn the former extent of the salt lake on the map of the valley of Tenochtitlán, using descriptions from the personal accounts of Cortés and a few of his contemporaries. In the year 1520, and for a long time thereafter, the villages of Iztapalapa, Coyohuacan (mistakenly called Cuyacan), Tacubaja, and Tacuba were very close to the shores of Lake Texcoco. Cortés expressly states[1] that most of the houses in Coyohuacan, Culhuacan, Chulubuzco, Mexicaltzingo, Iztapalapa, Cuitaguaca, and Mizqueque were built upon stilts in the water, so that canoes could enter them through a lower door. In Cortés's day, the small hill of Chapultepec, where the viceroy, the Count de Gálvez, had a castle built, was no longer an island in the lake. On this side, dry land came to within nearly 3,000 meters of the city of Tenochtitlán; the distance of two leagues that Cortés mentions in his letter to Charles V is therefore not entirely accurate. He should have reduced it by one-half, except for the part of the western side where the porphyritic hill of Chapultepec is located. It should be understood, however, that a few centuries earlier this hill was also a small island like the *Peñol del Márquez* and the island of

II.36

---

1. [Cortés, *Historia de Nueva-España,*] Lorenzana ed., pp. 229, 195, and 102.

Los Baños. Geological observations make it very likely that the lakes had been shrinking long before the Spanish arrived, and before the Huehuetoca canal was built.

Before founding the capital, which still exists, on a group of small islands in 1325, the Aztecs or Mexica had already lived for fifty-two years on another part of the lake farther to the south, whose exact location the Indians could not show me. The Mexica left Aztlan around 1160 and arrived in the Valley of Tenochtitlán only after migrating for fifty-six years, via Malinalco in the Toluca Cordillera and via Tula. They settled first in Sumpango and later on the southern slope of the Tepeyacac Mountains, where today the magnificent cathedral dedicated to Our Lady of Guadalupe is located. In 1245 (according to the Abbot Clavijero's chronology), they arrived in Chapultepec. Harassed by the lesser princes of Xalcotan, upon whom the Spanish historians bestow the title of kings, the Aztecs, to preserve their independence, took refuge on a group of small islands called Acocolco, near the southern edge of Lake Texcoco. They lived there in wretched misery for half a century, forced to feed on the roots of aquatic plants, insects, and a problematic reptile called the *Axolotl*, which Mr. Cuvier considers to be the tadpole of an unknown salamander.[1] Enslaved by the kings of Texcoco, or Acolhuacan [Culhuacan], the Mexica were forced to abandon their village in the middle of the water and take refuge on dry land in Tizapan. The services they provided to their masters during a war against the residents of Xochimilco won them back their freedom. They settled first in Acatzitzintlan, which they called Mexicalzingo after their god of war, Mexitli or Huitzilopochtli,[2] and then in Itzacalco. Carrying out an order given by the oracle Aztlan, they migrated from Iztacalco to the islands that arose at that time east-northeast of the Chapultepec hill, in the western part of Lake Texcoco. An ancient legend preserved among this horde maintained that the fateful end of their migration would be at the place where they would find an eagle sitting atop a nopal whose roots pierced through the crevices of a boulder. The Nopal (cactus) to which the oracle referred was revealed to the Aztecs in the year 1325 (which

II.37

II.38

---

1. Mr. Cuvier described it in my *Recueil d'observations zoologiques et d'anatomie comparée*, p. 119. ▼ Mr. Duméril believes that the axolotl, of which Mr. Bonpland and I brought back well-preserved specimens, is a new species of Proteus [cave-dwelling salamander]. *Zoologie analytique*, p. 93 [94–95].

2. Huitzilin refers to the hummingbird, and opochtli means "left"; for the god was depicted with the plumes of a hummingbird under his left foot. The Europeans garbled the name Huitzilopochtli into Huichilobos and Vizlipuzli. The brother of this god, who was particularly revered by the residents of Texcoco, was called Tlacahuepan-Cuexcotzin.

is the *second Calli*[1] of the Mexican era) on a small island that served as the foundation for the Teocalli, or Teopan, that is, the house of God, as the Spanish have called the great temple of Mextitli ever since.

The first *Teocalli*, around which the new city was built, was made of wood, like the most ancient Greek temple, that of Apollo in Delphi, described by ▼ Pausanias. The stone building, whose layout Cortés and Bernal Díaz admired, was built in the same location by King Ahuitzotl in 1486. It was a pyramidal monument at the center of a vast walled courtyard, thirty-seven meters high. There were five floors, as in several of the Sakkarah pyramids, particularly the *Medehun*. The base of the Teocalli of Tenochtitlán, which is exactly oriented like the Egyptian, Asian, and Mexican pyramids, is ninety-seven meters wide and forms such a truncated pyramid that from a distance, the monument appears to be an enormous cube, on the top of which rise small altars mounted with wooden cupolas. The highest point of the cupolas is fifty-four meters above the base of the building, or the cobblestones of the compound. These details show that the Teocalli was very similar in form to the ancient monument of Babylon, which ▼ Strabo calls the mausoleum of Belus, but which was simply a pyramid dedicated to Jupiter Belus.[2] Neither the Teocalli nor the Babylonian structure was a temple in our sense of the word, that is, according to the ideas that the Greeks and Romans have bequeathed to us. All the edifices dedicated to Mexican deities were truncated pyramids; the great monuments of Teotihuacán, Cholula, and Papantla, which have been preserved to this day, confirm this theory and give an idea of what the less considerable temples erected in the cities of Tenochtitlán and Texcoco were like. Covered altars were placed atop the Teocalli; in this respect, these buildings belong to the same category as the pyramidal monuments of Asia, traces of which were formerly found as far away as Arcadia: the conical mausoleum of Callistus,[3] an authentic *Tumulus* covered with fruit trees, formed the base of a small temple dedicated to Diana.

We know nothing about the construction materials of the Teocalli of Tenochtitlán. Historians merely report that the monument was covered

II.39

II.40

---

1. Since the *first Acatl* corresponds to the common year 1519, the *second Calli*, in the first half of the fourteenth century, can only be the year 1325, and not 1324, 1327, and 1342, the years when the interpreter of the *Raccolta di Mendoza*, as well as Sigüenza [y Góngora] (as cited by Boturini), and Betancourt (as cited by Torquemada) set the foundation of Mexico City. See the Abbot Clavigero's chronological dissertation, *Storia di Messico*, vol. 4, p. 54.

2. ▼ *Zoëga, De [usu] obelisorum*, p. 50.

3. Pausanias, [*Graeciae descriptio accvrata*,] book VIII, chap. 35.

with a hard, polished stone. The enormous fragments that one occasionally finds around the present cathedral are of porphyry with a base of greenstone filled with amphibole and vitreous feldspar. When the square around the cathedral was recently paved, sculpted rocks were found up to ten to twelve meters deep. There are few other peoples than the Mexica who have moved such large masses. The calendar stone and the sacrificial stones on view to the public in the great square are eight to ten cubic meters. The colossal statue of Teoyamiqui, covered with hieroglyphs and housed in one of the vestibules of the university, is two meters long and three meters wide. The canon, ▼ Mr. Gamboa, assured me that during the digging across from the Sagrario chapel, a sculpted rock seven meters long, six meters wide, and three meters high was found among a vast number of idols belonging to the Teocalli. They labored in vain to remove it.

The Teocalli was already in ruins[1] a few years after the siege of Tenochtitlán, which like that of Troy, ended with the almost total destruction of the city. I am thus inclined to believe that the exterior of the truncated pyramid was clay covered with a porous amygdaloid called *Tezontli*. In fact, quarries of this vesicular, spongy rock were first mined shortly before the temple was built, during the reign of King Ahuizotl. But nothing was easier to destroy than structures made of these light, porous materials, which resemble pumice. Although several accounts concur on the dimensions of the teocalli,[2] it is quite possible that these were slightly exaggerated. But the pyramidal shape of this edifice and its great similarity to the most ancient monuments of Asia should interest us much more than its sheer bulk and size.

II.41

II.42

The predecessor to Mexico City was connected to the mainland by three large causeways, the Tepejacac (Guadelupe), the Tlacopan (Tacuba), and the Iztapalapa. Cortés mentions four, no doubt because he is also

1. One of the most precious and oldest manuscripts preserved in Mexico City is the book of the Municipality (*libro de el Cabildo*). Father Pichardo, a respectable cleric well-versed in his country's history, showed me this manuscript, which begins on March 8, 1524, that is, three years after the siege. It mentions the place where the great temple had been ("la Plaza adonde estaba el templo mayor").

2. If the travelers who left us descriptions and drawings of the Teocalli merely related what the Indians had told them, instead of measuring it themselves, the fact that their accounts concur proves less than one might at first believe. In every country there are traditions and conventions governing the size of buildings, the height of towers, the diameter of craters, and the elevation of cataracts. National pride is only too willing to exaggerate these dimensions, and travelers' reports are in agreement insofar as they draw on the same source. In any case, in the particular case that concerns us, the exaggeration of height was probably fairly minor, since it was easy to determine the height of the monument according to the number of steps that led to the top.

counting the roadway that led to Chapultepec. The Calzada de Iztapala-pan had a branch that linked Choyohuacan with the small fort called *Xoloc*, the same one where the Mexican nobility greeted the Spanish when they first arrived. Robertson mentions a causeway that led to Texcoco, but such a causeway never existed, because of the distances involved and the great depth of the eastern side of the lake.

Seventeen years after the founding of Tenochtitlán in 1338, some of the residents separated from the others in the midst of civil discord. They settled on the small island northwest of the temple of Mexitli. The new city, which first took the name of Xaltilolco and then Tlatelolco, had a king who was independent of the ruler of Tenochtitlán. In the center of Anahuac, as in the Peloponnese, Latium, and everywhere human civilization had recently emerged, each city constituted a separate state for a long time. The Mexica king Axayacatl[1] conquered Tlatelolco, which was thenceforth connected to the city of Tenochtitlán by bridges. Among the hieroglyphic manuscripts of ancient Mexicans preserved in the viceroy's palace, I came across an inter-esting painting that depicts the last king of Tlatelolco, called Moquihuix, who was killed on top of a *house of God* (or truncated pyramid) and then thrown down the stairs that led to the sacrificial stone. After that calamity, the great Mexica market, previously held near the Teocalli of Mexitli, was moved to Tlatelolco. This is the very city that Cortés evoked in the descrip-tion of the Mexica market that we cited earlier.

What is now called the Barrio de Santiago takes up only a part of ancient Tlatelolco. One can walk for over an hour among the ruins of this ancient city on the road that leads to Tlalnepantla and Ahahuete. There, as on the road from Tacuba to Iztapalapa, one notes how much smaller Mexico City, as rebuilt by Cortés, is than Tenochtitlán was under the last of the Montezumas. The enormous size of the Tlatelolco market, whose boundaries are still dis-cernable, suggests just how large the population of the ancient city must have been. On this same square, the Indians point out a high point surrounded by walls. It is the site of one of the theaters of the war with the Mexica, where Cortés, a few days before the end of the siege, had established the famous catapult (*trabuco de palo*)[2] that towered over the besieged; but the contraption itself proved ineffectual, because of the artillerymen's clumsiness. This eleva-tion is now part of the porch of the chapel of Santiago.

II.43

II.44

1. Clavijero, [*Storia antica del Messico,*] vol. I, p. 251. Axayacatl reigned from 1464 to 1477 (IV, p. 58).

2. [Cortés, *Historia de Nueva-España,*] Lorenzana ed., p. 289.

The city of Tenochtitlán was divided into four districts, called Teopan or Xochimilco, Atzacualco, Moyotla, and Tlaquechiuhcan (or Cuepopan). This old division has been preserved until now in the borders of the districts of San Pablo, San Sebastián, San Juan, and Santa María. The present streets run largely in the same direction as they did before, more or less from north to south and east to west.[1] But as we observed above, what gives the new city a special and distinctive character is that it is almost entirely on dry land, between the edges of the two lakes, Texcoco and Xochimilco, and that it receives freshwater only via navigable canals from the latter.

Several circumstances contributed to this new order of affairs. The part of the salt lake between the southern and western causeways has always been the shallowest. Cortés himself complained that, despite the openings in the causeways, his fleet, the brigantines that he had built in Texcoco, could not sail all the way around the besieged city. These shallow pools of water gradually became marshlands. The ones that were intersected by trenches or small outflow canals were converted into *chinampas* and arable land. Lake Texcoco, which ▼ Valmont de Bomare[2] presumed to be connected with the Ocean—although, according to my measurements, it lies at an elevation of 2,277 meters—has no specific sources, as Lake Chalco does. If, on the one hand, one considers the small volume of water that small rivers contribute to this lake during dry years, and, on the other hand, the rapid rate of evaporation on the Mexican plateau, where I have conducted repeated experiments, one must admit—and geological experiments seem to confirm this—that the imbalance between the evaporated water and the affluent water progressively shrunk Lake Texcoco over the centuries. We learn from the Mexica annals[3] that, during the reign of King Ahuizotl, the water levels in this salt lake were already so low that navigation was disrupted; to obviate this problem and increase the number of tributaries to the lake, an aqueduct was hastily built from Coyohuacan to Tenochtitlán. This aqueduct brought the waters of Huitzilopochco to several of the city's canals that had dried up.

II.45

II.46

1. Specifically, from south 16° west to north 74° east, at least in the direction of the convent of Saint Augustine, where I took my azimuths. The direction of the former streets was probably determined by that of the main causeways. According to the location of the places where these causeways seem to have ended, it is highly unlikely that they represented meridians and parallels exactly.

2. [Valmont-Bomare,] *Dictionnaire [raisonné universel] d'histoire naturelle*, [vol. 7], LAC.

3. Paintings preserved in the Vatican library and Father Acosta's testimony.

This decrease in water, which had already been felt before the Spanish arrived, would probably have been very slow and barely noticeable had the human intervention not reversed the natural order since the time of the conquest. Those who have traveled across the Peninsula know how much the Spanish, even in Europe, hate plantings that offer shade around towns and villages. It seems that the first conquerors wanted the beautiful valley of Tenochtitlán to resemble in every way the soil of Castile, arid and devoid of vegetation. Since the sixteenth century, the plateau and the surrounding mountains have been clear-cut. The construction of the new city, which began in 1524, required a large quantity of lumber and stilts. People felled trees and still fell them daily without replanting, except around the capital, where the most recent viceroys have memorialized themselves through promenades[1] (*Paseos, Alamedas*) that bear their names. The lack of vegetation exposes the soil to direct sunlight, and whatever moisture remains, after the water has filtered through the spongy, basaltic amygdaloid rock, evapo-

II.47   rates quickly. It dissolves in the air wherever tree foliage or dense grasses do not protect the soil from the effects of the sun and the dry southerly winds.

Since this situation is the same throughout the valley, the abundance and circulation of water has noticeably diminished everywhere. Lake Texcoco, the most beautiful of the five lakes, which Cortés usually calls an interior *sea* in his letters, receives much less water through infiltration today than it did in the sixteenth century; the clearing of woods and destruction of forests have had the same effect everywhere. In his classic work on the Canal du Midi, ▼ General Andréossy proved that the springs around the Saint Ferréol reservoir have diminished simply because of a faulty system of forest management. In the province of Caracas, the picturesque Lake Tacarigua[2] has been slowly drying up ever since the sun has freely beamed on the deforested soil of the Aragua valley.

But the circumstance that has contributed the most to diminishing Lake Texcoco is the famous *open drain*, known as the *Desagüe real de Huehuetoca*, which we shall discuss later in this work. This *cut in the mountain*,

II.48   which was begun in 1607 in the form of an *underground tunnel,* has not only drastically shrunk the two lakes located in the northern part of the valley, Sumpango (*Tzompango*) and San Cristóbal, but has also prevented

---

1. *Paseo de Buccarely, de Revillagigedo, de Gálvez, de Asanza.*
2. The reduction of the waters also occasionally produces new islands (*las aparecidas*). Lake Tacarigua (Nueva Valencia) is 474 meters above sea level (see my *Tableaux de la nature,* vol. I, p. 72).

them from emptying their waters into the basin of Lake Texcoco during the rainy season. These waters used to flood the plains and wash over soil with a high content of carbonate and muriate of soda. Today, without standing still in pools and thereby increasing the humidity of the Mexican atmosphere, the waters run off into the Panuco River through an artificial canal and empty from there into the Atlantic Ocean.

This state of affairs resulted from the desire to convert the precursor to Mexico City into a capital that would be suitable for carriage traffic and less exposed to the danger of flooding. Water and vegetation have, in fact, diminished with the same rapidity with which Tequesquite (or carbonate of soda) has increased. In Montezuma's time and for a long time thereafter, the district of Tlatelolco, the barrios of San Sebastián, San Juan, and Santa Cruz were famous for the beautiful verdure of their gardens. Today, these same places, and especially the plains of San Lázaro, offer no more than a crust of efflorescent salts. Although still quite fertile in the southern part, the plateau is no longer as rich as when the city rose in the middle of the lake. The wise use of water, especially small irrigation canals, might restore the former fertility to the soil, and richness to a valley that nature seems to have chosen as the capital of a great empire.

II.49

The present edges of Lake Texcoco are somewhat vague; the soil is so clayey and so smooth that there is hardly a difference in level of two decimeters over the length of a mile. When the easterly winds blow strong, the water recedes toward the western edge of the lake, occasionally exposing an area over 600 meters long. It was perhaps a periodic sequence of these winds that gave Cortés the notion of regular tides,[1] the existence of which has not been verified by more recent observations. Lake Texcoco is usually only three to five meters deep. In some places, the bottom of the lake is less than one meter deep. As a result, the inhabitants of the small town of Texcoco have great difficulty conducting trade in the very dry months of January and February. The lack of water prevents them from taking their canoes to the capital. This inconvenience does not occur on Lake Xochimilco, because one can navigate with no interruption from Chalco, Mesquic, and Tlahuac, and Mexico City receives an abundance of vegetables, fruit, and flowers daily via the Iztapalapan canal.

Of the five lakes in the Valley of Mexico, the waters of Lake Texcoco are the most saturated with muriate and carbonate of soda. The presence

II.50

---

1. *Journal des sçavans* for the year 1676, p. 34. Lake Geneva also manifests a rather regular movement of its waters, which ▼ Saussure attributes to periodical winds.

of barium nitrate shows that this water contains no dissolved sulfates. Lake Xochimilco has the purest, clearest water; I found its specific weight to be 1.0009, while the specific weight of distilled water at a temperature of eighteen degrees Centigrade is 1.000, and that of the water of Lake Texcoco is 1.0215. The water of this lake is thus heavier than the water of the Baltic Sea; it is less heavy than water from the Ocean, which has been found to be between 1.0269 and 1.0285 in different latitudes. The amount of sulfurated hydrogen that emerges from the surface of all the Mexican lakes, and which the presence of lead acetate indicates to be in great abundance in Lakes Texcoco and Chalco, probably contributes to the seasonal insalubrity of the valley air. But there is also the curious fact that sporadic fevers are rare on the shores of these lakes, whose surface is partially hidden by rushes and aquatic grasses.

The first conquerors' accounts suggest that ancient Tenochtitlán—adorned with numerous teocalli that rise up like minarets; surrounded by water and causeways; built on small islands covered with verdure; and receiving thousands of boats that enlivened the lake hourly in its streets—must have resembled certain towns in Holland, China, or the flooded Delta of Lower Egypt. The capital, reconstructed by the Spanish, has perhaps a less cheerful appearance, but it is all the more imposing and majestic. Mexico City is certainly one of the most beautiful cities that Europeans have founded in either hemisphere. With the exception of St. Petersburg, Berlin, Philadelphia, and certain quarters of Westminster, there is hardly a city of the same size which can be compared with the capital of New Spain for the evenness of the ground on which it stands, the regularity and width of its streets, and the grandeur of its public squares. The architecture there is of a generally pure style; some buildings even bear an exquisite ordonnance. The exteriors of the houses are not overly decorated. Two types of cut stone—a porous amygdaloid called *tezontli* and, especially, a porphyry of vitreous, quartzless feldspar—give an air of solidity and sometimes even magnificence to Mexican constructions. The wooden balconies and galleries that disfigure all the European cities in both of the Indies are unknown here. The balustrades and grillwork are made of Biscay iron and adorned with bronze. The houses have terraces instead of roofs, like the houses in Italy and all southern countries.

Mexico City has been particularly beautified since the Abbot Chappe's stay in 1769. The building where the School of Mining is housed, for which the wealthiest individuals in the country provided over three million

II.51

francs,[1] could adorn the main squares of Paris and London. Mexican archi-   II.52
tects, students at the Academy of Fine Arts, have recently built two grand
residences, one of which, in the *Traspana* district, offers a very fine oval
peristyle of coupled columns in its inner courtyard. In the center of the
Plaza Mayor in Mexico City, opposite the cathedral and the viceroys' palace, the traveler will justly admire the vast enclosure paved with porphyry
tiles, bordered with grillwork richly ornamented in bronze and graced with
an equestrian statue[2] of King Charles IV on a pedestal of Mexican marble.
One must agree, however, that despite the progress that the arts have made
in the past thirty years, the capital of New Spain impresses Europeans
much less for the size and beauty of its monuments than for the breadth and
alignment of its streets, less for its buildings than for its complete regularity,
extent, and position. Thanks to an unusual set of circumstances, I have visited Lima, Mexico City, Philadelphia, Washington,[3] Paris, Rome, Naples,   II.53
and the major cities of Germany one after the other in a very short period of
time. By comparing impressions that follow in rapid succession, one is able
to revise an opinion that one may have adopted too easily. Despite these
comparisons, many of which might have been unfavorable for the capital of
Mexico, that city has left me with a memory of its splendor that I attribute to
the impressive character of its site and the surrounding scenery.

In fact, there is no richer and more varied sight than the view that the
valley presents on a beautiful summer morning, beneath a cloudless sky
and the deep blue characteristic of the dry and rarefied air of the high

1. See above, chap. VII, vol. I, p. 438 [p. 281 in this edition].

2. This colossal statue, which has been discussed earlier, was commissioned by the
Marquis of Branciforte, a former viceroy of Mexico and the brother-in-law of the ▼ Prince de
la Paz [Manuel Godoy]. It weighs 450 quintals. It was modeled, cast, and put in place by the
same artist, Mr. *Tolsa*, whose name deserves a place of honor in the history of Spanish sculpture. The merits of this man of genius can only be properly appreciated by those who know
the difficulties that the execution of such large works of art present, even in civilized Europe.

3. According to the plan drawn up for the city of Washington and the magnificence of
its Capitol (of which I have only seen a part completed), the *Federal City* will certainly one
day be more beautiful even than Mexico City. Philadelphia also has the same regularity of
construction. The boulevards of plane trees [plantains], acacia, and *populus heterophilia* that
adorn its streets give it an almost rustic beauty. The vegetation on the banks of the Potomac
and the Delaware is richer than what is found at an elevation of over 2,300 meters on the ridge
of the Mexican Cordilleras. But Washington and Philadelphia will always resemble beautiful
European cities. They will not strike the traveler's eyes with the special—might I say exotic—
character of Mexico City, Santa Fe, Bogotá, Quito, and all the other capitals in the tropics that
are built at the same altitude as the Great Saint Bernard Pass, or at even higher elevations.

mountains, when one climbs up one of the towers of the cathedral of Mexico
City or ascends Chapultepec Hill. Beautiful vegetation surrounds this hill.
Ancient cypress trunks,[1] over fifteen to sixteen meters in circumference,
raise their bare tips above the tops of the Schinus [molle], whose bearing is
reminiscent of the weeping willows of the Orient. In this solitude, from the
summit of the porphyritic rock of Chapultepec, the eye surveys a vast plain,
carefully tended fields that extend as far as the foot of the colossal, glacier-
covered mountains. The city appears as if bathed in the waters of Lake Texcoco, whose basin, surrounded by villages and hamlets, recalls the most
beautiful mountain lakes in Switzerland. Broad avenues of elms and poplars lead to the capital from all directions; two aqueducts built on very high
arches cross the plain and present a sight that is as pleasant as it is interesting. The magnificent convent of Our Lady of Guadalupe appears to the
north, with the Tepeyacac Mountains in the background, between ravines
that shelter a few date trees and tree-like yucca. To the south, the entire
terrain from San Ángel, Tacubaya, and San Agustín de las Cuevas appears
to be a vast garden of orange, peach, apple, cherry, and other European
fruit trees. These beautiful orchards contrast with the wild appearance of
the mountains that enclose the valley, among which one can distinguish the
famous volcanoes of La Puebla, Popocatepetl, and Iztaccihuatl. The first of
these forms an enormous cone whose crater, always aflame, spewing smoke
and ashes, gapes open in the midst of the eternal snows.

Mexico City is also remarkable for the order that is maintained there.
Most of the streets have wide sidewalks; they are tidy and well-lit by lamps
with flat, ribbon-shaped wicks. These innovations are the work of Count de
Revillagigedo, who found the capital extremely dirty when he arrived there.

Water is everywhere in the Mexican soil, just below the surface; but it is
brackish, like the water of Lake Texcoco. The two aqueducts that bring the
city its freshwater (which we have mentioned above) are modern constructions worthy of the traveler's attention. The sources of drinking water are
east of the city, one on the isolated hillock of Chapultepec, the other in the
Cerros de Santa Fe near the Cordillera that separates the valley of Tenochtitlán from those of Lerma and Toluca. The arches of the Chapultepec aqueduct are over 3,330 meters long. The water from Chapultepec enters the
southern part of the city at the *Salto del Agua*; it is not very pure and is
only drunk in the suburbs of Mexico City. The water least saturated with
carbonate of lime comes from the aqueduct of Santa Fe, which runs along

---

1. *Los Ahuahuetes, Cupressus disticha* I.

the Alameda and ends at La Transpana, at the Marescala bridge. This aqueduct is nearly 10,200 meters long, but the slope of the terrain only allowed   II.56
for the water to be transported on arches for one-third of this length. The former city of Tenochtitlán had aqueducts that were no less sizable.[1] At the beginning of the siege, the two captains, ▼ Alvarado and Olid, destroyed the Chapultepec aqueduct. In his first letter to Charles V, Cortés also describes the Amilco spring near Churubusco, whose waters were brought to the city by way of terracotta pipes. This spring is near the Santa Fe spring. The ruins of this great aqueduct, built with double pipes, can still be identified; one pipe held the water while the other was being cleaned.[2] This water was sold from canoes that crossed the streets of Tenochtitlán. The springs of San Agustín de las Cuevas are the most beautiful and also the purest; and I thought I saw traces of an ancient aqueduct on the road that leads from this charming village to Mexico City.

We mentioned above (on p. 324) the three main causeways that con-   II.57
nected the ancient city with terra firma. These causeways still partially exist, and their number has increased. Today, they are the great paved roads that cross marshy terrain, and which, being quite elevated, have the double advantages of serving as a roadway for carriages and accommodating the floodwaters of the lakes. The calzada of Iztapalapa is built on the same ancient embankment where Cortés performed such prodigious feats of valor in his encounters with the besieged. The calzada of San Antón is still identifiable today because of the large number of small bridges that the Spanish and the Tlaxcalteca found there when Cortés's companion-at-arms, ▼ Sandoval, was wounded near Coyohuacan.[3] The calzadas of San Antonio Abad, La Piedad, San Cristóbal, and La Guadalupe (formerly called the Tepeyacac causeway) were completely rebuilt after the great flood of 1604, under the viceroy ▼ Don Juan de Mendoza y Luna, Marquis of

1. Clavijero, [*Storia antica del Messico,*] vol. III, p. 195; Solís, [*Historia de la conquista de México,*] vol. I, p. 406.

2. [Cortés, *Historia de Nueva-España,*] Lorenzana ed., p. 108. "The largest and finest construction that the local residents built is the aqueduct of the city of Texcoco. One can still admire the vestiges of a large embankment that was constructed to raise the water level. Generally, how can one not but admire the industry and activity of the ancient Mexicans in irrigating such arid lands? In the coastal part of Peru, I saw the remnants of walls across which water was transported over a distance of more than five to six thousand meters, from the foot of the Cordillera to the coast. The sixteenth-century conquerors destroyed the aqueducts, and that part of Peru, like Persia, became a desert devoid of vegetation. Such was the civilization that the Europeans brought to the nations whom they deemed to be barbarous."

3. [Cortés, *Historia de Nueva-España,*] Lorenzana ed., p. 229, 243.

Montesclaros. The only learned people at that time, Fathers Torquemada and ▼ Gerónimo de Zarate, undertook to level and realign the roadways. It was also in that period that Mexico City was first paved; no other viceroy before Count Revillagigedo had been more successful than the Marquis of Montesclaros in imposing order on the city.

II.58   The sites that generally attract the traveler's attention are: (1) the *Cathedral*, of which a small section is built in the style commonly called Gothic; the main building, with its two towers decorated with pilasters and statues, has a beautiful ordonnance and is of recent construction; (2) the *Mint*, attached to the viceroys' palace, a building from which over six and one-half million gold and silver coins have been issued since the beginning of the sixteenth century; (3) the *convents*, among which the important convent of San Francisco stands out, with an annual revenue of half a million francs in alms alone. This huge edifice was first supposed to have been built on the ruins of the temple of Huitzilopochtli; but since these ruins themselves were to provide the foundation of the cathedral, the convent was begun in its present location in 1531. It owes its existence to the important activity of a servant brother, or lay monk, ▼ Fray Pedro de Gante, an extraordinary man who was reputed to be the illegitimate son of Charles V, and who became the Indians' benefactor; he was the first to teach them the most useful European mechanical arts; (4) the *Almshouse*, or, rather, two hospices together, of which the first cares for six hundred and the other eight hundred children and old people. This establishment, where relative order and cleanliness reign but little industry, benefits from a revenue of 250,000 francs. A rich merchant recently bequeathed six million francs to the almshouse in his will; the capital was assumed by the royal treasury, with the guarantee of

II.59   interest payments at 5 percent; (5) the *Acordada*, a beautiful building whose prisons are generally spacious and well-ventilated. In this establishment and in the other prisons of the Acordada under its control, there are over twelve hundred persons, among them a large number of smugglers and the unfortunate Indian prisoners who were brought to Mexico City from the provincias internas (Indios Mecos); they were discussed earlier, in chapters VI and VII[1]; (6) the *School of Mines*, its newly begun building and the old temporary establishment, with its fine collections in physics, mechanics, and mineralogy[2]; (7) the *Botanical Garden* in one of the courtyards of the

---

1. See vol. I, p. 383, chap. VI, and p. 446, chap. VII [p. 250 and p. 285 in this edition].

2. Two other remarkable oryctognostic and geological collections belong to Professor Cervantes and the ▼ *Oídor* Carvajal. This respectable magistrate also has a superb cabinet of

viceroy's palace, very small but extremely rich in plants that are both rare and interesting for industry and trade; (8) *the buildings of the University and the Public Library*, which is unworthy of such an old and important establishment; (9) the *Academy of Fine Arts*, with its collection of antique plaster casts;[1] (10) *the equestrian statue of King Charles IV* on the Plaza mayor, and the sepulchral monument that the Duke of Monteleone dedicated to the great Cortés in a chapel of the hospital of Los Naturales; it is a simple family monument, adorned with a bronze bust cast by Mr. Tolsa, representing the hero at a mature age. Nowhere else in Spanish America, from Buenos Aires to Monterrey, from Trinidad and Puerto Rico to Panama and Veragua, does one find a national monument in public recognition of the glory of Christopher Columbus or Hernán Cortés!

Avid students of the history and antiquities of the Americas will not find within the walls of the capital the great ruins of structures that one finds in the environs of Cuzco and Guamachuco, in Pachacamac near Lima, or in Mansiche near Trujillo, in Peru; in Cañar and Cayo, in the province of Quito; and near Milta and Cholula in the intendancies of Oaxaca and Puebla, in Mexico. It seems that the only Aztec monuments were the Teocalli, whose odd shape we have already mentioned. It was not only Christian fanaticism that led to their destruction; the conquerors' safety also made it necessary to raze them. Some of this destruction occurred during the siege itself, because the terraced, truncated pyramids served as a refuge for the combatants, as the temple of Baal-Berith had for the people of Canaan; each one was like a castle from which the enemy had to be dislodged.

As for private residences, which the Spanish historians describe as being squat, we should not be surprised to find only foundations or lowlying ruins, like the ones discovered in the Barrio of Tlatelolco and near the Itztacalco canal. Even in most of our European cities, what small number of houses is there whose construction dates back to the sixteenth century? Nonetheless, the Mexican buildings did not become ruins because of their decrepitude. Animated by the same destructive spirit that the Romans displayed in Syracuse, Carthage, and in Greece, the Spanish conquerors believed that the siege of a Mexican town was only complete when they had razed the buildings to the ground. In his third letter[2] to Emperor Charles V,

II.61

---

shells, collected during his stay in the Philippines, where he had already demonstrated the same enthusiasm for the natural sciences that distinguishes him so honorably in Mexico City.

1. See vol. I, chap. VII, p. 424 [p. 274 in this edition].
2. [Cortés, *Historia de Nueva-España*,] Lorenzana ed., p. 278.

Cortés himself divulges the horrifying system he used in his military operations. "Despite our having claimed the upper hand," he writes,

> I saw clearly that the residents of Temixtitlan [Tenochtitlán] were so rebellious and stubborn that they all preferred to die rather than surrender. I no longer knew what means to use to spare ourselves so much danger and hardship, and without completely destroying the capital, which was the most beautiful thing in the world (*a la ciudad, porque era la más hermosa cosa del Mundo*). It was useless to tell them that I would not break camp nor withdraw my fleet of brigantines, and that I would not desist from making war on them on land and by water until I was master of Temixtitlan. I pointed out to them in vain that they could expect no help, and that there was no corner of land where they might hope to get corn, meat, fruit, and water. The more we exhorted them, the more they showed us that they were far from discouraged. They only wanted to fight. In this state of affairs, considering that more than forty to fifty days had already elapsed since our siege had begun, I finally resolved to take a measure by which, in providing for our safety, we would be able to press our enemies more closely. *I devised the plan of demolishing all the houses on both sides of the streets as the latter came under our control, so that we would not advance one foot without having destroyed and brought down everything behind us, converting anything that had been water into firm ground, no matter how slow the work and the delay to which we would expose ourselves.*[1] For this purpose, I gathered the lords and the chiefs of our allies and explained my decision to them. I enjoined them to send a large number of workers with their *coas*, which are like the hoes used for excavating in Spain. Our allies and friends approved of my plan, since they hoped that the city would be entirely destroyed, which they had ardently desired for a long time. Three or four days went by without combat, because we awaited the men from the countryside who would help us with the demolition.

After reading the chief general's naïve account in his third letter to his sovereign, we should not be surprised to find almost no trace of the ancient

II.62

II.63

---

1. "Accordé de tomar un medio para nuestra seguridad y para poder más estrechar a los enemigos; y fue como fuessemos ganando por las calles de la ciudad, que fuessen derrocando todas las casas de ellas, de un lado y del otro; por manera que no fuessemos un passo adelante sin la dejar todo asolado, y que lo que era agua hacerlo tierra firme; aunque hubiesse toda la dilación que se pudiesse seguir" ([Cortés, *Historia de Nueva-España,*] Lorenzana ed., number XXXIV).

Mexica buildings. Cortés related that as soon as they learned that the capital was being destroyed, the indigenous peoples swarmed there from the most far-flung provinces to avenge themselves for the hardships they had endured under the rule of the Aztec kings. The rubble from the demolished houses was used to fill the canals. The streets were dried to allow the Spanish cavalry to take action. The low houses, like those in Peking in China, were partly built of wood and partly from tezontli, a spongy stone that is light and easily smashed. Cortés writes:

> Over fifty thousand Indians helped us on the day when, after climbing over heaps of cadavers, we at last reached the great street of Tacuba and burned the house of king Guatimucin.[1] Consequently, all we did was burn and raze the houses to the ground. The townspeople told our allies [the Tlaxcalteca] that they were wrong for helping us destroy the town, because one day, they would have to restore these very buildings with their own hands, either for the besieged, if they were victorious, or for us Spanish, who were indeed already forcing them to rebuild what had been demolished.[2]

II.64

Leafing through the Libro del Cabildo, an unpublished work that we already mentioned [on p. 325n], and which contains the history of the new city of Mexico from 1524 to 1529, the only thing I found on every page were the names of those who appeared before the alguacils "to ask the site (*solar*) where the house of such and such Mexica lord formerly stood." Even now, they are busy filling in and drying out the old canals that run through many streets in the capital. The number of these canals has particularly decreased since the Count of Gálvez's regime, although, because of the extreme width

II.65

1. The true name of this unfortunate king, the last of the Aztec dynasty, is *Quauhte-motzin*. He is the same man whose soles Cortés ordered slowly burned after having his feet dipped in oil. This torture did not induce the king to reveal the place where his treasures were hidden. He met the same end as the king of Acolhuacan (Texcoco) and Tetlepanguet-zaltzin, the king of Tlacopan (Tacuba). The three princes were hung from a tree, and, as I have seen represented in a hieroglyphic painting that Father Pichardo possesses (in the convent of San Felipe Neri), they were hanged by their feet to prolong their torment. Cortés's act of cruelty, which recent historians have had the cowardice to depict as the result of a far-sighted policy, elicited murmurs in the army itself. "The death of the young king," said Bernal Díaz del Castillo (an old soldier whose words were full of rectitude and sincerity), "was most unjust. Therefore, we were all blamed, since we were in the captain's retinue during his march on Comayagua."

2. [Cortés, *Historia de Nueva-España*,] Lorenzana ed., p. 286.

of the streets of Mexico City, the canals there are less of a hindrance to carriage traffic than in most cities in Holland.

Among the meager remnants of Mexican antiquities that may interest the educated traveler, either within the walls of Mexico City or in its environs, one may count the ruins of the causeways (albaradones) and the Aztec aqueducts; the sacrificial stone, adorned with a relief depicting the triumph of a Mexica king; the great calendar monument (unearthed with the stone in the Plaza Mayor); the colossal statue of the goddess Teoyamiqui, reclining on her back in one of the galleries of the university building, and normally covered with three or four inches of dirt; Aztec hieroglyphic paintings on agave paper, deerskin, and cotton canvas: a valuable collection wrongly taken away from the Chevalier Boturini[1] and very poorly preserved in the archives of the viceroys' palace, and showing in every figure the wanton imagination of a people who delighted in seeing the beating hearts of human victims offered up to gigantic and monstrous idols; the foundation of the palace of the kings of Acolhuacan in Texcoco; the colossal relief carved on the western face of the porphyritic crag called the Peñol de los Baños, and several other objects that will remind the educated traveler of the institutions and works of the people of the Mongol race, descriptions and drawings of which will appear in the personal narrative of my *Voyage to the Equinoctial Regions of the New Continent.*

II.66

The only ancient monuments in the Valley of Mexico whose grandeur and mass may impress Europeans are the remains of the two pyramids of San Juan de Teotihuacán located northeast of Lake Texcoco and dedicated to the sun and the moon; the locals call them Tonatiuh Yztaqual, the house of the Sun, and Meztli Ytzaqual, the house of the Moon. According to the measurements that a young Mexican scientist, Doctor Oteiza,[2] made in 1803, the first pyramid, which is also the southernmost, now has a base that is 208 meters (645 feet) long and has 55 meters (Mexican varas,[3] or 171 feet) of perpendicular elevation. The second, the pyramid of the moon, is

II.67

1. The author of the ingenious work *Idea de una nueva historia general de la América septentrional.*

2. Mr. Bullock, who recently visited the Otumba plains, has confirmed Mr. Oteiza's description. He believes that the large pyramid is even higher (*Six Months Residence*, pp. 408 and 418). It is quite unusual that the people from whom Mr. Bullock requested information about the monuments—whose position I indicated in 1805 on my map of the Valley of Mexico—denied their existence in 1822. (Bullock, *[A Descriptive Catalogue of the] Exhibition*, p. 44).

3. Velázquez found that the Mexican vara is exactly thirty-one inches, of the former pied du roi (of Paris). The northern façade of the Hôtel des Invalides in Paris is only 600 feet long.

eleven meters (thirty-four feet) shorter, and its base is much smaller. Both the accounts of the first travelers and the form that these monuments still exhibit suggest that they were the model for the Aztec Teocalli. The people whom the Spanish found settled in New Spain attributed the pyramids of Teotihuacán[1] to the Toltec people; their construction thus goes back to the eighth or ninth century, given that Tollan's kingdom lasted from 667 to 1031. These buildings face exactly north-south and east-west, within fifty-two minutes in arc. Their interiors are made of clay mixed with small stones. This core is covered with a thick wall of porous amygdaloid. Furthermore, traces of a layer of lime covering the stones (the tezontli) are noticeable on the outside. Some sixteenth-century authors claim that, according to an Indian legend, the inside of these pyramids is hollow. The Chevalier Boturini says that the Mexican geometer Sigüenza tried in vain to pierce a gallery into these buildings. There were four levels, but the ravages of time, as well as cactus and agave plants, have had a destructive effect on the exterior of these monuments. A staircase built of large cut stones once led to the top of each pyramid; it was there, according to the accounts of the earliest travelers, that one found statues covered with very thin sheets of gold. Each of the four main levels was subdivided into smaller tiers one meter high, the edges of which are still distinguishable. These tiers were covered with shards of obsidian, which were probably the sharp instruments with which the Toltec and Aztec priests (*Papahua Tlamacazque or Teopixqui*) cut open the chests of their human victims during their barbaric sacrifices. We know that obsidian (*itztli*) was excavated in the large mines whose traces can still be seen in the innumerable shafts between the Moran mines and the village of Totonilco el Grande, in the porphyritic mountains of Oyamel and the Jacal, a region that the Spanish call the "mountain of knives," the Cerro de la Navajas.[2]

II.68

One would, of course, like to solve the riddle of whether these strange buildings, one of which (the Tonatiuh Ytzaqual), according to the precise measurements of my friend, Mr. Oteiza, has a volume of 128,970

II.69

1. In his unpublished notes, however, Sigüenza claims that they are the work of the Olmec people who lived near the Sierra de Tlaxcala and were called the Matlalcueje. If this theory, whose historical foundation is unknown to us, were true, then these monuments would be even older, since the Olmec are among the first people whom the Aztec chronology places in New Spain. It is even claimed that they were the only people to have come not from the north or northwest (Mongol Asia?) but from the east (Europe?).

2. I found an elevation of 3,124 meters for the summit of Jacal; the Roca de las Ventanas, at the foot of the Cerro de las Navajas, is 2,950 meters above sea level.

cubic toises, were completely manmade, or if the Toltec took advantage of some natural hills that they then covered with stones and lime. The same question has recently come up with respect to several of the pyramids of Giza and Sakkara; it has become doubly interesting because of the fantastic theories that ▼ Mr. Witte has ventured on the origin of the colossal monuments of Egypt, Persepolis, and Palmyra. ▼ Since neither the pyramids of Teotihuacán nor of Cholula (which we shall discuss later) have been penetrated diametrically, it is not possible to speak with certainty of their internal structure. The Indian legends that maintain that they are hollow are vague and contradictory. Their location on otherwise hill-less plains makes it rather improbable that there is any natural rock at the core of these monuments. What is also very remarkable—especially if we recall ▼ Pococke's assertions about the symmetrical position of the small Egyptian pyramids—is that, surrounding the houses of the Sun and the Moon in Teotihuacán, one finds a group, I would venture to say a system

II.70 of pyramids, which are barely nine to ten meters high. These monuments, of which there are hundreds, are arranged in wide streets that follow exactly the direction of parallels and meridians, and which end at the four faces of the two great pyramids. The small pyramids are more frequent toward the southern side of the temple of the Moon than toward the temple of the Sun; according to the local legend, they, too, were dedicated to the stars. It is quite certain that they were the burial place of the tribal chiefs. The entire plain, which the Spanish call *Llano de los Cues*, after a word from the language of the island of Cuba, was formerly called *Mixcoatl*, the road of the dead, in the Aztec and Toltec languages. How many similarities there are to the monuments of the Old Continent! And when the Toltec people, arriving on Mexican soil in the seventh century, built the colossal forms of these monuments according to a uniform plan, the truncated pyramids that were divided into terraces, like the temple of Belus in Babylonia, where did they find the model for these buildings? Were they of the Mongol race? Or descended from a common stock[1] with the Chinese, the Hiong-nu, and the Japanese?

Another ancient monument very worthy of the traveler's attention is the military entrenchment of Xochicalco, located south-southwest of the town

II.71 of Cuernavaca near Tetlama, which belongs to the parish of Xochitepeque.

1. See ▼ Mr. Herder's *Idée d'une histoire philosophique de l'espèce humaine*, vol. II, p. 59 [24–25]; vol. III, p. 11 (in German); [Churchill: 140, 292], and ▼ Mr. Gatterer's *Essai d'une histoire universelle*, p. 489 (in German).

It is a solitary hill, 117 meters high and surrounded by ditches, divided by men's labor into five levels or terraces covered with masonry. Together, they form a truncated pyramid whose four faces are precisely arranged in accordance with the four cardinal points. The basaltic porphyry stones have a very regular cut and are decorated with hieroglyphic figures, among which one can make out crocodiles spouting water and—what is very odd indeed—men sitting cross-legged in the Asian manner. The platform of this extraordinary monument[1] is almost 9,000 square meters, and we find there the ruins of a small square building that must have served as the final retreat of the besieged.

I shall conclude this quick overview of Aztec antiquities by mentioning a few places that may be called classic sites because of the interest they evoke in those who have studied the history of the Spanish conquest of Mexico.

Montezuma's palace was located on the same site where the residence of the Duke de Monteleone now stands, commonly called the Casa del Estado, on the Plaza Mayor, southwest of the cathedral. Like those of the emperor of China, this palace—of which ▼ Sir George Staunton and Mr. Barrow have provided precise descriptions—was composed of several spacious but squat houses. They occupied all the land between the Empedradillo, the broad street of Tacuba and the convent of La Profesa. After taking the city, Cortés set up his residence across from the ruins of this palace of the Aztec kings, where the viceroys' palace is situated today. But it was soon decided that Cortés's house was better suited for the assemblies of the Audiencia. Consequently, the government took possession of the Casa del Estado, the old residence that belonged to Cortés's family. In exchange, this family, which bears the title of the Marquesado del Valle de Oaxaca, received the site formerly occupied by Montezuma's palace. It was there that they constructed the fine building where the archives del Estado are found, and which was passed on to the Neapolitan Duke of Monteleone together with the rest of the inheritance.

When Cortés made his initial entry into Tenochtitlán on November 8, 1519, he and his army were housed not in Montezuma's palace but in a building where King Axayacatl had lived. It was in this building that the Spanish and their allies the Tlaxcalteca sustained the Mexicas' assault. The unfortunate king Montezuma[2] died there, from a wound that he had

II.72

II.73

1. José Antonio Alzate, *Descripción de las antigüedades de Xochicalco dedicada a los Señores de la Expedición marítima bajo las ordenes de Don Alessandro Malaspina*, 1791, p. 12.

2. It is from one of his sons, initially called *Tohualicahuatzin* and *Don Pedro Montezuma* after his baptism, that the counts of Montezuma y Tula in Spain are descended. The Cano Montezuma, the Andrade Montezuma, and, if I am not mistaken, even the counts de

received while haranguing his people. The meager remains of this Spanish quarter are still recognizable[1] in the houses behind the convent of Santa Teresa, on the corner of the streets of Tacuba and Indio Triste.

A small bridge near Bonavista preserves the name Alvarado's Leap (*salto de Alvarado*) in memory of the prodigious leap that Pedro de Alvarado made during the famous *sad night*[2] when the Mexicans, having cut the Tlacopan causeway in several places, forced the Spanish to retreat from the town to the mountains of Tepeyacac. It seems that the historical truth of this event, which has been handed down as a popular legend to all classes of the Mexican population, was disputed even in Cortés's own day. Bernal Díaz considers the account of the leap a mere boast by his companion-at-arms, whose courage and presence of mind he nonetheless praises. He claims that the ditch was much too wide to be jumped. I must note, however, that this anecdote is related in great detail in the manuscript of a noble Mestizo from the republic of Tlaxcala, ▾ Diego Muños Camargo, which I consulted at the convent of San Felipe Neri, and with which Father Torquemada[3] also seems to have been familiar. This Mestizo historian was a contemporary of Fernando Cortés's. He tells the story of Alvarado's leap with great simplicity, with no apparent exaggeration, and without mentioning the width of the ditch. In his simple tale, we seem to recognize a hero of antiquity who, in supporting his shoulder and arm on his lance, makes an enormous bound to save himself

II.74

Miravalle in Mexico City trace their origin back to the beautiful princess *Tecuichpotzin*, the youngest daughter of the last king, Montezuma II, or *Motecuzoma Xocoyotzin*. This king's descendents did not mingle their blood with whites until the second generation.

1. Proof of this assertion is found in Mr. Gama's unpublished manuscripts, which are preserved in the convent of San Felipe Neri in Father Pichardo's hands. The palace of Axayacatl was probably a large enclosure containing several buildings, because almost seven thousand men were quartered there (Clavijero, [*Storia antica del Messico*,] III, p. 79.) The ruins of the city of Mansiche in Peru give us a very clear idea of this type of American construction. Every great lord's residence formed a separate district where courtyards, streets, walls, and trenches were distinguishable.

2. *Noche triste*, July 1, 1520.

3. [Torquemada,] *Monarquía indiana*, book IV, chap. 80; Clavijero, [*Storia antica del Messico*,] I, p. 10. In Mexico and in Spain there are still many unpublished historical works written in the sixteenth century, whose publication by way of excerpts would greatly clarify Anahuac history. These unpublished works are by Sahagún, Motolinia, ▾ Andrés del Olmos, Zurita, J[uan de] Tovar, Fernando Pimentel Ixtlilxochitl, Antonio Montezuma, Antonio Pimentel Ixtlilxochitl, Tadeo de Niza, Gabriel Ayala, Zapata, Ponce, Cristóbal de Castillo, Fernando Alva Ixtlilxochitl, Pomar, Chimalpahin, Alvarado Tezozomoc, and Gutiérrez. All these authors, with the exception of the first five, were baptized Indians, indigenous peoples of Tlaxcala, Texcoco, Cholula, and Mexico City. The Ixtlilxochitls were descended from the royal family of Alcohuacan.

from the hands of his enemies. Camargo adds that other Spaniards wanted to follow Alvarado's example but fell into the ditch (*Azequia)* because they lacked his agility. The Mexicans, he relates, were so amazed by Alvarado's   II.75 skill that when they saw him safe, they ate dirt—a figurative expression that the Tlaxcala author borrows from his own language and which means "to be stupefied in admiration." "The children of Alvarado, who was called *the Captain of the leap*, called on witnesses to offer proof of their father's prowess before the judges of Texcoco. They were forced to resort to this because of a trial, in the course of which they presented the exploits of *Alvarado de el Salto*, their father, during the conquest of Mexico."

Foreigners are shown the Clerigo Bridge near the Plaza Mayor of Tlatelolco as the memorable site where the last Aztec king, Quauhtemotzin—the nephew of his predecessor, King Cuitlahuatzin,[1] and Montezuma II's son-in-law—was taken. The careful studies that I have conducted with Father Pichardo suggest, however, that the young king fell into the hands of ▼Garci Holguín[2] in a large pool of water that once existed between the Garita del   II.76 Paralvillo, the square of Santiago de Tlatelolco, and the Amaxac Bridge. Cortés was on the terrace of a house in Tlatelolco when the captured king was brought to him: "I made him sit," writes the conqueror in his third letter to Emperor Charles V, "and spoke trustingly with him, but the young man put his hand on a dagger that I wore on my belt and begged me to kill him, because after having done his duty to himself and his people, he had no other desire but to die." This trait is worthy of the finest days of Greece and Rome. In all places, regardless of the color of men's skin, the language of valorous souls is the same when they confront adversity. We have seen above what a tragic end this unfortunate Quauhtemotzin met with!

After completely destroying ancient Tenochtitlán, Cortés and his people stayed in Cojohuacan[3] for four or five months, a place for which he always showed a great fondness. At first, he was unsure if he should rebuild the capital in a different place around the lakes. He decided on the ancient

---

1. This king Cuitlahuatzin (whom Solis and other European historians, who confuse all Mexican names, call Quetlabaca) was the brother and successor of Montezuma II. This same prince showed a great fondness for gardens and, according to Cortés, had a collection of rare plants which was admired long after his death in Itzapalapa.

2. On August 31, 1521, the seventy-fifth day of the siege of Tenochtitlán, Saint Hippolytus' day. The same day is still celebrated every year by a procession that the viceroy and the *Oidores* make through the city on horseback, following the standard of Cortés's victorious army, now carried by the second lieutenant-major of the *very noble city of Mexico*.

3. [Cortés, *Historia de Nueva-España,*] Lorenzana ed., p. 307.

site because "the town of Temixtitlan had become famous, because it is in a marvelous location and was always considered the seat of the Mexican

II.77 provinces (*como principal y señora de todas estas provincias*)." There is no doubt, however, that because of the frequent floods that both the old and the new Mexico City have suffered, it would have been better to situate the city east of Texcoco or on the heights between Tacuba and Tacubaya.[1] In fact, it was to these very heights that King Philip III ordered the new capital to be relocated during the great flood of 1607. The *Ayuntamiento* (city magistrate) informed the court that the value of the houses ordered destroyed was 105 million francs. Madrid seemed unaware of the fact that the capital of a kingdom that was built eighty-eight years earlier is not a mobile camp that can change location at will!

It is impossible to determine the population of the former Tenochtitlán

II.78 with any degree of certainty. Judging by the fragments of ruined houses, by the first conquerors' accounts, and, especially, by the number of combatants whom the kings Ciutlahuatzin and Quauhtimotzin sent against the Tlaxcalteca and the Spanish, the population of Tenochtitlán seems to have been at least three times larger than the present population of Mexico City. Cortés insists that after the siege, so many Mexican artisans worked for the Spanish as carpenters, masons, weavers, and founders that, in 1524, the new city of Mexico already counted thirty thousand residents. Modern authors have put forth the most contradictory ideas about the population of the capital. In his excellent work on the ancient history of New Spain, the Abbot Clavijero shows that estimates range from sixty thousand to one and a half million residents.[2] These contradictions should not surprise us, if we consider how new statistical research is, even in the most civilized part of Europe.

According to the most recent and the least unreliable information, the present population of the capital of Mexico appears to be between 135,000

---

1. [Diego de] Cisneros, *Sitio [naturaleza y propriedades, de la ciudad de] Mexico.* Alzate, "[Descripción] topográfica de México" (*Gazeta de Literatura* [de Mexico], 1790, p. 32). However new they may appear, most of the large cities in the Spanish colonies are in disadvantageous locations. I am not speaking here of the settings of Caracas, Quito, Pasto, and several other cities in South America but only of Mexican cities: for example, Valladolid, which could have been built in the beautiful Tepare valley; or Guadalajara, which is quite near the delightful plain of the Río Chiconahuatenco, or San Pedro; or Patzcuaro, which one would have liked to have seen built in Tzintzontza. It would seem that everywhere, the new colonists chose of the neighboring sites whichever one was more mountainous or more exposed to flooding. But the Spanish also built almost no new cities; they only inhabited or enlarged those that had been founded by the indigenous peoples.

2. Clavijero, [*Storia antica del Messico,*] IV, p. 278, note *p.*

and 140,000 persons (including troops). The 1790 census, taken by order of the Count of Revillagigedo, only yielded a result[1] of 112,926 inhabitants, but we know that this number is over one-sixth too low. The regular troops and the militia in garrison comprise between five and six thousand men at arms. It is highly probable that the present population includes:

II.79

2,500	European Whites
65,000	Creole Whites
33,000	Natives (copper-colored Indians)
26,500	Mestizos, mixture of Whites and Indians
10,000	Mulattos
137,000	inhabitants

As a result, there are in Mexico City 69,500 persons of color and 67,500 whites, but some *Mestizos* are nearly as white as the Europeans and the Spanish Creoles!

In the twenty-three monasteries in the capital, there are nearly 1,200 persons, among them nearly 580 priests and choristers. In the fifteen convents, there are 2,100 persons, among them nearly 900 professed nuns.

The clergy in Mexico City is extremely large, although it is one-quarter smaller than the clergy in Madrid. The 1790 census shows the following:

In the monasteries	573 priests and choristers	
	59 novices	867
	235 lay brothers	
In the nuns' convents	888 professional nuns	
	35 novices	923
Prebends		26
Total		1,816 individuals
Curates		16
Vicars		43
Secular clergy		517
	Total	2,392 individuals

II.80

According to Mr. de Laborde's excellent work, the clergy in Madrid is composed of 3,470 persons; the ratio of the clergy to the entire population is

1. See note C at the end of this work.

thus one and a half to one hundred in Mexico City and two to one hundred in Madrid.

We have presented above the table of the revenue of the Mexican clergy (vol. I, p. 440 [p. 282 in this edition]). The archbishop of Mexico City has an income of 682,500 livres tournois. This sum is somewhat less than the income of the convent of the Jeronimites in the Escorial. The archbishop of Mexico City is thus less wealthy than the archbishops of Toledo, Valencia, Seville, and Santiago. The archbishop of Toledo, however, has an income of three million livres tournois. Nonetheless, Mr. de Laborde has shown—and this fact is but little known—that before the Revolution, the French clergy was larger, compared to the whole population, and richer as a body than the Spanish clergy. The revenue of the inquisition tribunal of Mexico City, a tribunal that extends throughout the kingdom of New Spain, Guatemala, and the Philippine Islands, is 200,000 livres tournois.

Using an average period of one hundred years, the number of births in II.81 Mexico City is 5,930; the number of deaths is 5,050. In 1802, there were as many as 6,155 births and 5,166 deaths, which, assuming a population of 137,000 persons, would mean one birth for every twenty-two and a half persons, and one death for every twenty-six and a half persons. We have seen earlier in chapter four (vol. I, p. 311 [p. 207 in this edition]) that in the countryside of New Spain, the ratio of births to the population[1] is one to seventeen, and the ratio of deaths to the population is one to thirty. Consequently, there appears to be a very high mortality rate and a very low number of births in the capital. There is a profusion of sick people, not only from the most destitute class of society, which seeks relief from the hospitals, where the number of beds totals 1,100, but also from the wealthy individuals who are brought to Mexico City because they can find neither doctors nor remedies in the countryside. This circumstance explains why the numbers of deaths in parish registers are so high. On the other hand, the convents, the celibacy of the secular clergy, advances in luxury, the militia, and the destitution of the Indian *saragates* who live II.82 in idleness like the lazaroni in Naples, are the main causes for the unfavorable birthrate.

---

1. In France, the birth-to-death rate is such that, in relation to the entire population, only one-thirtieth dies annually, whereas one-eighth is born. Peuchet, *Statistique*, p. 251. In cities, this ratio depends on a combination of local and variable circumstances. In London, there were 18,119 births and 20,454 deaths in 1787; in Paris, there were 21,818 births and 20,390 deaths in 1802.

By comparing the parish registers in Mexico City with those of several European cities, Mr. Alzate and Mr. Clavijero[1] tried to demonstrate that the capital of New Spain must have over 200,000 inhabitants. But how are we to believe that an error of 87,000 souls was made in the 1790 census, given that this represents over two-fifths of the entire population? Furthermore, by their very nature the comparisons made by the two learned Mexicans can hardly yield reliable results, because the towns whose mortuary registers they present are located at different elevations and in very different climates, and because the levels of civilization and comfort of the bulk of their inhabitants exhibit the most striking contrasts. In Madrid, there is one birth for every thirty-four persons; in Berlin, one for every twenty-eight persons. Neither of these ratios is useful for calculating the population of the cities in equinoctial America. Moreover, the difference between them is so great that it alone would increase or diminish the population of Mexico City by 36,000, if one assumes an annual birth rate of 6,000. Determining the number of inhabitants in a particular district or a province by using the number of deaths or births is perhaps the best method when, *for a given country*, political arithmetic has painstakingly arrived at figures that express the ratios of births and deaths to the entire population. But these same figures, which result from a long process of induction, cannot be applied to countries with radically different physical and moral situations. They indicate the average level of prosperity of a population, the bulk of which lives in the country. One cannot, therefore, use these same ratios to determine the number of inhabitants in a capital city.

II.83

Mexico City is the most populous city on the New Continent; it has nearly 40,000 fewer residents than Madrid.[2] Because it is shaped like a large square, each side of which is almost 2,750 meters long, its population is spread over a broad terrain. Because the streets are very wide, they often appear to be rather deserted, all the more so since, in a climate that dwellers of the tropics find cold, the people venture less into the open air than

II.84

1. The Abbot Clavijero is mistaken when he states that a census yielded a population of over 200,000 for Mexico City. Otherwise, he suggests (and with good reason) that this city usually counts one quarter more births and deaths than Madrid. In fact, in Madrid, the number of births was 4,897, the number of deaths was 5,915 in 1788; in 1797, there were 4,441 deaths, and 4,911 births (Alexandre de Laborde, [*Itinéraire descriptif de l'Espagne,*] II, p. 102).

2. According to Mr. de Laborde, the population of Madrid is 156,272; if, however, one includes the garrison, the foreigners, and the Spaniards who flock there from the provinces, the population may be increased to 200,000. The longest length of Mexico City is 3,900 meters; in Paris, it is 8,000 meters.

in the towns located at the foot of the Cordillera. Consequently, these cities (*ciudades de tierra caliente*) always seem to be more populous than cities in more temperate or cold regions (*ciudades de tierra fría*). Although Mexico City has more inhabitants than the cities of Great Britain and France—with the exception of London, Dublin, and Paris—its population is also smaller than the large cities in the Levant and the East Indies. Calcutta, Surate, Madras, Haleb, and Damascus all have over two, four, and even six hundred thousand inhabitants.

The Count de Revillagigedo commissioned precise studies on food consumption in Mexico City. The following table for 1791 will be of some interest to those familiar with Mr. Lavoisier and Mr. Arnould's important work on food consumption in Paris and in France as a whole.

CONSUMPTION IN MEXICO CITY	
I. Meat	
Steers	16,300
Calves	450
Sheep	278,923
Pigs	50,676
Goats and rabbits	24,000
Chickens	1,255,340
Ducks	125,000
Turkeys	205,000
Pigeons	65,300
Partridges	140,000
II. Cereals	
Corn or Turkish wheat, cargas of three fanegas	117,224
Barley, cargas	40,219
Wheat flour, cargas of twelve arrobas	130,000
III. Liquids	
Pulque, fermented agave juice, cargas	294,790
Wine and vinegar, barrels at four and a half arrobas	4,507
Brandy, barrels	12,000
Spanish oil, arrobas of twenty-five pounds	5,585

II.85

If one assumes, as does Mr. Peuchet, that the population of Paris is four times larger than that of Mexico City, one observes that the consumption of beef is almost proportional to the number of inhabitants, but that the consumption of mutton and pork is much greater in Mexico City. Here is the difference:

	CONSUMPTION		QUADRUPLE OF MEXICO CITY'S CONSUMPTION
	OF MEXICO CITY	OF PARIS	
Steers	16,300	70,000	65,200
Sheep	273,000	350,000	1,116,000
Pigs	50,100	35,000	200,400

Through his calculations, Mr. Lavoisier found that, in his day, the residents of Paris consumed annually ninety million pounds of all types of meat, which amounts to 163 pounds (seventy-nine and seven-tenth kilograms) per person. If one estimates the amount of edible meat provided by the animals listed in the previous table, using Mr. Lavoisier's principles (adjusted for the location), the consumption of meat in general in Mexico City is twenty-six million pounds, or 189 pounds (ninety-two and four-tenth kilograms) per person. This difference is even more striking in that Mexico City includes 33,000 Indians who eat little or no meat. II.86

The consumption of wine has greatly increased since 1791, especially since the introduction of the ▼ Brownian system to Mexican medical practices. The general enthusiasm with which this system was embraced in a country where asthenic or debilitating remedies were overprescribed for centuries, has had—as the accounts of all the merchants of Veracruz suggest—a most striking effect on the consumption of Spanish sweet wines. But these wines are only drunk by the upper classes. The Indians, Mestizos, Mulattos, and even a great many white Creoles prefer the fermented juice of the agave called *pulque,* of which the enormous quantity of forty-four million bottles (each containing forty-eight cubic inches) is consumed annually. In Mr. Lavoisier's time, the annual consumption of the large population of Paris was only 281,000 hogsheads of wine, eau-de-vie, cider, and beer, which amounts to 80,928,000 bottles. II.87

Bread consumption in Mexico City is equal to that of European cities. This fact is even more striking, since in Caracas, Cumaná, Cartagena de Indias, and all other American cities located in the Torrid Zone, but at sea level or at low elevations, the Creole residents live on almost nothing but cornmeal bread and jatropha manihot [sweet cassava]. If one assumes, as does Mr. Arnoud, that 325 pounds of flour produce 416 pounds of bread, one finds that the 130,000 loads of flour consumed in Mexico City could produce 49,900,000 pounds of bread, which would amount to a consumption of 363 pounds per person, regardless of age. If one estimates the normal population of Paris to be 547,000 inhabitants, and bread consumption to be 206,788,000 pounds, one finds 377 pounds per person for Paris. In Mexico City, the consumption of corn is almost equal to that of wheat. Furthermore, Turkish wheat is the most sought-after food among the Indians. One could refer to it by the name that Pliny gave to barley (Homer's χριθη [chrithe]) *antiquissimum frumentum*, since zea [Indian corn] is the only farinaceous-grained gramen that the Americans grew before the Europeans arrived.

II.88   The market of Mexico City is richly stocked with food, especially vegetables and fruits of all kinds. It is an interesting spectacle that can be enjoyed every morning at sunrise, watching these provisions and huge numbers of flowers arrive on flatboats steered by Indians descending the canals of Itztacalco and Chalco. Most of the vegetables are grown on the *chinampas*, which the Europeans call floating gardens. There are two types—some are mobile and are blown here and there by the wind, others are stationary and attached to the shore. Only the first deserve the name of floating gardens, but their number decreases every day.

The resourceful invention of the chinampas seems to go back to the fourteenth century. It is related to the extraordinary situation of a people who, surrounded by enemies and forced to live in the middle of a lake with few fish, refined the methods of providing subsistence for themselves. It is likely that nature herself originally suggested the idea of floating gardens to the Aztecs. On the marshy shores of lakes Xochimilco and Chalco, the turbulent waters of the flood season carry away grass-covered clumps of earth interlaced with roots. These clumps floated here and there for a long time at the mercy of the winds and sometimes joined together to form small islands. A tribe of people, too weak to provide for themselves on the continent, believed that they could take advantage of these bits of earth that chance offered them, and whose ownership no enemy disputed. The oldest chinampas were only clumps of grass artificially joined together, which

II.89   the Aztecs dug up and seeded. These floating islands exist in all zones. I have seen some in the Kingdom of Quito, in the Guayaquil River, eight to nine meters long, drifting in the middle of the current and carrying young bamboo stems, pistia stratiotes, pontederia, and a riot of other plants whose roots are easily entwined. I have also found them in Italy, in the small *lago di acqua solfa* of Tivoli, near Agrippa's baths, small islands composed of sulfur, carbonate of lime, and ulva thermalis leaves, which move with the slightest gust of wind.

Simple clumps of earth broken off from the shore gave rise to the invention of the chinampas, but the industriousness of the Aztec people gradually perfected this system of cultivation. The prolific floating gardens that the Spanish found, of which several subsist in Lake Chalco, were rafts made of reeds (totora), rushes, roots, and branches of brushwood. The Indians cover these light, interwoven materials with black loam naturally impregnated with muriate of soda. This salt is gradually removed as the soil is sprinkled with water from the lake; the terrain becomes all the more fertile the more frequently this leaching is repeated. This process even works with salt water from Lake Texcoco; since it is far from the point of saturation, the water can still dissolve the salt as it filters through the loam. The chinampas sometimes even hold the hut   II.90
of the Indian who serves as guard for a group of the floating islands. They are moved from one shore to the other through towing or pushing.

As the freshwater lake moved farther from the salt lake, the chinampas became stationary. This second type can be seen all along the Viga canal in the marshland between Lake Chalco and Lake Texcoco. Every chinampa forms a parallelogram that is one hundred meters long and five to six meters wide. Narrow ditches that run symmetrically between them separate the squares. The loam fit for cultivation, which has been de-salted through frequent irrigation, is almost one meter higher than the surface of the surrounding water. Broad beans, peas, peppers (chile, capsicum), potatoes, artichokes, cauliflower, and a wide variety of other vegetables are grown on these chinampas. The borders of these squares are usually decorated with flowers and sometimes with a hedge of rosebushes. The boat excursion that one can take around the chinampas of Itztacalco is one of the most enjoyable things one can experience on the outskirts of Mexico City. Vegetation is quite vigorous in soil that is constantly watered.

The valley of Tenochtitlán presents to the scrutiny of physicists two sources of thermal springs, those of Our Lady of Guadeloupe and those of the Peñón de los Baños (rock of baths). These springs contain carbonic   II.91
acid, sulfate of lime and soda, and muriate of soda. The Peñón spring has a rather high temperature. Very healthful and comfortable baths have been established there. The Indians harvest salt near the baths of the Peñón de los Baños. They leach muriate of soda from clayey earth and condense water that has only twelve to thirteen parts salt per one hundred parts. The poorly built cauldrons have a surface area of only six square feet and a depth of two to three inches. The only fuels used are mule and cow dung. The fire

is so poorly managed that to produce twelve pounds of salt, which sells for thirty-five sous (in French currency), twelve sous of fuel are needed! This salt pit dates back to at least Montezuma's time, and the only change in the technical procedure has been the substitution of beaten copper cauldrons for the old earthen vats.

The young viceroy Gálvez chose Chapultepec Hill for the construction of a pleasure palace for himself and his successors. The exterior of the palace is finished, but the apartments have not yet been furnished. This building cost the king almost one and a half million livres tournois. As usual, the court in Madrid did not approve this expense until after it had been made. The layout of the building is very unusual. It is fortified on the side facing

II.92   Mexico City, where one finds protruding walls and parapets fitted out for cannons, although these parts appear to be mere architectural decorations. On the north side, there are moats and vast underground chambers where provisions can be stored for several months. Popular opinion in Mexico City has it that the viceroys' residence on Chapultepec is a disguised fortress. Count Bernardo de Gálvez was accused of plotting to make New Spain independent of the Peninsula. It is supposed that the Chapultepec rock was intended as his refuge and defense in case of an attack by European troops. I have known respectable men in high places who share this suspicion of the young viceroy. It is the duty of a historian not to be swayed by such grave accusations. The Count of Gálvez belonged to a family whom King Charles III had quickly elevated to an extraordinary level of wealth and power. Young, likable, fond of pleasure and luxury, he obtained through his sovereign's generosity one of the foremost positions to which an individual may rise. For this reason, it did not seem worthwhile for him to break the ties that had bound the colonies to the metropole for three centuries. Despite his conduct, which won him the approval of the populace of Mexico City, and despite the influence of a vicereine who was as beautiful as she was widely loved, the Count of Gálvez would have shared the same fate as any European

II.93   viceroy[1] who leaned toward independence. In a great revolutionary movement, he would not have been forgiven for not being an American!

1. Among the fifty viceroys who have governed Mexico from 1535 to 1808, only one was born in America: ▼ the Peruvian Don Juan de Acuña, Marquis of Casa Fuerte (1722–1734), an impartial man and a good administrator. Some of my readers will perhaps be interested to know that descendants of *Christopher Columbus* and King *Montezuma* have been viceroys of New Spain. ▼ Don Pedro Nuño Colón, Duke of Veraguas, entered Mexico City in 1673 and died six days later. ▼ Viceroy Don José Sarmiento Valladares, Count of Moctezuma, governed from 1697 to 1701.

The Chapultepec palace must be sold, to the government's benefit. Since it is difficult in any country to find buyers for a fortified residence, some of the ministers of the *Real Hacienda* have begun selling window-panes and frames at auction. This vandalism, which passes for thrift, has already contributed greatly to the degradation of a building that is 2,325 meters high, and which, in a harsh climate, is exposed to the full fury of the winds. It would perhaps be wise to preserve this residence as the only place where archives could be housed, silver bars from the mint might be placed, and the viceroy's person protected during a popular uprising. In Mexico City, the memories of the uprisings (*motinos*) of January 15, 1624, and June 8, 1692, have not been lost. In the latter uprising, the Indians, in need of corn, burned the palace of viceroy ▼ Don Gaspar de Sandoval, Count of Galve, who took refuge with the guardian of the convent of San Francisco. But it was only at that time that the monks' protection was equal to the safety of a fortified palace. II.94

To conclude the description of the Valley of Mexico, we must still quickly present the hydrographic tableau of this region interlaced with lakes and small rivers. I daresay that this overview will be of equal interest to the physicist and to the civil engineer. We mentioned above that the surface area of the four major lakes takes up nearly one-tenth of the valley, or twenty-two square leagues. In fact, Lake Xochimilco (and Chalco) covers six and a half square leagues; Lake Texcoco ten and one-tenth; Lake San Cristóbal three and six-tenth; and Lake Sumpango one and one-tenth (at twenty-five to an equatorial degree). The valley of Tenochtitlán or Mexico City is a basin surrounded by a circular wall of very high porphyritic mountains. This basin, whose bottom is at an elevation of 2,277 meters above sea level, resembles a small version of the vast Bohemian Basin and—if the comparison is not too random—the valleys of the mountains of the moon described by ▼ Mr. Herschel and Mr. Schroeter. All the humidity that comes from the Cordilleras surrounding the Tenochtitlán plateau collects in the valley. No river flows from there, except the small Tequisquiac stream (arroyo), which crosses the northern mountain chain via a narrow ravine before emptying into the Río de Tula or de Montezuma. II.95

The main affluents in the Tenochtitlán valley are: (1) the Papalotla, Texcoco, Teotihuacán, and Tepeyacac (Guadalupe) Rivers, which flow into into Lake Texcoco; (2) the Pachuca and Cuautitlan (*Quauhtitlan*) Rivers, which empty into Lake Sumpango. The last of these rivers (the Río de Cuautitlan) has the longest course; the volume of its water is greater than all the other affluents put together.

The Mexican lakes, which are all natural recipients into which torrents deposit the waters from the surrounding mountains, rise incrementally in proportion to their distance from the center of the valley or the site of the capital. After Lake Texcoco, Mexico City is the lowest point in the entire valley. According to Mr. Velázquez and Mr. Castera's very accurate survey, the *Plaza mayor* of Mexico City, on the south corner of the viceroy's pal-

II.96 ace, is one Mexican vara, one foot, and one inch[1] higher than the average level of the waters of Lake Texcoco.[2] This lake is four varas zero feet eight inches lower than Lake San Cristóbal, whose northern part is called Lake Xaltocan. The villages of Xaltocan and Tananitla are found on two islands in that part of the lake. Lake San Cristóbal itself is separated from Lake Xaltocan by a very old embankment that leads to the villages of San Pablo and San Tomás de Chiconautla. The northernmost lake in the Valley of Mexico, Lake Sumpango (Tzompango) is ten varas one foot six inches higher than the average level of the waters of Lake Texcoco. Another causeway (*la*

II.97 *Calzada de la Cruz del Rey*) divides Lake Sumpango into two basins, the westernmost of which is called the Laguna de Zitlaltepec, and the easternmost the Laguna de Coyotepec. Lake Chalco is at the southern edge of the valley. It surrounds the pretty little village of Xico, which is built on an island and is separated from Lake Xochimilco by la Calzada de San Pedro de

1. According to Mr. Ciscar's classic work (*[Memoria elemental] sobre los nuevos pesos y medidas decimales*), the ratio of the Castilian vara to the toise is 0.5130 to 1.1963, and one toise is 2.3316 varas. Don Jorge Juan estimated a Castilian vara to be three Burgos feet, and each Burgos foot comprises 123 lines, two-thirds of a *pied du roi*. In 1783, the court of Madrid ordered the naval artillery corps to take measurements in varas and the land artillery corps to do the same in French toises; it would be difficult to justify the usefulness of this difference. ▼ Don Francisco Javier Rovira, *Compendio de matemáticas*, vol. IV, pp. 57 and 63. The Mexican vara is the equivalent of 0.839 meters.

2. The unpublished material I used in drafting this notice on the *Desagüe* consists of (1) detailed plans drawn up in 1802 by order of the dean of the high court of justice (*Decano de la Real Audiencia de Mexico*), ▼ Don Cosme de Mier y Trespalacios; (2) the report that ▼ Don Juan Díez de la Calle, second officer of the State secretariat in Madrid, presented to King Philip IV in 1646; 3) the instructions which the venerable ▼ Palafox, bishop of La Puebla and viceroy of New Spain, communicated to his successor, the viceroy ▼ Count of Salvatierra (Marquis of Sobroso); 4) a report which cardinal Lorenzana, then archbishop of Mexico City, presented to the viceroy Bucarely; 5) a notice drafted by the tribunal of the Cuentas de México; 6) a report written by order of the Count of Revillagigedo; and 7) the *Informe de Velásquez*. I must also mention ▼ Zepeda's [Cepeda] curious work, *Historia del Desagüe*, printed in Mexico City. I myself examined the Huehuetoca Canal twice, once in the month of August 1803, and the second time from June 9, 1803, to January 12, 1804, in the company of the viceroy ▼ Don José de Iturrigaray, whose thoughtfulness and loyalty in my regard I cannot praise highly enough. (See note *D* at the end of this work.)

Tlahua, a narrow embankment that leads from Tuliagualco to San Francisco Tlaltengo. The level of the freshwater lakes of Chalco and Xochimilco is only one vara, eleven inches higher than the *Plaza Mayor* in the capital. I thought that these details would be of interest to hydrographic engineers seeking an accurate idea of the great canal (*Desagüe*) de Huehuetoca.

In the Tenochtitlán valley, the difference in elevation of the four main water reservoirs has been felt during each of the great floods to which Mexico City has been exposed over the course of several centuries. In all of them, the sequence of phenomena has always been the same. Swollen by the extraordinary rise in the water level of the Río de Cuautitlan and the tributaries of the Pachuca, Lake Sumpango pours its waters into Lake San Cristóbal, into which the *Ciénagas* of Tepejuelo and Tlapanahuiloya also flow. Lake San Cristóbal breaks the embankment that separates it from Lake Texcoco. Finally, the level of the overflow from this basin rises over one meter and flows back impetuously, crossing the salty terrain of San Lázaro into the streets of Mexico City. This is the usual path of the floods: they arise in the north and northwest. The outflow canal, which is called the Desagüe Real de Huehuetoca, is supposed to alleviate danger; it is certain, however, that by a set of various circumstances, the southern tributaries (*avenidas del Sur*) on which the Desagüe unfortunately has no impact, could become an equal menace to the capital. Lakes Chalco and Xochimilco would overflow if, during a strong eruption of the Popocatepetl volcano, this colossal mountain suddenly shed its snow. During my stay in Guayaquil, on the coast of the province of Quito in 1802, the cone of Cotopaxi became so hot because of volcanic fire that it lost its enormous snowcap in a single night. On the New Continent, eruptions and large earthquakes are often followed by heavy showers that last entire months. What new dangers would threaten the capital if these phenomena took place in the Valley of Mexico, in an area where there is up to fifteen centimeters of rainfall in relatively dry years?[1]

The inhabitants of New Spain believe that there is a constant period of years that intervene between the great floods. In fact, experience proves that the extraordinarily high waters have occurred in the Valley of Mexico almost every twenty-five years.[2] Since the Spanish arrived, the capital has witnessed five great floods, viz. in 1533, under viceroy ▼ Don Luis de

II.98

II.99

---

1. See above, vol. 1, chap. III, p. 284 [p. 190 in this edition].

2. ▼ Toaldo insists that he can conclude from several observations that especially rainy years, therefore years of great floods, recur every nineteen years, according to the terms of a saros cycle. Rozier, *Journal de physique*, 1783.

Velasco (el Viejo), constable of Castile; in 1580, under the viceroy Don Martín Enríquez de Almanza; in 1604, under the viceroy Marquis of Montesclaro; in 1607, under the viceroy Don Luis de Velasco (el Segundo), Marquis of Salinas; and in 1629, under the viceroy ▼ Marquis of Cerralvo. This last flood is the only one that occurred subsequent to the opening of the Huehuetoca draining canal, and we shall see later what the circumstances were that brought it about. Since 1629, there have been seven very alarming floods in the Valley of Mexico, but the city was saved by the *Desagüe*. These seven very rainy years were 1648, 1675, 1707, 1732, 1748, 1772, and 1795. When we compare the eleven periods just listed, we find for the occurrence of the fatal event the numbers twenty-seven, twenty-four, three, twenty-six, nineteen, twenty-seven, thirty-two, twenty-five, sixteen, twenty-four, and twenty-three years, a number series that probably shows somewhat more regularity than the one that is supposedly observed in Lima with respect to the occurrence of the great earthquakes.

II.100   The situation of the capital of Mexico is even more dangerous because the difference in elevation between the surface of Lake Texcoco and the ground upon which the houses are built shrinks from year to year. This ground is a level plane, especially since the streets of Mexico City were all paved under the Count of Revillagigedo's government. On the contrary, the bottom of Lake Texcoco is rising little by little, from the *mud* that is stirred up by small torrents and that creates deposits of soil in the reservoirs into which the latter flow. To avoid a similar problem, the Venetians deviated the Brenta, the Piave, the Livenza, and other rivers from the lagunas where they formed deposits.[1] If all the results of the leveling surveys conducted in the sixteenth century were reliable, one would likely find that the Plaza Mayor in Mexico City was once more than eleven decimeters above the level of Lake Texcoco, and that the average level of the lake varies from year to year. While, on the one hand, the felling of forests has led to a decrease in atmospheric humidity and in the water level of springs in the mountains that surround the valley, on the other hand, the clearing of the land has also amplified the effect of soil deposits and the swiftness of floods. In his excellent work on the Languedoc canal, General Andréossy has empha-

II.101   sized these causes, which are the same for all climates. The waters that glide over sward-covered slopes carry away much less soil than those that flow over loose earth. This sward, whether it is composed of graminaceae (as in Europe) or small alpine plants (as in Mexico), can be preserved only in the

---

1. See Andréossy on the *canal du Midi* [*Histoire du Canal du Midi*], p. 14.

shade of a forest. Shrubs and brushwood also provide obstacles to the melted snow running down the mountain slopes. When these slopes are stripped of vegetation, the streams of water meet with less resistance and more rapidly join the torrents that rise and swell the lakes in the vicinity of Mexico City.

It is quite natural that in the order of hydraulic works undertaken to protect the capital from the danger of floods, the system of *embankments* preceded the system of *outflow canals*. In 1446, when the city of Tenochtitlán was so flooded that not one of its streets was dry, Montezuma I (*Huehue Motecuzoma*), guided by the counsel of Netzahualcoyotl, the king of Texcoco, had an embankment built that was over 12,000 meters long and twenty meters wide. This embankment, partly raised up from the lake, consisted of a wall made of stone and clay and drilled on each side with a row of palisades, large remnants of which may still be seen on the plains of San Lázaro. Montezuma I's embankment was enlarged and repaired after the great flood of the year 1498, caused by King Ahuizotl's imprudence. As we have mentioned above, this prince had the abundant headsprings of Huitzilopochco routed to Lake Texcoco. He forgot that that lake, which has no water in dry spells, becomes more dangerous in rainy years as the number of its tributaries increases. Ahuizotl had Tzotzomatzin, a citizen of Coyohuacan, put to death because he dared to predict the danger to which the capital was exposed by the new Huitzilopochco acqueduct. Shortly thereafter, the young Mexican king nearly drowned in his palace. The swell of water came so swiftly that the prince was seriously injured as he escaped through a door that led from the ground floor chambers to the street.

II.102

The Aztecs had thus constructed the causeways (calzadas) of Tlalma, Mexicaltzingo, and the Albaradón, which extend from Iztapalapa to Tepeyacac (Guadalupe), and whose ruins, in their present state, are eternally useful to Mexico City. This system of embankments, which the Spanish continued through the beginning of the seventeenth century, offered a means of defense that, if not quite secure, was at least adequate in a period when the residents of Tenochtitlán sailed in canoes and were less concerned by the effects of small floods. The abundance of forests and plantings made it easier to build on stilts. This frugal people lived well enough on the produce of the floating gardens (*chinampas*). They required only a small parcel of cultivable land. The overflowing of Lake Texcoco was less of a threat to people who lived in houses, many of which were intersected by canals.

II.103

When the new city of Mexico, rebuilt by Hernán Cortés, experienced its first flood in 1553, the viceroy Velasco I ordered the Albaradón de San Lázaro to be built. Constructed following the model of the Indian causeways, this work suffered greatly during the second flood in 1580. In the third flood, in 1604, it had to be rebuilt completely. At that time, the viceroy Montesclaro added the water-supply point (*presa*) of Oculma and the three *calzadas* of Our Lady of Guadalupe, San Cristóbal, and San Antonio Abad for the safety of the capital.

No sooner were these important constructions completed than—through an extraordinary set of circumstances—the capital was flooded again in 1607. Never before had two floods followed each other so closely; never since has the fatal cycle of these calamities been shorter than sixteen or seventeen years. Weary of building causeways (*Albaradones*) that the waters periodically destroyed, the people finally decided that it was time to abandon the Indians' old hydraulic system and adopt a system of overflow canals. This change was even more necessary because the city where the Spanish lived no longer resembled the capital of the Aztec empire. The

II.104 ground floors of houses were now inhabited, and few streets could be navigated by boat. Both the inconveniences and the real losses brought about by the floods thus became greater than they had been in Montezuma's time.

Since the exceptional high waters of the Cuautitlan River and its tributaries were considered the main cause of the floods, the idea naturally emerged of preventing this river from flowing into Lake Sumpango, the mean level of whose waters is seven and one-half meters higher than the grounds of the main square of Mexico City. In a circular valley surrounded by high mountains, the Río de Cuautitlan could only be routed through an underground tunnel or an open canal dug across the same mountains. In fact, as early as 1580, at the time of the great flood, two intelligent men, the ▼ *licenciado Obregón* and the *maestro Arciniega*, had proposed to the government to have a tunnel bored between the Cerro de Sincoque and the Loma de Nochistongo. This spot, more than any other, was to draw the attention of those who had studied the topography of Mexico. It was nearest to the Río de Cuautitlan, justly considered the most dangerous enemy of the capital. The mountains that surround the plateau are lowest there and are no less massive than they are north-northwest of Huehuetoca, near the Nochistongo Hills. On close examination of this marly soil, the horizontal strata of which occupy

II.105 a porphyritic gorge, one might say that this is the place where the valley of Tenochtitlán was formerly connected with the Tula valley.

In 1607, the viceroy the Marquis of Salinas, put ▼ *Enrico* (Henri) *Martínez* in charge of the artificial draining of the Mexican lakes. It is generally believed in New Spain that this famous engineer, the author of the *Desagüe de Huehuetoca*, was Dutch or German. His name certainly suggests that he was of foreign descent, though he was raised in Spain. The king had granted him the title of cosmographer; a treatise that he authored on trigonometry, printed in Mexico City, exists, but has now become very rare. Enrico Martínez, Alonso Martínez, ▼ Damián Dávila, and Juan de Isla conducted a general survey of the valley, and the precision of their survey was proven by the work carried out in 1774 by the learned geometer Don Joaquin Velázquez. The royal cosmographer, Enrico Martínez, presented two projects for canals, one for draining the three lakes of Texcoco, Sumpango, and San Cristóbal, the other only for Lake Sumpango. According to both projects, the outflow of water was to take place through the underground tunnel of Nochistongo, which Obregón and Arciniega had proposed in 1580. But since the distance from Lake Texcoco to the mouth of the Río Cuautitlan was nearly 32,000 meters, the government decided to restrict its efforts to the Sumpango canal. This canal was begun in such a way as to receive the waters of the eponymous lake at the same time as the waters of the Cuautitlan River. Consequently, it is not true that the conception of the *Desagüe* projected by Martínez was *negative*, and that it was intended only to prevent the Río de Cuautitlan from flowing into Lake Sumpango. The branch of the canal that carried the waters from the lake to the tunnel became filled with soil deposits; from then on, the *desagüe* only served the Cuautitlan River, which was diverted from its course. Thus, when Mr. Mier recently took charge of the direct draining of the lakes of San Cristóbal and Sumpango, people only barely remembered that 188 years earlier the same work had already been executed for the first [second?] of these great basins.

II.106

The famous underground gallery of Nochistongo was begun on November 28, 1607. In the presence of the *Audiencia*, the viceroy struck the first blow of the pickaxe. Fifteen thousand Indians were employed for this project, which was accomplished at an extraordinary pace because work was done in several pits at the same time. The unfortunate indigenous peoples, who were given only a pickaxe and a shovel to pierce the loose and *crumbling* earth, were treated with extreme harshness. After eleven months of continuous work, the tunnel (*el socavón*) was finished, measuring over 6,600 meters (or 1.48 common leagues) in length, 3.5 meters in width, and 4.2 meters in height. In December 1608, the engineer Martínez invited the viceroy and the archbishop of Mexico to go to Huehuetoca to see the waters[1] of Lake Sumpango and the Río de Cuautitlan flow through the tunnel. According to Zepeda, the viceroy, the Marquis de Salinas, rode more

II.107

---

1. The water flowed for the first time on September 17, 1608.

than 2,000 meters on horseback inside this underground tunnel. The Río de Moctezuma (or de Tula), which flows into the Panuco River, is on the opposite side of the Nochistongo Hill. From the northern end of the Socabón, called the Boca de San Gregorio, Martínez had dug an open channel that led the water from the tunnel to the small cascade (*Salto*) of the Río de Tula, over a distance of 8,600 meters. According to my measurements, the same water must then still descend from this cascade to the Gulf of Mexico near the Tampico sandbar, a height of 2.135 meters, which translates into a mean fall of six and three-fifth meters per thousand meters over a total length of 323,000 meters.

An underground passage that serves as a drainage canal, completed in less than a year, 6,600 meters long, and with a cross-section opening of ten and a half square meters, is a hydraulic work that would draw the attention of engineers in today's Europe. In fact, it has only been since the end of the seventeenth century, since the illustrious example that François Andréossy provided for the *canal du midi* with the Malpaís passage, that these underground openings have become more common. The canal near Saperton, which connects the Thames with the Severn, leads through a chain of very high mountains over the course of more than 4,000 meters. The great underground canal of Bridgewater, near Worsley on the outskirts of Manchester, which serves to transport coal, is 19,200 meters (or four and three-tenth common leagues) long, including its various ramifications. According to the preliminary plan, the Picardy canal, which is now being built, should have a navigable underground passage 13,700 meters long, seven meters wide, and eight meters high.[1]

As soon as some of the water from the Valley of Mexico began to flow toward the Atlantic Ocean, Enrico Martínez was reprimanded for not having dug a gallery that was large, durable, and deep enough to accommodate the high floodwaters. The chief engineer (*Maestro del Desagüe*) replied that he had presented several projects, but that the government had chosen the solution that could be carried out with the greatest haste. In fact, the filtrations and erosions caused by the alternating periods of humidity and drought make the loose earth crumble frequently. They were soon

---

1. ▼ [James] Millar and [William] Vazie, [*Observations on the Advantages and Practicability of Making Tunnels Under Navigable Rivers Millar and Vazie on Canals*], 1807. The Georg-Stollen in the Harz, a tunnel begun in 1777 and completed in 1800, is 10,438 meters long and cost 1,600,000 francs. Men work in the coal mines near Forth at more than 3,000 meters below sea level without being exposed to infiltrations. The length of the underground canal of Bridgewater is equal to two-thirds of the width of the Pas-de-Calais.

compelled to prop up the ceiling, which was entirely formed of alternate   II.109
layers of marl and a hardened clay called *tepetate*. First, *timber work* was
used, placing corniced beams on pillars. But because resinous wood was
not common in that part of the valley, Martínez substituted *masonry stack*
for the timber. Judging from its remains that have been discovered in the
obra del consulado, this masonry stack was very well built, but the principle itself was flawed. Instead of covering the gallery from the ceiling to the
channel in the floor with a complete, elliptically shaped vault—the kind
used in mines whenever a transverse gallery is dug in loose sand—the engineer merely built arches on shaky ground. The waters, which had been
allowed to drop down too steeply, gradually undermined the lateral walls,
depositing a large quantity of earth and gravel in the gallery channel, as
no method of filtering the water had been adopted, such as sending it first
through a mesh of fabric from *petate*, which the Indians wove from palm-
tree leaf filaments. To overcome these obstacles, Martínez built a sort of cof-
fer dam, or small sluices, spaced throughout the gallery, which, by opening
rapidly, would serve to clear the passageway. This strategy also proved in-
effective, and the gallery became stopped up by continual deposits.

Since 1608, Mexican engineers have argued about whether to enlarge   II.110
the *socabón* of Nochistongo, whether to finish the masonry stack or make
an exposed aperture by removing the top of the vault, or, finally, whether
to build a new draining tunnel at a lower point, one capable of receiving the
waters of Lake Texcoco in addition to those of the Río de Cuautitlan. In
1611, the viceroy and archbishop ▼ Don Garcia Guerra, a Dominican clergyman, commissioned new surveys made by Alonso de Arias, the superintendent of the king's arsenal (*Armero mayor*) and inspector of fortifications
(*Maestro mayor de fortificationes*), a man of probity who enjoyed a great
reputation at the time. Arias seems to have approved of Martínez's work,
but the viceroy made no definitive decisions. Annoyed by the engineers' arguments, the court of Madrid sent a Dutchman, ▼ Adrien Boot, to Mexico
City in 1614. His knowledge of hydraulic architecture is praised in the papers of the time, preserved in the archives of the viceroyalty. This foreigner,
who was recommended to ▼ Philip III by his ambassador to the court of
France, spoke out once again in favor of the Indian system; he advised that
large embankments and reinforced earthen levees be built around the capital. It was only in 1623, however, that he succeeded in convincing people to
abandon the Nochistongo gallery. A new viceroy, the ▼ Marquis of Gelves,
had recently arrived in Mexico. He had not, therefore, witnessed the floods   II.111
caused by the overflowing of the Cuautitlan River. He had the audacity to

order the engineer Martínez to block the underground tunnel and to make the water of Sumpango and San Cristóbal flow into Lake Texcoco to see if the danger was actually as great as had been described to him. That lake swelled to an extraordinary level, and the orders were revoked. Martínez resumed work on the gallery until June 20, 1629,[1] when an event occurred of which the true causes remain shrouded in mystery.

The rains had been plentiful; the engineer blocked the underground passage. The next morning, Mexico City was covered in water one meter deep. Only the Plaza Mayor, the Plaza del Volador, and the suburb of Santiago de Tlatelolca remained dry. People went by boat in the other streets. Martínez was thrown in jail. It was purported that he had closed the outflow tunnel to provide the unbelievers with obvious and negative proof of the utility of his work. The engineer declared that, on the contrary, having seen that the volume of water was much too large to be admitted into the underground gallery, he preferred to expose the capital to the passing danger of a flood rather than see the work of so many years destroyed in a single day. Contrary to all expectations, Mexico City remained flooded for five years, from 1629 to 1634.[2] People crossed the streets in boats, as they had done before the conquest of old Tenochtitlán. Wooden bridges that served as platforms for pedestrians were built alongside the houses.

II.112

During this time, four different projects were presented and discussed by the viceroy, the Marquis de Ceralvo. A resident of Valladolid de Michoacan, ▼ Simón Méndez, explained in a report that the ground of the Tenochtitlán plateau rose significantly on the northwest side, toward Huehuetoca and the Nochistongo Hill; that the point where Martínez had opened the chain of mountains that encircle the valley corresponded to the average level of the highest lake (Sumpango) and not to the level of the lowest lake, Texcoco; that, on the contrary, the level of the valley is considerably lower north of the village of Carpio and east of the lakes of Sumpango and San Cristóbal. Méndez suggested draining Lake Texcoco through an outflow tunnel that would run between Xaltocan and Santa Lucía, flowing into the Tequisquiac creek (*Arroyo*), which, as we have seen above, empties into the Río de Moctezuma (or de Tula). Méndez began the projected *desagüe* at its lowest point; four air wells (*lumbreras*) were already finished when the government, forever irresolute and wavering, abandoned the

1. According to some unpublished manuscripts, September 20.
2. Several accounts record that the flood only lasted until 1631, but that it began again toward the end of the year 1633.

project for being too long and costly. In 1630, ▼ Antonio Román and Juan Alvarez de Toledo proposed drying out the valley through a mid point, II.113 Lake San Cristóbal, by conducting the waters to the ravine (*barranca*) of Huiputztla, north of the village of San Mateo and four leagues west of the small town of Pachuca. The viceroy and the *Audiencia* paid as little attention to this project as to the one presented by the mayor of Oculma, ▼ Cristóbal de Padilla, who, having discovered three perpendicular caverns or natural chasms (*boquerones*) within the precinct of the small town of Oculma itself, wanted to use these openings to drain the lakes. The small river of Teotihuacán disappeared into these *boquerones*. Padilla also proposed introducing the water of Lake Texcoco into them by bringing it to Oculma via the farm at Tezquititlan.

The idea of using the natural caverns formed by layers of porous amygdaloid inspired a similar and no less gigantic project in the mind of the ▼ Jesuit Francisco Calderón. This priest insisted that at the bottom of Lake Texcoco, very near the Peñol de los Baños, there was a hole (*sumidero*) that, if enlarged, could engulf all the waters. He tried to support this claim with testimony from the most intelligent indigenous peoples and with old Indian maps. The viceroy ordered the prelates of all the religious orders (who were probably the best informed about hydraulic matters) to examine the project. The monks and the Jesuit searched in vain for three months, from September to December 1635, but the *sumidero* was not found, although even today, many Indians believe with the same obstinacy as Father Calderón II.114 that it does exist. Whatever opinion one may form of the volcanic or neptunian origin of the porous amygdaloids (*blasiger Mandelstein*) in the Valley of Mexico, it is hardly likely that this problematic rock could contain holes large enough to accommodate the waters of Lake Texcoco, which must be estimated at more than 251,700,000 cubic meters even in times of drought. Only in sections of secondary gypsum, like those found in Thuringia, can one sometimes risk bringing relatively small masses of water into natural caverns (*Gipsschlotten*), where outflow tunnels that begin inside a gallery of coppery schist are allowed to terminate with no concern for where the waters that impede the mining works ultimately go. But how could one rely on this local method in a major hydraulic operation?

During the Mexico City flood, which lasted five consecutive years, the misery of the poor was amplified. Trade ceased, many houses collapsed, and others became uninhabitable. In these unfortunate times, the archbishop ▼ Francisco Manzo y Zúñiga distinguished himself by his kindliness. He went out daily in a canoe to distribute bread to the poor in the flooded

II.115 streets. In 1635, the court in Madrid ordered for the second time that the town be moved to the plains between Tacuba and Tacubaya, but the magistrate (*Cabildo*) argued that the value of the buildings (*fincas*) that the court had proposed to abandon—estimated in 1607 at 150 million livres tournois—amounted at that time to over 200 million livres. In the midst of these misfortunes, the viceroy had the statue of the Holy Virgin of Guadalupe[1] brought to Mexico City. The statue remained in the flooded city for a long time. But the waters only receded in 1634 when, amid very strong and frequent earth-

II.116 quakes, the ground cracked open, a phenomenon that (according to the skeptics) was of great benefit to the miracle attributed to the revered image.

The viceroy, Marquis de Cerralvo, freed the engineer Martínez. He had the *Calzada* (causeway) of San Cristóbal built, more or less as we know it today. Sluices (*compertuas*) allowed for a connection between Lake San Cristóbal and Lake Texcoco whose water level is generally thirty to thirty-two centimeters lower. Martínez had already begun (in 1609) converting a small part of the underground gallery of Nochistongo into an open well. After the flood of 1634, he was ordered to suspend this project, for it was too long and too costly, and to complete the Desagüe by enlarging the earlier gallery. The Marquis de Salinas had intended the revenue from a special tax on the consumption of commodities (*derecho de Sisas*) for the maintenance of Martínez's hydraulic projects. ▼ The Marquis of Cadereyta increased the revenue of the *Desagüe* fund with a new tax of twenty-five piasters on each large cask of wine imported from Spain. Both the *Sisa* duty and the duty on drink are still in place today, but only a small part of the taxes goes to the *Desagüe*. At the beginning of the eighteenth century, the court allotted

---

1. During public calamities, the residents of Mexico City have recourse to two famous images of Our Lady, that of *Guadalupe* and *Remedios*. The first one is considered indigenous, having appeared first among the flowers in an Indian's kerchief; the second was brought from Spain at the time of the conquest. The difference in allegiance between the Creoles and the Europeans (*Gachupines)* lends a special devotional nuance. During the great droughts, the Creole and Indian lower classes saw with dismay that the archbishop preferred to have the image of the Virgin of Remedios brought to Mexico City. Hence the proverb that characterizes the mutual hatred of the castes: everything, even our water, must come to us from Europe (*hasta el água nos debe venir de la Gachupina!*). If the drought continues despite the presence of the Holy Virgin of *los Remedios* (as people claim to have happened, in rare instances), then the archbishop allows the Indians to go and seek the image of Our Lady of Guadalupe. Such indulgence spreads happiness among the Mexican people, especially when long droughts end (like everywhere else) in abundant rains. I have seen works of trigonometry printed in New Spain and dedicated to the Holy Virgin of Guadalupe. On Tepejacac Hill, at the foot of which her elaborate sanctuary is built, formerly stood the temple of the Mexican Ceres, called *Tonantzin* (our mother), *Cen-teotl* (the goddess of corn), or *Tzin-teotl* (the goddess of reproduction).

half of the excise tax on wine for the upkeep of the great fortifications of the castle of San Juan de Ulúa. Since 1770, the fund for hydraulic projects in the Valley of Mexico has not even collected more than five francs of the duty paid on each barrel of European wine imported through Veracruz.   II.117

Work on the *Desagüe* was carried on with little zeal from 1634 until 1637, when the viceroy, the ▼ Marquis of Villena (Duke of Escalona) asked Father Luis Flores, the commissioner-general of the order of St. Francis, to administer it. This priest's work was widely praised, and the drying system was changed for the third time under his administration. The definitive decision was made to abandon the gallery (*socavón*), to remove the top of the vault, and to make an enormous *cut in the mountain* (*tajo abierto*), the former underground passage of which was to become merely a ditch.

The monks of St. Francis were able to retain the management of the hydraulic projects. They were even more successful in this endeavor since, at that time,[1] the viceroyalty was held, virtually in succession, by the bishop of Puebla, Palafox; the bishop of Yucatán, ▼ Torres; a certain Count of Baños, who ended a brilliant career by becoming a barefooted Carmelite; and the archbishop of Mexico City, a monk of St. Augustine, ▼ Enriquez de Ribera. Annoyed by monastic ignorance and delays, a lawyer, the fiscal [public prosecutor] ▼ Martín de Solis, obtained the administration of the *Desagüe* in 1675 from the court in Madrid. He promised to cut through the mountain chain in two months. His project succeeded so well that eighty years were hardly enough to repair the damage he caused in just a few days.   II.118 Advised by the engineer ▼ Francisco Pozuelo Espinoa, the fiscal had more earth thrown into the ditch that even the impact of the waters could carry away. The passage was blocked. In 1760, traces of the fallen earth caused by Solis's imprudence were still identifiable. The viceroy, the ▼ Count of Monclova, rightly believed that the slowness of the monks of St. Francis was less detrimental than the legal adviser's temerity. ▼ Father Manuel Cabrera was reinstated in 1687 in his position as superintendent (*super-intendente de la Real obra del Desagüe de Huehuetoca*). He took his revenge on the fiscal by publishing a book with the curious title: *Illuminated Truths or Defeated Deceptions by Which a Powerful and Poisoned Pen Attempted to Prove, in an Ill-Conceived Report, that the Desagüe Project Was Completed in 1675.*[2]

---

1. From June 9, 1641, to December 13, 1673.

2. *Verdad aclarada y desvanecidas imposturas, con que lo ardiente y envenenado de una pluma poderosa en esta Nueva España, en un dictamen mal instruido, quisó persuadir averse acabado y perfeccionado el año de 1675, la fábrica del Real Desagüe de México.*

The underground passage was dug and lined with masonry in only a few years. It took two centuries to finish the open cut, in loose earth and in cross-sections that were eighty to one hundred meters wide and forty to fifty meters in perpendicular depth. The work was neglected in years of drought but resumed with extraordinary energy during the few months that

II.119 followed the period of very high water levels or an overflowing of the Cuautitlan River. The flood that threatened the capital in 1747 compelled the Count of Guemes to attend to the Desagüe. But there was a new delay until 1762 when, after a very rainy winter, there were strong signs of overflowing. There were still 2,310 Mexican varas (or 1,938 meters) at the northern end of Martínez's underground opening that had not been converted into an open trench (*tajo abierto*). Because this tunnel was too narrow, it often happened that the waters from the valley could not flow freely toward the Salto de Tula.

Finally, in 1767, under the administration of a Flemish viceroy, the Marquis de Croix, the body of Mexican merchants who formed the board of the *Consulado* in the capital took over the completion of the *Desagüe,* on the condition that they could levy the Sisa duty and the duty on wine to compensate for advance outlay. The engineers estimated the cost of the project at six million francs. In fact, the Consulado carried out the work at a cost of four million, but instead of completing the cut in five years (as had been stipulated), and instead of making the channel eight meters wide, the canal was not finished until 1789 and even preserved the width of Martínez's gallery. Since then, they have made continual improvements to the work, widening the bottom of the cut and, especially, making the slope gentler.

II.120 Much work still needs to be done, however, for the present canal to be in a state where there is no longer any worry about the walls crumbling. This is even more dangerous because lateral erosions increase in proportion to the obstacles that impede water flow.

When studying the history of the hydraulic projects of Nochistongo in the archives of Mexico City, one notices a continuous indecisiveness on the part of governors, a wavering of opinions and ideas, which increases the danger instead of removing it. One finds visits by the viceroy, accompanied by the *Audiencia* and the canons; papers drawn up by the fiscal and other lawyers; *juntas*; advice given by the monks of St. Francis; impetuous activity every fifteen to twenty years, whenever the lakes threatened to overflow; and, conversely, slowness and a blameworthy carelessness once the danger was past. Twenty-five million livres tournois were spent, because no one ever had the courage to follow the same plan, because they vacillated for two centuries between the Indian system of embankments and the plan for overflow canals, between the project for an underground gallery (*socavón*) and that for an open cut in the mountain (*tajo abierto*). Martínez's tunnel was left to deteriorate because they wanted to dig a larger and deeper one; completion of the Nochistongo cut (*tajo*) was neglected because they were arguing over the project for a canal at Texcoco, which was never carried out.

In its present state, the *Desagüe* undoubtedly belongs to the most co- II.121
lossal hydraulic projects ever undertaken. One regards it with a certain admiration, especially if one considers the nature of the terrain, and the enormous width, depth, and length of the pit. If the pit were filled with water to a depth of ten meters, the largest warships could pass between the mountain range that borders the plateau of Mexico City to the northeast. The admiration that this work inspires is mixed, however, with troubling ideas. At the sight of the Nochistongo cut, one remembers how many Indians perished there, either through the engineers' neglect or because of the excessive strain to which they were subjected during centuries of barbarism and cruelty. One asks whether it was really necessary to adopt such slow and costly methods to drain off a relatively small volume of water from a valley enclosed on all sides. It is deplorable that so much collective strength was not used for a greater and more useful goal, for instance, to open not a canal but a *passage* through some isthmus that blocks navigation.

Enrico Martínez's project was wisely conceived and carried out with astonishing speed. The nature of the land and the shape of the valley made an underground opening necessary. The problem could have been solved in a complete and long-lasting way, if (1) the gallery had begun at a lower point, one, in other words, that corresponded to the level of the lower lake, and (2) this gallery had been cut in an elliptical form and completely reinforced by II.122
a solid wall with a similarly elliptical vault. The underground passage that Martínez built was only fifteen square meters in profile, as we have seen above. To determine in which dimensions a gallery of efflux should have had been constructed, one would have to know the exact amount of water that the Cuautitlan River and Lake Sumpango carry during the high-water periods. I have found no such studies in papers written by Zepeda, Cabrera, Velázquez, and Mr. Castera. But, according to my own on-site research at that part of the cut in the mountain (*el corte o tajo*) called *la obra del consulado*, it seemed to me that, during the time of ordinary rains, the waters have a profile of eight to ten square meters, and that this amount increases

to thirty to forty square meters[1] during extraordinary swells of the Guautitlan River. The Indians insisted that when this happens, the channel that forms the bottom of the *tajo* becomes so full that the ruins of Martínez's former vault are completely submerged beneath the water. Had the engineers found great difficulties in making an elliptical gallery four or five meters wide, it would certainly have been better either to support the vault with a central pillar or to dig two galleries at the same time, instead of making an open trench. Such trenches are useful only when the hills are low, not very broad, and contain layers less subject to erosion. It was believed that what was necessary to pass a volume of water that is usually eight square meters and sometimes fifteen to twenty square meters in profile through the Nochistongo mountain was to dig a trench with a profile between 1,800 and 3,000 square meters over long stretches.

II.123

According to Mr. Velázquez' measurements,[2] the outflow canal (*Desagüe*) of Huehuetoca in its present state measures:

	MEXICAN VARAS	METERS
From the lock at Vertideros to the bridge of Huehuetoca	4,870 or	4,087
From the bridge of Huehuetoca to the lock at Santa María	2,660	2,232
From the Compuerta de Santa María to the lock at Valderas	1,400	1,175
From the Compuerta de Valderas to the Boveda Real	3,290	2,761
From the Boveda Real to the remains of the old subterranean gallery called Techo Bajo	650	545
From Techo Bajo to the viceroys' gallery	1,270	1,066
From the Cañon de los Virreyes to the Boca de San Gregorio	610	512
From the Boca de San Gregorio to the demolished lock	1,400	1,175
From the Presa demolida to the bridge at the Waterfall	7,950	6,671
From the Puente del salto to the waterfall itself (Salto del Río de Tula)	430	361
Length of the canal from Vertideros to Salto	24,530 or	20,585

1. The engineer *Iniesta* even suggested that, during great floods, the water rises to a height of twenty to twenty-five meters in the canal near the Boveda Real. But Velázquez insists that these estimates are greatly exaggerated (*Declaración del Maestro Iniesta* and *Informe de Velásquez*, both unpublished papers).

2. *Informe y exposición de las operaciones hechas para examinar la posibilidad del Desagüe general de la Laguna de México y otros fines á el conducientes*, 1774 (unpublished report, folio 5).

For one-quarter of this distance of four and three-fifth common leagues,   II.124
the chain of the Nochistongo hills (east of the Cerro de Sincoque) has been
cut to an extraordinary depth. At the point where the ridge is the highest,
near the old well of Juan García, the cut in the mountain presents a perpendicular depth of forty to sixty meters. From one talus to the next toward the
summit, its width is from eighty-five to 110 meters.[1] Over a length of more
than 3,500 meters, the depth of the cut is from thirty to fifty meters. The   II.125
channel into which the water flows is usually only three to four meters wide,
but for a large part of the Desagüe, as seen in the cross sections that I have
appended to plate 15 of my Mexican atlas, the upper part of the cross section
is not proportional to the lower part, such that the sides, instead of having
an inclination of 40° or 45°, are much too steep and are constantly subject
to landslides. Especially in the *Obra del Consulado*, we see the enormous
accumulation of *drift deposits* that nature has left on the basaltic porphy-
ries in the Valley of Mexico. Descending the *Viceroys' Staircase*, I counted
twenty-five layers of hardened clay, alternating with the same number of
marly layers containing fibrous calcareous pellets with a vesicular surface.
It was in digging the pit for the desagüe that fossil elephant bones were discovered, which I have discussed in another work.[2]

Large hills formed of rubble are visible on both sides of the cut in the
mountain; these are gradually overgrowing with vegetation. Since the re-
moval of this debris was an infinitely painful and slow task, more recently
people have resorted to Enrique Martínez's former method. The water level   II.126
was raised by small sluices, so that the strength of the current would carry
away the rubble cast into the channel. In the course of this work, twenty to
thirty Indians sometimes perished at once. They were attached to ropes
and forced to work while suspended, gathering debris in the middle of the
current; and it often happened that they were crushed when hurled by the
unpredictable current that threw them against detached masses of rocks.

We have mentioned above that the branch of Martínez's canal that
stretched toward Lake Sumpango was blocked in 1623, and so—to use the
expression of the Mexican engineers today—the *Desagüe* had simply become *negative*; that is, it impeded the Cuautitlan River from emptying into
the lake. During high waters, one felt the drawbacks that this state of affairs

1. To form a clearer idea of the enormous breadth of the pit in the *Obra del Consulado*,
one has only to recall that the breadth of the Seine in Paris at the port of Orsay is 102 meters;
at the Pont-Royal, 136 meters; and at the Pont d'Austerlitz, near the Jardin des Plantes, 175
meters.

2. In the *Recueil de mes observations de zoologie et d'anatomie comparée.*

produced for the city of Mexico. When it overflowed, the Cuautitlan River poured some of its water into the lake, which, swollen even more by the tributaries of San Mateo and Pachuca, joined Lake San Cristóbal. It would have been very costly to enlarge the riverbed of the Cuautitlan, cut off its meanders, and *correct* its course. Even this remedy would not have removed all danger of flooding. Thus, at the end of the last century, the very wise resolution was adopted, under the direction of Don Cosme de Mier y Trespalacios, superintendent-general of the Desagüe, to open two canals that direct the waters of Lakes Sumpango and San Cristóbal to the cut in the Nochistongo Mountain. The first of these canals was begun in 1796, the second in 1798. One is 8,900 meters long, and the other is 13,000 meters. The San Cristóbal drain canal joins the Sumpango canal southeast of Huehuetoca, at a distance of 5,000 meters from its entry into the Desagüe de Martínez. Both projects cost more than one million livres tournois. The water level in these channels is from eight to twelve meters lower than the nearby ground; they have the same defects as the main Nochistongo trench. Their sides are much too steep, and they are almost perpendicular in many places. The erosion of loose earth is, therefore, so frequent that the maintenance of Mr. Mier's two canals costs more than 16,000 to 20,000 francs annually. When the viceroys go to inspect (*la visita*) the Desagüe (a two-day journey that, in the past, earned them a gift of 3,000 piastras fuertes), they embark near their palace on the southern bank of Lake San Cristóbal and travel by boat even beyond Huehuetoca, over a distance of seven common leagues.[1]

II.127

According to an unpublished report by Don Ignacio Castera, the present inspector (*Maestro mayor*) of hydraulic works in the Valley of Mexico, the cost of the *Desagüe*, including repairs to the embankments (*Albaradones*) from the year 1607 to 1789 was 5,547,670 piastras fuertes. If we add to this enormous sum the 600,000 to 700,000 piasters spent in the subsequent fifteen years, we find that all these projects—the cut in the Nochistongo mountain, the embankments, and the two canals for the upper lakes—cost over *thirty-one million livres tournois*. The estimate for the cost of the canal du Midi, which is 238,648 meters long, was only 4,897,000 francs (despite the construction of sixty-two sluices and the magnificent reservoir of St. Ferréol). But the maintenance of this canal from 1686 to 1791 cost 22,999,000 francs.[2]

II.128

1. The so-called *Palacio de los Vireyes*, from which there is a magnificent view of Lake Texcoco and the glacier-covered Popocatepetl volcano, looks more like a large farmhouse than a palace.

2. Andréossy, *Histoire du canal du Midi*, p. 289.

To summarize what we have just stated about the hydraulic projects completed on the plains of Mexico City: we see that the safety of the capital currently rests on (1) the stone embankments that prevent the waters of Sumpango from pouring into Lake San Cristóbal and the waters of the latter from entering Lake Texcoco; (2) the embankments and sluices of Tlahuac and Mexicaltzingo, which counteract the overflow of Lakes Chalco and Xochimilco; (3) Enrico Martínez's Desagüe, through which the Cuautitlan River crosses the mountains before entering the Tula valley; and (4) Mr. Mier's two canals, by which one can drain Lakes Sumpango and San Cristóbal at will.

II.129

These numerous measures, however, do not secure the capital against the floods that come from the north and northwest. Despite all the expenses that have been paid, the city continues to run great risks as long as no canal leads directly to Lake Texcoco. The water of this lake can swell without the water of San Cristóbal, breaking the embankment that holds them back. The great flood of Mexico City during the reign of Ahuitzotl was only caused by frequent rains[1] and the overflowing of the southernmost lakes, Chalco and Xochimilco. The water rose to five or six meters above street level. In 1763 and at the beginning of 1764, the capital was in the greatest danger from the same cause. Flooded everywhere, it was effectively an island for several months, even without a drop of water from the Cuautitlan River emptying into Lake Texcoco. This flood was caused exclusively by the small tributaries that come from the east, west, and south. Water was seen seeping from the earth everywhere, probably because of the hydrostatic pressure it underwent while entering the surrounding mountains. On September 6, 1772, such a heavy and sudden rainstorm[2] fell on the Valley of Mexico that it had every appearance of a waterspout (*manga de agua*). Fortunately, this phenomenon took place in the northern and northwest parts of the valley. The Huehuetoca canal had the most beneficial effect at that time, though a large section of terrain between San Cristóbal, Ecatepec, San Mateo, Santa Inés, and Cuautitlan was so flooded that many buildings crumbled. If this rain cloud had burst above the basin of Lake Texcoco, the capital would have been exposed to the most imminent danger. These

II.130

---

1. The Indian historians relate that, at that time, great masses of water containing fish that are found only in rivers in the hot regions (*pescados de tierra caliente*) were seen emerging from the interior of the earth on the mountain slopes, a physical phenomenon that is difficult to explain because of the elevation of the Mexican plateau.

2. *Informe de Velasquez* (unpublished), folio 25.

circumstances, and many others that we have presented above,[1] provide sufficient evidence as to how indispensable it has become for the government to attend to draining the lakes closest to Mexico City. This necessity increases every day, because deposits of earth are raising the elevation of the bottom of the basins of Texcoco and Chalco.

In fact, during my stay in Huehuetoca in January 1804, viceroy Iturrigaray ordered the construction of the Texcoco canal, which Martínez had already projected, and Velázquez recently surveyed. This canal, whose cost is estimated at three million livres tournois, will begin at the northwest extremity of Lake Texcoco, at a point near the first sluice of the Calzada de San Cristóbal, 36° southeast, at a distance of 4,593 meters. It will first pass through the broad arid plain where the isolated peaks of *Las Cruces de Ecatepec* and *Chiconautla*[2] are located; then, it will lead through the Santa Iñés farm toward the Huehuetoca canal. Its total length, as far as the Vertideros sluice, is 37,978 Mexican varas, or 31,901 meters. But the need to deepen the channel of the old Desagüe from Vertideros to a point beyond the Boveda Real will make the completion of this project more expensive; the first of these two points is 9.078 meters higher, and the second 9.181 meters lower than the average level of Lake Texcoco.[3] The distance between them is almost 10,200 meters. To avoid deepening the bed of the present Desagüe over an even greater length, they intend to give the new canal a drop of only 0.2 meters over a course of 1,000 meters. The engineer Martínez's project was rejected in 1607 for the simple reason that it was assumed that the running waters had to have a drop of half a meter per hundred meters. Alonso de Arias then proved on ▼ Vitruvius's authority ([*De architectura*,] book VIII, chap. 7) that to make the water of Lake Texcoco

II.131

II.132

1. pp. 355ff.

2. According to Mr. Velázquez's geodesic measurements, the first of these peaks is 404 Mexican varas; the second, 378 Mexican varas (339 and 317 meters, respectively) in elevation above the average level of the waters of Lake Texcoco.

3. To complete the description of this great hydraulic work and make the plate that represents the cross section of the mountain cut more interesting, we shall record here the principal results of Mr. Velázquez's survey. Corrected for the refraction error and then converted from the apparent level to the true level, these results generally agree with those which Enrico Martínez and Arias obtained at the beginning of the seventeenth century. But they prove the inaccuracy of the surveys conducted in 1764 by ▼ Don Ildefonso de Iniesta, according to which the draining of Lake Texcoco was a much more difficult problem to resolve than it actually is. We shall designate by U the points that are higher and by – the points that are lower than the average level of Lake Texcoco in 1773 and 1774, where the signal placed near its bank at south 36° is at the first sluice of the Calzada de San Cristóbal at a distance of 5,475 Mexican varas.

enter the Río de Tula, the new canal would require a prodigious depth, and II.133
that even at the foot of the cascade, near the Hacienda del Salto, the level
of its waters would be two hundred meters lower than the level of the river.
Martínez had to capitulate to the power of biased opinions and the authority of the ancients! We believe that while it is prudent to give a gentle slope
to navigational canals, it is generally useful to cut drainage canals much
steeper. But there are specific cases where the nature of the terrain does not
allow for hydraulic works that combine all the dictates of theory.

When one considers the expenses that excavations for the Río del Desagüe would require, from the Vertideros or the Valderas sluices to the
Boveda Real, one is tempted to believe that it would perhaps be easier to
protect the capital from the dangers that Lake Texcoco still wields by returning to the project that Simón Méndez[1] had begun during the great flood
of 1629–1634. Mr. Velázquez reviewed this project in 1774. After surveying the terrain, that geometer claimed that twenty-eight air shafts and an II.134
underground gallery 13,000 meters long, which would lead the waters
of Texcoco through the Zitlaltepec Mountains to the Tequizquiac River,

		VARAS	PALMOS	DEDOS	GRANOS
The bottom of the Guautitlan River near the lock at Vertideros	+	10	3	2	3
The bottom of the Desague, below the port of Huehuetoca	+	8	0	2	1
Id. Near the lock at Santa Maria	+	4	3	8	3″
Id. Below the lock at Valderas	+	2	1	11	2
Id. Below the Boveda Real	−	10	3	9	3
Id. Below the Boveda de Techo Bajo	−	15	0	6	1
Id. Below the Boca de San Gregorio	−	23	1	11	2
Id. Above the Salto del Río	−	90	1	9	0
Id. Below the Salto del Río	−	107	2	9	0

We must note that the vara is divided into four palms, forty-eight dedos, and 192 granos;
that one toise is equivalent to 3.32258 Mexican varas, and one Mexican vara equals 0.839169
meters, according to the experiments done in varas and preserved in the *Casa del Cabildo* in
Mexico City since the time of ▼ King Philip II.

1. Since my departure, the great project of the *Desagüe directo* has begun, a canal that
begins at Lake Texcoco, crosses Lakes Cristóbal and Sumpango, and carries its waters to the
Huehuetoca trench, which must be dug as far as Lake Texcoco. Political revolutions have not
only interrupted this new work but also reduced the older projects to the most deplorable
state because of lack of maintenance.

could be completed more quickly and at lesser expense than enlarging the Desagüe pit, expanding its bottom over a length of more than 9,000 meters, and digging a canal from Lake Texcoco to the Vertideros sluice near Huehuetoca. I was present for the consultations in 1804 that preceded the decision to drain the latter lake through the old cut in Nochistongo mountain. The advantages and disadvantages of Méndez's project were not discussed at those meetings.

One hopes that when the new Texcoco canal is dug, the fate of the Indians will be attended to more seriously than it has been to date, even during the digging of the Sumpango and San Cristóbal channels in 1796 and 1798. The Indians have the bitterest hatred of the Desagüe de Huehuetoca. They consider a hydraulic project to be a public calamity, not only because serious accidents that occurred during Martínez's cut in the mountain led to the deaths of a large number of individuals, but especially because they were plunged into dire poverty when forced to work and to neglect their own affairs, during the draining of the lakes. Several thousand Indian laborers II.135 have been occupied almost continuously for two centuries. The Desagüe may be considered a primary factor in the misery of the indigenous peoples in the Valley of Mexico. They were ravaged by fatal illnesses stemming from dampness to which they were exposed in the Nochistongo trench. Only a few years ago, the Indians were cruelly tied to ropes and forced to work like slaves, sometimes sick and dying as they worked. Through an abuse of the law, especially of the principles introduced since the intendancies were established, work on the Desagüe de Huehuetoca is looked upon as an extraordinary corvée. This physical labor demanded of the Indian is a vestige of the *mita*[1], which one would not expect to find in a country where mining work is now completely voluntary, and where the Indian has more personal freedom than a peasant in northeastern Europe. In bringing these important considerations to the viceroy's attention, I was able to rely on numerous testimonies found in the *Informe de Zepeda*. There, one reads on every page "that the Desagüe led to a decrease in both the population and the well-being of the Indians, and that no one dares to undertake such and such II.136 hydraulic project because the engineers no longer have as many Indians at their disposal as they did in the time of viceroy Don Luis de Velasco II." It

1. See vol. 1, chap. V, p. 338 [p. 227 in this edition]. At the Desagüe, the Indian is paid two reals of *plata* [silver] or twenty-five soles per day. In the seventeenth century, during Martínez's rule, Indians were paid only five reals, or three francs, per week; but they were also given a certain amount of corn for sustenance in addition.

is of some consolation to observe, as we tried to explain at the beginning of chapter four, that this progressive depopulation only occurs in the central part of the former Anahuac.

In all of the hydraulic projects in the Valley of Mexico, water was considered solely as an enemy against which it was necessary to defend oneself, either with embankments or by means of drainage canals. We have proven above (pp. 327 and following) that this plan of action, and especially the European system of artificial desiccation, has destroyed the basis of the fertility of a large part of the Tenochtitlán plateau. Efflorescences of carbonate of soda (*Tequesquite*) have increased as the atmospheric humidity and the mass of running water have decreased. Beautiful savannahs have gradually assumed the appearance of arid steppes. In large open spaces, the valley's soil is merely a crust of hardened clay (*Tepetate*) devoid of vegetation and cracked through contact with the air. It would have been quite easy, however, to put the natural advantages of the soil to good use by employing the same canals not only to *draw* water from the lakes but also for *irrigating* arid plains and for internal *navigation*. Large basins of water stacked on top of each other would facilitate the layout of irrigation canals. Southeast of Huehuetoca are three sluices called *los Vertideros*, which are opened whenever it is felt necessary to discharge the Cuautitlan River into Lake Sumpango or to dry out the *Río del Desagüe* (the cut in the mountain) to clear or deepen the channel. Since the course of the old mouth of the Río de Cuautitlan (the one that existed in 1607) had gradually disappeared, a new canal was dug from Vertideros to Lake Sumpango. Rather than continuously drawing the water from both this lake and Lake San Cristóbal out of the valley toward the Atlantic Ocean, they might have distributed the waters of the Desagüe for the benefit of agriculture in the lowest parts of the valley, during the eighteen- to twenty-year periods in which the water levels do not rise significantly. Water reservoirs for times of drought might have been built. But they preferred to blindly follow the order formerly dispatched from Madrid, which stated "that not a single drop of water from Lake San Cristóbal must enter Lake Texcoco, other than once a year when the sluices (*las compuertas de la calzada*) are opened for fishing[1] in the first of these basins." The trade among the Indians of Texcoco languishes for

II.137

II.138

---

[1]. This fishing season is one of the most important outdoor festivals for the inhabitants of the capital. The Indians would build huts on the banks of Lake Cristóbal, which is almost too dried out for fishing. This recalls the fishing that ▼ Herodotus relates the Egyptians did twice a year in Lake Mœris when the irrigation sluices were opened.

months on end because of the lack of water in the salt lake that separates them from the capital, and dry land lies below the mean level of the Cu-autitlan's waters and those of the northern lakes. It has nevertheless been centuries since anyone has thought of providing for agricultural needs or for internal navigation. For a long time, there was a small canal (*Sanja*) lead-ing from Lake Texcoco to Lake San Cristóbal. A sluice lock with a fall of four meters could have allowed canoes to sail back from the capital to the latter lake. Mr. Mier's canals might even have led them to the village of Hue-huetoca. A water connection would thus have been established between the southern bank of Lake Chalco and the northern edge of the valley, across a distance of over 80,000 meters. Learned men motivated by noble patriotic zeal have dared to raise their voices[1] in favor of these ideas. But the gov-ernment, having steadfastly rejected the best-conceived projects, refused to consider the Mexican lakes as anything but a harmful presence of which the outskirts of the capital should be rid, and from which the only outflow should be toward the Ocean coasts.

II.139　　　Now that the viceroy Don José de Iturrigarray has ordered the Tex-coco canal opened, there will be no obstacle to free navigation thoughout the large and beautiful valley of Tenochtitlán. Wheat and other products from the districts of Tula and Cuautitlan will be brought to the capital by water. The price for a mule load, calculated at a weight of 300 pounds from Huehuetoca to Mexico City, is five reales,[2] or four francs. It is estimated that when navigation is established, the freight charge for an Indian canoe carrying 15,000 pounds of cargo will be only four or five piasters, so that it will only cost nine sous to transport 300 pounds (which make one *carga*). Mexico City will get lime, for instance, at a cost of six or seven piasters per cartload (*carretada*), whereas today it costs from ten to twelve piasters.

But the most beneficial effect of a navigable canal from Chalco to Hue-huetoca will be the improvement to trade in the interior of New Spain, which is called *comercio de tierra adentro* [inland trade] and which runs in a straight line from the capital to Durango, Chihuahua, and Santa Fe in New Mexico. Huehuetoca may become the storehouse for this important trade route, on which more than fifty to sixty thousand beasts of burden (*re-cuas*) are employed. Over a route of five hundred leagues, the mule drivers

---

1. Mr. Velázquez, for example, at the end of his *Informe sobre el Desagüe* (unpublished).

2. A double piaster is eight silver reals (*de Plata*); in works on the Spanish colonies in America, *pesos fuertes and reales de Plata* are always understood (see vol. I, the note to p. 441 [p. 282 in this edition]).

(*arrieros*) in Nueva Vizcaya and Santa Fe fear no day of travel more than the   II.140
one between Huehuetoca and Mexico City. The roads in the northwest part
of the valley, where the basaltic amygdaloid is covered with a thick layer
of clay, become almost impassable in the rainy season. Many mules perish
there. The others are unable to recover from their exhaustion on the outskirts of the capital, which offer neither good pastures nor the large com-
mons (*exidos*) that they would find in Huehuetoca. Only after living in a
country where all trade is conducted by camel- or mule-driven caravans can
one appreciate the impact of the topics we have just discussed on the inhabitants' well-being.

    The lakes in the southern part of the valley of Tenochtitlán release from
their surface clouds of sulfuric hydrogen that one can smell on the streets in
Mexico City whenever the south wind blows. For this reason, people in this
country regard this wind as very unhealthy. In their hieroglyphic script, the
Aztecs represented it by a skull. Lake Xochimilco is partly filled with plants
from the Juncaceae and Cyperoides families that float near the surface, beneath a layer of stagnant water. It was recently proposed[1] to the government
that a navigable canal be dug in a straight line from the small town of Chalco
to Mexico City, a canal that would be one-third shorter than the existing   II.141
one. It was proposed at the same time that the basins of Lake Xochimilco
and Lake Chalco be dried out and that these lands, which had been washed
by fresh water for centuries and had become very fertile, be sold. Since the
center of Lake Chalco is slightly deeper than Lake Texcoco, it would not be
fully drained. Mr. Castera's project would be of benefit both to agriculture
and to air quality, since the southern extremity of the valley generally offers
the soil that is best suited for cultivation. Carbonate and muriate of soda are
less abundant there because of the continuous filtrations fed by the rivulets
of water that descend from the heights of the Cerro de Axusco, the Guarda,
and the Volcanoes. We must not forget, however, that draining the two lakes
will likely exacerbate the dryness of the atmosphere in a valley where ▼ De-
Luc's hygrometer[2] often descends to fifteen degrees. This disaster is inevitable if no attempt is made to link these hydraulic projects to a general
system, and, simultaneously, to increase the number of irrigation canals, to
form large reservoirs for periods of drought, and to build sluices capable of

---

    1. *Informe de Ignacio Castera* (manuscript), leaf 14.

    2. With the air temperature at twenty-three degrees Centigrade, fifteen degrees on
Deluc's whalebone hygrometer are equivalent to forty-two degrees on Saussure's hair tension
hygrometer. I have discussed the causes of this extreme aridity in the physical Tableau of the
equinoctial regions appended to my *Essai sur la géographie des plantes*, p. 98.

II.142

counterbalancing the different pressures of unequal reaches and that would open to receive and retain the high water from the rivers. Placed at appropriate elevations, these water reservoirs could even serve periodically to clean and wash the streets of the capital.

At the dawn of a civilization, bold concepts and colossal projects are more seductive than are the simplest and most easily realized ideas. Instead of setting up a system of small canals for inland navigation of the lake, at the time of the viceroy Count de Revillagigedo, people were distracted by vague speculations about the possibility of a water connection between the capital and the port of Tampico. Seeing the lake water descend through the Nochistongo Mountains via the Río de Tula (also called the Río de Moctezuma) and the Río de Panuco to the Gulf of Mexico, they entertained the hope of opening up the same route to the Veracruz trade. Goods valued at more than one hundred million livres tournois are transported annually on mule-back from the coastline facing Europe to the interior plateau. Flour, leather, and valuable metals move the other way, from the central plateau to Veracruz. The capital is the storehouse of this immense trade. The overland road that, in the absence of a canal, must be built from the coast to Perote will cost several million piasters. To this point, the air of the port of Tampico has seemed less harmful to Europeans

II.143

and to the inhabitants of the cold regions of Mexico than the climate of Veracruz. Although its sandbar prevents the former port from receiving ships that require a depth of forty-five to sixty decimeters of water, this might even be preferable to the dangerous anchorage in the shallows of Veracruz. Because of this set of circumstances, some means of navigation from the capital to Tampico would be desirable, regardless of the expense that such a bold project might entail.

But such expense is not to be feared in a country where one private individual, Count de la Valenciana, has dug three shafts in a single mine[1] that cost him over eight and a half million francs. The possibility of digging a canal from the valley of Tenochtitlán to Tampico should also not be ruled out. Recent advances in hydraulic architecture enable ships to sail across high mountain chains wherever nature offers dividing ridges that link different drainage basins. General Andréossy has indicated several of these points in the Vosges and other parts of France.[2] ▼ Mr. de Prony has calculated the time it would take a boat to cross the Alps, if, by taking advantage of the

1. Near Guanajuato.
2. Andréossy, *Sur le canal du Midi* [*Histoire du canal du Midi*], p. 45.

lakes near the hospice of Mont Cenis, a water connection were established between Lans-le-Bourg and the Susa Valley. Using his own calculations, this II.144 distinguished engineer proved that, in this particular case, overland transport was preferable to time-consuming travel via locks. Inclined planes, invented by ▼ Mr. Reynolds and perfected by Fulton, and Mr. Huddleston and Mr. Betancourt's plunger sluices—two ideas that are also applicable to the system of small canals—have advantageously increased the number of means that human ingenuity has produced for navigation in mountainous country. But no matter how much water or time one might succeed in saving, there are certain maximum elevations of a canal's highest point above which canals become less efficient than roads. The waters of Lake Texcoco to the east of the capital of Mexico are 2,276 meters higher than the sea near the port of Tampico! Even if the locks were placed side by side, it would take almost two hundred sluices to raise boats to such an enormous height. If the reaches of the Mexican canal were distributed as they are in the canal du Midi, whose dividing ridge (at Narouse) has a perpendicular elevation of only 189 meters, the number of locks would increase to 330 or 340. I am not familiar with the riverbed of the Moctezuma beyond the Tula Valley (the former Tollan); nor do I know what its partial drop is to the outskirts of Zimapán and the Doctor; but I recall that without sluices, and across distances of 180 leagues, pirogues [large canoes] on the great rivers of South America ascend to elevations of three hundred meters, either by II.145 being towed or by rowing against the current. Despite this similarity and those presented by the great works carried out in Europe, I have difficulty persuading myself that a navigation canal from the Anahuac plateau to the Antillean Sea coasts would be an advisable hydraulic project!

: :

The notable towns (*Ciudades y villas*) in the intendancy of Mexico City are as follows:

MEXICO CITY, capital of the kingdom of New Spain.
Elevation: 2,277 meters; 137,000.
TEXCOCO, with its once considerable cotton factories, which have suffered greatly from competition from producers in Querétaro; 5,000.
CUYOACAN, with its convent of nuns founded by Hernán Cortés. According to his will, the great captain wished to be buried in this convent, "wherever in the world he might end his days." We have seen above that this clause of the will was not honored.

TACUBAYA, west of the capital, with its archbishop's palace and a beautiful plantation of European olive trees.

TACUBA, formerly Tlacopan, capital of a small kingdom of Tepanecs.

CUERNAVACA, formerly Quauhnahuac, on the southern slope of the Cordillera of Guchilaque, in a temperate climate, one of the most delightful and most suitable for growing European fruit trees. Elevation:[1] 1,655 meters.

CHILPANSINGO (Chilpantzinco), surrounded by fertile wheat fields. Elevation: 1,080 meters.

TAXCO (Tlachco), with a beautiful parochial church built and endowed around the mid-eighteenth century by a Frenchman, ▼ Joseph de Laborde, who, in a short time, acquired immense wealth by exploiting the Mexican mines. The construction of the church alone cost this individual over two million francs. Reduced to dire poverty toward the end of his life, he obtained permission from the archbishop of Mexico City to profit personally from selling to the capital's metropolitan church the magnificent sun (*Custodia*) embellished with diamonds, which, in happier times, he had gifted to the tabernacle of the parochial church of Taxco out of devotion. Elevation of the town: 783 meters.

ACAPULCO (Acapolco), which backs upon a chain of granite mountains that, due to the reflection of the sun's rays, increases the stifling heat of the climate. The famous mountain cut (*abra de San Nicolás*) near the bay of La Langosta, intended to let in the sea winds, was recently finished. The population of this miserable town, inhabited almost exclusively by people of color, rises to 9,000 when the galleon from Manila (*Nao de China*) arrives. Its normal population is only 4,000.

ZACATULA, a small port on the South Sea bordering of the intendancy of Valladolid, between the ports of Siguantanejo and Colima.

II.146

II.147

---

1. In the literary gazette published in Mexico City ([*Gacetas de literatura de México*], 1760 [1791], p. 221), Mr. Alzate [y Ramírez] confirms that in New Spain the absolute elevation of localities has very little influence on their temperature. For example, he cites the town of Cuernavaca, which, according to him, is at the same elevation above sea level as the capital of Mexico, and which owes its delicious climate only to its position south of a high chain of mountains. But Mr. Alzate made an error of more than 600 meters for the elevation of Cuernavaca! Cortés, who changed all Aztec names, calls this town *Coadnabaced*, a word in which it is difficult to recognize Quauhnahuac (*Carta de relación al Emperador don Carlos*, paragraph 19).

LERMA, at the entrance to the Toluca Valley, on marshy terrain.

TOLUCA (Tolocan), at the foot of the porphyritic mountain of San Miguel de Tutucuitlalpilco, in a valley where corn and maguey (agave) abound. Elevation: 2,687 meters.

PACHUCA, the oldest mining location in the kingdom, next to Tasco. Like the nearby village Pachuquillo, Pachuco is assumed to be the first Christian village founded by the Spanish. Elevation: 2,482 meters.

CADEREITA, with beautiful quarries of porphyry on a base of clay (*Thonporphyr*).

SAN JUAN DEL RÍO, surrounded by gardens adorned with vines and annona shrubs. Elevation: 1,978 meters.

QUERÉTARO, famous for the beauty of its buildings, its aqueduct, and its cloth factories. Elevation: 1,940 meters. Normal population: 35,000.   II.148

The town has 11,600 Indians, eighty-five secular ecclesiastics, 181 monks, and 143 nuns. In 1793, the food supply of Querétaro amounted to 13,618 *cargas* of wheat flour, 69,445 *fanegas* of corn, 656 *cargas* of chili (capsicum), 1,770 barrels of eau-de-vie, 1,682 cattle and cows, 14,949 sheep, and 8,869 pigs.[1]

The most important mines in this intendancy, with no consideration of their present wealth, are:

La Veta Vizcaína de Real del Monte near Pachuca; Zimapán, El Doctor, and Tehulolitepec near Tasco.

: :

II. The Intendancy of Puebla

POPULATION (IN 1803): 813,300
SURFACE AREA IN SQUARE LEAGUES: 2,696
INHABITANTS PER SQUARE LEAGUE: 301

This intendancy, which the waves of the Great Ocean lap across a coastline of only twenty-six leagues, extends from 16°57' to 20°40' northern latitude. It is thus entirely situated in the Torrid Zone, bordering the intendancy of Veracruz to the northeast, the intendancy of Oaxaca to the east, the Ocean to the south, and the intendancy of Mexico City to the west. Its greatest   II.149

---

1. ▼ *Noticia del Doctor don Juan Ignacio Briones* (unpublished manuscript).

length, from the mouth of the small Tecoyame River to the vicinity of Mextitlan, is 118 leagues, and its greatest width, from Techuacan to Mecameca, is fifty leagues.

The bulk of the intendancy of Puebla is taken up by the high Cordillera of Anahuac. Above 18° latitude, the entire country is comprised by a plateau eminently rich in wheat, corn, agave, and fruit trees; this plateau is 1,800 to 2,000 meters above sea level. The highest mountain in all of New Spain, Popocatepetl, is also found in this intendancy. This volcano, which I was the first to measure, is always ablaze; but for centuries it has emitted only smoke and ashes from its crater. It is six hundred meters higher than any of the tall peaks on the Old Continent. From the Isthmus of Panama to the Bering Straits, which separate Asia from Africa [sic: America], we know of only one elevation, Mount St. Elias, which is higher than the great volcano of Puebla.

The population of this intendancy is still more unevenly distributed than in the intendancy of Mexico City. It is concentrated on the plateau that extends from the eastern slope of the *Nevados*[1] to the environs of Perote, especially on the beautiful, high plains between Cholula, Puebla, and Tlaxcala. Almost all of the country stretching from the central plateau toward San Luis and Ygualapa, near the South Sea coasts, is desert, although very suitable for growing sugar, cotton, and other valuable products of the tropics.

The plateau of Puebla offers remarkable vestiges of the most ancient Mexican civilization. The fortifications at Tlaxcallan were built after the great pyramid of Cholula, a curious monument whose design and detailed description I shall give in the *Personal Narrative* of my travels in the interior of the New Continent. Suffice it to say here that this pyramid, from the top of which I made a great number of astronomical observations, consists of four levels; that in its present state it has only fifty-four meters of perpendicular elevation but 439 meters of horizontal width at its base; that its sides are very precisely positioned to align with the meridians and the parallels; and that (judging from the hole bored a few years ago on the north side) it is built of alternating layers of brick and clay. This information is enough to identify the shape of this edifice as being the same as the pyramids at

II.150

II.151

---

1. The Spanish words *Nevado* and *Sierra Nevada* do not refer to mountains occasionally covered with snow during the summer but to that rise to the region of perpetual snows. I prefer this foreign word to lengthy sentences or the inappropriate expression "*snowy* mountains," which is sometimes used by the academicians sent to Peru. Besides, the word *Nevado*, when it accompanies the name of a mountain, gives an idea of the minimum elevation attributable to a peak (see the *Recueil d'observations astronomiques*, vol. I, p. 134).

Teotihuacán, which we have described above. It also provides sufficient evidence of the important similarity[1] between these brick monuments erected by the oldest inhabitants of Anahuac, the temple of Belus in Babylon, and the pyramids of Menschich-Dahshur, near Sakkara in Egypt.

The platform of the truncated pyramid of Cholula has a surface area of 4,200 square meters. In the middle of the platform, there is a church dedicated to Our Lady of los Remedios that is surrounded by cypress trees, and where mass is celebrated every morning by an Indian ecclesiastic whose regular abode is the top of this monument. From this platform, one has a delightful and impressive view of the Puebla volcano, Orizaba Peak, and the small Cordillera de Matlalcueje,[2] which formerly separated the territory of the Cholulans from that of the Tlaxcalteca republicans.

The pyramid or Teocalli of Cholula has exactly the same height as the Tonatiuh Yztaqual of Teotihuacán that we have described above (pp. 338–39); it is three meters higher than the Mycerinus, the third of the great Egyptian pyramids in the Giza group. The visible length of its base exceeds that of all buildings of this type that travelers have found on the Old Continent. This base is almost double the size of the great pyramid known by the name of Cheops. Those who wish to form a clear idea of the considerable mass of this Mexica monument should imagine a square four times larger than Place Vendôme, covered with a mound of bricks twice the height of the Louvre! Perhaps the whole interior of the pyramid of Cholula is not made of brick; perhaps, as Mr. Zoëga, a famous antiquarian in Rome, already suspected, bricks are merely the veneer covering a heap of stones and cement, like several of the Sakkara pyramids visited by Pococke and, more recently, by ▼ Mr. Grobert.[3] But the road from Puebla to Mecameca, which was dug through part of the first level of the Teocalli, contradicts this supposition.

We do not know the former height of this extraordinary monument. In its present state, the ratio of the length of its base[4] to its perpendicular height is eight to one, whereas for the three great pyramids of Giza, this proportion is

II.152

II.153

---

1. Zoëga, *De origine et usu obeliscorum*, p. 380. Pococke, *Voyages* (Neufchâtel edition), 1752, vol. I, pp. 156 and 167. ▼ Denon, *Voyage*, quarto edition, pp. 86, 194, and 237. Grobert, *Description des pyramides*, pp. 6 and 12.

2. Also call the *Sierra Malinche* or *Doña Maria*. Malinche seems to be derived from Malintzin, a word that (I know not why) today designates the name of the Holy Virgin.

3. See note *E* at the end of this work.

4. I record here the actual dimensions of the three great pyramids of Giza, according to Mr. Grobert's interesting work. Next to them, I place the dimensions of the pyramidal monuments in brick of Sakkara in Egypt and Teotihuacán and Cholula in Mexico. The numbers are in pieds de roi.

II.154

more like one and six-tenth and one and seven-tenth to one, or roughly eight to five. We observed above that the houses of the sun and the moon, or the pyramidal monuments of Teotihuacán northeast of Mexico City, are surrounded by a system of small pyramids, symmetrically arranged. Mr. Grobert has published a very interesting drawing of the equally regular arrangement of the small pyramids that surround Cheops and Mycerinus at Giza. The Teocalli of Cholula—if one may compare it to the great monuments of Egypt—appears to have been built according to a similar plan. On its western side, opposite the Cerros de Tecaxete and Zapoteca, one finds two perfectly prismatic masses. One of these masses is now called Alcosac or Istenenetl, the other the Cerro de la Cruz. The latter, built of rammed earth is only fifteen meters high.

In the intendancy of Puebla, the curious traveler also finds one of the oldest monuments of vegetation. The famous Ahahuete,[1] or cypress, of the village of Atlixco is twenty-three meters (or seventy-three) feet in circumference: measured internally (since its trunk is hollow), it is fifteen feet in diameter. The Atlixco cypress is thus within a few feet of the same size[2] as the baobab (Adansonia digitata) of Senegal.

	STONE PYRAMIDS				BRICK PYRAMIDS	
	Cheops	Chephren	Mycerinus	with five layers in Egypt near Sakkara	with four layers, in Mexico	
					Teotihuacan	Cholula
Height	448 feet	398 f.	162 f.	150 f.	171 f.	172 f.
Length of base	728	655	280	210	645	1,355

It is curious to observe that (1) the people of Anahuac had intended to give the pyramid of Cholula the same height and double the base of Tonatiuh Yztaqual; and (2) that the largest of all the Egyptian pyramids, Asychis, whose base is 800 feet long, is not made of stone but of brick (Grobert, [*Description des pyramides de Ghizé,*] p. 6). The cathedral of Strasbourg is eight feet lower, and the cross of St. Peter's in Rome is forty-one feet lower than Cheops. In Mexico, there are pyramids several levels high in the forest of Papantla, slightly above sea level, on the plateaus of Cholula and Teotihuacán, at elevations that surpass those of our Alpine passes. We are astonished to see that in regions that are the farthest from each other, in the most different of climes, people follow the same model of construction, decoration, and customs, even in the form of their political institutions.

1. Cupressus disticha. Linnaeus.

2. On the antiquity of plant species, see my *Mémoire sur la physiognomie des plantes* in my *Tableaux de la nature*, vol. II, pp. 108 and 137.

The district of the former republic of Tlaxcala, inhabited by Indians   II.155
very protective of their privileges and most inclined to civil conflicts, has
long constituted a separate government. In my general map of New Spain,
I indicate that it still belongs to the intendancy of Puebla. Because of a recent change in financial administration, however, Tlaxcala and Guautla de
la Hamilpas have been adjoined to the intendancy of Mexico City, while
Tlapa and Ygualapa have been removed from it.

In 1793, in the intendancy of Puebla there were (not counting the four
districts of Tlaxcala, Guautla, Ygualapa, and Tlapa):

Indian men		187,531 souls
Indian women		186,221souls
Spaniards or Whites	male	25,617
	female	29,393
People of mixed race	male	37,318
	female	40,590
Secular clergymen		585
Monks		446
Nuns		427
Total census result		508,028 souls

These were distributed among six towns, 133 parishes, 607 villages, 425
farms (*Haciendas*), 886 outlying homes (*ranchos*), and thirty-three convents, of which two-thirds house monks.

In 1793, the government of Tlaxcala had a population of 59,177 souls,
among them 21,849 Indian men and 21,029 Indian women, distributed   II.156
among twenty-two parishes, 110 villages, and 139 farms. The vaunted privileges of the citizens of Tlaxcallan may be reduced to the following three
points: (1) the town is governed by a cacique and four Indian Alcaldes [may-
ors] who represent the former heads of the four districts that are still called
Teopectipac, Ocotelolco, Quiahutztláan, and Tizatlán. These alcaldes answer to an Indian governor who is himself subject to the Spanish intendant;
(2) in accordance with a royal *cedula* of April 16, 1585, whites may not occupy any seat in the municipality of Tlaxcala; and (3) the cacique, or Indian
governor, enjoys the stature of an *Alférez real* [royal second lieutenant].

In 1793, the district of Cholula had a population of 22,423 souls; it had
forty-two villages and forty-five farms. Cholula, Tlaxcala, and Huetzocingo
are the three republics that resisted the Mexica empire for centuries,

although the wretched aristocracy of their society would have given the lower classes hardly any more freedoms than they would have had under the Aztec kings' feudal regime.

Advances in national industry and in the well-being of the inhabitants of this province have been very slow, despite the active zeal of an intendant as enlightened as he is respectable, ▼ Don Manuel de Flon, who has recently inherited the title of Count de la Cadena. The flour trade, formerly thriving, has suffered greatly from the enormous cost of transportation from II.157 the Mexican plateau to Havana, due especially to the lack of beasts of burden. The trade in hats and ceramics that the town of Puebla conducted with Peru until 1710 has completely ceased. But the greatest hindrance to public prosperity is that four-fifths of all properties (*fincas*) belong to mortmain owners, in other words, to communities of monks, chapter houses, brotherhoods, and hospitals.

The intendancy of Puebla has rather large salt marshes near Chila, Xicotlan, and Ocotlan (in the district of Chiautla), as well as near Zapotitlan. The beautiful marble known as Puebla marble—preferable to stone from Bizaru, Real del Doctor—is quarried in Totamehuacan and Tecali, two to seven leagues from the capital of the intendancy. The carbonate of lime from Tecali is transparent, like the gypsum alabaster from Volterra [Italy] and the phengite of the ancients.

The indigenous peoples of this province speak three completely different languages—Mexica [Nahuatl], Totonac, and Tlapanec. The first is peculiar to the residents of Puebla, Cholula, and Tlaxcala; the second, to the residents of Zacatula; the third has been preserved on the outskirts of Tlapa.

: :

The most noteworthy towns in the intendancy of Puebla are:

II.158   LA PUEBLA DE LOS ANGELES, capital of the intendancy, more populous than Lima, Quito, Santa Fe, and Caracas: after Mexico City, Guanajuato, and Havana, it is the largest city in the Spanish colonies on the New Continent. La Puebla belongs to a very small number of American cities that were founded by European colonists; at the beginning of the sixteenth century, there were only a few huts inhabited by the Indians of Cholula on the plain of Acaxete (or Cuitlaxcoapan) on the site where the provincial capital is now found. The privileges of the town of Puebla dates from September 28, 1531. The residents' consumption of wheat flour in 1802 was 52,951 cargas (each weighing 300 pounds) and 36,000 cargas of corn. The ground elevation of the Plaza Mayor is 2,196 meters. Population: 67,800.

TLAXCALA is so far removed from its former grandeur that it has no more than 3,400 residents, among whom there are only 900 pure-blooded Indians. Nonetheless, Hernán Cortés found a population there that seemed larger to him than the population of Granada: 3,400.

CHOLULA, which Cortés[1] called Churultecal, surrounded by beautiful agave plantations. Population: 16,000.   II.159

ATLIXCO, justly famous for the beauty of its climate, the exceptional fertility of its fields, and the abundance of delicious fruits, especially annona cherimolia, Lin. (*chilimoya*) [cherimoya] and several kinds of passion-flower (*parchas*) grown in the environs.

TEHUACAN DE LAS GRANADAS, the former Teohuacan de la Mixteca, one of the sanctuaries most visited by the Mexica before the arrival of the Spanish. TEPEACA, or Tepeyacac, which belongs to Cortés's marquisat. At the beginning of the conquest, this town was called Segura de la Frontera (*Cartas de Hernán Cortés*, p. 155). In the district of Tepeaca, one finds the pretty Indian village now called   II.160 Huacachula (the former Quauhquechollan) located in a valley rich in fruit trees.

HUAJOCONGO, or Huexotzinco, formerly the administrative center of a small eponymous republic, the enemy of the republics of Tlaxcala and Cholula.

---

1. This great *Conquistador,* with the simplicity of style that characterizes his writing, paints a curious picture of the ancient town of Cholula. "The inhabitants of this town," he writes in his third letter to Emperor Charles V, "are better dressed than those we have seen until now. The upper class wears coats (*albornoces*) over their clothes. These coats are different from those in Africa because they have pockets, although their cut, fabric, and fringe are the same. The environs of the town are very fertile and well-cultivated. Almost all the fields can be watered, and the town is more beautiful than all those in Spain because it is well-fortified and built on very level ground. I can assure Your Highness that from the top of a mosque (*mezquita* is the word Cortés uses to designate the *Teocalli*), I counted over four-hundred towers, all mosques. The number of inhabitants is so large that there is not an inch of uncultivated land; in many places, however, the Indians feel the effects of famine, and there are many poor people who ask alms from the rich in the streets, in houses, and in the market place, like the beggars in Spain and other civilized countries" (*Cartas de Cortés,* p. 69). It is rather odd that the Spanish general considers begging in the streets a sign of civilization. As he writes, "*Gente que piden como hay en España y en otras partes que hay gente de razón*" [Beggars in Spain and other civilized countries].

However unpopulated the intendancy of Puebla may be, its *relative population*[1] is nevertheless four times greater than that of the kingdom of Sweden, and almost the same as that of the kingdom of Aragon.

The industry of the inhabitants of this province is only barely channeled into gold and silver mining; the mines at Yxtacmaztitlan, Temeztla, and Alatlauquitepec in the Partido de San Juan de los Llanos, those of La Cañada near Tetla de Xonotla, and those at San Miguel Tenango near Zacatlan are nearly abandoned or relatively unworked.

: :

III. Intendancy of Guanajuato

POPULATION (IN 1803): 517,300
SURFACE AREA IN SQUARE LEAGUES: 911
INHABITANTS PER SQUARE LEAGUE: 586

II.161

This province, entirely situated on the ridge of the high Cordillera of Anahuac, is the most populous in New Spain; it also has the most evenly distributed population of any province. From Lake Chapala to northeast of San Felipe, its length is fifty-two leagues; it width from Villa de León to Celaya is thirty-one leagues. Its territorial extent is almost the same as that of the kingdom of Murcia, and its relative population exceeds that of the kingdom of Asturias. This population is even greater than the relative population of the departments of Hautes-Alpes, Basses-Alpes, Pyrénées-Orientales, and the Landes. The highest point in this mountainous country seems to be the mountain of Los Llanitos in the Sierra de Santa Rosa. I found its elevation above sea level to be 2,815 meters.

The cultivation of this beautiful province, which is part of the ancient kingdom of Michoacan, is almost entirely due to the Europeans who introduced the first seeds of civilization there in the sixteenth century. It was in these northern regions, on the banks of the Río de Lerma (formerly called Tololotlan), that they fought against the nomadic hunter tribes to whom historians refer by the vague name of Chichimecs, and who belonged to the Indian tribes of the Pames, Capuces, Samues, Mayolias, Guamanes, and Guachichiles. As the country was gradually abandoned by these wandering, warlike peoples, the Spanish conquerors transplanted colonies of Mexica (or Aztec) Indians there. For a long time, agriculture made greater

---

1. See above, pp. 314–15.

advances there than mining. The local mines, little known at the beginning of the conquest, were almost entirely abandoned during the seventeenth and eighteenth centuries. Their wealth has only surpassed the mines of Pachuca, Zacatecas, and Bolaños in the last thirty to forty years. As we shall discuss later, their metal production is now greater than the production of Potosí or any other mine on the two continents has ever been.

II.162

In the intendancy of Guanajuato, there are three *ciudades* (viz. Guanajuato, Celaya, and Salvatierra); four *villas* (viz. San Miguel el Grande, León, San Felipe, and Salamanca); thirty-seven villages or pueblos, thirty-three parishes (*parroquias*), 448 farms or *haciendas*, 225 individuals in the secular clergy, 170 monks, thirty nuns, and, out of a population of 180,000 Indians, 52,000 subject to tribute.

The most noteworthy cities in this intendancy are the following:

GUANAJUATO, or Santa Fe de Guanajuato. The construction of this city was begun by the Spanish in 1554. It received the royal privilege of *villa* in 1619, and that of *ciudad* on December 8, 1741. Its current population is:

Within the city confines (*en el casco de la ciudad*)	41,000
In the mines that surround the city, and whose buildings adjoin the latter, at Marfil, Santa Ana, Santa Rosa, [II.163] Valencia, Rayas, and Mellado	29,600

among whom are 4,500 Indians. The elevation of the city at the Plaza Mayor is 2,084 meters. The elevation of Valenciana at the edge of the new pit (*tiro nuevo*) is 2,313 meters. The elevation of Rayas at the mouth of the tunnel is 2,157 meters.

II.163

SALAMANCA, a pretty town, located on a plain the rises unnoticeably through

Temascatio, Burras, and Cuevas, toward Guanajuato. Elevation: 1,757 meters.

CELAYA. Splendid edifices were recently built in Celaya, Querétaro, and Guanajuato.

The church of the Carmelites at Celaya is well-proportioned and adorned with

Corinthian and Ionic columns. Elevation: 1,835 meters.

VILLA DE LEON, on a plain that is highly fertile in wheat. One finds the finest fields of wheat, barley, and corn between this town and San Juan del Río.

SAN MIGUEL EL GRANDE, famous for the industry of its inhabitants, who make cotton cloth.

The hot springs of San José de Comangillas are found in this province, spewing forth from a basaltic crack whose temperature (according to the measurements I made jointly with ▼ Mr. Rojas) is 96.3° on the Centigrade thermometer.

: :

IV. The Intendancy of Valladolid

POPULATION (IN 1803): 376,400
SURFACE AREA IN SQUARE LEAGUES: 3,446
INHABITANTS PER SQUARE LEAGUE: 109

At the time of the Spanish conquest, this intendancy was part of the kingdom of Michoacán (Mechoacan), which stretched from the Río de Zacatula to the port of La Navidad, and from the mountains of Xala and Colima to the Lerma River and Lake Chapala. The capital of the kingdom of Michoacan, which (like the republics of Tlaxcala, Huexocingo, and Cholollan) had always been independent of the Mexica Empire, was Tzintontzan, a town situated on the shores of an infinitely picturesque lake, called Lake Patzcuaro. Tzintzontzan, which the Aztec inhabitants of Tenochtitlán called Huitzitzila, is now only a poor Indian village, although it has kept the ostentatious title of city (*ciudad*).

The intendancy of Valladolid, which is popularly known as Michoacan, is bordered to the north by the Río de Lerma, which, farther to east, takes the name of Río Grande de Santiago. It borders the intendancy of Mexico City to the east and northeast; to the north, the intendancy of Guanajuato; to the west, that of Guadalajara. The greatest length of the province of Valladolid is seventy-eight leagues, from the port of Zacatula to the basaltic mountains of Palangeo; therefore in the direction of south-southeast to north-northeast. It is washed by the South Sea across a coastal area of more than thirty-eight leagues.

Situated on the eastern slope of the Cordillera of Anahuac, intersected by charming hills and valleys, and offering to the traveler's eye an appearance uncommon in the Torrid Zone—that of wide prairies watered by streams—the province of Valladolid usually enjoys a mild, temperate climate that is very favorable to its inhabitants' health. Only when one

II.165

descends the Ario plateau and approaches the coast does one find terrain where the new colonists and often even the indigenous peoples are exposed to the scourge of intermittent and putrid fevers.

The highest mountain peak in the intendancy of Valladolid is the *Tancitaro* peak, to the east of Tuspan. I was unable to view it closely enough to take a precise measurement, but it is certainly higher than the Colima volcano and is more often covered in snow. East of the Tancitaro peak, the *volcán de Jorullo* (Xorullo or Juruyo) was formed during the night of September 29, 1759, which we discussed above,[1] and in whose crater Mr. Bonpland and I arrived on September 19, 1803. The great catastrophe during   II.166 which this mountain rose up from the earth and through which a considerable extent of terrain totally changed its appearance is perhaps one of the most extraordinary physical revolutions in the annals of the history of our planet.[2] Geology identifies the areas in the ocean where, in recent times during the past two thousand years, volcanic islands rose above the surface of the water near the Azores, in the Aegean Sea, and south of Iceland. But it offers us no example in which, in the interior of a continent, thirty-six leagues from the coast, and over forty-two leagues from any other active volcano, a mountain of scoria and ashes 517 meters high (only in comparison   II.167 to the former level of the nearby plains) suddenly formed in the center of a thousand small flaming cones. This remarkable phenomenon was sung in Latin hexameter by a Jesuit father, ▼ Rafael Landivar, an indigenous of Guatemala. The Abbot Clavijero[3] mentions it in the ancient history of his country; but it remains unknown to mineralogists and physicists in Europe, though it happened only fifty years ago and took place six days' journey

---

1. Vol. I, chap. III, p. 284 [p. 190 in this edition], and *Géographie des plantes*, p. 130. The elevations I now indicate are based on Mr. Laplace's barometric formula. They are the result of Mr. Oltmanns' latest work and sometimes vary by twenty to thirty meters from those I recorded in the *Géographie des plantes*, written a few months after I returned to Europe, at a time when it was impossible to give such a large number of calculations all the precision they deserve. (See the note written in the ▼ month of nivose in the year 13 at the end of the *Géographie des plantes*, p. 147.)

2. Strabo relates ([*Strabonis Rerum geographicarum*] ed. Alm[eloveen], vol. I, p. 102) that on the plains near Methona, on the shore of the Gulf of Hermione, a volcanic explosion produced a mountain of scoria (a *monte novo*) to which he gave the tremendous height of seven stadia which, if we assume this to mean Olympic stadia (Vincent, *Voyage of Nearchus*, p. 56), would be 1,249 meters! However exaggerated this claim may be, the geological fact certainly deserves the attention of travelers.

3. [Clavijero,] *Storia antica di Messico*, vol. I, p. 42, and *Rusticatio Mexicana* (a poem by Father Landivar, the second edition of which was published in Bologna in 1782), p. 17.

from the capital of Mexico, if one descends from the central plateau toward the South Sea coasts!

A vast plain extends from the Aguasarco hills as far as the villages of Teipa and Petatlán, also famous for their cotton fields. Between the *Picachos del Mortero*, the *Cerros de la Cuevas* and those of *Cuiche*, this plain is only 750 to 800 meters above sea level. Basaltic cones rise in the middle of a terrain in which porphyry with a grünstein base predominates. Their peaks are covered with evergreen oaks, foliage of laurel and olive trees, intermingled with small palm trees with flabelliform leaves. This lush vegetation contrasts sharply with the aridity of the plain that was devastated by the effects of volcanic fire.

II.168   Until the mid-eighteenth century, fields planted with sugarcane and indigo stretched between the two streams called Cuitimba and San Pedro. They were bordered by basaltic mountains whose structure seems to indicate that, in remote times, this entire country had already been struck by volcanoes on several occasions. These artificially watered fields belong to the plantation (*Hacienda*) of San Pedro de Jorullo, one of the largest and wealthiest in the country. In June 1759, an underground noise was heard. Successive earthquakes that lasted for fifty to sixty days and plunged the residents of the *Hacienda* into the greatest consternation, accompanied frightful groans (*bramidos*). From the beginning of September, everything seemed to indicate perfect calm when, during the night from the twenty-eighth to the twenty-ninth, a horrible underground fracas was heard once again. The terrified Indians fled to the Aguasarco Mountains. A three- to four-square-mile piece of land that goes by name of *Malpaís* rose up in the shape of a bladder. The edges of this upheaval are still distinguishable today in the fractured strata. Near its boundaries, the *Malpaís* is only twelve meters higher than the former level of the plain called *Las playas de Jorullo* [the beaches of Jorullo]. But the convexity of the raised terrain increases progressively toward its center up to an elevation of 160 meters.

II.169   Those who witnessed this great cataclysm from the top of Aguasarco claim that flames were seen for more than half a square league, that fragments of incandescent rock were thrown to prodigious heights, and that through a thick cloud of ash, lit up by the volcanic fire and resembling an agitated sea, one had the impression of seeing the softened crust of the earth swell up. As a result, the Cuitimba and San Pedro rivers plummeted into fiery crevices. The decomposition of the water helped stoke the flames, which were visible from the town of Patzcuaro, even though it is situated on a broad plateau at an elevation of 1,400 meters above the plains of *las playas* of Jorullo. Eruptions of mud, especially from the layers of clay that covered the disintegrated basaltic rounds in concentric layers, seem to indicate that the underground waters played an important role in this extraordinary upheaval. Thousands of small cones only two to three meters wide, which the indigenous peoples call *ovens* (*hornitos*), jut out from the raised mound of the *Malpaís*. Although in the past fifteen years, according to the Indians' testimony, the heat of these volcanic ovens has greatly decreased, I have seen the thermometer rise there to ninety-five degrees when I plunged it into some of the vents that exhale water vapor. Each small cone is a *fumarole* from which thick smoke rises to a height of ten to fifteen meters. In many of them, one hears an underground noise that seems to announce the proximity of a boiling liquid.

In the midst of the ovens, six large hillocks, each four to five hundred meters above the former level of the plain, sprang up from a chasm that runs from north-northeast to south-southeast. This is the same phenomenon as the Monte Nuovo in Naples, repeated several times in a range of volcanic hills. The tallest of these enormous bluffs, which recall the *puys* of the Auvergne, is the great volcano of Jorullo. It is always aflame and on its north side, it ejected an enormous amount of scorified and basaltic lava containing fragments of primitive rocks. These great eruptions of the central volcano continued until February 1760. In the years that followed, the Indians, who were very frightened by the horrible fracas caused by the new volcano, had first abandoned their villages located seven or eight leagues from the playas de Jorullo. In a few months, they became accustomed to the terrifying spectacle; once returned to their huts, they descended toward the Aguasarco and Santa Inés mountains to behold the sprays of fire thrown up by an infinite number of large and small volcanic openings. At that time, ashes covered the roofs of the houses in Querétaro at a distance of more than forty-eight leagues in a straight line from the site of the explosion. Although the underground fire seems fairly inactive[1] now, and although the Malpaís and

II.170

II.171

---

1. We found that the air at the bottom of the crater was forty-seven degrees C, in some places fifty-eight and sixty degrees C. We had to cross chasms exhaling sulfurous vapors where the thermometer rose to eighty-five degrees C. Passage over these crevices and the masses of scoria that cover considerable hollows make the descent into the crater rather dangerous. I shall reserve the details of my geological research on the volcano of Jorullo for the personal narrative of my travels. The atlas accompanying this account will contain three plates: (1) the picturesque view of the new volcano, which is three times higher than the Monte Novo of Puzzole, which emerged from the earth in 1538, almost on the shore of the Mediterranean; (2) the vertical crosssection, or Profile, of the Malpaís and the entire raised part; (3) the geographical Map of the plains of Jorullo, drawn by means of a sextant, using the method of

the great volcano are beginning to be covered with vegetation, we found nonetheless that the ambient air is so heated by the action of the small *ovens* (*hornitos*) that the thermometer reached forty-three degrees [Centigrade] even quite far from the ground and in the shade. This fact seems to prove that there is no exaggeration in the accounts of a few elderly Indians who relate that several years after the first eruption, even at great distances from the upraised terrain, the plains of Jorullo were uninhabitable because of the excessive heat that prevailed there.

Near the Cerro de Santa Inés, the traveler can also see the Cuitimba and San Pedro Rivers, whose clear waters once irrigated the sugarcane that was grown on ▼ Don Andrés Pimentel's plantation. These springs vanished

II.172   during the night of September 29, 1759; but farther west, at a distance of 2,000 meters, on the raised ground itself, one now sees two small rivers that break through the clayey mound of the *hornitos*, resembling hot springs in which the thermometer rises to 52.7°. The Indians have kept the names San Pedro and Cuitimba because, in many parts of the *Malpaís*, one seems to hear great masses of water flowing from east to west, from the mountains of Santa Inés toward the *Hacienda de la Presentación*. Near this property, there is a stream that releases sulfurous hydrogen. At more than seven meters wide, it is the most abundant hydrosulfurous spring that I have ever seen.

According to the indigenous peoples, the extraordinary transformations that we have just described—the earthen crust raised up and cracked by volcanic fire and the mountains of scoria and piled ashes—are the work of the monks, certainly the greatest that they ever produced in either hemisphere! In the hut where we were living at the *Playas* de Jorullo, our Indian host told us that, in 1759, Capuchin missionaries preached at the San Pedro property, but that when they did not encounter a favorable reception (perhaps not having dined as well as expected), they hurled at the then-beautiful and fertile plain the most horrible and elaborate curses; they prophesied that

II.173   the property would first be engulfed in flames that would rise up from the earth, and that, later on, the surrounding air would become so cold that the nearby mountains would remain eternally covered with snow and ice. Since the first of these curses had such a disastrous outcome, the lower-class Indians already saw in the progressive cooling of the volcano the sinister

---

perpendicular bases and angles of elevation. Volcanic samples from this convulsed terrain are to be found in the collection of the School of Mining in Berlin. Plants collected in the vicinity are included in the herbaria I consigned to the Museum of Natural History in Paris.

sign of perpetual winter. I thought it necessary to mention this popular legend, worthy of a place in the Jesuit Landivar's epic poem because it adds a
colorful aspect to the portrait of the customs and prejudices of these faraway lands. It demonstrates the ingenuity of a class of men who, by repeatedly taking advantage of the people's credulity and pretending to suspend
the immutable laws of nature through their influence, are able to use any
means to base their empire on the fear of bodily harm.

The position of the new volcano of Jorullo gives rise to an intriguing
geological observation. We have already remarked above, in the third chapter, that there is in New Spain a *parallel of great elevations*, or a narrow zone
between 18°59' and 19°12' latitude, where one finds all the Anahuac peaks
that rise above the eternal snow line. These summits are either volcanoes that
are still active or mountains that, by their shape as well as the type of their
rocks, make it highly likely that they once contained an underground fire.
Moving from the shores of the Sea of the Antilles, we find, from east to
west, Orizaba Peak, the two volcanoes of Puebla, the Nevado de Toluca,
Tancitaro Peak, and the Colima volcano. Instead of forming the crest of the
Anahuac Cordillera and following its direction (which is from southeast to
northwest), these high elevations are located on a line that runs perpendicular to the axis of the great mountain chain. It is doubtless worth noting that,
in 1759, the new Jorullo volcano rose up in the extension of this line, on the
same parallel as the old Mexican volcanoes!

II.174

A glance at my map of the outskirts of Jorullo reveals that the six large
buttes arose from the earth on a line that traverses the plain from the Cerro
de las Cuevas to the Pichado del Mortero; the *bocche nuove* [new craters]
of Vesuvius are also arranged along the extension of a crevice. Do these
similarities not give one the right to suppose that, in this part of Mexico,
deep in the earth's interior, a crevice runs from east to west over a length
of 137 leagues, a crack through which the volcanic fire broke through the
outer crust of the porphyritic rocks in different epochs and rose to the surface, from the shoreline of the Gulf of Mexico to the South Sea? Could this
crevice extend as far as the small group of islands that Mr. Collnet calls the
Archipelago of Revillagigedo, around which pumice stone has been seen
floating on the very *parallel of the Mexican volcanoes*? Naturalists who distinguish between facts presented by descriptive geology, on the one hand,
and theoretical reveries about the original state of our planet, on the other,
will forgive us for having recorded these observations on the general map
of New Spain in the Mexican Atlas. Furthermore, across an area of forty
square leagues, from Lake Cuiseo, which is saturated with muriate of soda

II.175

and gives off sulfurous hydrogen, to the town of Valladolid, there are many hot springs that generally contain only muriatic acid with no traces of terreous sulfates or metallic salts. Such are the thermal waters of Cuchandiro, Cuinche, San Sebastián, and San Juan Tararamco.

The surface area of the intendancy of Valladolid is one-fifth smaller than that of Ireland; but its relative population is twice as large as that of Finland. There are three towns in this province (Valladolid, Tzintzontzan, and Patzcuaro); three *villas* (Citaquaro, Zamora, and Charo), 263 villages, 205 parishes, and 326 farms. The inaccurate census of 1793 gave a total population of 289,314, among them 40,399 white males, 39,081 white females, 61,352 Indian men, 58,016 Indian women, 154 monks, 138 nuns, and 293 persons in the secular clergy.

II.176
The Indians living in the province of Valladolid form three tribes of different origins: the Tarasca, famous in the sixteenth century for their mild-manneredness, their industry in the mechanical arts, and the musicality of their vowel-rich language; the Otomi, a tribe whose civilization is backward even now and who speak a language full of nasal and guttural aspirates; and the Chichimecs, who, like the Tlaxcalteca, the Nahuatlaca, and the Aztecs, have preserved the Mexica language [Nahuatl]. Indians inhabit the entire southern part of the intendancy of Valladolid. The only white face one meets in the villages there is the priest, who is often Indian or Mulatto himself. Earnings are so low there that the bishop of Michoacan has the greatest difficulty finding clerics who might want to settle in a country where Spanish is almost never heard, and where, along the coastline of the Great Ocean, priests affected by the contagious miasmas of malignant fevers often die after a stay of only seven or eight months.

The population of the intendancy of Valladolid decreased during the famine years of 1786 and 1790. It would have suffered even more if the respectable bishop whom we mentioned in the sixth chapter had not made extraordinary sacrifices to ease the lot of the Indians; he willingly lost 230,000 francs in a few months by buying 50,000 fanegas of corn which he resold at a very reduced price to combat the disgusting greed of several rich proprietors who tried to take advantage of the people's misery at a time of public calamity.

II.177
The most remarkable places in the province of Valladolid are the following:

VALLADOLID de Michoacan, the capital of the intendancy and bishop's see, blessed with a lovely climate. Its elevation above sea level is 1,950 meters; yet, even at this relatively low altitude, and below 19°42′ latitude, snow has been seen to fall on the streets of Valladolid. This example of a sudden cooling[1] of the atmosphere, caused most likely by a north wind, is much more striking than snow falling on the streets of Mexico City the day before the Jesuit fathers were removed! The new aqueduct, which brings potable water to the city, was built at the expense of the last bishop, Fray Antonio de San Miguel; it cost him almost half a million francs. *Population*: 18,000 inhabitants.

PATZCUARO, on the banks of the picturesque lake of the same name, opposite the Indian village of Janicho, located a short league away on the charming little island in the middle of the lake. The ashes of a quite remarkable man, whose memory the Indians have honored for two centuries, the famous ▼ Vasco de Quiroga, the first bishop of Michoacan, who died in the village of Uruapa in 1556, are at rest in Patzcuaro. This zealous prelate, whom the indigenous peoples still call their father (*Tata don Vasco*), was more successful in protecting the unfortunate inhabitants of Mexico than the virtuous bishop of Chiapas, ▼ Bartolomé de las Casas. Quiroga became the special benefactor of the Tarasca Indians, whose industry he encouraged, designating a specific branch of trade for each village. These useful institutions have been largely preserved until today. Elevation of Patzcuaro: 2,200 meters. *Population*: 6,000.

TZINTZONTZAN, or Huitzitzilla, the former capital of the kingdom of Michoacan, which we have discussed above. *Population*: 2,500.

The intendancy of Valladolid is home to the mines of Zitaquaro, Angangueo, Tlapuxahua, Real de Oro, and Ynguaran.

: :

V. Intendancy of Guadalajara

POPULATION (IN 1830): 630,500
SURFACE AREA IN SQUARE LEAGUES: 9,612
INHABITANTS PER SQUARE LEAGUE: 66

This province, part of the kingdom of Nueva Galicia, is almost twice the size of Portugal, with a population that is five times smaller. It borders the

---

1. See above, vol. I, p. 281 [p. 188 in this edition], and my *Géographie des plantes*, p. 133.

II.179   intendancies of Sonora and Durango to the north, those of Zacatecas and Guanajuato to the east, the province of Valladolid to the south, and the Pacific Ocean to the west, with a coastal length of 123 leagues. Its greatest width is one hundred leagues from the port of San Blas to the town of Lagos; its greatest length, from south to north, is one hundred and eighteen leagues from the Colima volcano to San Andrés Teul.

From east to west, the intendancy of Guadalajara is crossed by the Río de Santiago, a good-sized river that is connected to Lake Chapala and which, one day (once civilization has become more developed in this country) may be of interest for inland navigation from Salamanca and Zelaya to the port of San Blas.

The entire eastern part of this province occupies the western plateau and slope of the Anahuac Cordillera. The coastal regions, especially those that stretch toward the great bay of Bayona, are covered with forests and furnish superb timber for construction. But the inhabitants there are exposed to unhealthy, excessively hot air. The interior of the country enjoys a healthful, temperate climate.

The Colima volcano, whose position has not yet been determined by astronomical observations, is the farthest west of the volcanoes of New Spain, which are on the same line and follow a parallel course. Colima ejects ashes

II.180   and smoke. *Don Manuel Abad [y Queipo]*, vicar general of the diocese of Michoacán, had taken several very precise barometric measurements there long before I arrived in Mexico. He had estimated the elevation above sea level of the Colima volcano at 2,800 meters. "This isolated mountain," observed Mr. Abad, "appears to be only of average height when its summit is compared to the ground level of Zapotilti and Zapotlan, two villages that are 2,000 varas above the coastline. The volcano is visible in all its grandeur from the small town of Colima. It is covered with snow only when, through the effect of the north winds, there is snowfall on the nearby mountain chains. On December 8, 1788, the volcano was covered with snow up to nearly two-thirds of its height[1]; but this snow lasted for two months only on the northern slope of the mountain, toward Zapotlan. At the beginning of 1791, I traveled around the volcano through Sayula, Tuxpan, and Colima, and there was not the slightest trace of snow on its peak."

1. Let us suppose that the snow covered only half the height of the volcano. Snow sometimes falls on the western part of New Spain, below 18° to 20° latitude, at 1,600 meters elevation. These meteorological considerations would give roughly 3,200 meters for the height of the Colima volcano.

According to an unpublished report sent to the court of the Consulado de Veracruz by the intendant of Guadalajara, the value of the agricultural products from this intendancy amounted to 2,599,000 piasters (almost three million francs) in 1802, which included a stock of 1,657,000 fanegas of corn, 43,000 cargas of wheat, 17,000 *tercios* of cotton (at five piasters per tercio), and 20,000 pounds of Autlan cochineal. The value of the manufacturing industry was estimated at 3,302,200 piasters or sixteen-and-a-half million francs.

II.181

The province of Guadalajara has two ciudades, six villas, and 322 villages. Its most famous mines are those of Bolaños, Asientos de Ibarra, Histotipaquillo, Copala, and Guichichila, near Tepic.

: :

The most notable cities are:

GUADALAJARA, on the left back of the Río de Santiago, residence of the intendant and seat of the high court of justice (Audiencia). Population: 19,500.

SAN BLAS, a port, residence of the *Departamento de Marina*, at the mouth of the Río de Santiago. Its officials (*Officiales reales*) are in Tepic, a small town whose climate is less hot and more salubrious. The question of whether it would be useful to move the dockyards, warehouses, and the entire naval department from San Blas to Acapulco has been debated for the past ten years. Acapulco is lacking in lumber for construction. The air there is probably as unhealthy there as it is in San Blas, but by favoring the concentration of naval forces, the projected change would make it easier for the government to recognize the needs of the navy and the means of providing them.

COMPOSTELA, south of Tepic. Northwest of Compostela, as in the partidos of Autlan, Ahuxcatlan, and Acaponeta, tobacco of superior quality was once grown.

II.182

AGUAS CALIENTES, south of the mines of Los Asientos de Ibarra, a small, densely populated town.

VILLA DE LA PURIFICACIÓN, northwest of the port of Guatlan, once called Santiago de Buena Esperanza, and celebrated in Diego Hurtado de Mendoza's voyage of discoveries from 1532.

LAGOS, north of the city of León, on a fertile plateau where wheat is grown, on the border of the intendancy of Guanajuato.

COLIMA, two leagues south of the Colima volcano.

: :

VI. The Intendancy of Zacatecas

POPULATION (IN 1803): 153,300
SURFACE AREA IN SQUARE LEAGUES: 2,355
INHABITANTS PER SQUARE LEAGUE: 65

This strikingly uninhabited province is situated on mountainous, arid terrain continually exposed to inclement weather. To the north, it borders on the intendancy of Durango; to the east on that of San Luis Potosí; to the south on the province of Guanajuato; and to the west on the province of Guadalajara. Its greatest length is eighty-five leagues, its maximum width, from Sombrerete to Real de Ramos, fifty-one leagues.

II.183   The intendancy of Zacatecas has approximately the same surface area as Switzerland, with which it shares many geological similarities. Its relative population barely equals that of Sweden.

The plateau that forms the center of the intendancy of Zacatecas, and that rises to an elevation of over 2,000 meters, is made of syenite, a rock that is, according to Mr. Valencia's[1] excellent observations, found beneath layers of shale and foliated chlorite (*Chlorithschiefer*). Slate forms the base of the mountains of *Grauwacke* and trappean porphyry. North of the town of Zacatecas are nine small lakes with heavy amounts of muriate and especially carbonate of soda.[2] This carbonate, that is called tequesquite [rock salt] from the old Mexican word *tequixquilit*, is of great use in dissolving muriates and silver sulfides. A lawyer from Zacatecas, ▼ Mr. *Garcés*, recently drew his countrymen's attention to the tequesquite that is also found in Zacualco, between Valladolid and Guadalajara, in the San Francisco valley near San Luis Potosí, in Acusquilco near the Bolaños mines, in Chorro

II.184   near Durango, and in five lakes around the town of Chihuahua. The central plateau of Asia is no richer in soda than is Mexico.

: :

The most remarkable places in this province are:

1. *Don Vicente Valencia*, a student of the learned and well-respected Don Andrés del Río and the School of Mining in Mexico City, has written a very interesting description of the mines of Zacatecas ([del Río, "Continuación del Discurso sobre las formaciones de las montañas de algunos Reales de Minas,"] *Gazeta de Mexico*, vol. XI [1803], p. 417).

2. Joseph Garcés y Eguía, *Del beneficio de los metales de oro y plata*. Mexico City, 1802, pp. 11 and 40 (a work that presents very solid knowledge of chemistry).

ZACATECAS, at present, after Guanajuato, the most famous mining site in New Spain. Its population is at least 33,000 inhabitants.

FRESNILLO, on the road from Zacatecas to Durango.

SOMBRERETE, the administrative center, seat of a *Diputación de minería*.

Besides the three places listed, the intendancy of Zacatecas offers interesting metal-bearing veins near the Sierra de Pinos, Chalchiguitec, San Miguel del Mezquitas, and Mazapil. This is also the province where the *Veta Negra de Sombrerete* demonstrated the greatest mining wealth ever extracted in either hemisphere.

: :

VII. The Intendancy of Oaxaca

POPULATION (IN 1803): 534,800
SURFACE AREA IN SQUARE LEAGUES: 4,447
INHABITANTS PER SQUARE LEAGUE: 120

The name of this province, which other geographers incorrectly call *Guax-aca*, is derived from the Mexica name of the town and the valley of *Huaxy-* II.185
acac, one of the principal towns in Zapoteca country, which was almost as large as the capital, Teotzapotlan. The intendancy of Oaxaca is one of the most delightful regions in this part of the globe. Its beautiful and healthy climate, fertile soil, and the richness and variety of its products all contribute to the well-being of its inhabitants. This province has been the center of an advanced civilization since the remotest times.

To the north it borders on the intendancy of Veracruz; to the east on the kingdom of Guatemala; to the west on the province of Puebla; and to the south, for a coastal length of 111 leagues, on the Great Ocean. Its surface area exceeds that of Bohemia and Moravia combined, while its absolute population is nine times smaller. As a result, its relative population is equal to that of European Russia.

The mountainous terrain of the intendancy of Oaxaca contrasts singularly with that of the provinces of Puebla, Mexico City, and Valladolid. Instead of the layers of basalt, amygdaloids, and porphyry with a base of greenstone [Grünstein] that cover Anahuac from 18° to 22° latitude, only granite and gneiss are found in the mountains of Mixteca and Zapoteca. The chain of mountains of trappean formation only resumes to the southeast, on the western coastline of the kingdom of Guatemala. We do not know the

II.186   elevation of any of the granite summits in the intendancy of Oaxaca. The inhabitants of this beautiful country believe that the Cerro de Senpualtepec near Villalta, from which the two seas are visible, is one of the highest. But this extent of horizon would indicate an elevation of only 350 meters.[1] It is said that the same impressive spectacle may be enjoyed from *La Gineta*, on the border of the dioceses of Oaxaca and Chiapas, twelve leagues from the port of Tehuantepec, on the main road leading from Guatemala to Mexico City.

    The vegetation is beautiful and vigorous throughout the province of Oaxaca, especially halfway down the coast in the temperate region where there are abundant rains from May until October. In the village of Santa María del Tule, three leagues east from the capital, between Santa Lucía and Tlacochiguaya, there is an enormous cupressa disticha (sabino) trunk that is thirty-six meters in circumference. This ancient tree is thus larger than the Atlixco cypress that we have mentioned above, larger than the dragon tree in the Canary Islands and all the baobabs (Adansoniae) in Af-

II.187   rica. But on close inspection, Mr. Anza observed that what inspires travelers' admiration was not a single trunk but the three joined trunks that form the famous sabino of Santa María del Tule.

    The intendancy of Oaxaca includes two mountainous lands that have been called *Mixteca* and *Zapoteca* since time immemorial. These names, which have been preserved until today, indicate a great difference of origin among the indigenous population. The former Mixtecapan is now divided into upper and lower Mixteca (*Mixteca alta y baja*). The eastern border of Mixteca alta, which neighbors the intendancy of Puebla, runs from Ticomabacca, on Quaxiniquilapa, toward the South Sea, passing between Colotepeque and Tamasulapa. The Indians of Mixteca are an active, intelligent, and industrious people.

    Although the province of Oaxaca contains no monuments of ancient Aztec architecture whose size is as astonishing as the god-dwellings (*Teocalli*) of Cholula, Papantla, and Teotihuacán, it presents ruins of buildings that are more remarkable for their ordonnance and the elegance of their adornments. The walls of the *Mitla* palace are decorated with Greek-key

---

1. The visual horizon of a mountain whose elevation is 2,350 meters has a diameter of 3°20′. It has been debated whether the two seas might be visible from the summit of the Nevado de Toluca. The visual horizon of this mountain has a radius of 2°21′, or fifty-eight leagues, assuming only ordinary refraction. The two Mexican coasts closest to the Nevado are those of Coyuca and Tuxpan, at a distance of fifty-four and sixty-four leagues, respectively.

patterns and labyrinths made of mosaics of small porphyritic stones. This is the same design that one admires on the vases erroneously called Etruscan, or on the frieze of ▼ the old temple of *Deus Redicolus* near the grotto   II.188
of the nymph Egeria in Rome. I have commissioned engravings of some of these American ruins, which have been carefully drawn by Colonel Don Pedro de la Laguna and by the skilled architect Don Luis Martín. If one is rightly struck by the great similarity of the adornments of the Mitla palace with those used by the Greeks and Romans, one must not for this reason engage glibly in historical theories about ancient connections that might have existed between the two continents. It is necessary to bear in mind that in all climes, humankind takes pleasure in the rhythmical repetition of the same forms, and that this repetition constitutes the main characteristic of what we call the Greek-key pattern,[1] meanders, labyrinths, and arabesques.

The village of Mitla was formerly called *Miguitlan*, a word that, in the Zapoteca language, signifies a gloomy place, a place of sadness. The Zapoteca Indians call it *Leoba*, which means "tomb." Indeed, according to the indigenous peoples' tradition, and as the layout of all its parts indicates, the palace of Mitla, whose age is unknown, was a palace built above the kings' tombs. It was a building where the sovereign would retire for a time when a son, wife, or mother died. Comparing the magnificence of these tombs   II.189
with the compactness of the houses that served as abodes for the living, one would agree with Diodorus of Sicily ([*Bibliothecæ historicæ*,] book I, chap. 51) that some peoples erect splendid monuments for the dead because, considering this life as short-lived and fleeting, they think that it is not worth building such dwellings for the living.

The palace, or, rather, the tombs of Mitla, form three symmetrically arranged buildings in an extremely romantic setting. The main building is the best preserved and is nearly forty meters long. A staircase built in a shaft leads to an underground chamber that is twenty-seven meters long and eight meters wide. This doleful apartment, intended for the tombs, is covered with the same Greek-key pattern that decorates the exterior walls of the building.

What distinguishes the ruins of Mitla from all other remains of Mexica architecture, however, are the six porphyry columns placed in the middle of an immense hall, which support the ceiling. These columns—almost the only ones on the New Continent—are a testament to the infancy of art.

---

1. The greatest connoisseur of Egyptian antiquities, Mr. Zoëga, has made the curious observation that the Egyptians never used this ornamental style.

They have neither bases nor capitals; one notices only a short tapering of their upper section. Their total height is five meters; but their shaft is a solid piece of amphibolic porphyry. Rubble that has heaped up for centuries hides more than a third of the height of these columns. When he discovered

II.190 them, Mr. Martín found that this height is equivalent to six diameters, or twelve modules. This would suggest that their ordonnance is even stockier than the Tuscan order, were the lower diameter of the Mitla columns not in a proportion of three to two to their upper diameter.

The distribution of the apartments throughout this unusual building offers striking comparisons with what is found in the monuments of Upper Egypt, as drawn by Mr. Denon and the scholars who make up the Cairo Institute. Among the ruins of Mitla, Mr. de [la] Laguna found curious paintings representing war trophies and sacrifices. I shall have occasion elsewhere (in the Personal Narrative of my journey) to return to these remains of an ancient civilization. The further south one journeys from Mexico City, the more one encounters vestiges of buildings and sculptures indicative of a more advanced civilization. It is especially in the southeast of the intendancy of Oaxaca that one may admire the ruins of the great cities of Palenque (or Culhuacan) and Utatlan, commonly called Quiche, after the name of the ▼ Toltec king Nima Quiche. The first set of ruins is located in the province of Tzendales (*partido de Ciudad real*, in the diocese of Chiapas), where a large parish still bears the name of Santo Domingo Palenque. The second set surrounds the village of Santa Cruz del Quiche (in the province of Solola). Someone recently had the fortuitous idea of publishing in Great Britain the drawings that captain Don Antonio del Río made in ▼ Palenque, whose figures, with their enormous aquiline noses,[1] look

II.191 very strange, as do the crosses to which offerings are made and the poses reminiscent of the deities of Hindustan. ([Antonio del Río,] *Description of the Ruins of an Ancient City Discovered in the Kingdom of Guatemala*, 1822. Juarros [y Montúfar], *Compendio de la historia de la Guatemala*, vol. I, pp. 14 and 64.)

The intendancy of Oaxaca is the only one that has preserved the cultivation of cochineal (coccus cacti), a branch of industry that it formerly shared with the provinces of Puebla and Nueva Galicia.

1. These large noses are found in unpublished accounts or Mexica hieroglyphic painting. See my *Vues des Cordillères*, vol. II, p. 200. The triumphal pose that I represent in plate XI is a sculpture from Palenque, as I have recalled in the addition, vol. II, p. 392 (octavo edition).

The family of Hernán Cortés carries the title of Marquis of the Valley of Oaxaca. Its majorat covers the four *villas del Marquesado* and forty-nine villages, with a population of 17,700.

: :

The most remarkable places in this province are:

OAXACA, or GUAXACA, the ancient Huaxyacac, called *Antequera* at the beginning of the conquest. ▼ Thierry de Menonville attributes only 6,000 inhabitants to it, but the 1792 census found 24,400.

TEHUANTEPEC, or Teguantepeque, a port located at the base of a cove   II.192
that the ocean forms between the small villages of San Francisco, San Dionisio, and Santa María de la Mar. This port, protected by rather dangerous shallows, will one day become important when navigation in general, and especially the transport of indigo from Guatemala, becomes more frequent on the Río Guasacualco.

SAN ANTONIO DE LOS CUES, a very populous place on the road from Orizaba to Oaxaca, famous for the remnants of old Mexica fortifications.

The mines in this intendancy that are worked most attentively are Villalta, Zolaga, Yxtepexi, and Totmostla.

: :

VIII. The Intendancy of Mérida

POPULATION (IN 1803): 465,800
SURFACE AREA IN SQUARE LEAGUES: 5,977
INHABITANTS PER SQUARE LEAGUE: 81

This intendancy, about which ▼ Mr. Gilbert[1] has provided us with valu-   II.193
able information, includes the great Yucatán peninsula, situated between the bays of Campeche and Honduras. Before the irruption of the Antillean

---

1. This enlightened observer has traveled widely in the Spanish colonies. He had the misfortune of losing the statistical data he had collected in a shipwreck south of the island of Cuba between the depths of the *King's Garden*, whose astronomical position I determined. It is useful to mention here that without knowledge of the information that I procured, Mr. Gilbert, by himself evaluating the number of villages and their population, found that the Yucatán must have contained nearly half a million inhabitants of all castes and colors in 1801.

Sea, Mexico appears to have been connected to the island of Cuba by Cabo Catoche, fifty-one leagues from the calcareous hills of Cabo San Antonio.

To the south, the province of Mérida borders on the kingdom of Guatemala, to the east on the intendancy of Veracruz, from which it is separated by the Río Baraderas, also called the river of Crocodiles (*Lagartos*); to the west, the British settlements extend as far as the mouth of the Río Honda [Hondo] north of the bay of Hannover [Bahia de Chetumal], across from the island of Ubero (Ambergris Caye). In this part of the intendancy, Salamanca, or the small fort of *San Felipe de Bacalar*, is the southernmost point on the coastline inhabited by the Spanish.

The Yucatán Peninsula, whose northern coastline from Cabo Catoche, near the island of Contoy, to Punta de Piedras (over a distance of eighty-one leagues) follows the direction of the *rotation current* exactly, is a vast plain whose interior is crossed by a chain of low hills from northwest to southwest. The country that extends east of these hills toward the bays of
II.194 Ascensión and Espíritu Santo seems to be the most fertile, and so was once the most inhabited. The ruins of European buildings that one discovers on the island of Cozumel, in the middle of a grove of palm trees, demonstrate that from the very outset of the conquest, Spanish colonists populated this island, which is now deserted. Since the British have settled between Omo and Río Hondo, the government, to reduce smuggling, concentrated the Spanish and Indian population in that part of the peninsula west of the Yucatán Mountains. Colonists are not allowed to settle on the west coast, on the banks of the Río Bacalar and on the Río Honda. The entire vast region remains depopulated: there is only the military fort (*presidio*) of Salamanca.

The intendancy of Mérida is one of the hottest, but also one of the healthiest countries in equinoctial America. The salubrity of the climate in the Yucatán, as in Coro, Cumaná, and on Margarita Island, must clearly be attributed to the extreme aridity of the soil and of the atmosphere. Along the entire coastline, from either Campeche or the mouth of the Río de San Francisco as far as Cabo Catoche, sailors will not find a single freshwater spring. Near this cape, nature has repeated the same phenomenon found on the southern side of the island of Cuba, in the bay of Jagua, which I have described elsewhere.[1]
II.195 On the northern coastline of the Yucatán, at the mouth of the Río Lagartos, 400 meters from the shore, fresh water springs spout up amid

1. In my *Tableaux de la nature*, vol. II, pp. 174 and 235. [See also Humboldt's *Political Essay on the Island of Cuba.*]

the salt water. These remarkable springs are called *mouths* (*bocas*) *of Conil*. Because of strong hydrostatic pressure, it is likely that the fresh water, after breaking through the beds of calcareous rock between whose vents it has flowed, rises above the level of the salt water.

The Indians in this intendancy speak the *Maya* language, which is very guttural, and of which there are four rather complete dictionaries, compiled by ▼ Pedro Beltrán, Andrés de Avendaño, Fray Antonio de Ciudad Real, and Luis de Villalpando. The Yucatán Peninsula was never ruled by the Mexica or Aztec kings. The first conquerors, Bernal Díaz, ▼ Hernández de Córdova, and the valorous Juan de Grijalva, were nevertheless struck by the advanced civilization of the inhabitants of this peninsula. They found houses built of stone cemented with lime, pyramidal buildings (teocalli) that they compared to Moorish mosques, fields enclosed by hedges, and a clothed, civilized people very different from the indigenous peoples of the island of Cuba. To the east of the small central chain of mountains, one can still find many ruins, especially of sepulchral monuments (*guacas*). Some Indian tribes have preserved their independence in the southern part of this hilly terrain, which the dense forests and lush vegetation make almost inaccessible.

Like all lands in the Torrid Zone, the province of Mérida, whose surface does not rise more than 1,300 meters above sea level, produces only corn, jatropha, and dioscorea roots to feed its population, but no European wheat. The trees that furnish the famous Campeche wood (*Hæmatoxilon campechianum*, L.) grow abundantly in many districts of this intendancy. *Cutting* (*Cortes de palo Campeche*) is done annually on the banks of the Río Champotón, the mouth of which lies south of the town of Campeche, four leagues from the small village of Lerma. It is only with special permission from the intendant of Mérida, who holds the title of *Governor-Captain-General*, that merchants may occasionally cut Campeche wood east of the mountains, near the bays of Ascensión, Todos los Santos, and Espíritu Santo. In these coves on the eastern coastline, the British engage in smuggling, as extensive as it is lucrative. After being cut, Campeche wood must be dried for a year before being sent to Veracruz, Havana, or Cádiz. In Campeche, one quintal of this dried wood (*palo de tinta*) fetches a price of two to two and a half piasters (ten francs, fifty centimes to twelve francs, eighty-eight centimes). Haematoxilon, which is very abundant in the Yucatán and on the coast of Honduras, is also scattered across all the forests of equinoctial America, wherever the average air temperature is not below twenty-two degrees on the Centigrade thermometer. The coastline of Paria, in the province of

II.196

II.197

Nueva Andalucía, may one day carry on significant trade in Campeche and Brazilwood (*Caesalpinia*), which it produces in abundance.

: :

The most remarkable places in the intendancy of Mérida are:

MÉRIDA DE YUCATÁN, capital, ten leagues into the interior, on an arid plain. The small port of Mérida is called *Sizal*, west of Chaboana and facing a sand bar nearly twelve leagues long. *Population*: 10,000.

CAMPECHE, on the Río de San Francisco, with a rather unsafe port. Ships are forced to anchor far from the shore. In the Maya language, *cam* means snake and *peche* the small insect (acarus) that the Spanish call *garrapata* [tick] that burrows into the skin and causes burning pain. Between Campeche and Mérida, there are two sizable Indian villages called Xampolan and Equetchecan. Exports of Yucatán wax form one of the most lucrative branches of trade. The regular population of the town is 6,000.

VALLADOLID, a small town whose environs produce large amounts of high-quality cotton. This cotton brings a low price, however, because it has the serious flaw of sticking firmly to the seeds. No one in the area knows how to clean it (*despepitar* or *desmotar*). Hauling absorbs two-thirds of its value due to the weight of the seeds.

II.198

: :

IX. The Intendancy of Veracruz

POPULATION (IN 1803): 156,000
SURFACE AREA IN SQUARE LEAGUES: 4,141
INHABITANTS PER SQUARE LEAGUE: 38

Situated beneath the blazing sun of the tropics, this province extends along the Gulf of Mexico from the Río Baraderas (or *de los Lagartos*) to the great Panuco River, whose source lies in the metalliferous mountains of San Luis Potosí. It thus includes a large part of the eastern coast of New Spain. Its length, from the bay of Términos near Carmen Island to the small port of Tampico, is 210 leagues, whereas its width is generally only between twenty-five and twenty-eight leagues. To the east, it borders on the peninsula of Mérida; to the west, on the intendancies of Oaxaca, Puebla, and Mexico City; and to the north, on the colony of Nuevo Santander.

A glance at the ninth and twelfth plates of my Mexican atlas will show II.199
the extraordinary shape of this land, which was once called *Guetlachtlan*.
There are few regions on the New Continent where the contrast between
two adjacent climes is more striking to the traveler. The entire western part
of the intendancy of Veracruz is on the slope of the Cordilleras of Anahuac.
In the space of a single day, the inhabitants descend from the zone of eternal snows to seaside plains where suffocating heat prevails. Nowhere can
one better observe the admirable order with which different types of plant
life succeed one another, as if in layers, than by ascending from the port of
Veracruz toward the Perote plateau. At every step, one sees changes in the
physiognomy of the land, in the appearance of the sky, in the bearing of the
plants, the shape of the animals, the customs of the inhabitants and the type
of cultivation they practice.

As one ascends, nature becomes less animated, the beauty of the forms
of plant life diminishes, stalks are less lush, flowers smaller and less colorful. The Mexican oak reassures the traveler who has landed at Veracruz. Its
presence tells him that he has left the zone that people in the north justly fear,
where yellow fever wreaks its ravages in New Spain. This same lower growth II.200
line of the oaks warns the settler living on the central plateau to what point he
may descend toward the coastline without fearing the deadly sickness of the
vómito. Near Jalapa, the fresh foliage of forests of liquidambar [Sweetgum]
announces that this is the elevation where the clouds suspended above the
Ocean graze the basaltic summits of the Cordillera. Even higher, near the
Bandillera, the nourishing fruit of the banana tree does not even ripen. In this
foggy, cold region, necessity thus motivates the Indian to work harder and
kindles his industry. At the elevation of San Miguel, pine trees intermingle
with the oaks, and the traveler finds them as far as the high plains of Perote,
which present him with the cheerful sight of wheat-planted fields. Eight hundred meters higher, the climate already becomes too cold for oaks to grow.
Only firs cover the rocks whose peaks enter the zone of perpetual snows. In
only a few hours in this enchanted country, the naturalist can cover the gamut
of vegetation, from heliconia and the banana tree, whose shiny leaves grow to
such extraordinary size, to the shriveled parenchyma of the resinous trees!

Nature has enriched the province of Veracruz with its most precious
creations. At the foot of the Cordillera, in the evergreen forests of Papantla,
Nautla, and San Andrés Tuxtla, grows the liana (epidendrum vanilla), II.201
whose pleasant-smelling fruit is used to flavor chocolate. Near the Indian
villages of Colipa and Misantla, one finds the beautiful convolvulacea (con-
volvus jalapae [Jalapa Bindweed]) whose tuberous root provides jalap,

one of the most powerful and beneficial purgatives. In the eastern part of the intendancy of Veracruz, the forests that extend to the Baraderas river produce myrtle (myrtus pimenta) whose seeds are a pleasant spice known commercially as *pimienta de Tabasco* [allspice]. Acayucan cacao would be more sought-after if the indigenous peoples were more industrious in cultivating their cacao trees. On the eastern and southern slopes of the Orizaba Peak, in the valleys that extend toward the small town of Córdoba, excellent tobacco is grown, which provides the crown with annual revenue of more than eighteen million francs. The smilax [greenbrier], whose root is true sarsaparilla, grows in the humid, shady crevices of the Cordillera. Cotton from the coasts of Veracruz is famous for its softness and whiteness. Sugarcane there yields almost as much sugar as it does on the island of Cuba, and more than on the plantations of Saint-Domingue.

II.202

This intendancy alone would be enough to invigorate trade in the port of Veracruz, if there were more settlers, and if their sloth—the effect of nature's bounty and the ease of providing for the basic necessities of life without effort—did not impede the progress of industry. The original population of Mexico was concentrated in the interior of the country, especially on the plateau. The Mexica peoples, originally from northern regions (as we have explained above), preferred the ridges of the Cordillera as they migrated, because these places offered them a climate similar to that of their native country. When the Spanish first arrived on the beach at Chalchiuhcuecan (Veracruz), the entire coastline from the Papaloapan (Alvarado) River to Huaxtecapan must have been more populous and better cultivated than it is today. As the conquerors climbed the plateau, however, they found that villages were closer together, the fields divided into smaller parcels, and the people more civilized. The Spanish, who believed that they were founding new cities when they gave European names to cities built by the Aztecs, followed the traces of the indigenous civilization. They had strong reasons for wanting to live on the Anahuac plateau, fearing the heat and the diseases of the plains. The quest for precious metals, the cultivation of wheat and European fruit trees, the similarity with the climate of Castile, and other reasons mentioned in the fourth chapter of this work, compelled them to settle on the ridge of the Cordillera. As long as the *Encomenderos*, abusing the

II.203

rights that they were granted by law, treated the Indians like serfs, many Indians were transplanted from coastal regions to the interior plateau, either to work in the mines or simply to bring them closer to their masters' residences. For two centuries, there was virtually no trade in American indigo, sugar, and cotton. Nothing enticed the whites to settle on plains that had the true climate of the Indies. One might say that the Europeans ventured below the tropics only to live in the temperate zone.

Since the consumption of sugar has increased substantially and trade with the New Continent has provided many products that Europe had formerly imported only from Asia and Africa, the plains (*tierras calientes*) have certainly become more attractive for colonization. Sugarcane and cotton plantations have multiplied in the province of Veracruz, especially since the dire events in Saint-Domingue, which gave a great boost to industry in the Spanish colonies. These advances, however, are not very noticeable on the Mexican coasts. It will take centuries to repopulate these deserts. Today, plots several square leagues wide have only two or three shacks on them (*hatos de ganado* [cattle ranches]) around which half-wild cattle roam. A few powerful families who live on the central plateau own most of the coastline of the intendancies of Veracruz and San Luis Potosí. No agrarian law forces these rich landowners to sell their majorats (*mayorazgos*) if they themselves persist in refusing to cultivate the vast lands that belong to them. They harass their farmers and chase them away at will.

II.204

Other causes of depopulation may be added to this difficulty that the coast of the Gulf of Mexico shares with Andalusia and a large part of Spain. The intendancy of Veracruz has a militia far too large for such an underpopulated country. Military service is a burden on the laborer, who flees from the coast so as not to be forced to enter the corps of *lanceros* and *milicianos*. The drafts conducted to provide the royal navy with sailors are too often repeated and are carried out in too arbitrary a fashion. Until now, the government has neglected all means of increasing the population of this desolate coastline. Due to this state of affairs, there is a lack of manpower and a scarcity of provisions, which contrasts with the great fertility of the country. An ordinary worker in the port of Veracruz earns from five to six francs for a day's work. A master mason or any man who has a specific trade can earn from fifteen to twenty francs per day there; in other words, three to four times as much as on the central plateau.

There are two gigantic peaks within the borders of the intendancy of Veracruz, the first of which, the *Orizaba volcano*, is the highest mountain in New Spain after Popocatepetl. The tip of this truncated cone faces southeast. The cone's indentation makes its crater visible from very far away, even from the town of Jalapa. According to my measurements, the second summit, the *Cofre de Perote*, is almost four hundred meters higher than the Peak of Tenerife and serves as a signal to mariners when they put in at Veracruz. Since this circumstance makes it very important to determine its

II.205

astronomical position, I observed circum-meridion altitudes of the sun on the *Cofre* itself. A thick layer of pumice stone surrounds this porphyritic mountain. Nothing prepares us for a crater at the summit of the mountain, but the currents of lava that one sees between the small villages of Viga and Hoya appear to have stemmed from a very old lateral explosion. The small *volcano of Tuxtla*, which backs onto the Sierra de San Martín, is located four leagues from the coast, to the southeast of the port of Veracruz near the Indian village of Santiago de Tuxtla. It is thus outside the line that we mentioned above as the parallel of the active volcanoes of Mexico. Its last major eruption took place on March 2, 1793. Volcanic ashes covered the roofs of houses in Oaxaca, Veracruz, and Perote. In the latter site, which is fifty-seven leagues[1] from the Tuxtla volcano in a straight line, the underground noise sounded like the discharge of heavy artillery.

II.206

In the northern part of the intendancy of Veracruz, west of the mouth of the Río Tecolutla and two leagues from the large Indian village of Papantla, there is a pyramidal building of great antiquity. The pyramid of Papantla remained unknown to the first conquerors. It is located in the middle of a thick forest, called *Tajin* in the Totonac language. For centuries, the indigenous peoples hid this monument, an object of ancient worship, from the Spanish. Some hunters discovered it by chance thirty years ago. Mr. Dupé,[2] a modest and learned observer, who has long devoted himself to very interesting research on Mexica architecture and idols, visited the pyramid of Papantla. He carefully examined the cut of the stones of which it is built and sketched the hieroglyphs that adorn these enormous stones. One hopes that he will decide to publish a description of this interesting monument. The figure[3] printed in the *Gazette of Mexico* [*Gazeta de literatura de México*] in 1785 is very inaccurate.

II.207

Unlike the pyramids of Cholula and Teotihuacán, the pyramid of Papantla is built neither of bricks nor of clay mixed with gravel and covered

---

1. This distance is greater than that from Naples to Rome, but Vesuvius cannot even be heard beyond Gaeta. Mr. Bonpland and I distinctly heard the rumblings of Cotopaxi when it exploded in 1802 in the South Sea west of the island of La Puna, at a distance of seventy-two leagues from the crater. The same volcano was heard in 1744 in Honda and in Mompox on the banks of the Magdalena River. See my *Géographie des plantes*, p. 53.

2. Captain in the service of the King of Spain. Mr. Dupé has in his possession the basaltic bust of a Mexica priestess that I had engraved by ▼ Mr. Massard, and which shows a strong resemblance with the *calantica* of the heads of Isis. Thanks to the efforts of the enlightened persons who represent the present government of the Mexican Confederation, Mr. Dupé's drawings have been gathered together into a collection that is open to the public.

3. See also *Monumenti di architettura messicana di Pietro Márquez*, Rome, 1804, table I.

with a layer of amygdaloids. The only material used was immense porphyritic cut stones. Mortar is discernable in the seams. But the building is less remarkable for its size than for its structure, the smoothness of its stones, and the great regularity of their cut. The base of the pyramid is perfectly square; each side is twenty-five meters long. Its vertical elevation is barely sixteen to twenty meters. Like all the Mexica Teocalli, this monument is composed of several levels. Six of them are still visible, and it is believed that the seventh is hidden beneath the vegetation that covers the entire flank of the pyramid. A main staircase of fifty-seven steps leads to the truncated top of the Teocalli, where human victims were sacrificed. The main staircase is flanked on both sides by small staircases. The facing of the levels is adorned with hieroglyphs, and one can make out serpents and crocodiles carved in relief. Each level has several symmetrically arranged square re-   II.208
cesses. There are twenty-four of them on each side of the first level, twenty on the second, and sixteen on the third. There are 366 of these recesses on the body of the pyramid, and twelve on the staircase that one can make out to the east. The ▼ Abbot Márquez assumes that these 378 recesses refer to the Mexica calendar system. He even thinks that each of them contained one of the twenty figures that in the Toltecs' hieroglyphic language symbolized the day of the common year and the intercalated days at the end of cycles. Indeed, the year was composed of eighteen months, each of which had twenty days, resulting in 360 to which, conforming to the Egyptian custom, were added the five complementary days called *nemonteni*. The intercalation was done every fifty-two years, increasing the cycle by thirteen days, which made $360 + 5 + 13 = 378$ simple or compound signs of the civil calendar called *cempohualilhuitl*, or *tonalpohualli*, to distinguish it from the *comilhuitlaohualliztli*, the ritual calendar that the priests used to mark the return of the sacrifices. I shall not examine here the Abbot Márquez's theory, which also recalls the astronomical explanations that a famous historian, Mr. Gatterer, gave of the number of chambers and stairs found in the great Egyptian labyrinth.

: :

The most remarkable cities in this province are: II.209

VERACRUZ, the intendant's residence, and the center of trade with Europe and the Antillean islands. The city is pretty and laid out very regularly, inhabited by intelligent, active merchants who desire only the good of their country; its internal security has been much improved in recent years. The beach at Veracruz was once called Chalchiuh-cuecan. The island where the fortress of San Juan de Ulúa was built at enormous expense—according to rumors, at a cost of two hundred million francs—was already visited by Juan de Grijalva in 1518. He gave it the name of Ulúa because when he found the remains of two unfortunate victims[1] there and, inquiring to the indigenous people why they sacrificed humans, they told him that it was by order of the kings of *Acolhua*, or Mexico. The Spaniards, whose only interpreters were Yucatán Indians, misunderstood the reply and thought that Ulúa was the name of the island. Peru, the coastline of Paria, and many other provinces owe their present names to similar misunder-

II.210 standings. The city of Veracruz is often called *Vera Cruz Nueva* to distinguish it from *Vera Cruz Vieja*, located near the mouth of the Río Antigua, which almost all historians consider to be the first colony founded by Cortés. The town founded in 1519 was called *Villarica*, or the Villa Rica de la Veracruz, and was located three leagues from Cempoalla, the Totanacs' main town, near the small port of *Chiahuit-zla*, which one barely recognizes in Robertson's work, where it bears the name Quiabislan. Three years later, Villarica was deserted, and the Spanish founded another town to the south, which has kept the name of *Antigua*. It is believed locally that this second colony was abandoned once again because of the vómito disease, which already at that time was cutting down two-thirds of the Europeans who disembarked during the hot season. The viceroy ▼ Count de Monterey, who governed Mexico at the end of the sixteenth century, had the foundation of Nueva Veracruz (the present city) laid on the beach of Chalchiuhcuecan, facing the small island of San Juan de Ulúa, on the very site where Cortés had landed on April 21, 1519. This third city of Veracruz received its city privileges only under King Philip III in 1615. It is located on a dry plain devoid of rivers and streams, where

II.211 the north winds that blow impetuously from October until April have created hills of shifting sand. These dunes (*Meganos de arena*) change their shape and location every year. They are between eight and twelve meters high and are largely responsible for the suffocating heat of the air in Veracruz, because of the reflection of the sun's rays and the high

---

1. Apparently, these sacrifices were made on many of the small islands that surround the port of Veracruz. One of these small islands that navigators dread is still called the *Isla de Sacrificios*.

temperature that they themselves reach during the summer months. Between the city and the Arroyo Gavilán, there are marshlands in the middle of the dunes, which are covered with mangrove trees and other brushwood. The stagnant water of the Bajio de la Tembladera and the small lagunas of Hormiga, Rancho de la Hortaliza, and Arjona causes intermittent fevers among the indigenous peoples. It probably also features prominently among the deadly causes of the *vómito prieto*, which we shall examine later on in this work. Since there are no rocks on the outskirts of the city, all of the buildings in Veracruz and the castle of Ulúa were built from material taken from the Ocean floor, formations of stony coral, or madrepore (*piedras de mucara*). Sand covers the secondary formations that rest on the porphyry of the Encero, which appears on the surface only near Acazonica, a Jesuit farm famous for its quarries of fine foliated gypsum. Digging into the sandy soil of Veracruz, one finds freshwater one meter down, but this water results II.212
from the filtration of the marshes or lagoons formed between the dunes. It is the rainwater that has been in contact with plant roots; its quality is very poor and it is used only for washing. The lower class— and this is an important fact in the medical topography of Veracruz— is compelled to use the water from a ditch (*zanja*) that comes from the *Meganos* and is slightly better than well water or the water from the Tenoya stream. By contrast, the upper class drinks rainwater collected in very poorly built cisterns; an exception are the beautiful cisterns (*algibes*) at the castle of San Juan de Ulúa, whose very pure and healthy water is distributed only to military personnel. For centuries, this lack of good potable water has been considered one of the many causes of disease among the inhabitants. In 1704, a project was developed to divert part of the beautiful Jamapa River to the port of Veracruz. ▼ King Philip V sent a French engineer to examine the terrain. The engineer, who was likely unhappy with his stay in such a hot and unpleasant land, declared that it was impossible to carry out the project. In 1756, the debates resumed among engineers, the municipality, the governor, the viceroy's assessor, and the tax office. Heretofore, 2,250,000 francs have been spent on experts' visits and legal fees (since everything becomes a lawsuit in the Spanish colonies!). Before the ground had II.213
been surveyed, an embankment (levee) was built 1,100 meters above the village of Jamapa; having cost one and a half million francs, it is already half destroyed. For more than twelve years, the government has imposed a flour tax on the public, which brings in annually more

than 150,000 francs. A stone aqueduct (*atarjea*) more than nine hundred meters long has already been built and can provide 116 square centimeters of water. But despite all these expenses, and the jumble of reports and information piled up in the archives, the waters of the Río Jamapa are still over 23,000 meters away from the city of Veracruz. In 1795, the project ended where it should have begun: the ground was surveyed, and it was discovered that the average water level of the Jamapa is 8.83 meters (ten Mexican varas and twenty-two and a half inches) above the level of the streets in Veracruz. It was acknowledged that the large embankment should have been placed in Medellín, and that through ignorance it had been built in a place that was not only too high but also 7,500 meters farther from the port than was required by the drop necessary for conducting the water. In the present state of things, the construction of the aqueduct from Río Xamapa to Veracruz is estimated at five or six million francs. In a country where there is an immense wealth of precious metals, it is not the size of the sum that frightens the government. The project is suspended because it has been recently calculated that ten public cisterns placed outside the city limits would cost only 700,000 francs all told and would suffice for a population of 16,000 if each cistern held 670 cubic meters of water. "Why," one reads in the report to the viceroy, "seek so far afield what nature affords us firsthand? Why not take advantage of these regular and plentiful rains that, according to Colonel Costansó's precise studies, annually provide three times as much water as falls in France and Germany?" The regular population of Veracruz, excluding the militia and seafarers, is 16,000.

II.214

JALAPA (Jalapa), a town at the foot of the basaltic mountain of Macultepec, in a very romantic location. From a distance, the convent of San Francisco, like all those that Cortés founded, resembles a fortress; in the early days of the conquest, convents and churches were build to serve as defenses in case of an insurrection on the part of the indigenous peoples. From this same convent of San Francisco in Jalapa, one has a magnificent view of the colossal summits of the Cofre and Orizaba Peak, the slope of the Cordillera (looking toward Encero, Otates, and Apazapa), the Antigua River, and even the Ocean. Dense forests of styrax, piper, melastoma, and tree ferns, especially those traversed by the Pacho and San Andrés roads; the shores of the small lake of Los Berrios; and the heights that lead to the village of Huastepec—all of these make for the most delightful walks. The

II.215

Jalapa sky, beautiful and serene in summer, inspires melancholy from December to February. Every time the north wind blows in Veracruz, a thick fog envelops the residents of Jalapa. The thermometer then drops to twelve or sixteen degrees [Celsius]; in that season (*estación de los Nortes*), one often goes two or three weeks without seeing the sun and the stars. The richest merchants of Veracruz have country houses in Jalapa, where they enjoy the pleasantly cool temperature, while the mosquitoes, the excessive heat, and yellow fever make the coast almost uninhabitable. There is an establishment in this small town whose existence confirms what I have suggested above about the advances of intellectual culture in Mexico: opened a few years ago, it is an excellent drawing school where the children of poor artisans are taught at the expense of more well-to-do citizens. The elevation above sea level of Jalapa is 1,320 meters. Its population is estimated at 13,000.

PEROTE (formerly Pinahuizapan). The small fortress of San Carlos de Perote is located north of the large town of Perote. It is more of an arms depot than a fortress. The surrounding plains are very barren and covered with pumice stone. There are no trees, except for a few odd cypress and molina trunks! The elevation of Perote is 2,353 meters. II.216

CÓRDOBA, a town on the eastern slop of Orizaba Peak, in a much warmer climate than that of Jalapa. The environs of Córdoba and Orizaba produce all the tobacco that is consumed in New Spain.

ORIZABA, east of Córdoba, slightly north of the Río Blanco, which empties into the Alvarado lagoon. It has long been debated whether the new road from Mexico City to Veracruz should go through Jalapa or Orizaba. Since these two towns both take great interest in the direction of this route, in their rivalry they have used every means available to win over the constituted authorities to their side. Consequently, the viceroys have alternatively taken either side, and during this time of uncertainty no road was built. Finally, in recent years, a beautiful road has been started from the fortress of Perote to Jalapa, and from Jalapa to the Encero.

TLACOTLALPAN, the administrative center of the former province of Tabasco. The small towns of Victoria and Villa Hermosa are farther north; the first of these is one of the oldest in New Spain. There is no metal mining of any importance in the intendancy of Veracruz. The mines of Zomelahuacan near Jalacingo are almost completely abandoned. II.217

: :

X. The Intendancy of San Luis Potosí

POPULATION (IN 1803): 334,900
SURFACE AREA IN SQUARE LEAGUES: 27,821
INHABITANTS PER SQUARE LEAGUE: 12

This intendancy includes the entire northeast part of New Spain. Since it borders on countries that are either deserted or inhabited by free and no-madic Indians, we may say that its northern borders are almost impossible to determine. The mountainous terrain called the *Bolsón de Mapimi* en-compasses more than 3,000 square leagues; it is from there that the Apache come out to attack settlers in Coahuila and Nueva Vizcaya. Hemmed in be-tween these two provinces and bordered on the north by the great Río del Norte, the Bolsón de Mapimi is considered either as a land that the Spanish did not conquer or as a part of the intendancy of Durango. I have drawn the borders of Coahuila and Texas near the mouth of the Río Puerco and toward the source of the Río de San Saba as I found them on the ▼ special maps preserved in the archives of the viceroyalty and drawn up by engineers in the service of the King of Spain. But how to determine the territorial bor-

II.218 ders in the immense savannahs where farms are fifteen to twenty leagues apart, and where there is almost no trace of cleared land or cultivation!

The intendancy of San Luis Potosí includes some very diverse areas, whose various names have caused many geographical errors. It is composed of provinces that belong, in some cases, to the *Provincias internas* and, in others, to the kingdom of New Spain proper. Among the first, there are two that are directly under the command of the *Provincias internas*; the other two are considered *Provincias internas del Virreinato*. The following table presents these complicated and rather unnatural divisions.

The intendant of San Luis Potosí governs

A) *In Mexico proper*:
>The *Province of San Luis*, which extends from the Río de Panuco to the Río de Santander and includes the important mines of Charcas, Potosí, Bamos, and Catorce.

B) *In the Provincias internas del Virreinato*:
1) *The new kingdom of León.*
2) *The colony of Nuevo Santander.*

C) *In the Provincias internas de la comandancia general oriental:*

1) *The province of Coahuila.*
2) *The province of Texas.*

It follows from what we have mentioned above about the latest changes that have been made in the organization of the *comandancia general* of Chihuahua that, besides the province of Potosí, the intendancy of San Luis now includes all the land called the *Provincias internas orientales*. As a result, there is only one intendant at the head of an administration that covers a greater surface area of the globe than all of European Spain. But this immense land, endowed with the most precious natural products and located under a serene sky in the temperate zone near the border of the tropics, is largely a wild desert that is even less populated than the lands of Asian Russia! Its position on the eastern border of New Spain, the proximity of the United States, frequent communications with settlers in Louisiana, and several other circumstances that I shall not discuss here, will probably soon favor civilizational advances and the prosperity of citizens in these vast and fertile regions. II.219

The intendancy of San Luis includes almost 230 leagues of coast, the same distance as that between Genoa and Reggio di Calabria. But with the exception of a few ships that come from the Antilles to load up on meats—either at the Tampico sandbar near Panuco or the mooring in New Santander—this entire coast is uninhabited and without trade. The section that extends from the mouth of the great Río del Norte to the Río Sabina is still virtually unknown and has never been studied by navigators. It would, however, be very important to discover a good port in this northern edge of the Gulf of Mexico. Unfortunately, the eastern coastline of New Spain presents the same obstacles everywhere: a lack of depth for ships needing a depth of over thirty-eight decimeters of water; sandbars at the mouths of rivers; necks of land and small, but long islands that lie parallel to the continent and prevent access to the interior basin. The coastline of the provinces of Santander and Texas, from 21° to 29° latitude, is strikingly scalloped and offers a series of interior basins that are four to five leagues wide and forty to fifty leagues long. They are called *lagunas*, salt lakes. Some (like the laguna de Tamiagua) are true *impasses*. Others, like the laguna Madre and the laguna de San Bernardo, are connected with the ocean by several canals. The canals encourage coastal trading, since the coasters are protected there from the sea troughs. It would be interesting for geologists to study these sites to determine whether currents have formed these *lagoons*, with their irruptions cutting deeply into the land, or whether these small, long, and narrow islands arranged parallel to the coast are sandbars that have gradually risen above the average water level. II.220

II.221   In the entire intendancy of San Luis Potosí, only the section that adjoins the province of Zacatecas, where the rich mines of Charcas, Guadalcazar, and Catorce are located, is cold and mountainous country. The diocese of Monterey, which bears the pompous title of the New Kingdom of León, Coahuila, Santander, and Texas are very low regions; the terrain is almost completely flat, and the land is covered with secondary formations and alluvial tracts. Their climate is rather uneven, excessively hot in summer, and exceedingly cold in winter, when the north winds chase columns of cold air from Canada toward the Torrid Zone.

Since the cession of Louisiana to the United States, the boundaries between the province of Texas and the county of Natchitoches (a county that is an integral part of the confederation of American republics) have become the object of a long and unfruitful political discussion. Several members of Congress in Washington thought that the Louisiana territory could be extended to the left bank of the Río Bravo del Norte. According to them, "all the country that the Mexicans call the province of Texas formerly belonged to Louisiana; now the United States must possess the latter province to the full extent of the rights by which it was possessed by France before its cession to Spain; and neither the new names introduced by the viceroys of Mexico nor

II.222   the eastward migration of the population of Texas can infringe on the lawful titles of Congress." During the course of these debates, the American government repeatedly cited the settlement that a Frenchman, ▼ Mr. de Lasalle, had established around 1685 near the bay of San Bernard, without having appeared to encroach on the rights of the crown of Spain.

But upon close examination of the general map that I have drawn of Mexico and the countries that border it to the east, one will see that there is a considerable distance between the bay of San Bernard and the mouth of the Río del Norte: thus, the Mexicans rightly allege in their favor that the Spanish population of Texas is very old, that it came here in the earliest days of the conquest from the interior of New Spain via Linares, Revilla, and Camargo, and that Mr. de Lasalle already found Spanish among the savages whom he tried to fight when he disembarked west of the Mississippi, the mouth of which he had missed. At present, the intendant of San Luis Potosí considers the Río Mermentas or Mexicana, which empties into the Gulf of Mexico east of the Río de la Sabina, as the eastern boundary of the province of Texas, and consequently of his entire intendancy.

It is useful to point out here that this dispute over the true boundaries of New Spain will become significant only when the land cleared by colonists in Louisiana directly borders on land inhabited by Mexican settlers, when

a village in the province of Texas is built near a village in the county of the   II.223
Opelousas. Fort Clayborne, located near the former Spanish mission of Los
Adayes (Adaes or Adaisses) on the Red River, is the settlement in Louisiana
that most closely resembles the military outposts (*presidios*) in the province
of Texas; but there are still nearly sixty-eight leagues between the Presidio
de Nacogdoches and Fort Clayborne. Vast grass-covered steppes serve as
shared borders between the territory of the American confederation and
Mexican territory. All of the country west of the Mississippi, from the Ox
River to the Río Colorado in Texas, is uninhabited. These steppes, partly
marshy, present obstacles that are easily overcome. They may be considered inlets that separate the adjoining coastline but that the industry of
new settlers will soon bridge. The Atlantic provinces in the United States
saw their population spread first toward Ohio and Tennessee, then toward
Louisiana. A portion of this mobile population will later migrate west. The
very name of the Mexican territory will lead them to imagine that mines
are nearby. On the banks of the Río Mermentas, the American settlers will
believe that they have already reached ground concealing metallic riches.
This error, which is widespread among the lower classes, will spawn new
emigrations; people will realize too late that the famous mines of Catorce,
which are the mines closest to Louisiana, are still almost 300 leagues away.

Many of my Mexican friends have taken the dirt road from New Orleans   II.224
to the capital of New Spain. This route, cleared by inhabitants of Louisiana,
who come to buy horses in the Provincias internas, is over 540 leagues long;
its length is thus almost equal to the distance from Madrid to Warsaw. The
road is said to be very punishing because of the lack of water and accom-
modations, but it is far from presenting the same natural obstacles that must
be overcome on the trails that run across the ridge of the Cordillera from
Santa Fe in New Granada to Quito, or from Quito to Cusco. ▼ Mr. Pagès,
captain of a vessel in the French Navy and an intrepid traveler, traveled this
Texan road from Louisiana to Acapulco in 1767. The details he gives about
the intendancy of San Luis Potosí and on the road from Querétaro to Acapulco—a journey that I made thirty years after him—show a fair-minded
spirit motivated by respect for the truth; yet, this traveler was regrettably
so inaccurate in transcribing Mexican and Spanish names that it is very
difficult to recognize from his descriptions the places through which he
passed.[1] The road from Louisiana to Mexico City presents very few obsta-

---

1. Mr. Pagès calls *Loredo* la Rheda; the fort of *la Bahia del Espíritu Santo*, Labadia;
Orquoquissas, Acoquissa; *Saltillo*, Sartille; *Coahuila*, Cuwilla.

cles until one reaches the Río del Norte, and it is only after Saltillo that one
begins to climb toward the Anahuac plateau. The slope of the Cordillera is
gentle there: considering the advances of civilization on the New Continent,
one hopes that overland connections between the United States and New
Spain will gradually become more frequent. One day, carriages will travel
from Philadelphia and Washington to Mexico City and Acapulco.[1]

 The three counties of the state of Louisiana, or New Orleans, that are
the closest to the deserted country that is considered the eastern boundary of the province of Texas are, ranging from south to north, the counties
of the Attakapas, the Opelousas, and the Natchitoches. The last settlements belonging to Louisiana are located on a meridian twenty-five leagues
east of the mouth of the Río Mermentas. The northernmost town is Fort
Clayborne in Natchitoches, seven leagues east of the former site of the Los
Adayes mission. Northwest of Clayborne is the *Spanish lake*, in the middle
of which rises a great rock covered with stalactites; following this lake south-
southeast, one finds at the edges of this beautiful land cultivated by settlers
of French origin first the small village of St. Landry, three leagues north of
the source of the Río Mermentas; then the settlement of St. Martin, and finally New Iberia on the Teche River near the Boutet Canal, which leads to
Lake Tase. Since there are no Mexican settlements beyond the east bank of
the Río Sabina, it follows that the uninhabited country that separates the
villages in Louisiana from the missions in Texas encompasses more than
1,500 square leagues. The southernmost part of these prairies, between the
bay of Carcusiu and the bay of Sabina, presents only impassable marshes.
Thus, the road that leads from Louisiana to Mexico City goes farther north
and follows the thirty-second parallel. From Natchez, travelers head north
from Lake Cataouillou toward Fort Clayborne in Natchitoches; from there,
they pass from the former settlement of Los Adayes to Chichi and to Father
Gama's fountain. Mr. Lafond, a skilled engineer whose map sheds a great
deal of light on these regions, observes that the coal-rich hills that rise eight
leagues north of the Chichi outpost emit an underground noise like cannon
fire, which can be heard from afar. Could this curious phenomenon indicate
a hydrogen emission produced by a burning coal bed? From Los Adayes,
the route from Mexico City goes through San Antonio de Bajar, Laredo (on
the banks of the Río Grande del Norte), Saltillo, Charcas, San Luis Potosí,
and Querétaro to the capital of New Spain. It takes two and a half months

II.225

II.226

---

1. This was written in 1803: today (August 1825), United States legislators are very seriously planning to set up a great rail connection from Philadelphia to Mexico City. E—R.

to cross this vast expanse where, from the left bank of the Río Grande del Norte to Natchitoches, one almost always sleeps beneath the stars.

The most remarkable places in the intendancy of San Luis are: II.227

SAN LUIS POTOSÍ, the intendant's residence, is located on the eastern slope of the Anahuac plateau, west of the source of the Río de Panuca. The typical population of this town is 12,000.

NUEVO SANTANDER, capital of the eponymous province. The Santander sandbar does not allow entry to ships drawing more than ten *palms* of water. The village of *Soto la Marina* east of Santander could become very interesting for trade along this coastline if the port were dredged. Today, the province of Santander is so deserted that ten to twelve square leagues of fertile land were sold for two to three francs in 1802.

CHARCAS, or SANTA MARÍA de las Charcas, a very considerable small town, the seat of a Diputación de Minas [mining council].

CATORCE, or LA PURÍSIMA CONCEPCIÓN de Alamos de Catorce, one of the richest mines in New Spain. ▼ The Real de Catorce has only existed since 1773, when Don Sebastian Coronado and Don Bernabé Antonio de Zepeda discovered the famous veins that produce a value of over eighteen to twenty million francs annually.

MONTEREY, seat of a diocese, in the small kingdom of León.

LINARES, in the same kingdom, between the Río Tigre and the great Río   II.228
Bravo del Norte.

MONCLOVA, a military outpost (*presidio*), capital of the province of Coa-huila, and residence of a governor.

SAN ANTONIO DE BEJAR, capital of the province of Texas between the Río de los Nogales and the Río de San Antonio.

: :

XI. The Intendancy of Durango

POPULATION (IN 1803): 159,700
SURFACE AREA IN SQUARE LEAGUES: 16,873
INHABITANTS PER SQUARE LEAGUE: 10

Like Sonora and Nuevo Mexico (which we have yet to describe), this intendancy, better known as Nueva Vizcaya, belongs to the *Provincias internas occidentales*. It occupies a larger land area than the three united kingdoms of Great Britain combined; but its total population scarcely exceeds that of the two cities of Birmingham and Manchester together. Its length from north to south, from the famous mines of Guarisamey to the Carcay Mountains located northwest of the Presidio de Yanos, is 232 leagues. Its width is very irregular and is barely fifty-eight leagues near Parral.

II.229   To the south, the province of Durango, or Nueva Vizcaya, borders on Nueva Galicia, in other words, the two intendancies of Zacatecas and Guadalajara; to the southeast, on a small part of the intendancy of San Luis Potosí; to the west, on the intendancy of Sonora. But for a stretch of over two hundred leagues to the north and especially the east, it borders on an uncultivated country inhabited by independent, warlike Indians. The Acoclames, Cocoyames, and Mescalero and Faraones Apaches occupy the Bolsón de Mapimi, the Chanate and Los Órganos mountains on the left bank of the Río Grande del Norte. The Mimbreños Apaches are found farther west in the wild ravines of the Sierra de Acha. The Comanche and the many tribes of Chichimecs, whom the Spanish group under the vague name of Mecos, are a source of anxiety to the inhabitants of Nueva Vizcaya, forcing them to travel only when well-armed and in caravans. The military outposts (*presidios*) that dot the vast borders of the *Provincias internas* are too far apart from each other to be able to prevent the incursions of these savages, who, like Bedouins in the desert, know all the tricks of petty warfare. The Comanche Indians, the Apaches' mortal enemies, several hordes of whom live peaceably with Spanish colonists, are the most fearsome for the inhabitants of Nueva Vizcaya and New Mexico. Like the Patagonians in the Straits of Magellan, they have learned how to tame the
II.230   horses that went feral in these regions after the Europeans' arrival. Well-informed travelers confirm that the Arabs are not more agile and nimble horsemen than the Comanche Indians. For centuries, these Indians have roamed plains dotted with mountains that give them the opportunity to ambush travelers. Like almost all the plains savages, the Comanches do not know where their original homeland is. They have tents of buffalo hide that they do not load on their horses but on large dogs that accompany the wandering tribe. This circumstance, already mentioned in the unpublished journal of ▼ Bishop Tamarón's journey,[1] is quite extraordinary; it recalls similar customs among several tribes in northern Asia. The Spanish are even more afraid of the Comanches because they kill all adult

1. *Diario de la visita diocesana del Ilustríssimo Señor Tamarón, Obispo de Durango, hecho en 1759 y 1760* (unpublished).

prisoners and allow only children to live, to be raised with care so they can be used as slaves.

The number of warlike wild Indians (*Indios bravos*) who overrun the borders of Nueva Vizcaya has decreased somewhat since the end of the last century. They make fewer attempts to penetrate into the interior of the inhabited country to pillage and destroy Spanish villages. Their hatred of the whites, however, has remained constant; it is the result of a war of extermination carried out under a barbarous policy and sustained with more II.231 bravery than success. The Indians are concentrated toward the north, in the Moqui [Hopi] and in the Nabajoa [Navajo] mountains, where they have reconquered a large area of land from the inhabitants of New Mexico. This state of things had dire consequences that will be felt for centuries and are definitely worthy of being examined. These wars have not destroyed but have at least removed the hope of bringing civility to these wild tribes through acts of kindness. A spirit of revenge and an inveterate hatred have raised almost insurmountable barriers between the Indians and the whites. Many Apaches, Moqui, and Yuta [Ute] tribes, referred to as peaceable Indians (*Indios de paz*), are agrarian, build their huts together communally, and grow corn. They might be less reluctant to join the Spanish settlers if there were Mexica Indians among them. Similar manners and customs, and the resemblance not only of the sounds but also of the mechanics and general structure of the American languages, can become powerful bonds among people of the same origin. Prudent legislation could perhaps succeed in erasing the memory of those barbarous times when a corporal or sergeant in the *Provincias internas* chased Indians with his men as one hunts down wild animals. It is more likely that a copper-skinned man would prefer to live in a village inhabited by people of his own race than mix with whites II.232 who condescend to him. But we have seen above, in chapter VI, that unfortunately, there are virtually no Indian farmers of the Aztec race in Nueva Vizcaya or New Mexico. In Nueva Vizcaya, there is not a single [Indian] tributary; all the inhabitants are *white*, or at least consider themselves so. They all believe that they have the right to put the title *Don* before their baptismal name, even if they represent what, in the French islands, through an aristocratic refinement that enriches language, were once called *petits blancs* or *messieurs passables*.

This centuries-long struggle against the indigenous peoples; the constant need of settlers who live on isolated farms or travel across arid deserts to be mindful of their own safety and to defend their flocks, homes, wives and even their children against the nomadic Indians' incursions—in short, the natural state preserved amidst the appearance of an ancient civilization—gives the character of people who live in the northern part of New Spain a peculiar energy and, dare I say, temperament. To these factors we must add the temperate climate, an eminently salubrious atmosphere, the need to work less rich and fertile land, and the complete absence of Indians and slaves whom the whites might employ to give themselves over to II.233   idleness and indolence with impunity. In the *Provincias internas*, a very active life, much of which is spent on horseback, favors the development of physical strength. This is especially the case because of the care required by the numerous cattle herds that roam almost wild on the savannahs. Strength of mind and a happy disposition of intellectual faculties complement the strength of a healthy, robust body. Directors of educational institutions in Mexico City have long observed that the youth who excelled by making rapid progress in the exact sciences have come mainly from the northernmost provinces of New Spain.

The intendancy of Durango occupies the northern edge of the great Anahuac plateau, which descends northeast toward the banks of the Río Grande del Norte. But, according to Don Juan José de Oteiza's barometric measurements, the environs of the town of Durango are still over 2,000 meters above sea level. The land seems to maintain this great elevation as far as Chihuahua, since it is the central chain of the Sierra Madre that (as we have shown in the general physical tableau of the country),[1] near San José del Parral, runs north-northwest toward the Sierra Verde and the Sierra de la Grulla.

II.234   In Nueva Vizcaya there is one city or *ciudad* (Durango), six *villas* (Chihuahua, San Juan del Río, Nombre de Dios, Papasquiaro, Saltillo, and Mapimi), 199 villages (or *pueblos*), seventy-five parishes (or *parroquias*), 152 farms (or *haciendas*), thirty-seven missions, and 400 shacks (or *ranchos*).

: :

The most remarkable places there are:

DURANGO or Guadiana, the residence of an intendant and a bishop, in the southernmost part of Nueva Vizcaya, 170 leagues from Mexico City as the crow flies; 298 leagues from the town of Santa Fe. The elevation

1. See vol. I, in the third chapter, p. 267 [p. 181 in this edition].

of the city is 2,087 meters. It often snows there, and the thermometer (below 24°25′ latitude) descends to eight degrees Centigrade below freezing. A group of scoria-covered boulders called the *Breña* rise in the middle of a very level plateau between the capital, the dwellings at Ojo and Chorro, and the small town of Nombre de Dios. This grotesquely shaped formation—twelve leagues long from north to south and six leagues wide from east to west—deserves special attention from mineralogists. The boulders that constitute the Breña are basaltic amygdaloid and appear to have been raised up by volcanic fire. Mr. Oteiza has studied the nearby mountains, especially the Frayle near the Hacienda del Ojo. He found a crater on its peak that is nearly one hundred meters in circumference and over thirty meters in perpendicular depth. There is also an enormous, solitary mass of malleable iron and nickel on the plain around Durango, whose composition is identical to that of the aerolith that fell on Hraschina, near Agram in Hungary, in 1751. Don Fausto d'Elhuyar, the learned director of the *tribunal de Minería de México*, sent me samples that I deposited in various European collections, and of which Mr. Vauquelin and Mr. Klaproth have published analyses. It has been confirmed that the Durango mass weighs almost 1,900 myriagrams, which is 400 myriagrams more than the aerolite that ▼ Mr. Rubin de Celis found in Olumpa in Tucumán [Argentina]. In 1792, a distinguished mineralogist, Mr. Friedrich Sonneschmidt,[1] who had traveled much more widely in Mexico than I did, also identified a mass of malleable iron weighing ninety-seven myriagrams inside the town of Zacatecas. He found its exterior and physical characteristics analogous to the malleable iron described by the celebrated ▼ Pallas. The population of Durango is 12,000.

CHIHUAHUA, residence of the captain-general of the Provincias internas, surrounded by extensive mines, east of the great Real de Santa Rosa de Cosiquiriachi. Population: 11,600.

SAN JUAN DEL RÍO, southwest of Lake Parras. This town should not be confused with the similarly named place in the intendancy of Mexico City, which is located east of Querétaro. Population: 10,200.

NOMBRE DE DIOS, a sizable town on the road to the famous mine of Sombrerete in Durango. Population: 6,800.

PAPASQUIARO, a small town, south of Río de Nasas. Population: 5,600.

II.235

II.236

---

1. [Sonneschmidt, "Carta al Autor de esta,"] *Gazeta de Mexico*, vol. V, p. [1]56.

SALTILLO, on the border of the province of Cohahuila and the small kingdom of León. This town is surrounded by arid plains, where the traveler suffers greatly from the lack of fresh water. The plateau where Saltillo is located descends toward Monclova, the Río del Norte, and the province of Texas where, instead of European wheat, one finds only fields covered with cactus. Population: 6,000.

MAPIMIS, with a military outpost (*presidio*), east of the Cerros de la Cadena, on the uncultivated border called the Bolsón de Mapimi. Population: 2,400.

PARRÁS, near the eponymous lake, west of Saltillo. A type of wild vine that grows at this beautiful site led the Spanish to call it *Parrás*. The conquerors transplanted Vitis vinifera from the Old Continent there, and this new branch of industry has succeeded so well that, despite the stiff resistance that the monopolists in Cádiz have shown for centuries to cultivating olives, grapevines, and mulberry trees in the provinces of Spanish America.

II.237

SAN PEDRO DE BATOPILAS, formerly very famous for the wealth of its mines west of the Río de Conchos. Population: 8,000.

SAN JOSÉ DE PARRAL, residence of a *Diputación de Minas*. The name of this *Real* is derived, like that of the town of Parrás, from the large number of wild vine stocks that covered the countryside when the Spanish first arrived. Population: 5,000.

SANTA ROSA DE COSIQUIRIACHI, surrounded by silver mines, at the foot of the Sierra de los Metates. I have seen a very recent report by the intendant of Durango in which the population of this Real was estimated at 10,700.

GARISAMEY, very old mines on the road from Durango to Copala. Population: 3,800.

: :

XII. The Intendancy of La Sonora

POPULATION (IN 1803): 121,400
SURFACE AREA IN SQUARE METERS: 19,143
INHABITANTS PER SQUARE LEAGUE: 6

This intendancy, which is even less populous than the intendancy of Durango, extends along the Gulf of California, also called the Sea of Cortés. Its coastline is more than two hundred and eighty leagues long, from the great

Bay of Bayona, or the Río del Rosario, to the mouth of the Río Colorado,   II.238
formerly called the Río de Balzas, on whose banks the missionary monks
Pedro Nadal and Marcos de Niza made astronomical observations in the
sixteenth century. The intendancy is of irregular width. From the tropic of
Cancer to almost 27° latitude, it is just over fifty leagues wide, but farther
north, toward the Río Gila, it stretches so far that on the parallel of Arispe,
it is more than 128 leagues wide.

The intendancy of La Sonora occupies an area of mountainous terrain
that is more than half the size of France. But its absolute population does not
amount to one quarter that of the most populous departments of that kingdom. Like his counterpart in San Luis Potosí, the intendant that lives in the
town of Arispe is responsible for the administration of several provinces
that have preserved the special names they had before their unification.
Consequently, the intendancy of Sonora comprises the three provinces of
Cinaloa (or Sinaloa), *Ostimuri*, and *Sonora proper.* The first extends from
the Río del Rosario to the Río del Fuerte, the second, from that river to the
Mayo; the province of Sonora, which old maps also designate by the name
of Nueva Navarra, occupies the entire northern extremity of this intendancy. The small district of Ostimuri is now considered to be contained entirely within the province of Cinaloa. To the west, the intendancy of Sonora   II.239
borders on the sea; to the south on the intendancy of Guadalajara; to the
east, on a very uncultivated part of Nueva Vizcaya. Its northern borders are
quite vague. The villages of the Pimería Alta are separated from the banks
of the Río Gila by a region inhabited by autonomous Indians whom neither
the soldiers stationed at the presidios nor the monks in the nearby missions
have succeeded in conquering.[1]

The three largest rivers in Sonora are the Culiacán, the Mayo, and the
Yaqui (or Sonora). Bearing government dispatches and public correspondence, the courier sails for California from the mouth of the Río Mayo at
the port of Guitivis, also called Santa Cruz de Mayo. This courier rides on
horseback from Guatemala to Mexico City, and from there through Guadalajara and Rosario to Guitivis. After crossing the Sea of Cortés in a *lancha*,
he disembarks at the village of Loreto in Old California. From that village,
letters are sent from mission to mission as far as Monterey and the port of

---

1. To go "*a la conquista*" and "to conquer" (*conquistar*) are the technical terms that
missionaries use in America to signify that they have planted crosses around which the
Indians have built a few huts; unfortunately for the indigenous peoples, however, the words
to conquer and *to civilize* are not synonymous.

San Francisco, located in New California below 37°48' northern latitude.
II.240 On this postal route, letters travel more than 920 leagues, in other words, a
distance equal to that from Lisbon to Kherson [Ukraine]. The course of the
Yaqui (or Sonora) River is quite long and originates on the western slope of
the Sierra Madre, whose low crest runs between Arispe and the Presidio de
Fronteras. Near its mouth is the small port of Guaymas.

The northernmost part of the intendancy of Sonora is called *Pimería*
because of a large tribe of Pima Indians who live there. For the most part,
these Indians live under the dominion of the missionary monks and practice the Catholic ritual. The Pimería *alta* is different from the Pimería *baja*.
The latter includes the Presidio de Buenavista. The former extends from
the military outpost (*presidio*) of Ternate toward the Río Gila. The mountainous terrain of the Pimería alta is the Chocó of North America. All of the
ravines and even some of the plains there contain washed gold dispersed
throughout alluvial terrain. *Nuggets* of pure gold weighing two or three
kilograms have been found there. But these *lavaderos* [washes] are barely
worked because of frequent raids by autonomous Indians, and especially
because of the high cost of provisions that must be hauled from afar into
this uncultivated country. Extremely warlike Indians, the *Seri*, live farther
north, on the right bank of the Río de la Ascensión. Many Mexican scholars
attribute Asian origins to this tribe because of the similarity of their name
II.241 with that of the Seri, whom ancient geographers placed at the foot of the Ot-
torocorras [Uttara-Kuru] mountains east of ▼ *Scythia extra Imaum*.

Until now, there has been no permanent connection between Sonora,
New Mexico, and New California, although the court in Madrid has often
ordered presidios and missions to be established between the Río Gila and
the Río Colorado. Don José Gálvez's extravagant military expedition did
nothing to expand permanently the northern borders of the intendancy of
Sonora. Nevertheless, by crossing country inhabited by autonomous Indians, two brave and enterprising monks, ▼ Fathers Garcés and Font managed to travel overland from the missions in Pimería to Monterrey and the
port of San Francisco without crossing the Sea of Cortés or setting foot on
the peninsula of Old California. This bold expedition, about which the
Propaganda [fide] college in Querétaro has published an interesting account, also provided new information about the ruins of the *Casa grande*
that Mexican historians[1] regard as having been the Aztecs' dwelling place
when they arrived at the Río Gila toward the end of the twelfth century.

1. Clavijero, [*Storia antica del Messico,*] vol. I, p. 159.

Charged with making latitudinal observations, Father Francisco Gar-   II.242
cés, accompanied by Father Font,[1] left from the Presidio of Horcasitas on
April 20, 1773. After traveling for eleven days, he arrived at a beautiful, wide
plain one league from the southern bank of the Río Gila. There he identified
the ruins of an old Aztec town, in the middle of which one finds a building
called the *Casa grande*. These ruins occupied a space of almost one square
league. The *main house* is laid out in exact accordance with the four cardinal
points and is 136 meters long from north to south and eighty-four meters
wide from east to west. It is built of adobe (*tapia*). The rammed-clay bricks
are of unequal size but symmetrically arranged. The walls are twelve decimeters thick. One can see that the building had three floors and a terrace.
There was an exterior staircase, probably of wood. The same type of construction is still found in all the villages of autonomous Moqui [Hopi] Indians west of New Mexico. There were five rooms in the *Casa grande*, each
8.3 meters long, 3.3 meters wide, and 3.5 meters high. A wall interrupted
by large towers surrounds the main building and seems to have served as a
means of defense. Father Garcés discovered the remnants of a manmade canal that brought water from the Río Gila to the town. The entire surrounding   II.243
plain is covered with broken clay pitchers and pots, prettily painted in white,
red, and blue. One also finds pieces of obsidian (*itztli*) among this debris of
Mexica earthenware, a rather interesting phenomenon since it proves that
the Aztecs had traveled through an unknown northern region that contained
this volcanic substance, and that it is not the profusion of obsidian found in
New Spain that inspired the idea of razors and weapons made of *itztli*. The
ruins of the town of Gila, the center of an ancient civilization of American
peoples, should not be confused with the *Casas grandes* in Nueva Vizcaya,
located between the presidio of Yanos and the presidio of San Buenaventura. The indigenous peoples call these places the third dwelling place of
the Aztecs, on the rather vague supposition that the Aztec nation, migrating
from Aztlan to Tula and to the valley of Tenochtitlán, made three stops: the
first near Lake Teguyo (south of the mythical town of ▼ Quivira, the Mexica
Dorado!); the second at the Río Gila, and the third near Yanos.

1. ▼ Fray Domingo Arricivita, *Chronica serifica de el Colegio de propaganda fide de
Querétaro*. Mexico, 1792, vol. II, pp. 396, 426 [427], and 462. This chronicle, which
comprises a large folio volume of six hundred pages, certainly deserves to be excerpted. It
contains very precise geographical information about the Indian tribes who live in California, Sonora, Moqui [Hopi], Nabajoa [Navajo], and on the banks of the Río Gila. I have not
been able to learn which astronomical instruments Father Font used during his excursions on
the Río Colorado from 1771 to 1776. I fear that it was a solar ring.

The Indians who live on the plains near the Casas grandes of the Río Gila, and who have never had the slightest contact with the inhabitants of Sonora, do not in any way deserve to be called *Indios bravos* [wild Indians]. Their social culture markedly contrasts with the state of the savages who roam on the banks of the Missouri and other parts of Canada. Fathers Garcés and Font found the Indians south of the Río Gila clothed and peacefully farming fields planted with corn, cotton, and calabazas [gourds], two or three thousand of them gathered together in villages that they called Uturicut and Sutaquisan. In their efforts to convert the Indians, the missionaries showed them a large picture painted on a piece of cotton cloth and representing a sinner condemned to the flames of hell. This picture frightened the Indians so much that they begged Father Garcés not to unroll it again and not to speak to them about what they believed would happen to them when they were dead. These indigenous peoples are by nature gentle and loyal. Father Font had his interpreters explain to them the safety that prevailed in the Christian missions where an Indian alcalde [mayor] administered justice. The chief of Uturicut replied to him: "This order of things may be necessary for the rest of you. We do not steal, we rarely argue, so of what use would an alcalde be among us?" The civilization that one finds among the indigenous peoples as one approaches the northwest coast of the Americas, from the 33° to 54° latitude, is a very striking phenomenon that cannot but shed light on the history of the earliest migrations of the Mexica peoples.

II.244

There is one city (*ciudad*), Arispe, in the province of Sonora, two towns (*villas*), Sonora and Hostimuri; forty-six villages (*pueblos*), fifteen parishes (*parroquias*), forty-three missions, twenty smallholdings (*haciendas*), and twenty-five farms (*ranchos*).

II.245

The province of Sinaloa encompasses five towns (Culiacán, Sinaloa, Rosario, Fuerte, and Los Álamos), nintety-two villages, thirty parishes, fourteen *haciendas*, and 450 *ranchos*.

In 1793, the number of tributary Indians in the province of Sonora was only 251, whereas in the province of Sinaloa it rose to 1,851. The latter province was thus populated at an earlier time than the former.

: :

The most remarkable places in the intendancy of Sonora are:

ARISPE, residence of the intendant, south and west of the presidios of Bacuachi and Bavispe. Those who accompanied Mr. Gálvez on his expedition from Sonora confirm that the mission of Ures near Pitci would have been more suitable than Arispe as the capital of the intendancy. *Population*: 7,600.

SONORA, south of Arispe, northwest of the presidio of Horcasitas. *Population*: 6,400.

HOSTIMURI, a small, very populous town surrounded by considerable mines.

CULIACÁN, famous in Mexican history under the name of Hueicolhuacan. The population is estimated at 10,800.

SINALOA, also called the *villa de San Felipe y Santiago*, east of the port of Santa María de Aome. *Population*: 9,500. | II.246

EL ROSARIO, near the rich mines of Copala, *Population*: 5,600.

VILLA DEL FUERTE or Montesclaros, north of Cinaloa. *Population*: 7,900.

LOS ÁLAMOS, between the Río del Fuerte and the Río Mayo, residence of a *Diputación de Minería*. *Population*: 7,900.

: :

XIII. The Intendancy of New Mexico

POPULATION (IN 1803): 40,200
SURFACE AREA IN SQUARE LEAGUES: 5,709
INHABITANTS PER SQUARE LEAGUE: 7

Many geographers seem to confuse New Mexico with the *Provincias internas*: they refer to it as a rich mining country with a vast surface area. The celebrated author of the *Histoire philosophique des établissements européens dans les deux Indes* contributed to propagating this error. What he calls the empire of New Mexico is only a shoreline inhabited by poor colonists. It is a fertile but unpopulated land devoid—as far we know—of all metallic wealth, extending along the Río del Norte from 31° to 38° northern latitude. This province is 175 leagues from north to south, and thirty to fifty leagues from east to west. The size of the territory is thus far smaller than | II.247 what those who live in the country itself, but have but little knowledge of geography, imagine it to be. National vanity is content to enlarge spaces and expand—if not in reality then at least in the imagination—the borders of the lands occupied by the Spanish. In the reports I was given on the position of the Mexican mines, the distance from Arispe to Rosario is estimated at three hundred nautical leagues, and at four hundred leagues from Arispe to Copala, without taking into account that the entire intendancy of Sonora is only 280 leagues long. For the same reason, and especially to curry favor with the court, the *conquistadores*, missionary monks, and early colonists gave grand names to small things. Above, we have described a kingdom—the kingdom of León—whose entire population does not equal the number of Franciscan monks in Spain. A few huts grouped together often assume the pompous title of town. A cross raised in the forests of Guiana appears on the maps of missions that are sent to Madrid and Rome as a village inhabited by Indians. Only after having lived for some time in the Spanish colonies, after having unmasked firsthand these fictions involving kingdoms, towns, and villages, can a traveler apply an accurate scale for reducing these objects to their true value.

A few years after the destruction of the Aztec empire, the Spanish conquerors built permanent settlements in the northern part of Anahuac. The town of Durango was founded during the administration of the second viceroy of New Spain, *Velasco el Primero*, in 1559. It was a military outpost to defend against raids by the Chichimec Indians. Toward the end of the sixteenth century, the viceroy, Count of Monterey, sent the valorous ▼ *Juan de Oñate* to New Mexico. After running off the tribes of itinerant indigenous peoples, this general settled the banks of the great Río del Norte.

One can travel by coach from the town of Chihuahua to Santa Fe in New Mexico. In doing so, one usually rides in a type of open carriage, or calash, that the Catalans call *volants*. The road is both beautiful and level, running along the eastern bank of *the Great River* (*Río Grande*), which is crossed at the Paso del Norte. The banks of the river are very picturesque, adorned with beautiful poplars and other trees common to the temperate zone.

It is rather striking to note that, after two centuries of *colonization*, the province of New Mexico still does not border on the intendancy of Nueva Vizcaya. The two provinces are separated by a desert where travelers are sometimes attacked by Comanche Indians. It extends from the Paso del Norte toward the town of Albuquerque. But before 1680, a period marked by a ▼ general revolt among the Indians of New Mexico, this expanse of uncultivated and uninhabited land was less considerable than it is today. There were then three villages—San Pascual, Semillete, and Socorro—that were located between the swamp of Muerto and the town of Santa Fe . The bishop Tamarón was still able to see their ruins in 1760. He found apricot trees growing wild in the fields, an indication of earlier plantings there. The two most dangerous places for travelers are the Robledo gorge, west of the

II.248

II.249

Río del Norte and opposite the Sierra de Doña Ana, and the ▼ Muerto desert, where many whites have been killed by roving Indians.[1]

The Muerto desert is a waterless plain thirty leagues long. The entire country is alarmingly arid, since there is not a single stream in the Los Mansos Mountains located east of the road that leads from Durango to Santa Fe. Despite the mildness of the climate and the progress of trade, a large part of this country, as well as Old California and several districts of Nueva Vizcaya and the intendancy of Guadalajara, will never be able to sustain a significant population.

Although it is at the same latitude as Syria and central Persia, New Mexico has a climate that is eminently cold. It freezes there in the month of May. Near Santa Fe and slightly to the north (at the same parallel of Morea), the Río del Norte is sometimes covered in several consecutive years with ice so thick that a horse and carriage can cross it. We do not know the ground elevation of the province of New Mexico. But I doubt that the riverbed below the thirty-seventh parallel is more than 700 to 800 meters above sea level. The mountains that frame the valley of the Río del Norte, even those at whose foot the village of Taos is located, already lose their snow cover toward the beginning of June. $\qquad$ II.250

The *Great Northern River*, as we have mentioned above, begins in the Sierra Verde, which is the watershed that divides the rivers that flow into the Gulf of Mexico from those that flow into the South Sea. Like the Orinoco, the Mississippi, and several rivers on the two Continents, it has periodic high waters (*crecientes*). The waters of the Río del Norte begin to rise in April. High water reaches its *maximum* level at the beginning of May and drops significantly at the end of June. Only during the summer droughts, when the current is very weak, can inhabitants ford the river, on unusually tall horses. In Peru, these horses are called *caballos chimbadores*. Several persons can ride them at the same time, and if the animal occasionally takes footing as it swims, this way of crossing the river is called *pasar el río a volapié*.

The waters of the Río del Norte, like those of the Orinoco and all the great rivers of South America, are extremely muddy. In Nueva Vizcaya, the cause of this phenomenon is considered to be a small river called the Río Puerco (*dirty* II.251 *river*), the mouth of which is south of the town of Albuquerque, near Valencia.

1. Between the Missouri and the Arkansas; indigo and cotton can only be grown up to 36° latitude, sugar below 37° and a half (Long, *[Narrative of an] expedit[ion to the source of St. Peter's River]*, II, 348).

Mr. Tamarón has observed, however, that the water is muddy well above Santa Fe and the town of Taos. The inhabitants of Paso del Norte still recall an unusual event that took place in 1752. They suddenly saw the entire riverbed go dry, thirty leagues above and more than twenty leagues below Paso: the river water poured into a newly formed crevice and only reemerged near the Presidio of San Eleazario. This *loss of the Río del Norte* lasted for a rather long time. The lovely countryside around the Paso, which is intersected by small irrigation canals, remained unirrigated; the inhabitants dug wells in the sand that fills the riverbed. Finally, after several weeks, the water resumed its former course, probably because the crevice and the underground channels had been filled. The phenomenon I have just described bears a certain resemblance to a fact that the Indians in the province of Jaén de Bracamoros related to me during my stay in Tomependa. At the beginning of the eighteenth century, the inhabitants of the village of Puyaya saw, to their horror and astonishment, the riverbed of the Amazon go almost completely dry for several hours. Near the cataract (*Pongo*) de Rentema, a section of the sandstone boulders had col-

II.252 lapsed due to an earthquake, and the course of the Marañón was blocked until the water was able to flow over the barrier that had been formed. Near Taos, in the northern part of New Mexico rivers originate whose waters blend with those of the Mississippi. The Río Pecos is probably identical to the red river of Natchitoches, and the Río Napestla is perhaps the same river that is called the Arkansas farther east.

Known for their energetic character, the colonists in this province live in a state of perpetual war with the Indians. Because of the lack of safety in the countryside, towns here are more populous than one would expect. The situation of the inhabitants of New Mexico resembles in many ways that of the Europeans in the Middle Ages. As long as isolation exposes men to physical danger, there can be no equilibrium between urban and rural populations.

The Indians who live in hostility with the Spanish settlers, however, are far from being all equally aggressive. The ones in the east are nomads and warriors. Although they trade with the whites, this is often done without seeing each other face-to-face, and according to principles that may be traced back to several African nations. During their excursions north of the

II.253 Bolsón de Mapimi, the Indians set up small crosses along the road that leads from Chihuahua to Santa Fe, on which they hang a leather pouch filled with a little venison. A buffalo hide is spread beneath the cross. The Indians use these signs to show that they want to barter with those who worship the cross. They offer Christian travelers a hide in exchange for food, though the amount of the latter is not specified. The soldiers from the *presidios* who understand the Indians' hieroglyphic language take the buffalo hide and leave salted meat at the foot of the cross.[1] This system of trading requires an extraordinary mixture of good faith and suspicion.

The Indians west of the Río del Norte, between the Gila and Colorado Rivers, contrast with the nomadic and suspicious Indians who roam the savannas east of New Mexico. In 1773, Father Garcés is one of the last missionaries to visit *Moqui* [Moki or Hopi] country, which is traversed by the Río de Yaquesila. He was astonished to find an Indian town there with two great squares, houses several stories tall, and well laid-out streets parallel to each other. The locals gathered every evening on terraces formed by the rooftops of houses. The construction of the Moqui dwellings is the same as that of the *Casas grandes* on the banks of the Río Gila, which we have mentioned above. The Indians who live in the northern part of New Mexico also build their houses quite tall so as to see their enemies approach. Everything in these regions appears to point to traces of the ancient Mexica culture. Indian legends even teach us that, after their departure from Axtlan, the Aztecs first settled on the banks of the Nabajoa River twenty leagues north of the Moqui, near the mouth of the Río Zaguananas. When one compares the civilization that exists at various sites on the northwest coastline of the Americas, among the Moqui, and on the banks of the Gila, one is tempted to believe—and I venture to repeat it here—that during the migrations of the Toltecs, Acolhua, and Aztecs, several tribes were separated from the larger group of people and settled in these northern regions. Nevertheless, the language spoken by the Moqui Indians, the Yabipais [Navajo], who have long beards, and those who live on the plains near the Río Colorado differs fundamentally[2] from the Mexica language.

In the seventeenth century, many Franciscan missionaries settled among the Moqui and Navajo Indians. They were massacred in the great Indian revolt that took place in 1680. I have seen the name *Provincia de Moqui* on unpublished maps drawn up before that period.

The province of New Mexico has three *villas* (Santa Fe, Santa Cruz de la Cañada y Taos, Albuquerque y Alameda), twenty-six *pueblos* (villages), three *Parroquias* (parishes), nineteen missions, and no isolated farms (*ranchos*).

II.254

II.255

---

1. *Diario del Illmo Señor Tamarón* (unpublished).
2. See the testimony of several missionary monks who were well-versed in the knowledge of the Aztec language ([Arricivita,] *Crónica seráfica [y apostólica] del colegio [de propaganda fide de la Santa Cruz] de Querétaro*, p. 408.)

SANTA FE, the capital, east of the Gran Río del Norte. *Population*: 3,600.

ALBUQUERQUE, opposite the village of Atrisco, west of the Sierra Oscura. *Population*: 6,000.

TAOS, which old maps place sixty-two leagues too far to the north, at 40° latitude. *Population*: 8,900.

PASO DEL NORTE, a *presidio*, or military post, on the right bank of the Río del Norte, separated from the town of Santa Fe by over sixty leagues of uncultivated land. This small town, which some unpublished maps preserved in archives in Mexico City consider to be under the control of Nueva Vizcaya, is not to be confused with the *Presidio del Norte* (or de las Juntas) situated farther south, at the mouth of the Río Conchos. Travelers stop at the Paso del Norte to pick up the supplies needed before continuing on to Santa Fe. Around the Paso is delightful countryside that resembles the most beautiful parts of Andalusia. The fields are planted with corn and wheat. The vineyards produce excellent sweet wines that are preferable even to the wines from Parrás in Nueva Vizcaya. All kinds of European fruit trees abound in the gardens—fig, peach, apple, and pear. Since the country is very dry, an irrigation canal brings the water of the Río del Norte to the Paso. At times when the rivers are very low, it is very difficult for the inhabitants of the Presidio to preserve the batardeau that forces their waters into the irrigation ditch (*Azequia*). During the high-water season of the Río del Norte, the force of the current destroys this batardeau almost every year, in the months of May and June. The means of rebuilding and reinforcing the dam is rather ingenious. The locals construct baskets of stakes held together by tree branches and filled with earth and rocks. These gabion baskets (*cestones*) are released into the swirling current that deposits them at the point where the canal separates from the river.

II.256

: :

XIV. The Intendancy of Old California

POPULATION (IN 1803): 9,000
SURFACE AREA IN SQUARE LEAGUES: 7,295
INHABITANTS PER SQUARE LEAGUE: 1

The history of geography offers several examples of countries whose position was known to the earliest mariners but which were long regarded as having only been discovered in recent times. Among them are the

II.257

Sandwich Islands, the western coast of New Holland, the great Cyclades (which ▼ Quirós had once called the archipelago *del Espíritu Santo*), the land of Arsacides [Solomon Islands] as seen by ▼ [Álvaro] Mendaña, and, above all, the coastline of California. The latter was recognized as a peninsula before the year 1541; however, 160 years later, Father *Kühn* (Kino) received the honor of having been the first to prove that California was not an island but was attached to the Mexican mainland.

After astonishing the world with his exploits on the mainland, Cortés demonstrated no less admirable fortitude of character in his naval endeavors. Restless, ambitious, and tormented by the idea that the country that his valor had conquered might be administrated either by a corregidor from Toledo, a president of the Audiencia, or by a bishop of Santo Domingo,[1] he devoted himself entirely to expeditions of discovery in the South Sea. He appeared to have forgotten that the powerful enemies he had at the court had been created by the grandeur and rapidity of his successes, and he took pride in forcing them into silence with the brilliant new career upon which he embarked. The government, on the other hand, was mistrustful of such   II.258 an extraordinary man and encouraged him in his plan to journey across the Ocean. The emperor, believing that he no longer needed Cortés's military prowess since Mexico had been conquered, was happy to see the latter thrown into dangerous undertakings. Above all, the emperor wished to distance the hero from the stage where his courage and audacity had shone so brightly.

As early as 1523, Charles V, in a letter dated from Valladolid, had recommended that Cortés search on the eastern and western coastline of New Spain for *the secret of a strait* (*el secreto del estrecho*) that would shorten by two-thirds the navigations from Cádiz to the East Indies, then called the *land of spices*. In his reply to the emperor, Cortés spoke most enthusiastically about the likelihood of this discovery, "which (he added) will make Your Majesty master of so many kingdoms that he may consider himself the monarch of the entire world."[2] It was during one of these navigations, undertaken at Cortés's own expense, that Hernando de Grijalva discovered the coastline of California in February 1534.[3] His pilot, Fortún Ximénez,   II.259

1. The corregidor Luis Ponce de León; ▼ president Nuño de Guzmán, and the bishop Sebastian Ramírez de Fuenleal.

2. *Cartas de Cortés*, pp. 374, 382, 385.

3. In an unpublished paper preserved in the archives of the viceroyalty of Mexico City I found that California had been discovered in 1526. I do not know on what authority this assertion is based. In his letters to the emperor written up to 1524, Cortés often mentions

was killed by the Californians in the bay of Santa Cruz, later called the port of La Paz or of the Marquis del Valle. Frustrated by the sluggishness and lack of success of the discoveries in the South Sea, Cortés himself set out for the port of Chiametlan (*Chametla*) in 1535, with four hundred Spaniards and *three hundred black slaves* [*nègres esclaves*]. He sailed along both coastlines of the gulf that was then named the *Sea of Cortés,* which the historian Gómara, in 1557, very sensibly compared to the Adriatic Sea. During his stay in the bay of Santa Cruz, Cortés received the troublesome news that the first viceroy had just arrived in New Spain. The great conqueror relentlessly continued his discoveries in California when the news of his death spread throughout Mexico. His wife, Juana de Zuñiga, fitted out two ships and a *caravelle* to find out whether this alarming story was true. After risking a thousand dangers, Cortés dropped anchor safely in the port of Acapulco. Covering all expenses, he had Francisco de Ulloa pursue the career that he himself had begun so gloriously. During the course of a two-year navigation, Ulloa himself scouted out the coasts of the Gulf of California almost to the mouth of the Río Colorado.

II.260   The map that the pilot Castillo drew up in Mexico City in 1541, to which we have referred several times, represents the direction of the coastline of the California peninsula more or less as we now know it. Despite the advances made in geography that we owe to Cortés's genius and his deeds, several writers during the ineffectual reign of ▼ King Charles II began to consider California an archipelago of large islands, called the *Islas Carolinas.* Pearl fishing only occasionally drew a few ships, sent from the ports of Jalisco, Acapulco, and Chacala; when three Jesuits—Fathers Kühn [Kino], ▼ Salvatierra, and Ugarte—explored in great detail the coastline around the Sea of Cortés (*mar rojo o vermejo*) between 1701 and 1721, Europeans believed to have discovered for the first time that California is a peninsula.

The less known a country is and the further removed it is from the most populous European colonies, the more easily it acquires a reputation for great wealth in metals. People's imaginations delight in the tales of wonders that the gullibility and cunning of the early travelers enabled to spread in such a mysterious and beguiling way. In Cumaná and Caracas, people are enraptured by the wealth of the lands between the Orinoco and the Río

---

pearls found near the islands in the South Sea; however, the excerpts that the author [Espinosa y Tello] of the *Relación del Viaje al Estrecho de Fuca* (pp. VII–XXII) took from the valuable manuscripts preserved in the Academy of History in Madrid appear to prove that California had not even been sighted during Diego Hurtado de Mendoza's expedition in 1532.

Negro; in Santa Fe, the missions of the Andaqui are endlessly extolled; the same is true in Quito for the provinces of the Macas and Maynas. For a long time, the California peninsula was the *El Dorado* of New Spain. According to popular reasoning, a country rich in pearls should also produce gold, diamonds, and other precious stones in abundance. A traveling monk, Fray Marcos de Niza, enthralled the Mexicans with the fantastic tales he told them about the beauty of the land north of the Gulf of California, the magnificence of the city of Cibola,[1] its immense population, its orderliness, and the civilization of its inhabitants. Cortés and the viceroy Mendoza had previously disputed the conquest of this Mexica *Timbuktu*. The settlements that Jesuits created in Old California beginning in 1683 have led to a general acknowledgment of how arid and difficult to cultivate this land is. The scant success of the mines worked at Santa Ana, north of Cabo Pulmo, dampened the enthusiasm with which people had imagined the metallic wealth of the peninsula. But the malice and hatred toward the Jesuits gave rise to the suspicion that the Order was hiding from the government the treasures of once-vaunted land. Such considerations led the Visitador Don José de Gálvez—whose knightly spirit had compelled him to join an expedition against the Indians of the Sonora—to travel to California. There, he found bare mountains without arable land and water, a few Barbary fig and mimosa trees growing from cracks in the rocks. There were no signs of the gold and silver that the Jesuits were accused of having extracted from the bosom of the earth. But there were traces everywhere of their industry and the admirable zeal with which they strove to cultivate a desert-like and parched country. On this California expedition, the Visitador Gálvez was accompanied by a man who was as remarkable for his talent as for the great vicissitudes of fortune he experienced: the Chevalier de Asanza, who served as

II.261

II.262

1. Castillo's old unpublished map situates the legendary town of Cibola, or Cibora, below 37° latitude. But if we reset its position to that of the mouth of the Río Colorado, we are tempted to think that the ruins of the *Casas grandes* in Gila, which we mentioned in the description of the intendancy of Sonora, might have given wings to the good Father Marcos de Niza's tales. The great civilization that this man of the church claims to have found among the inhabitants of these northern regions, however, appears to me to be a rather important fact, one that is related to what we presented in our discussion of the Indians of the Río Gila and Río Moqui. Sixteenth-century authors placed a second El Dorado north of Cibora, below 41° latitude. According to them, this was the location of the kingdom of Tatarrax and an enormous town called *Quivira*, on the shores of the Lake Teguayo, close to the Río Aguilar. Although based on the Anahuac Indians' claims, this lore is rather remarkable, because Aztec historians mention the shores of Lake Teguayo—which is perhaps identical to Lake Timpanogos [Lake Utah]—as the country of the Mexica.

Mr. Gálvez's secretary. He stated plainly what the maneuvers of the small army demonstrated much better than did the doctors of Pitic: he dared to say that the Visitador's mind was deranged. Mr. de Asanza was arrested and imprisoned for five months in the village of Tepozotlan, where, thirty years later, he made his formal entrance as viceroy of New Spain.

II.263

With a surface area equal to that of England but with a population smaller than the small towns of Ipswich and Deptford, the California peninsula is located on the same parallel as Bengal and the Canary Islands. The sky there is constantly clear, dark blue, and cloudless; although clouds may appear momentarily at sunset, they radiate the most beautiful shades of violet, purple, and green. Everyone who has traveled in California (and I have met several such persons in New Spain) remembers the extraordinary beauty of this phenomenon, which depends on a special composition of the vesicular mist and the purity of the air in those climes. An astronomer could not find more propitious sites than Cumaná, Coro, Pampatar on the island of Margarita, and the coastline of California. But, unfortunately, the sky is more beautiful than the earth on this peninsula. A dusty, dry soil, like that of Provence, can scarcely sustain a few plants.

A chain of mountains runs through the center of the peninsula, the highest of which, the Cerro de la Giganta, is fourteen or fifteen hundred meters high and appears to be of volcanic origin. This Cordillera is inhabited by animals that, in shape and behavior, resemble the Sardinian *mountain sheep* or *mouflon* (ovis ammon); ▼ Father Consag gave a rather inaccurate account of them. The Spanish call them wild sheep (*carneros cimarrones*). They jump like ibex, with their heads downward. Their horns are curved spirals. According to Mr. Costansó's[1] observations, this animal is fundamentally different from the *wild goats*, which are ashen white, much larger, and native to New California, especially the Sierra de Santa Lucia near Monterrey. Furthermore, the locals refer to those goats, which belong perhaps to the genus of the antelope, as *Berendos*. Their horns, like those of the chamois, are curved backwards.

II.264

At the foot of the mountains of California, there is only sand or a stony stratum where cylindrical cacti (*Órganos del Tunal*) grow to extraordinary

1. *Diario Histórico de los viajes de mar, tierra hechos al norte de la California*, written in 1769 (unpublished). This interesting paper had already been published in Mexico City when, by the minister's order, all copies were confiscated. For the sake of the advancement of zoology, one hopes that attentive travelers will soon make it possible to distinguish the true specific characteristics of the *Carneros cimarrones* of Old California from the *Berendos* of Monterrey.

heights. There are very few springs, and one notices with regret that the rock is bare where the water springs up, whereas there is no water in places where the rock is covered with humus. Wherever springs and humus are found together, the earth is immensely fertile. In these rare spots blessed by   II.265
nature, the Jesuits established their first missions. Corn, jatropha, and di-oscorea [yams] grow vigorously there; vines produce excellent grapes with wine reminiscent of Canary Islands' wine. But in general, due to the arid nature of its soil and the lack of water and humus in the interior of the country, Old California will never be able to sustain a large population, no more than the northernmost part of the Sonora, which is nearly as dry and sandy.

Of all the natural products of California, pearls are the thing that, since the sixteenth century, has most drawn mariners to the coastline of this desert country. Pearls are especially abundant in the southern part of the peninsula. Since pearl fishing has ceased near the island of Margarita, opposite the coast of Araya, the Gulfs of Panama and California are the only areas in the Spanish colonies that supply pearls for the European market. California waters have beautiful pearls; they are large but are often irregularly shaped and unpleasing to the eye. The shells that produce the pearls are found mainly in the Bay of Ceralvo and around the islands of Santa Cruz and San José. The most precious pearls that the court of Spain possesses were found in 1615 and 1665 during the expeditions of ▼ Juan Iturbi and Bernal de Pi-ñadero. During the Visitador Gálvez's stay in California in 1768 and 1769, a   II.266
common soldier from the presidio in Loreto, *Manuel Ocio*, made a large fortune in short time by fishing for pearls along the coast of Ceralvo. Since that time, the number of California pearls that have come on the market annually has been reduced to almost nil. The Indians and Blacks who devote themselves to the strenuous occupation of divers are so poorly paid by whites that such pearl fishing has been virtually abandoned. This branch of industry is languishing for the same reasons that have escalated the price of vicuna skins, rubber, and even febrifugal quinquina bark in South America.

Although Hernán Cortés spent over two hundred thousand ducats of his own wealth on his California expeditions, and although ▼ Sebastian Vizcaíno—who deserves to be placed in the highest ranks of the navigators of his century—took formal possession of the peninsula, it was not until 1642 that Jesuits were able to establish permanent settlements there. Intent on protecting their power, they successfully rebuffed the incursions of the Franciscans, who sought from time to time to introduce themselves among the Indians. They had more difficult enemies to overcome, namely the soldiers at military outposts, since in the farthest reaches of the Spanish possessions on the New Continent, at the edge of European civilization, legislative and executive powers were allocated in a very strange way. The poor Indian knows no other master there than a corporal or a missionary.

II.267 In California, the Jesuits triumphed over the military posted at the presidios. By royal *warrant*, the court decided that everyone, even the captain of the Loreto detachment, would submit to the orders of the father at the head of the missions. The interesting voyages of the three Jesuits—Eusebius Kino, Maria Salvatierra, and Juan Ugarte—shed light upon the physical situation of the country. The village of Loreto had been founded as early as 1697 under the name of Presidio of San Dionisio. During the reign of Philip V, especially after 1744, the Spanish settlements in California became quite large. The Jesuit fathers engaged in the same commercial ventures there to which they had owed so much success but which had also exposed them to so much defamation in both Indies. In a few short years, they built sixteen villages in the interior of the peninsula. Since the Jesuits' expulsion in 1767, the administration of California has been entrusted to the Dominican monks in Mexico City. It appears that they have been less successful in the settlements in Old California than the Franciscans have been on the coastline of New California.

The indigenous peoples of the peninsula who do not live in the missions are among all savages those who are the closest to that state commonly called the natural state. They spend entire days lying on their stomachs, stretched

II.268 out on the sand when it is warmed by the reflections of the sun's rays. Like many of the tribes that we saw on the Orinoco, they abhor clothing. Father Venegas relates that a dressed-up monkey would appear less laughable to people in Europe than a clothed man does to the California Indians. Despite this state of apparent imbecility, the first missionaries were able to distinguish different religious sects among these indigenous peoples. Three deities who waged a war of extermination against each other were objects of terror for three Californian tribes. The Pericú feared the power of Niparajá; the Mengwe and the Vehidi feared the power of Wactupuran and Sumongo. I say that these hordes dreaded rather than adored these invisible beings, since the savage's religion is only tantamount to being possessed by fear: the feeling of a secretive and religious horror.

According to the information that I obtained from the monks who now govern the two Californias, the population of Old California has decreased so sharply in the past thirty years that there are no more than four to five thousand indigenous farmers (*Indios reducidos*) in the missionary villages. The number of these missions has also been reduced to sixteen. The Santiago and Guadalupe missions remain deserted for lack of inhabitants.

Smallpox, and another disease that the European peoples convinced themselves that they contracted in the very continent where they themselves introduced it, and which wreaks havoc on the South Sea Islands, are cited as the main causes of the depopulation of California. It must be assumed that II.269
there are other reasons having to do with the political institutions themselves; and it may be time for the Mexican government to concern itself seriously with removing the obstacles to the well-being of the inhabitants of the peninsula. The number of savages there is scarcely four thousand. One notes that those who live in northern California are somewhat more civilized and docile than the indigenous peoples in the southern part.

: :

The main villages in this province are:

LORETO, presidio and administrative center for all the missions in Old California, founded at the end of the seventeenth century by Father Kühn, the astronomer from Ingolstadt.
SANTA ANA, mission and *Real de minas*, famous for Velásquez's astronomical observations.
SAN JOSÉ, the mission where the Abbot Chappe died, a victim of his own zeal and devotion to the sciences.[1]

XV. The Intendancy of New California II.270

POPULATION (IN 1803): 15,600
SURFACE AREA IN SQUARE LEAGUES: 2,125
INHABITANTS PER SQUARE LEAGUE: 7

On Spanish maps, the part of the coastline of the Great Ocean that extends from the isthmus of Old California, or from the Bay of Todos los Santos (south of the port of San Diego) to Cape Mendocino, bears the name of *New California* (Nueva California). It is a long and narrow expanse of

---

1. People who have lived in California for a long time have assured me that Father Venegas's *Noticia*, against which the enemies of the suppressed order and even Cardinal Lorenzana, raised doubts, is quite accurate (*Cartas de Cortés*, p. 327). In the archives in Mexico City, one also finds the following manuscripts that ▼ Father *Barco* did not use in his *Storia di California*, printed in Rome: (1) *Chronica histórica de la provincia de Mechoacan con varios mapas de la California*. (2) *Cartas originales del Padre Juan Maria de Salvatierra*. (3) *Diario del* ▼ *Capitán Juan Mateo Mangi que acompañó a los padres apostólicos Kino y Kappus*.

land where the Mexican government has established missions and military posts in the past forty years. No village or farm north of the port of San Francisco is more than seventy-eight leagues from Cabo Mendocino. In its present state, the province of New California is only 197 leagues long and nine to ten leagues wide. Mexico City is equidistant, as the crow flies, from Philadelphia and Monterrey, the administrative center of the missions in New California, and its latitude is the same as Cádiz's to nearly four minutes in arc.

II.271   We mentioned above the journeys of several missionaries who, traveling by land from the peninsula of Old California to Sonora at the beginning of the past century, walked around the Sea of Cortés. At the time of Mr. Gálvez's expedition, military detachments went from Loreto to San Diego. Even now, mail goes out from this port to San Francisco along the northwest coast. The northernmost of all the Spanish possessions on the new Continent, San Francisco is situated almost on the same parallel[1] as the small town of Taos in New Mexico. The latter is only three hundred leagues away, and although Father Escalante ventured as far as the western bank of the Zaguananas River near the mountains of *Los Guacaros* during the apostolic excursions he made in 1777, no traveler has yet journeyed from New Mexico to the coast of New California. This must be a stunning fact to those who know, from the history of the conquest of the Americas, the enterprising spirit and the admirable courage that animated the Spanish in the sixteenth century. Hernán Cortés first landed on the coast of Mexico at Chalchiuhcuecan in 1519, and four years later he already had ships built on the shores of the South Sea, at Zacatula and Tehuatepec. In 1537, ▼ Alvar Núñez Cabeza de Vaca and two of his companions, exhausted, naked, and covered with wounds, reached the coastline of Cu-

II.272   liacán opposite the California peninsula. He had landed in Florida with Pánfilo Narváez, and after two years of excursions, after having traversed all of Louisiana and the northern part of Mexico, he reached the edge of the Great Ocean in Sonora. The distance that Núñez covered is almost as great as the route Captain Lewis followed from the shores of the Mississippi to Nootka and the mouth of the Columbia River.[2] When one considers the

---

1. See the first chapter of this work.
2. Captain Lewis's admirable voyage was undertaken under the auspices of Mr. Jefferson who, by this important service rendered to science, has added new claims to the gratitude he is owed by scientists of all nations.

first Spanish conquerors' bold ventures throughout Mexico, Peru, and the Amazon River, one is astonished to find that, for the past two centuries, that same nation has been unable to find an overland path from Taos to the port of Monterrey in New Spain; from Santa Fe to Cartagena in New Granada, or from Quito to Panama; from Esmeralda to Saint Thomas de Angostura in Guiana!

In keeping with English maps, many geographers call New California by the name of *New Albion*. This designation is based on the rather inaccurate theory that, in 1578, the navigator ▼ Drake had been the first to discover the northwest coast of America between 38° and 48° latitude. Sebastián Vizcaíno's famous voyage took place, of course, twenty-four years after Francis Drake's discoveries. But ▼ Knox[1] and other historians seem to forget that Cabrillo had already explored the coastline of New California as far as the forty-third parallel, the end of his journey, in 1542; this follows from a comparison of the old observations of latitude with those taken today. According to reliable historical information, the name *New Albion* should be restricted to the part of the coastline that extends from 43° to 48°, that is, from *Martín de Aguilar's Cabo Blanco* to the *Strait of Juan de Fuca*.[2] Furthermore, over one thousand leagues of coastline, inhabited by free men and populated by groups of otters and seals, stretch from the Catholic priests' missions to those of the Greek priests; in other words, from the Spanish village of San Francisco in New California to the Russian settlements on the Cook River, to Prince William Bay and the Kodiak and Unalaska Islands! Discussions about the surface area of Drake's New Albion and about the so-called rights that Europeans believe they obtain when they erect small crosses, leave inscriptions on tree trunks, or bury bottles may therefore be considered futile.

Although the great mariner Sebastián Vizcaíno carefully explored the entire coastline of New California (as proven by the maps he himself drew up in 1602), it was another 167 years before the Spanish occupied this beautiful country. Fearing that other European naval powers might establish settlements on the northwest coast of the Americas that could become dangerous to the old Spanish colonies, the court of Madrid ordered the viceroy, Chevalier de Croix, and the *Visitador* Gálvez to found missions

II.273

II.274

1. Knox, *Collection of Voyages*, vol. III, p. 18.
2. See the scientific research in the Introduction to [Espinosa y Tello's] *Viaje de las Goletas Sútil y Mexicana*, 1802, pp. XXXIV, XXXVI, LVII.

and presidios at the ports of San Diego and Monterrey. To that end, two packet boats set out from the port of San Blas and anchored in San Diego in April 1763. Another expedition took the land route across Old California. No European since Vizcaíno had disembarked on these remote coasts. The Indians appeared astonished to see men wearing clothes, even though they knew that men who were not copper-colored lived farther east. They even had in their possession a few pieces of silver, which had probably come to them from New Mexico. The first Spanish settlers suffered greatly from the lack of provisions and from an epidemic disease that was the result of poor fare, fatigue, and the lack of shelter; almost all became sick, and only eight individuals remained standing. Among them were two respectable men,

II.275 Fray Junípero Serra, a monk renowned for his travels, and Mr. Costansó, the chief engineer, of whom we have often spoken highly in the course of this work. With their own hands they dug the graves where their companions' corpses would be laid. The overland expedition was too late in bringing assistance to this unfortunate new colony. To announce the arrival of the Spanish, the Indians climbed up on barrels, waving their arms in the air to indicate that they had seen white men on horseback.

The soil of New California is as well watered and fertile as that of Old California is arid and stony. It is one of the most picturesque lands that one might ever see. The climate there is much milder than at the same latitude on the eastern coastline of the New Continent. The sky is cloudy, but the frequent fog that makes landing difficult on the coastline of Monterrey and San Francisco encourages vegetation and nourishes the soil, which is covered with spongy black humus. The eighteen *missions* that now exist in New California grow wheat, corn, and beans (*frijoles*) in abundance. Barley, fava beans, lentils, and chickpeas (*garbanzos*) grow well in most of the province. Since the thirty-six monks of St. Francis who administer these missions are all Europeans, they have carefully introduced to the Indians'

II.276 gardens most of the vegetables and fruit trees that are grown in Spain. The first settlers, who arrived in 1769, already found wild rootstock that produced rather large, but very sour clusters of grapes. Those were perhaps one of the numerous types of *Vitis* native to Canada, Louisiana, and Nueva Vizcaya, about which botanists still know little. The missionaries introduced the vines (*Vitis vinifera*) that the Greeks and Romans had spread throughout Europe, and which are certainly foreign to the New Continent. Good wine is made in the villages of San Diego, San Juan Capistrano, San Gabriel, San Buenaventura, Santa Barbara, San Luis Obispo, Santa Clara, and San José, that is, all along the coastline north and south of Monterrey as far as 37° latitude. The European olive tree is successfully grown near the Santa Barbara Channel, especially near San Diego, where an oil is pressed that is every bit as good as that of the Valley of Mexico or the oils of Andalusia. The very cold winds that blow impetuously from the north and northwest sometimes prevent fruit along the coast from ripening. As a result, the small village of Santa Clara, located nine leagues from Santa Cruz and protected by a chain of mountains, has better-planted orchards and more abundant fruit harvests than the *presidio* in Monterey. It is with great satisfaction that the monks in Monterrey show travelers several use-   II.277 ful vegetables grown from the seeds that ▼ Mr. Thouin entrusted to the unfortunate Lapérouse.

Of all the missions in New Spain, those on the northwest coast show the most rapid and noticeable progress. Since the public has read with great interest the details that Lapérouse, Vancouver, and, more recently, two Spanish navigators, Mr. de Galiano and Mr. Valdés,[1] have published on the state of these distant regions, during my stay in Mexico City I tried to obtain the statistical tables created in 1802 in those very places (San Carlos de Monterrey) by ▼ Father Fermin Lasuén,[2] the current head of the missions of New California. The comparison I made among the official documents preserved in the archives of the archdiocese of Mexico City suggests that there were only eight villages in 1776 and eleven in 1790; whereas by 1802, they numbered eighteen. The population of New California, counting only the Indians who, attached to the soil, began to devote themselves to farming, was:

In 1790:	7,748 souls
In 1801:	13,668
In 1802:	15,562

The number of inhabitants has thus doubled in the space of twelve   II.278 years. According to parish registers, since the missions were founded—or from 1769 to 1802—there have been 33,717 baptisms, 8,009 marriages, and 16,984 deaths in all. One should not attempt to deduce the ratio of births to deaths from this information, because adult Indians (*los neofitos*) are mixed with infants in the number of baptisms.

Assessments of soil yields, or harvest estimates, also offer convincing proof of the growth of industry and prosperity in New California. In 1791,

1. [Espinosa y Tello,] *Viaje de la[s Goletas] Sútil [y Mexicana]*, p. 167.
2. See the excerpt of these tables that I give in note *D* at the end of this work.

according to the tables Mr. de Galiano published, the Indians planted only 874 *fanegas* of wheat in the entire province, which yielded a harvest of 15,197 *fanegas*. By 1802, planting had doubled, because the amount of wheat sown was 2,089 *fanegas* and the harvest 33,576 *fanegas*.

The following table shows the number of animals in 1802.

STEERS	EWES	PIGS	HORSES	MULES
67,782	107,172	1,040	2,187	877

In 1791, there were only 24,958 head of livestock (*ganado mayor*) in all the Indian villages.

These advances in agriculture, these peaceful conquests of industry, are all the more interesting since the indigenous peoples of this coastline, so different from those of Nootka and Norfolk Bay, were still a nomadic people II.279 only thirty years ago, subsisting by fishing and hunting, and cultivating no vegetables whatsoever. The Indians of San Francisco Bay were then as destitute as the inhabitants of Diemen's Island [Tasmania]. Only in the Santa Barbara Channel did one find indigenous peoples who were somewhat more advanced agriculturally in 1769. They built large, pyramid-shaped houses near one another. Kind and hospitable, they gave the Spaniards containers artfully woven from reed stalks. The insides of these baskets, of which Mr. Bonpland has several in his collections, are lined with a thin coating of asphalt that makes them impenetrable to water and the fermented liquors that they may hold.

The northern part of New California is inhabited by both the Rumsen [Ohlone] and the Esselen peoples.[1] They speak completely different languages and they make up the population of the *presidio* and the town of Monterrey. In San Francisco Bay one finds the tribes of the Matalan, Ssalson, and Quiroste, whose languages share a common origin. In my view, many travelers whom I heard speak of the similarity between the Mexica, or Aztec, language and the languages that one finds on the northwest coast of the New Continent have exaggerated the resemblance between these American languages. When I examined closely the vocabularies from Nootka II.280 and Monterrey, I was struck by the homophony and the Mexica endings of several words in the language of the Nootka Indians: for example, *apquixitl* (to embrace), *temextixitl* (to kiss), *coctl* (otter), *hitltzitl* (to sigh), *tzitzimitz*

---

1. *Manuscrit du Père [Fermín Francisco de] Lasuén.* Mr. Galiano calls them Rumsien and Eslen.

(earth), *inicoatzimitl* (the name of a month). In general, however, the languages of New California and Quadra Island [British Columbia] are fundamentally different from the Aztec language, as we shall see in the cardinal numbers that I have brought together in the following table:

	MEXICAN LANGUAGE	ESSELEN LANGUAGE	RUMSEN LANGUAGE	NOOTKA LANGUAGE
1)	Ce	Pek	Enjala	Sahuac
2)	Ome	Ulhai	Ultis	Atla
3)	Jei	Julep	Kappes	Catza
4)	Nahui	Jamajus	Ultizim	Nu
5)	Macuilli	Pamajala	Haliizu	Sutcha
6)	Chicuace	Pegualanai	Halishakem	Nupu
7)	Chicome	Julajualanai	Kapkamaishakem	Atlipu
8)	Chicuei	Julepjualanai	Ultumaishakem	Atlacual
9)	Chiucnahui	Jamajusjualanai	Pakke	Tzahuacuatl
10)	Matlactli	Tomoila	Tamchaigt	Ayo

The Nootka words are taken from Mr. *Moziño*'s unpublished manuscript and not from Cook's vocabulary, in which *ayo* is confused with *haecoo*, *nu* with *mo*, etc.

Father Lasuén observed that on the coastline of New California, across a stretch of 180 leagues from San Diego to San Francisco, one hears seventeen languages that can hardly be considered dialects of a small number of mother tongues. This assertion will not surprise those who know the ▼ fascinating studies that Mr. Jefferson, Mr. Volney, Mr. Bartón, Mr. Hervas, William von Humboldt, Mr. Vater, and Friedrich Schlegel[1] have conducted on the American languages. II.281

The population of New California would have increased even more rapidly if the laws by which the Spanish *presidios* have been governed for centuries were not diametrically opposed to the real interests of the mother country and the colonies. According to these laws, soldiers stationed in Monterey are not allowed to live away from their barracks and settle like colonists. The monks generally oppose the settlement of colonists of the

---

1. See Mr. Schlegel's classic work on Hindu language, philosophy, and poetry, where one finds important views of the mechanism, I dare say, of the organization of languages in the two continents.

white caste because the latter, *as rational [or civilized] people* (gente de razón)[1] do not allow themselves to be subjected to such blind obedience as do the Indians. "It is most troublesome," wrote an educated and progressive navigator,[2] "that soldiers who spend a difficult and laborious life may

II.282 not settle in the country in their old age and become farmers. Forbidding them to build houses in the area surrounding the presidio is contrary to all the dictates of sound policy. If whites were allowed to practice agriculture and animal husbandry; if soldiers, by establishing their wives and children on independent farms, could prepare a refuge for themselves against the indigence to which they are too often subjected in their old age, New California would become a flourishing colony in no time, a most useful sanctuary for Spanish seafarers who engage in trade between Peru, Mexico, and the Philippine Islands." If the obstacles we have just mentioned were removed, the Malvinas Islands [Falkland Islands], the missions on the Río Negro, and the coastline of San Francisco and Monterrey would become populated by a great number of whites. But what a striking contrast there is between the principles of *colonization* followed by the Spanish and those by which Great Britain established villages on the eastern coast of New Holland [Dutch Brazil] in only a few years.

The Rumsen and Esselen Indians share a special predilection for hot baths with the people of the Aztec race and several tribes of Northern Asia. The Temazcalli that one still finds in Mexico City, and of which the Abbot Clavijero gave an exact description, are true steam baths. The Aztec

II.283 Indian lies outstretched in a hot oven whose floor is constantly sprinkled with water. The indigenous peoples of New California, on the contrary, take a bath that the famous Franklin once recommended, referring to it as a *hot-air bath*. A small vaulted building in the shape of a temazcalli is therefore found next to each hut in the missions. Returning from their labor, the Indians enter the oven where the fire was extinguished a few moments before. They stay there for a quarter of an hour, and when they feel completely soaked in sweat, they jump into the cold water of a nearby stream or else roll around in the sand. The rapid transition from hot to cold, the sudden suppression of perspiration, which a European would rightfully dread,

---

1. In Indian villages, indigenous peoples are distinguished from *gente de razón*. Whites, mulattos, blacks [les blancs, les mulâtres, les nègres], all the non-Indian castes, are designated by the name of *rational people*, a humiliating expression for the indigenous peoples whose origin goes back to centuries of barbarism.

2. [Espinosa y Tello,] *Journal de Don Dionisio Galiano.*

provokes pleasant sensations in the savage, who delights in whatever grips or excites him strongly or causes violent reactions in his nervous system.

For some years now, the Indians who live in the villages of New California have woven coarse woolen cloth (*frisadas*). But their main occupation, which might become an important branch of trade, is curing deerskins. I record here the information I was able to collect from Colonel Costansó's unpublished journals about the animals that live in the mountains between San Diego and Monterrey and the special skill with which the Indians are able to catch deer.

One finds neither buffalo nor elk in the low Cordillera that runs along the coast or in the nearby savannahs. Only *berendos* with small chamois horns, which we have mentioned above, graze on the crest of the mountains, which are covered with snow in the month of November. But all of the forests and the grass-covered plains are filled with herds of gigantic deer with extremely large, round antlers. One sometimes sees forty or fifty at a time; they are solid brown in color and without spots. Their antlers, the top of which is not flat, are almost fifteen decimeters (four and a half feet) long. All travelers agree that this large deer is one of the most beautiful animals in Spanish America. It is probably different from Mr. Hearne's *Wewakish* or the *Elk* known to the inhabitants of the United States, of which naturalists have misconstrued the two species of Cervus Canadensis and C. strongyloceros.[1] The deer in New California, which are not found in Old California, had already struck Sebastián Vizcaíno when he set off from the port of Monterrey on December 15, 1602. He claims to "have seen deer whose antlers were three meters (almost nine feet) long." These *venados* run extraordinarily fast, throwing their necks back and pressing their antlers on their backs. The horses of Nueva Vizcaya, reputed to be excellent runners, are unable to follow them closely. They can only keep up with them when the animal, who drinks only rarely, has just quenched his thirst. Then, when he is too heavy to exert all the energy of his muscular strength, he can be easily overtaken. The horseman following him can bring him down by throwing a lasso around him, as one does in all the Spanish colonies with wild horses and cattle. The Indians also use another clever ruse to get close enough to deer to kill them. They cut off the head of a *venado* with very

II.284

II.285

---

1. Much uncertainty still reigns about the specific characteristics that distinguish large and small deer (*venados*) on the New Continent. See Mr. Cuvier's interesting research in his paper on the fossilized bones of ruminants. [Cuvier, "Sur les os fossiles de Ruminans trouvés dans les terrains meubles,"] *Annales du Museum*, year VI [VII], p. 353.

long antlers; then they clean out the neck and put it over their own heads. Thus disguised, but also armed with bow and arrows, they hide in a thicket or in the dense, tall grass. By imitating the movement of the grazing deer, they entice the herd, which is duped by human ingenuity. Mr. Costansó saw this extraordinary hunt on the coast of the Santa Barbara Channel; the officers sailing on the schooners Sútil and Mexicana witnessed it twenty-four years later on the savannahs that surround Monterrey.[1] The enormous deer antlers that Montezuma showed to Cortés's companions as objects of curiosity came perhaps from *venados* in New California. I myself have seen two of them, found in the ancient monument of Xochicalco preserved in the viceroy's palace. Although there was little internal communication in the kingdom of Anahuac in the fifteenth century, it would not be unthinkable for these deer antlers to have been passed from hand to hand from 35° to 20° latitude, just as we find beautiful nephritic jade from Brazil (*piedras de Mahagua*) among the Caribs near the mouth of the Orinoco.

II.286

Since the Russian and Spanish settlements were, until now, the only European colonies that existed on the northwest coastline of the Americas, I think it will be useful to list all the missions in New California that have been founded up to the beginning of 1803. This information is particularly interesting now, when the United States is eager to move westward toward the coasts of the Great Ocean, facing China, which abound in fine sea otter pelts.

The missions of New California are located from south to north in the order in which they are presented here.

SAN DIEGO, a village founded in 1769, fifteen leagues from the northernmost mission in Old California. *Population* in 1802: 1,560.

II.287

SAN LUIS REY DE FRANCIA, a village founded in 1798. *Population*: 600.

SAN JUAN CAPISTRANO, a village founded in 1776. *Population*: 1,000.

SAN GABRIEL, a village founded in 1771. *Population*: 1,050.

SAN FERNANDO, a village founded in 1797. *Population*: 600.

SAN BUENAVENTURA, a village founded in 1782. *Population*: 950.

SANTA BARBARA, a village founded in 1786. *Population*: 1,100.

LA PURISSIMA CONCEPCIÓN, a village founded in 1787. *Population*, 1,000.

SAN LUIS OBISPO, a village founded in 1772. *Population*: 700.

SAN MIGUEL, a village founded in 1797. *Population*: 600.

SOLEDAD, a village founded in 1791. *Population*: 570.

1. [Espinosa y Tello,] *Viaje a Fuca*, p. 164.

SAN ANTONIO DE PADUA, a village founded in 1771. *Population*: 1,050.

SAN CARLOS DE MONTERREY, capital of New California, founded in 1770 at the foot of the Cordillera of Santa Lucia, which is covered with oaks, pines (*foliis ternis*), and rose bushes. The village is two leagues from the eponymous *Presidio*. It seems that Cabrillo already saw Monterrey Bay on November 15, 1542, and called it the *Bahía de los Pinos* because of the beautiful pines that crown the nearby mountains. *Vizcaíno* gave it its present name sixty years later, in honor of the viceroy of Mexico, Gaspardo de Zuñiga, Count of Monterrey, an energetic man to whom we are indebted for the great seagoing expeditions, and who enlisted Juan de Oñate for the conquest of New Mexico. The coastline near San Carlos produces the famous Monterrey *aurum merum* shell, sought after by the inhabitants of Nootka and used in the otter-skin trade. The *Population* of the village of San Carlos is 700.

II.288

SAN JUAN BAUTISTA, a village founded in 1797. *Population*: 960.

SANTA CRUZ, a village founded in 1794. *Population*: 440.

SANTA CLARA, a village founded in 1777. *Population*: 1,300.

SAN JOSÉ, a village founded in 1797. *Population*: 630.

SAN FRANCISCO, a village founded in 1776, with a beautiful port. Geographers often confuse this port with *Drake's Port*, which is farther north, at 38°10′ latitude, and which the Spanish call the *Puerto de Bodega*. The population of San Francisco is 820.

We do not know how many *whites*, *mestizos*, and *mulattos* [blancs, métis, et mulâtres] live in New California, either on the *presidios* or in the service of the Franciscan order. I believe that their number exceeded 1,300 in 1801 and 1802, for in the castes of whites and mixed-bloods [sang-mêlé], there were thirty-five marriages, 182 baptisms, and eighty-two deaths. This is the only part of the population on which the government can rely to defend the coastline in case of an attempted military attack by a European naval power!

II.289

The Population of New Spain in 1823		
Natives or Indians		3,700,000
Whites		1,230,000
Blacks, Africans		10,000
Mixed-blood castes		1,860,000
	Total	6,800,000

These numbers are the result of only an approximate calculation. The information on which they are based has been discussed above at the end of chapter four.

II.290   *Territories Northwest of Mexico*

After drawing a portrait of the provinces that comprise the vast empire of Mexico, we must now take a cursory glance at the coastline of the Great Ocean that extends from the port of San Francisco and Cabo Mendocino to the Russian settlements founded in the bay of Prince William (*Prince William's Sound*).

Spanish seafarers have visited this coastline since the end of the sixteenth century, but it was only in 1774 that the viceroys of New Spain began ordering in-depth explorations. Until 1792, many voyages of discovery were launched from the ports of Acapulco, San Blas, and Monterrey. The colony that the Spanish tried to create in Nootka attracted the attention of all European naval powers for some time. A few warehouses built on the beach, a pitiful fortress defended by swivel guns, and a few cabbages planted in a yard nearly set off a bloody war between Spain and Great Britain, and it was only through the destruction of the settlement on *the island of Quadra and Vancouver* that Macuina, the *Tays* (or Prince) of Nootka preserved his independence. Several European nations have visited the area since 1786 to trade in sea otter skins, but their rivalry had a detrimental outcome both for themselves and for the indigenous peoples of the country. As the price of skins rose on the coasts of the Americas, it dropped dramatically in China. The Indian ways became more corrupt; using the same strategy that has caused so much bloodshed on the African coasts, Europeans tried to take advantage of the discord among the *Tays*. Several of the most depraved sailors deserted their ships to settle among the indigenous peoples of the country. In Nootka, as in the Sandwich Islands, one already finds a frightful mixture of primitive barbarism with the vices of civilized Europe. It is difficult to believe that a few varieties of vegetables from the Old Continent, which travelers have transplanted to these fertile regions, and which appear in the list of good deeds that Europeans pride themselves on having showered upon the inhabitants of the island of the Great Ocean, were adequate compensation for these real evils.

In the sixteenth century, that glorious period when the Spanish nation, favored by a set of extraordinary circumstances, freely expended the resources of its genius and its strength of character, the *problem of a northwest*

II.291

passage—a direct route to the East Indies—occupied the minds of the Castilians with the same fervor as it had concerned the minds of other nations   II.292
for thirty or forty years. We are not referring to the apocryphal voyages
of ▼ *Ferrer Maldonado*, *Juan de Fuca*, and *Bartolomé Fonte*, the importance of which has long been exaggerated. Most of the fraud perpetuated in
the name of these three navigators has been undone through the painstaking studies and scholarly discussions conducted by several officers of the
Spanish navy.[1] Rather than invoking names that are quasi-mythical and
losing ourselves in the uncertainty of theories, we shall be content with indicating what has been incontestably proven by historical documents. The
following information, which is taken in part from the unpublished reports
of Don Antonio Bonilla and Mr. Casasola preserved in the archives of the
viceroyalty in Mexico City, presents facts whose connections may capture
the attention of readers. Painting, so to speak, a varied portrait of national
activity, which is sometimes alert and sometimes lethargic, these reports
will be of interest even to those who do not believe that a land inhabited by
free men belongs to the European nation that saw it first.

The names *Cabrillo* and *Gali* have not become as prominent as those of   II.293
Fuca and Fonte. The truth within the narrative of a modest mariner lacks
both the charm and the power of the fiction it contains. *Juan Rodríguez
Cabrillo* explored the coastline of New California as far as 37°10′, that is,
up to *Punta del Año Nuevo* north of Monterrey. He perished[2] on January 3,
1543, on the island of San Bernardo near the Santa Barbara Channel, but
his pilot, ▼ Bartolomé Ferrelo, continued his journey northward as far as
43° latitude, where he saw the coastline of Cabo Blanco, which Vancouver
calls Cape Orford.

During his voyage from Macao to Acapulco in 1582, *Francisco Gali* discovered the northwest coast of the Americas[3] below 57°30′. Like all those

---

1. *Mémoire de Don Ciriaco Cevallos. Recherches faites dans les archives de Séville
par ▼ Don Agustin Ceán. Introduction historique au voyage de Galiano et Valdes.* Pp. 49–56
and pp. 76–83. In all of my research, I have not been able to discover a single document in
New Spain which names either the pilot Fuca or Admiral Fonte.

2. According to the unpublished paper preserved in the *Archivo general de Indias* in
Madrid.

3. The author of this *Political Essay* adopts the opinion given in the historical introduction to the voyage of the *Goletas Sútil* and *Mexicana* and in the French translation of the *Relation de Linschot*. ▼ Mr. Eyriès, however, in his learned biography of Francisco Gali, reminds
us that this navigator probably saw only the coastlines of San Francisco and Monterrey, since
Hakluyt, as well as Linschot's original edition, mention 37° and a half latitude instead of 57°
and a half. E.—R.

who explored *New Cornwall* after him, he admired the beauty of the colossal mountains, whose summits are covered with eternal snows while their base is decorated with lush plantlife. Correcting[1] old observations with new information about places whose locations are now knowable, one finds that Gali sailed along a part of the Prince of Wales Archipelago [Alexander Archipelago] or the King George Archipelago. In 1578, Sir Francis Drake only reached up to 48° latitude north of Cape Grenville in New Georgia.

Of the two expeditions that Sebastián Vizcaíno undertook in 1596 and 1602, only the last one steered toward the coastline of New California. The thirty-two maps that the cosmographer Enrico Martínez[2] drew up in Mexico City prove that Vizcaíno surveyed this coastline with more attention and intelligence that any pilot had ever done before him. His crew's illnesses, the lack of provisions, and the extreme harshness of the season prevented him, however, from sailing beyond Cabo San Sebastián, located at 42° latitude, slightly north of Trinity Bay. Only one ship from Vizcaíno's expedition, the frigate Antonio Florez commanded, sailed beyond Cabo Mendocino. It reached 43° latitude at the mouth of a river that Cabrillo appears to have already discovered in 1543 and that ▼ ensign Martín de Aguilar believed to be the western extremity of the Strait of Anian.[3] This inlet, the Aguilar River—which has not been rediscovered in recent times—should not be confused with the mouth of the Columbia River (latitude 46°15′), made famous by the voyages of Vancouver, ▼ Gray, and Captain Lewis.

The brilliant era of discoveries that the Spanish had launched on the northwest coasts of the Americas ended with Gali and Vizcaíno. This history of the voyages carried out throughout the seventeenth century and the first half of the eighteenth does not present a single expedition that embarked from the coasts of Mexico to the immense seaboard stretching from Cabo Mendocino to the edges of eastern Asia. Instead of the Spanish flag, one sees the Russian flag flying in this area, as it was hoisted in 1741 on the ships commanded by two intrepid seafarers, ▼ Bering and Chirikov.

II.294

II.295

1. These corrections have already been made in this work wherever latitudes are given upon which early navigators relied. [Espinosa y Tello,] *Viaje de la[s Goletas] Sútil [y Mexicana]*, p. XXXI.

2. The same person whom we mentioned above (p. 359) in relating the history of the *Desagüe Real de Huehuetoca*.

3. In the sixteenth century, the Strait of Anian, which several geographers confuse with the Bering Strait, was synonymous with the Hudson Strait. It took its name from one of the two brothers who sailed on ▼ Gaspard de Corte Real's ship. See the scholarly research that Mr. de Fleurieu recorded in the historical introduction to the *Voyage de Marchand* [Marchand, *Voyage autour du monde*], vol. I, p. 5.

After an interruption of nearly 170 years, the court of Madrid finally focused its attention again to the coasts of the Great Ocean. But it was not merely the desire to make discoveries useful to the sciences that awoke the government from its slumber. Rather, it was the worry of being attacked in its northernmost possessions in New Spain and the alarm at seeing new European settlements crop up near those in California. Of all the Spanish expeditions from 1774 to 1792, only the last two bear the mark of real expe-   II.296
ditions of discovery. Naval officers whose work exhibits broad knowledge of nautical astronomy led these expeditions. The names of Alessandro Mala-spina, Galiano, Espinosa, Valdés, and ▼ Vernaci will always have a place of honor in the list of intelligent and bold seafarers to whom we owe precise information about the northwest coasts of the New Continent. If their pre-decessors were unable to provide measurements with the same degree of perfection, this was because, after having left the ports of San Blas or Mon-terrey, they found themselves lacking the instruments and other means that civilized Europe provides.

The first important expedition after Vizcaíno's voyage was carried out by ▼ *Juan Pérez*, who commanded the corvette *Santiago*. Since neither Cook nor Barrington nor Mr. de Fleurieu seem to have had any knowledge of this important voyage, I shall record several facts here, taken from an un-published journal[1] that I owe to the kindness of ▼ Don Guillermo Aguirre, a member of the audiencia of Mexico City.

Pérez and his pilot, ▼ Esteban José Martínez, left the port of San Blas on January 24, 1774. Their assignment was to explore the entire coastline from the port of San Carlos de Monterrey as far as 60° latitude. Having   II.297
reached Monterey, they set sail again on June 7. On July 20, they discovered Margarita Island [Langara Island] (the northwest tip of Queen Charlotte Island) and the strait[2] that separates that island from the Prince of Wales Strait. On August 9, they became *the first of all European navigators* to drop anchor in Nootka harbor, which they called the port of *San Lorenzo* and which the illustrious Cook, four years later, called *King George's Sound*. They bartered with the Indians, among whom they saw iron and copper, giving them hatchets and knives in exchange for furs and otter pelts. Pérez was unable to land: rough weather and a high, thrashing sea prevented it;

1. This journal was kept by two monks, Father Juan Crespi and Father Tomás de la Peña, who had sailed on the corvette Santiago. These details complement what has already been published in [Espinosa y Tello's] *Voyage de la Sútil*, p. XCII.
2. The *Entrada de Pérez* on Spanish maps.

his skiff nearly sank in attempting to reach land. The corvette had to cut its ropes and abandon its anchors to regain the open sea. The indigenous peoples stole several objects that belonged to Mr. Pérez and his crew. This fact, related in ▼ Father Crespi's journal, serves to resolve the famous problem of the silver spoons made in Europe that Captain Cook found in the hands of the Nootka Indians in 1778. The corvette *Santiago* returned to Monterrey on August 27, 1774, after a voyage of eight months.

II.298   A second expedition left San Blas the following year, under the orders of ▼ *Don Bruno de Heceta, Don Juan de Ayala*, and *Don Juan de la Bodega y Quadra*. We know of this voyage, which singularly advanced the discovery of the northwest, thanks to the ▼ pilot *Mourelle*'s journal, published by Mr. Barrington and appended to the instructions that the unfortunate Lapérouse received. Quadra discovered the mouth of the Río Colombia [Columbia River], which was called *Heceta's inlet, San Jacinto* peak (Mount Edgecombe) near Norfolk Bay, and the beautiful *port of Bucarely* (latitude 55°24′) which, as we know from Vancouver's research, belongs to the western coast of the Great Island of the Prince of Wales Archipelago. This port is surrounded by seven volcanoes, whose perpetually snow-covered peaks spew flames and ashes. Mr. Quadra found a great number of dogs that the Indians used for hunting. I have in my possession two very interesting small maps[1] engraved in Mexico City in 1788, which give the coastal bearings

II.299 from 17° to 58° latitude, as they were plotted during Quadra's expedition.

     In 1766, the court of Madrid ordered the viceroy of Mexico to prepare a new expedition to explore the coastline of the Americas as far as 70° northern latitude. Two corvettes, the *Princesa* and the *Favorita*, were built to this end in Guayaquil, but construction was so delayed that the expedition, headed by Quadra and ▼ Don Ignacio Arteaga, was only able to set sail from the port of San Blas on February 11, 1779. It was during this period that Cook visited the same coastline. Quadra and his pilot, Don Francisco Maurelle, carefully explored the port of Bucarely, Mount Saint Elias,

---

1. *Carta geográfica de la costa occidental de la California situada al Norte de la línea sobre el mar Asiático que se descubrió en los años de 1769 y 1775 por el Teniente de Navío, Don Juan Francisco de Bodega y Quadra y por el Alférez de Fragata, Don José Cañizares, desde los 17 hasta los 58 grados.* The coastline on this map appears to be without inlets or islands. The Ensenada de Heceta (Río Colombia) is noticeable, as is the inlet of Juan Pérez, but not the name of the port of San Lorenzo (Nootka) which was seen by the same Pérez in 1774. *Plan del gran Puerto de San Francisco descubierto por Don José de Cañizares en el mar Asiático.* Vancouver distinguishes the ports of Saint Francis, Sir Francis Drake, and La Bodega as three different ports. Mr. de Fleurieu finds them to be identical. *Voyage de Marchand*, vol. I, p. 54. As we have seen above, Quadra believes that Drake dropped anchor in the port of La Bodega.

the island of Magdalena, which Vancouver called Hinchinbrook (latitude 60°25′), located at the entrance to Prince William's Bay, and the island of Regla, which is one of the barren islands in the Cook River. The expedition returned to San Blas on November 21, 1779. In an unpublished work that I acquired in Mexico City, I discovered that the schistose rocks near the port of Bucarely on Prince of Wales Island contained metallic veins.

The memorable war that brought freedom to a large part of North America prevented the Mexican viceroys from continuing their expeditions of discovery north of Cabo Mendocino. The court in Madrid ordered that expeditions be suspended for the duration of the hostilities that had broken out between Spain and England. The suspension was prolonged even long after the peace of Versailles; it was not until 1788 that two Spanish ships, the frigate *La Princesa* and the pack boat *San Carlos*, commanded by Don Esteban Martínez and ▼ Don Gonzalo López de Haro, left the port of San Blas with the intention of exploring the position and the state of the Russian settlements on the northwest coast of the Americas. The Spanish government was deeply troubled by the existence of these settlements, of which Madrid appears to have had no knowledge until after the publication of the illustrious Cook's third voyage. It pained the Spanish to see that the fur trade drew British, French, and American vessels to a coastline that, before ▼ Lieutenant King's return to London, had been as unfrequented by Europeans as Nuyts's Land or D'Endracht's Land in New Holland.

Martínez and *Haro*'s expedition lasted from March 8 to December 5, 1788. These mariners sailed directly from San Blas to the Prince William Sound, which the Russians call *Gulf Chugatskaia*. They visited the Cook River, the *Kichtak* (Kodiak) Islands, *Schumagin*, *Unimak*, and *Unalaschka* (Onalaksa). They were treated warmly at the various trading posts that they found on the Cook River and in Unalaschka, and they even had the use of several maps that the Russians had drawn up of these areas. In the archives of the viceroyalty of Mexico City, I found a large folio volume bearing the title *Reconocimiento de los cuatro establecimientos Russos al Norte de la California, hecho en 1788*. The historical summary of Martínez's voyage presented in this unpublished paper, however, has but little information about the Russian colonies on the New Continent. No crewmember spoke a single word of Russian, and they could make themselves understood only by sign language. In undertaking this distant expedition, they had forgotten to have an interpreter brought from Europe. There was no way to remedy the ensuing dilemma. Besides, Mr. Martínez would have had as much difficulty

II.300

II.301

finding a Russian in the whole of Spanish America as Mr. George Staunton did in discovering a Chinaman in England or France.

Since the voyages of Cook, Dixon, ▼ Portlock, Mears, and Duncan, Europeans have begun to consider the port of Nootka as the primary fur market on the northwest coastline of the Americas. This consideration persuaded the court of Madrid in 1789 to do what it could have done more easily fifteen years earlier, immediately after Juan Pérez's voyage. Mr. *Martínez*, who had just visited the Russian trading posts, received the order to establish a permanent settlement in Nootka and explore carefully the part of the coastline between 50° and 55° latitude, which Captain Cook was unable to chart during his voyage.

II.302

The port of Nootka is situated on the eastern coast of an island that, according to Mr. Espinosa and Mr. Cevallos's survey in 1791, is twenty nautical miles wide and is separated from the large island, now called *Quadra and Vancouver Island*, by the Tasis Channel. It is, therefore, as false to claim that the port of Nootka, which the indigenous inhabitants call *Yucuatl*, belongs to the large island of Quadra as it is incorrect to maintain that Cape Horn is the tip of Tierra del Fuego. We do not know through what misunderstanding Cook transformed the name Yucuatl into Nootka, the latter word being unknown to the indigenous peoples of the country [Nuuchahnulth] and without any correspondence to other words in their language, except for the word *Noutchi*, which means *mountain*.[1]

1. *Mémoire de Don Francisco Moziño.* The esteemed author was one of the botanists on Mr. Sessé's expedition and stayed in Nootka with Mr. Quadra in 1792. In my attempt to procure as much information as possible on the northwest coastline of North America, I took excerpts from Mr. Moziño's unpublished paper in 1803, for which I am indebted to the friendship of professor Cervantes, the director of the botanical garden in Mexico City. Since then, I have seen that the same unpublished paper provided material to the learned editor [Espinoza y Tello] of the *Viaje de la Sútil*, p. 123. Despite the precise information provided by the English and French navigators, it would still be of great interest to publish Mr. Moziño's observations on Nootka customs. These observations include a great many interesting topics: the union of civil and sacerdotal power in the person of the princes or Tays; the struggle between the principles of good and evil that govern the world, between Quautz and Matlox; the origin of humankind at a time when deer had no antlers, birds were without wings, and dogs had no tails; the Eve of the Nootka people, who lived alone in a flowery bower in Yucuatl, when the god Quautz came to visit her in a beautiful copper dugout canoe; the education of the first man who, as he grew, lived in increasingly larger shells; the genealogy of the Nootka nobles who descend from the oldest son of the man raised in a shell, while the people (who have their own paradise, called *Pinpula*, even in the world beyond) do not dare trace their origin beyond the youngest sons in the family; the calendar system of the Nootka, which is based on beginning the year with the summer solstice, a division of the year

Don Esteban Martínez, the commander of the frigate *La Princesa* and II.303
the pack boat *San Carlos*, dropped anchor in the port of Nootka on May 5,
1789. He was very warmly received by the chief, ▼ Maquinna, who recalled
having seen him with Mr. Pérez in 1774, and who even displayed the beautiful shells from Monterrey that he had been given as gift at that time. Macuina,
the *Tays* of the island of Yucuatl, has absolute power; he is the Montezuma
of these regions, and his name has become famous among all nations that
engage in trading sea otter pelts. I do not know if Macuina is still alive, but
in Mexico City at the end of 1803 we learned from letters written from Mon- II.304
terrey that, being more protective of his independence than the king of the
Sandwich Islands—who declared himself a vassal of Great Britain—he had
attempted to acquire firearms and gunpowder to defend himself against the
abuses that European seafarers often visited upon him.

The port of *Santa Cruz de Nootka* (called the *Puerto de San Lorenzo* by
Pérez and *Friendly Cove* by Cook) is seven or eight fathoms deep. To the
southeast, it is almost entirely hemmed in by small islands; Martínez established the battery of San Miguel on one of them. The mountains in the interior of the island appear to be composed of *Thonschiefer* [English slate] and
other primary rocks. Mr. Moziño discovered veins of copper and sulfurated
lead there. He thought to have identified the products of volcanic fire in a
porous amygdaloid near a lake a quarter-league from the port. The climate
of Nootka is so mild that even at a more northerly latitude than Québec or
Paris, the smallest streams do not freeze before January. This curious phenomenon confirms Mackenzie's observations; he claims that the temperature
of the northwest coast of the New Continent is much higher than that of the II.305
eastern coasts of the Americas and Asia, which are located below the same
parallels. The inhabitants of Nootka, like those who live on the northern
coast of Norway, are almost completely unfamiliar with the sound of thunder. Electrical storms there are infinitely rare. The hills are covered with
pines, oaks, cypresses, beautiful rosebushes, vaccinium, and andromeda. It
was only at the most northerly latitudes that botanists on Vancouver's expedition discovered the pretty shrub named after Linnaeus. John Mears, and
particularly ▼ Don Pedro Alberni, a Spanish officer, succeeded in growing
all of the European vegetables in Nootka, except for corn and wheat, which
never yielded ripe grains there. This phenomenon seemed to have been the
result of excessively vigorous plant growth. True hummingbirds have been

---

into fourteen months of twenty days each, and many intercalated days that are added to the
end of several months, etc. etc.

seen among the birds on Quadra and Vancouver Island. Those unaware that Mr. Mackenzie saw hummingbirds at the source of the Peace River at 54°24' latitude, and that Mr. Galiano also found them at almost the same southern parallel in the Straits of Magellan, must be struck by this fact, which is so important in the geography of animals![1]

Martínez could not pursue his exploration beyond 50° latitude. Two months after entering the port of Nootka, he saw an English vessel, the

II.306 *Argonaut*, commanded by ▼ James Colnett, who is known for the observations he made in the Galapagos Islands. Colnett showed the Spanish navigator the order that his government had given him to set up a trading post in Nootka, build a frigate and a schooner, and prevent any other European nation from taking part in the pelt trade.[2] Martínez replied in vain that Juan Pérez had anchored in this area long before Cook's time. The dispute that arose between the commanders of the *Argonaut* and the *Princesa* nearly caused a rupture between the courts of London and Madrid. To ensure that his rights had precedence, Martínez took a violent and rather illegitimate step: he had Mr. Colnett arrested and sent to Mexico City via San Blas. The real proprietor of the Nootka territory, the Tays Macuina, wisely allied himself with the victor's side; but the viceroy, believing that he should recall Martínez more quickly, directed three other ships toward the northwest coast of the Americas at the beginning of 1791.

▼ *Don Francisco Eliza* and *Don Salvador Fidalgo*, the brother of the astronomer who surveyed the coastline of South America from the Dragon's Mouth to Portobelo, commanded the new expedition. Mr. Fidalgo visited

II.307 Cook's inlet and Prince William's Bay; he completed the survey of the area that the courageous Vancouver explored later. Below 60°54', at the northernmost edge of *Prince William's Sound*, Mr. Fidalgo witnessed a most extraordinary phenomenon, which was most likely of volcanic origin. The locals led him to a snow-covered plain where he saw great masses of ice and rock hurled upward to prodigious heights, accompanied by a terrifying sound. Don Francisco Elisa remained in Nootka to enlarge and fortify the settlement that Martínez had founded the previous year. It was not known in this part of the world that on October 28, 1790, Spain had renounced its claims to Nootka and the Cox Canal through a treaty signed at the Escorial in favor of the court of London. Thus the frigate *Dedalus*, which brought

1. Vol. II, p. 338.
2. A company called *The King George's Sound Company* was created in England in 1785; there was a project to establish an English colony in Nootka similar to the one in New Holland.

the order to Vancouver to preside over the execution of the treaty, did not arrive in the port of Nootka until August of 1792, at a time when Fidalgo was busying himself in creating a second Spanish settlement southeast of Quadra Island on the mainland itself, in the port of *Nuñez Gaona*, or *Quinicamet*, located at 48°20′ latitude at the strait of Juan de Fuca.

Captain Elisa's expedition was followed by two others that, in terms of the importance of the astronomical work that arose from them and the excellence of the instruments with which they were equipped, may be compared to the voyages of Cook, Lapérouse, and Vancouver. I am referring to the voyage of the illustrious *Malaspina* in 1791 and of *Galiano* and *Valdés's* II.308 expedition of 1792.

Malaspina's measurements and those made by the officers working under his orders encompass a stretch of coastline from the mouth of the Río de la Plata to Prince William's Inlet [the entrance to Prince William Sound]. But this talented navigator became even more famous for his misfortunes than for his discoveries. After traveling throughout both hemispheres and escaping all the dangers of a stormtossed sea, he found even greater perils at a court whose favor had turned treacherous. The victim of political intrigue, he wasted away in prison for six years. The French government obtained his freedom, and Alessandro Malaspina returned to his country. In his solitude on the banks of the *Arno*, he takes pleasure in the profound impressions that the contemplation of nature and the study of men in various climes have made on a sensitive soul tested by misfortune.

Malaspina's work has remained buried in archives, not because the government was reluctant to reveal secrets that it might have found useful to conceal, but because the name of this valiant navigator was meant to be consigned to eternal oblivion. Fortunately, the Hydrographic Office (*Depósito hidrográfico de Madrid*)[1] shared with the public the main results that came out of the astronomical observations of Malaspina's expedition. The nauti- II.309 cal maps that have been published in Madrid since 1799 are, in great measure, based on these important results; but instead of bearing the name of the leader, one finds on them only the names of the two corvettes, *La Descubierta* and *La Atrevida*, which Malaspina commanded.

His expedition,[2] which left Cádiz on July 30, 1789, only arrived in the port of Acapulco on February 2, 1791. At that time, the court of Madrid

1. This office was established by royal order on August 6, 1797.
2. *Extrait d'un journal tenu à bord de la Atrevida*, manuscript preserved in the archives of Mexico City. [Espinoza y Tello,] *Viaje de la Sútil*, pp. cxiii–cxxiii. Before the 1789

turned its attention once more to a subject of debate at the beginning of the seventeenth century: the so-called strait through which Lorenzo Ferrer Maldonado insisted he had sailed in 1588 from the coast of Labrador to the Great Ocean. A paper that ▼ Mr. Buache had read at the Academy of Sciences rekindled hope in the existence of this passage. The corvettes La Descubierta and the Atrevida received the order to sail up to high latitudes on the northwest coastline of the Americas and examine all the passes and inlets that interrupt the continuity of the seaboard between 58° and 60° latitude. Accompanied by the botanists ▼ Haenke and Née, *Malaspina* set sail from Acapulco on May 1, 1791. After sailing for three weeks, he landed

II.310 at Cape Saint Bartholomew, which had already been explored by Quadra in 1775, by Cook in 1778, and by Dixon in 1786. Malaspina surveyed the coast from Mount San Jacinto, near Cape Edgecumbe (*Cabo Engaño*, latitude 57°1′30″) to Montagu Island opposite the Prince William Sound. In the course of this expedition, the length of the pendulum and the inclination and declination of the magnetic needle were determined at several points on the coastline. The elevation of the mountains of Saint Elie and Beautemps (the *Cerro de Buen Tiempo* or *Mount Fairweather*), the principal summits of the Cordillera of New Norfolk, were very carefully measured. Knowledge of their elevation[1] and position may be of great help to navigators when bad weather prevents them from observing the sun for weeks on end; within sight of these peaks, at a distance of eighty or one hundred miles, they can plot the location of their vessels by using simple measurements and angles of elevation.

After a futile search for the strait mentioned in the Account of Maldonado's apocryphal voyage [*Relación del Descubrimiento del Estrecho de Anián*] and a stay in the port of Mulgrave in Bering Bay (latitude 59°34′20″),

II.311 Alessandro Malaspina made his way southward. He anchored in the port of Nootka on August 13, sounded the channels that surround the island of Yucuatl, and used purely astronomical observations to determine the positions of Nootka, Monterrey, the island of Guadalupe—where the galleon from the Philippines (the *Nao de China*) usually lands—and Cabo San

---

expedition, Mr. Malaspina, headed for Manila, had already sailed around the globe in the frigate *Astrée*.

1. Malaspina's expedition found the elevation of *Mount Saint Elie* to be 5,441 meters (6,507.6 varas) and that of *Mount Fairweather* 4,489 meters (5,368.3 varas). The elevation of the first of these two mountains is thus close to that of Cotopaxi, while the elevation of the second is almost equal to the height of Mount Rose. See above, vol. I, p. 266 [p. 180 in this edition] and my *Géographie des plantes*, p. 153.

Lucas. The corvette *La Atrevida* arrived in Acapulco and the corvette *La Descubierta* in San Blas in October of 1791.

A five-month voyage was probably not sufficient to explore and survey such an extensive coastline with the minute attention that we so admire in Vancouver's Voyage, which lasted three years. Malaspina's expedition, however, has one special merit that consists not only in the number of its astronomical observations but, above all, in the astute method used to obtain accurate results. The longitude and latitude of four points on the coastline—Cabo San Lucas, Monterrey, Nootka, and Port Mulgrave—were set as absolutes. The intermediate points were connected to these set points by means of four of ▼ Arnold's marine chronometers. This method, used by the officers who embarked on Mr. Malaspina's corvettes, Mr. Espinosa, Mr. Cevallos, and Mr. Vernaci, is highly preferable to the *partial* corrections that people, using lunar distance measurements, take the liberty of making to chronometric longitudes.

Dissatisfied with not having seen the coastline that extends from Nootka Island to Cabo Mendocino at close range, Alessandro Malaspina had just returned to the shores of Mexico when he entreated the viceroy, Count de Revillagigedo, to prepare a new expedition to the northwest coast of the Americas. Endowed with a lively and enterprising mind, the viceroy acquiesced even more easily to this wish since new information provided by the officers stationed in Nootka appeared to lend new credence to the existence of a channel whose discovery had been attributed to the Greek pilot Juan de Fuca since the end of the sixteenth century. Indeed, in 1774 Martínez had found a very wide inlet below 48°20′ latitude. In 1791, the pilot of the schooner *Gertrudis*, ▼ Ensign Don Manuel Quimper, who commanded the bilander *La Princesse Royale*, and Captain Elisa had successively visited this inlet; they even discovered secure and spacious ports there. To complete their reconnaissance, the schooners *Sútil* and *Mexicana* left Acapulco on March 8, 1792, under the command of Don Dionisio Galiano and Don Cayetano Valdés.

Accompanied by ▼ Mr. Salamanca and Mr. Vernaci, these able, experienced astronomers sailed around the large island that now bears the name *Quadra and Vancouver*, taking four months to complete this difficult and dangerous voyage. After passing through the straits of Fuca and Haro, they met the English navigators Vancouver and Broughton, who were busy conducting the same research that was the goal of their voyage, in the Rosario channel, which the English call the Gulf of Georgia. The two expeditions freely shared the results of their work and assisted each other in taking measurements. A clear mutual understanding and perfect harmony prevailed

II.312

II.313

among them until the moment they parted, the likes of which has not been demonstrated by any astronomer on the ridge of the Cordilleras in any another period.

On their return voyage from Nootka to Monterrey, Galiano and Valdés once again surveyed the inlet of la *Ascención*, which *Don Bruno Heceta* had discovered on August 17, 1775, and which the gifted American navigator, Mr. Gray, had called the Colombia River, after the sloop that he commanded. This survey was even more significant since Vancouver, who had already examined this coastline closely, had been unable to see any inlet from 45° latitude to the Fuca strait; even then, the learned navigator had doubted the existence of the Río de Colombia,[1] or the *Entrada de Eceta*.

II.314   In 1797, the Spanish government ordered that the maps drawn up during the expedition of Mr. Galiano and Mr. Valdés be published, "so that the public might have access to them before Vancouver's maps." They were not published until 1802, however, and geographers now have the advantage of being able to compare Vancouver's maps with those of the Spanish navigators, edited by the *Depósito hidrográfico* in Madrid, and the Russian map published in St. Petersburg in 1802 by the emperor's map office. This comparison is all the more necessary since the same capes, passes, and islands often have three or four different names, and their geographical synonymy has become, and for similar reasons, as confused as that of cryptogamic plants.

II.315   At the same time as the schooners *Sútil* and *Mexicana* were exploring the seaboard between the forty-fifth and fifty-first parallels in the greatest

---

1. I have already mentioned above (vol. I, p. 206 [p. 148 in this edition]) how easy it would be for Europeans to establish a colony on the fertile banks of the Colombia River and the doubts that have arisen about the identity that this river may share with the Tacoutché-Tessé or, as Mackenzie calls it, the Oregan. I do not know if this Oregan, or Oregon, flows into one of the great salt lakes that, using information given to me by Father Escalante, I placed on my map of Mexico at 39° and 41° latitude. I shall not decide if the Oregan, like several great rivers in South America, carves a passage through a chain of high mountains, and if its mouth is located in one of the little-known coves between the port of La Bodega and Cape Orford. But I would have hoped that a learned and prudent geographer would not have attempted to identify the name Oregan in that of Origen, which he believed to refer to a river on the map of Mexico published by Don Antonio Alzate (*Géographie mathématique, physique, et politique*, vol. XV, pp. 116 and 117). He has confused the Spanish word *orígen*, the origin or beginning of something, with the Indian word *Origan*. Alzate's map shows that only the Río Colorado receives the waters of the Río Gila. Near their juncture, one reads the following words: "Río Colorado, ó del Norte, *cuyo orígen se ignora*, whose origin is unknown." The sloppy engraving of these Spanish words (one reads "*Nortecuio*" and "*Seignora*") is most likely the source of such an extraordinary error.

For the true course of the Tacoutché-Tessé, or Fraser River, see the note that Mr. von Humboldt appended to this second edition, vol. I, p. 208 [p. 148 in this edition]. E.—R.

detail, the viceroy, Count de Revillagigedo, dispatched another expedition to higher latitudes. The mouth of *Martín de Aguilar*'s river had been sought in vain in the vicinity of Cape Orford and Cape Gregory. Instead of *Maldonado*'s famous channel, Alessandro Malaspina found only *culs-de-sac*, or *impasses*. Galiano and Valdés were sure that *Fuca*'s strait was but an arm of the sea that separated an island more than 1,700 square leagues in size,[1] *Quadra and Vancouver Island*, from the mountainous coastline of *New Georgia*. Doubts still remained about the existence of the strait supposedly below 53° latitude, whose discovery was attributed to Admiral *Fonte* (or *Fuentes*). Cook had regretted not having been able to explore the part of the continent called *New Hanover*; the claims of a capable navigator, Captain Colnet, made it likely that the continuity of the coastline was broken in this area. To resolve such an important question, the viceroy of New Spain ordered ▼ *Don Jacinto Caamaño*, the ship's lieutenant and commander of the frigate *Aranzazu*, to explore with the greatest possible care the seaboard stretching from 51° to 56° northern latitude. Mr. Caamaño, whom I had the pleasure of seeing often in Mexico City, set sail for a six-month voyage from the port of San Blas on March 20, 1792. He scrupulously explored the northern part of Queen Charlotte Island, the southern coastline of Prince of Wales Island, which he called *Isla de Ulloa*, the Revillagigedo Islands, Banks Island (or *Calamidad*), Ariztizábal Island, and the large *Inlet* of Moñino, the mouth of which faces Pitt's Archipelago. The great number of Spanish names that Vancouver preserved on his maps proves that the expeditions we have just summarized contributed in no small measure to the knowledge of a coastline that, from 45° latitude to Cape Douglas east of Cook's inlet, is now more accurately surveyed than most of the European coastline.

II.316

I have limited myself to bringing together at the end of this chapter all the information that I was able to gather on the voyages the Spanish made to the western coastline of New Spain in the northern part of New California from 1543 to this day. It seemed necessary to collect this material in a work that encompasses everything that has any bearing on the political and trade relations of Mexico.

Geographers eager to divide the world to facilitate the study of their scientific findings divide the northwest coastline into a British section, a neutral Spanish one, and a Russian one. These divisions were made without

II.317

1. The extent of *Quadra and Vancouver Island*, calculated according to Vancouver's maps, is 1,730 square leagues, at twenty-five to the sexagesimal degree. It is the largest island on the western coastline of the Americas.

consulting the chiefs of the various tribes that inhabit these regions! If the childish ceremonies that the Europeans call taking possession and the astronomical observations made on a recently discovered coastline could confer property rights, this portion of the New Continent would be singularly parceled out and distributed among the Spanish, English, Russians, French, and Americans of the United States. The same small island would sometimes be divided among two or three nations, because each one could prove that it had discovered a different cape. The great sinuosity of the coast between the fifty-fifth and sixtieth parallels encompasses discoveries made successively by Gali, Bering and Chirikov, Quadra, Cook, Lapérouse, Malaspina, and Vancouver!

Until now, no European nation has created a stable settlement on the vast littoral that extends from Cabo Mendocino to 59° latitude. The Russian trading posts begin beyond this boundary; most of them are scattered and far from each other, like the trading posts that European nations have established for three centuries on the shores of Africa. Most of these small Russian colonies are connected only by sea, and the new names of *Russian America* or *Russian Possessions on the New Continent* should not lead us to believe that the coastline of the *Bering Sea*, the *Alaska* Peninsula, or the country of the *Tschugatschi* have become Russian provinces, in the sense this word has when used to refer to the Spanish provinces of Sonora or Nueva Vizcaya.

II.318

The western coastline of the Americas offers a unique example of a seaboard 1,900 leagues long and inhabited by a single European people. As we have indicated at the beginning of this work,[1] the Spanish have created settlements from Fort Maullin in Chile to San Francisco in New California. Independent Indian tribes inhabit the area north of the 38° parallel. It is likely that these tribes will gradually be subjugated by Russian settlers, who have crossed over to the American continent since the end of the past century. The southward advances of these Siberian Russians should naturally be more rapid than the northward movement of the Mexican Spaniards. A hunting people accustomed to living in a foggy and excessively cold climate will find the temperature that dominates the coastline of New Cornwall agreeable. But this same coastline is like an uninhabitable country or a polar region for settlers from the temperate climate on the fertile, delightful plains of Sonora and New California.

II.319

Since 1788, the Spanish government has been troubled by the appearance of Russians on the northwest coastline of the New Continent.

1. See above, vol. I, p. 190 [p. 139 in this edition].

Considering every European nation as a dangerous neighbor, it has had the location of the Russian trading posts examined. The worry ceased as soon as Madrid ascertained that these posts did not extend eastward beyond *Cook's Inlet*. When ▼ Emperor Paul [of Russia] declared war on Spain in 1799, a bold plan was hatched for some time in Mexico to prepare a naval expedition against the Russian settlements in the Americas from the ports of San Blas and Monterrey. If the plan had been carried out, there would have been war between two nations that occupy opposite edges of Europe but find themselves in close proximity in the other hemisphere, on the easternmost and westernmost boundaries of their vast empires.

The space separating these borders is progressively shrinking, and it is in the political interest of New Spain to know exactly to which parallel the Russian nation has already advanced in the east and the south. An unpublished paper in the archives of the viceroyalty in Mexico City, which I have cited above, gave me only vague and incomplete information. It describes the state of the Russian settlements as they were twenty years ago. In his *Universal Geography*, Mr. Malte-Brun gives a very interesting account of the northwest coastline of the Americas. He was the first to draw attention to ▼ Billings's[1] travel narrative, published by Mr. *Sarytschew*, which is preferable to Mr. *Sauer*'s version. I am proud to be able to provide the positions of the Russian trading posts, based on very recent information taken from an official document;[2] they are, for the most part, merely a collection of storehouses and huts, but they serve as warehouses for the fur trade.

On the coastline closest to Asia, along the Bering Strait, from 67° to 64°10' latitude below the parallels of Lapland and Iceland, one finds many huts used by Siberian hunters. The main posts from north to south are:

II.320

---

1. *Account of the geographical and astronomical expedition undertaken for exploring the coast of the Icy sea, the land of the Tshutski and the islands between Asia and America, under the command of captain Billings, between the years 1785 and 1794. By Martin Sauer, secretary to the expedition.*—*Putetchestwie flota-kapitana Sarytschewa po severowostochnoï tschasti sibiri, ledowitawa mora, i wostochnogo okeana*, 1804 [*Putešestvie flota kapitana Saryčeva po sěverovostočnoj časti Sibiri, ledovitomu morju i vostočnomu okeanu, v prodolženie os'mi lět, pri Geografičeskoj i astronomičeskoj morskoj ěkspedicii, byvšej pod načal'stvom flota kapitana Billingsa, s 1785 po 1793 god*, 1802].

2. [Wilbrecht,] *Carte des découvertes faites successivement par des navigateurs russes dans l'Océan pacifique et dans la mer Glaciale, corrigée d'après les observations astronomiques les plus récentes de plusieurs navigateurs étrangers, gravée au dépôt de Sa Majesté l'Empereur de toutes les Russies, en* 1802. This fine map, which I owe to the kind generosity of ▼ Mr. de St. Aignan, is 1.231 meters long and 0.722 meters wide; it includes the extent of the coastline and the sea between 40° and 72° latitude and 125° and 224° western longitude of Paris. The names are written in Russian letters.

II.321   *Kigiltach, Leglelachtok, Tuguten, Netschich, Tchinkegriun,*], *Topar, Pin-tepata, Agulichan, Chavani,* and *Nugran* near *Cape Rodney* (Cape Parent). These dwellings of the indigenous peoples of *Russian America* are only

II.322   thirty to forty leagues[1] away from the Chukchi's huts in *Russian Asia*. The Bering Strait that separates them is filled with deserted islands, the northernmost of which is called Imaglin [Big Diomede Island]. The northeastern edge of Asia forms a peninsula attached to the great continental mass only by a narrow isthmus between the two gulfs of Mechigmenskiy and Kaltschin. The Asian coastline that borders on the Bering Strait is inhabited by a large number of cetacean mammals. The Chukchi, who live in constant war with the Americans, have groups of dwellings on this part of the coast; their small villages are called *Nukan, Tugulan,* and *Chigin.*

No Russian settlements are found along the coastline of the American continent from Cape Rodney and Norton Sound to Cape Malovodnoy ("the cape with little water"). But the indigenous peoples have assembled many of their huts together on the seaboard that extends between 63°20′ and 60°5′

1. Since it is more than likely that Asian and American peoples crossed the Ocean, it is interesting to examine the width of the arm of the sea that separates the two continents below 65°50′ northern latitude. According to most recent discoveries made by Russian navigators, America is closest to Siberia on a line that crosses the Bering Strait in a direction from southeast to northwest, *from Cape Prince of Wales to Cape Chukotsky.* The distance between these two capes is forty-four minutes in arc, or eighteen and three-tenth leagues at twenty-five to the degree. The island of Imaglin [Big Diomede Island] is almost in the middle of the channel; it is one-fifth closer to the Asian cape. Besides, it appears that to imagine how Asiatic tribes who settled on the plateau of Chinese Tartary were able to cross over from the Old to the New Continent, it is not necessary to take recourse to a transmigration made from such high latitudes. A chain of closely knit small islands extends from Korea and Japan to the southern cape of the Kamchatka Peninsula, between 33° and 51° latitude. The large island of Tchoka [Chukotka?], connected to the continent by an immense sand bank (below 52° latitude), facilitates travel between the mouth of the Amour River and the Kuril Islands. Another archipelago, enclosed to the south by the great Bering basin, extends from the Alaskan peninsula, four hundred leagues westward. The westernmost of the Aleutian Islands is only 144 leagues from the eastern coast of Kamchatka, and this distance is then divided into two almost equal parts by Bering Island and Mednoi Island [Copper Island] (located below 55° latitude). This rapid survey suffices to prove that Asian tribes were able to travel from one small island to the next, and from one continent to the other, *without going higher on the Asian continent than the fifty-fifth parallel,* without moving around the Okhotsk Sea to the west, and without making a crossing *in the open sea* of more than twenty-four to thirty-six hours. Northwest winds that blow in this area for most of the year favor navigation from Asia to the Americas between 50° and 60° latitude. The purpose of this note is to try and establish new historical theories or discuss those rejected forty years ago. Let it suffice to have presented precise information on the proximity of the two continents.

latitude. Their northernmost dwelling places are *Agibaniach* and *Chalmiagmi*, the southernmost are *Kuynegach* and *Kuymin*.

The Russian name for Bristol Bay, north of the Alaskan (or Aliaskan) peninsula, is the *gulf of Kamischezkaja*. They have not preserved on their maps any of the English names imposed by Captain Cook and Vancouver north of 55° latitude. They even prefer not to give names to the two large islands containing *Trubizin* peak (Vancouver's Mount Edgecumbe and Quadra's Cerro de San Jacinto) and Cape *Chirikov* (Cabo San Bartolomé) rather that adopt the names *King George's Archipelago* and *Prince of Wales Archipelago*.

II.323

Five tribes, who form again as many large territorial divisions in the colonies of Russian America, inhabit the coastline that extends from the Gulf of Kamischezkaja as far as New Cornwall. Their names are *Koniagi*, *Kenaizi*, *Chugachi*, *Ugalachmiuti*, and *Koliugi*.

The northernmost part of Alaska and the island of *Kightag* [Kodiak], which the Russians commonly call *Kikhtak* (although in the native language the word Kightak generally only means an island) belong to the *Koniagi* division. A large interior lake [Amanka Lake] over twenty-six leagues long and twelve leagues wide is connected with Bristol Bay via the Igtschiagik [Igushik] River. There are two forts and several trading posts on Kodiak (Kadiak) Island and the small adjacent islands. The forts established by ▼ Shelikhov are called *Karluk* and the fort *of the three saints* [Three Saints Bay]. Mr. Malte-Brun relates that according to the latest news, the archipelago of Kightak was supposed to be the site for the administrative center for all of the Russian settlements. ▼ Sarytschew claims that there is a bishop and a Russian monastery on Umanak (Umnak) Island. I do not know whether they have been established elsewhere, because the map published in 1802 shows no trading post in Umnak, Unimak, or Unalaska. I have read, however, in an unpublished journal of Martínez's voyage in Mexico City, that in 1788 the Spanish found several Russian houses and about one hundred small barks on Unalaska Island. The indigenous peoples of the Alaskan peninsula call themselves *men of the East* (Kagataya-Kung'n).

II.324

The *Kenaizi* inhabit the western coastline of Cook's Inlet, or the Gulf of Kenayskaja. The *Rada* trading post, which Vancouver visited, is located there, at 61°8'. The governor of Kodiak Island, the ▼ Greek Ivanitsch Delareff, assured Mr. Sauer that, despite the harsh climate, wheat would grow well on the banks of the Cook River. He had introduced cabbage and potato farming to the gardens established on Kodiak.

The *Chugatch* occupy the land stretching from the northern extremity of Cook's Inlet to east of Prince William Sound (the Gulf of Chugatskaya).

There are several trading posts and three small fortresses in this district: Fort Alexander, built near Port Chatham, and the forts on Tuk Island (Vancouver's Green Island) and Chalcha (Hinchinbrook Island).

II.325   The *Ugalachmiuti* are spread from Prince William Sound to *Yakutat Bay*, which Vancouver calls Bering Bay.[1] The Saint Simon trading post is near Cape Suckling (Cape Elie for the Russians). It appears that the central chain of the Cordillera of New Norfolk is quite far from the coastline at Saint Elie Peak; the indigenous peoples told Mr. Barrow, who traveled five hundred *werst* (120 leagues) up the Mednaya River (Copper River), that he would reach the high mountain chain only after a two-day journey northward.

The *Koliugi* inhabit the mountainous land of New Norfolk and the northern part of New Cornwall. The Russians, on their maps, show Burrough's Bay (latitude 55°50′) opposite Vancouver's Revillagigedo Island (the Isla de Gravina on Spanish maps) as the *southernmost and easternmost border* of the area over which they claim ownership: it thus appears that the large island in the King George Archipelago was explored more carefully and in greater detail by Russian navigators than by Vancouver. One is easily persuaded of

II.326   this when one compares attentively the western coast of this island, especially the environs of Cape Trubizin (Cape Edgecumbe) and the port of St. Michael the Archangel in Sitka Bay (Norfolk Sound for the English, Chinkitané Bay for Mr. Marchand), on the map published by the royal office in St. Petersburg and on Vancouver's own maps. The southernmost Russian settlement in the Koliugi district is a small fortress (*crepost*) built on Yakutat Bay at the foot of the Cordillera that connects Mount Beautemps and Mount Saint Elie, near port Mulgrave, at 59°27′ latitude. The proximity of mountains covered with perpetual snow and the size of the continent above 58° latitude give to this coastline of New Norfolk and to the land of the Ugalachmuiti an excessively cold climate that is unsuitable for plant cultivation.

When the schooners from Malaspina's expedition entered the interior of Yakutat Bay as far as the port of Desengaño, they found, in July, the northernmost edge of the port, below 59°59′ latitude, covered with a solid sheet of ice. One might think that this ice mass was part of a glacier[2] that culminates in high maritime Alps; but Mackenzie reported that, on a visit to

1. One should not mistake Vancouver's Bering Bay, located at the foot of Mount Saint Elie, for the Bering Bay on Spanish maps, which is found near Mount Fairweather (Nevado de Buentempo). Unless one has a precise knowledge of geographic synonymy, Spanish, English, Russian, and French works that discuss the northwestern coastline of America are almost unintelligible, and only a minute comparison of maps can stabilize this confusion.

2. Vancouver, [*Voyage de découvertes,*] vol. I, p. 67.

the banks of Slaves Lake 250 leagues eastward below 61° latitude, he found   II.327
the lake completely frozen over in the month of June. The temperature difference that one generally observes on the eastern and western shores of the
New Continent, and which we discussed above, appears to be very noticeable only south of the fifty-third parallel that passes through New Hanover
and the large island of Queen Charlotte.

There is approximately the same *absolute* distance between Saint Petersburg and the easternmost Russian trading post on the American continent
as there is between Madrid and the port of San Francisco in New California.
The breadth of the Russian empire at 60° latitude encompasses a stretch of
land measuring 2,400 leagues; the small fort in Yakutat Bay, however, is
still more than *six hundred leagues* away from the northern boundaries of
Mexican possessions. The indigenous peoples of these northern regions[1]
have been cruelly harassed for a long time by Siberian hunters: women and
children were taken as hostages to Russian trading posts. The instructions
that the Empress Catherine gave to Captain Billings, which were drafted
by the illustrious Pallas, exude philanthropy and a noble sensibility. The
current government is seriously concerned with reducing abuse and rep-
rimanding harassment; but it is difficult to prevent wrongdoing in the far   II.328
reaches of a vast empire, and Americans are conscious at every moment of
their distance from the capital.

Corrections and Supplementary Notes to the
Statistical Portrait of New Spain

Until now, the censuses taken during the tenure of Count de Revillagigedo have
been the only basis for estimating the population of Mexico at different times.
I believe that the changes I made in the original estimates in order to reduce them
to the period of 1803 have been justified (see vol. I, pp. 302 *passim* [p. 202 in this
edition]). It is more than likely that the *Confederation of Mexican States* today
includes at least 6,800,000 souls, of which there are 3,700,000 pure-blooded
Indians, 1,230,000 whites, 1,860,000 persons of mixed race [races mixtes],
and 10,000 blacks [nègres]. But it would be unwise to estimate population
growth in each state (formerly called an intendancy) and each state capital. "The

---

1. More recent information on the state of Russian America has been collected
in ▼ Hassel, *Vollst[ändiges] Handbuch der Erdbeschreibung*, 1822, vol. 16, pp. 548–78.

fundamental basis of good government," according to Mr. Alamán, the minister of the interior, "is a statistic or a precise knowledge of State resources: thus, from the earliest days of our independence, the provisory Junta ordered the provincial delegations and *Ayuntamientos* (municipal councils) to gather material that could contribute to a general work. Several delegations distributed forms with the most important questions among the municipalities. In spite of these efforts and despite the order of April 1, 1822, almost nothing came of it. The only province for which it is confirmed that a statistical report was officially drafted is Valladolid. Congress, however, has still not received information about the population of Valladolid, and to date no progress has been made in gathering detailed statistics about the allocation of taxes, the proper distribution of rights to national representation, and the knowledge of our means and strengths" (Lucas Alamán, *[Memoria que el] Secretario de Estado y del Despacho de Relaciones interiores [presenta al soberano Congreso constituyente], en el Informe al Congreso del 8 de nov. 1823*, p. 22). Here are a few corrections that I have largely drawn from information that was sent to me from Mexico:

II.329

Mexico City. The population, which was believed to be 140,000 souls in 1803 (see above, p. 344), was found in 1820 to be 168,846, of which there were 92,838 women and 76,008 men. The preponderance of women, which one sees in all censuses of large cities in Mexico and the United States ([▼ Sidney Edward Morse] and [Jedidiah] Morse, *[A New System of] Modern Geography*, p. 619), has been maintained, although new research done in Mexico City has confirmed the general principle that more boys than girls are born in Mexico City and in Bombay (*Trans[actions] of the Phil[osophical] Society of Bombay*, vol. I, p. 25), as is also true in the coldest regions, Siberia and Lapland. The census of births in thirty departments from 1817 to 1822 has disproven what we suggested above (vol. I, p. 455 [p. 290 in this edition]), following Mr. Peuchet, about the changes in the ratio of male to female births in the southern departments of France (*Annuaire du Bureau des longitudes, pour 1825*, p. 98). Only with respect to the birth of illegitimate children do female births come closer to the number of male births. The gender ratio for legitimate births is sixteen to fifteen, and for births of illegitimate children, it is twenty and a half to nineteen. Even today, Mexico City is the most populous of all capital cities on the New Continent. Rio de Janeiro has 135,000 inhabitants, of whom there are 105,000 Blacks; Havana, 130,000; New York City, 140,000; Philadelphia, 115,000; and Bahia, 100,000.

La Puebla de los Angeles. I doubt that the population has risen as much as some modern travelers, who increase it to 90,000, would have it. The 1820 census gives only 60,000.

II.330

Guadalajara, whose population was already underestimated in 1803, is today, the second largest city in the Confederation, after Mexico City. Its population is estimated at seventy thousand souls.

Guanajuato. According to the census taken in May 1822, only 15,379 souls remain in this town, and 20,354 live in the nearby mines and suburbs; the total is 35,733, or less than half of the population in 1805 (see above, p. 389). Since metal mining has started to thrive again, the number of inhabitants is now increasing ([J. M. "Minería,"] *Sol ó Gazeta de Mexico*, 1825, n. 597, p. 954, and [Poinsett,] *Notes on Mexico*, 1824, p. 162.)

Oaxaca. Although the statistical table of the intendancy of Oaxaca was sent to me in its original form in accordance with the order issued by the provincial delegation on September 27, 1820, I have no doubt that it is based entirely on the census taken in 1793 under Count de Revillagigedo's administration. That census gave 411,366, but how is one to believe that a province that has enjoyed the deepest peace for twenty-seven years, and that has a healthy and hard-working Indian population, has only increased by 9,000 souls from 1793 to 1822? During my stay in Mexico City, I had already found, in the archives of the viceroyalty, some population figures that the intendants had dated at the beginning of the nineteenth century and which were identical, to a person, to the lists from 1793. It is easier to copy an old table than to conduct a new census. The 420,973 souls that the *Estadistica general de la Intendencia y del Corregimiento de Oaxaca* (which we are recording here) presents are probably only one of the *variants* with which numerous copies of Count de Revillagigedo's *censo general* are riddled. What appears to confirm this supposition is the remarkable circumstance that, at the meeting of the first Mexican congress, the official number of deputies from Oaxaca was determined by assuming a population of 600,000 souls (vol. I, p. 319 [p. 213 in this edition]). Yet this decision would surely not have been made with such an assumption had the number of inhabitants still been believed in 1820 to have been only 420,973. The *Estatistica general* presents population and agriculture in relation to the different *subdelegaciones* by listing the number of parishes, farms, and villages for each of them.

II.331

CITIES AND PARTIDOS OF THE CORREGIMIENTO				
DE OAXACA	PARISHES	FARMS	VILLAGES	POPULATION
Ciudad de Oaxaca	1			15,624
Barrio de las Muertas		2	1	418
de Talatlaco	6		4	952
Partido de Ejutla	1	10	3	4,697
de Zauchila	1	1	6	5,172
de Ixtlán	1		5	1,462
de Sosola	1		6	686

(continued)

CITIES AND PARTIDOS OF THE CORREGIMIENTO			
DE OAXACA	PARISHES	FARMS	VILLAGES	POPULATION
de Atatlahuca	1	2	5	1,296
de Colotepeque		11	1	328
de Ocotlán	1		13	9,468
de Ayoguesco	3	2	13	6,263
de Talistaca	2	2	5	5,676
Total for the Corregimiento de Oaxaca	18	30	62	52,042
Subdelegaciones				
Villa alta	20		110	39,404
Marquesado	5	15	45	21,087
Miahuatlán	6	10	49	18,450
Muamelula	1	22	13	4,276
Tentitlan del Valle	3	8	23	12,862
Quiechapa	9	15	39	15,749
Muapiapam	15	13	79	33,765
Yxtepexe	2	11	9	4,871
Nochistlán	6	1	29	10,039
Huizo	2	9	13	8,211
Zimatlan	5	29	46	12,792
Teococuilco	5	4	31	11,077
Teposcolula	17	13	147	55,697
Tentila	7	13	35	23,353
Tehuantepec	6	171	28	24,922
Tentitlan del Camino	5	7	37	20,509
Tustlahuaca	6	6	26	8,171
Chontales	4	9	31	5,344
Iamiltepec	9	44	56	37,721
Villa de Xalapa del Estado	1	7	1	631
Total for the province of Oaxaca	152	437	909	120,973

[II.332]

San Blas. It is clear from the naval minister's recent report to Congress (December 16, 1824) that the transfer of military establishments from San Blas to Acapulco is a serious concern. There are complaints about the poor condition of the *estero* (roads) of San Blas, where three or four ships that require a depth of ten to fourteen feet of water can hardly be stationed at the same time, while the excessively unhealthy climate interrupts docking and shipbuilding for half the year. On the other hand, the superb port of Acapulco, as we have seen above (vol. I, p. 238 [p. 399 in this edition]), lacks the necessary timber for shipyards, and the government is deliberating at this time whether the port of Manzanillo (between Zacatula and Cabo Corrientes), which is closer to forests, should not be given preference over Acapulco. As for the port of San Blas, the position of which is very important for the merchant marine, it would retain the rights of a *puerto habilitado* [port of entry]. II.332

According to an official report, there were in the entire country of New Spain, in 1810, 1,073 parishes, 157 missions, 264 convents, 4,682 villages, 3,749 farms (*haciendas de campo*), 6,684 ranches (*ranchos* or *haciendas menores*), 1,195 dairy and sheep farms (*estancias* or *haciendas de cria de ganados*) (see *Miscellanea*, no. 200, p. 6). According to Don Fernando Navarro y Noriega's work, the allocation of parishes (*curatos*) in New Spain between the dioceses of nine bishops was as follows: II.333

DIOCESES	PARISHES
Mexico City	244
Puebla	241
Valladolid	116
Oaxaca	140
Guadalajara	120
Yucatán	85
Durango	46
Monterey	51
Sonora	30
Total	1,073

According to an official notice in my possession, only 2,300 ecclesiastics served these 1,073 parishes. According to these same documents, in 1813 there were eighteen missions in the diocese of Mexico City, five in Valladolid, forty-five in Durango, eighteen in Monterey, and sixty-six in Sonora—for a total of 157.

I. *Misiones del Arzobispado de México.*

Custodia del Salvador de Tampico—supervised by the Franciscan order in the province of Santo Evangelio.

II. *Misiones del Obispado de Valladolid.*

Custodia de Santa Catalina Mártir de Rioverde—supervised by Franciscans from the province of Los Santos Apostolos de Michoacan.

II.334

III. *Misiones del Osbipado de Durango.*

a) Custodia de la Conversión de San Pablo de Nuevo México—under the supervision of monks from the province of Santo Evangelio.

b) Custodia del Paso del Norte.

c) Custodia de la Taraumara Alta—supervised by the Colegio apostolico de N.S. de Guadalupe de Zacatecas.

d) Custodia del Parral—supervised by Franciscans from the province of Zacatecas.

IV. *Misiones del Obispado de Monterrey.*

a) Misiones de Gualapuiser in the Nuevo Reyno de León—supervised by monks from the province of Zacatecas.

b) Misiones de Cohahuila—supervised by the Colegio apostolico de Pachuca.

c) Misiones de Texas—supervised by the Colegio apostolico de Zacatecas.

V. *Misiones del Obispado de la Sonora.*

a) Misiones de Sonora and Arispe—supervised by the Colegio apostolico de la Santa Cruz de Querétaro.

b) Misiones de la Antigua (or Baja) California—supervised by the province of Santiago del Orden de Predicadores.

c) Misiones de la Nueva (or Alta) California—supervised by the Colegio apostolico de San Fernando de México.

In the report that the interior minister of the ecclesiastical department made to the Mexican Congress in 1824, one finds the following table that lists the Colegios de Propaganda Fide and the number of monks and missions that belong to these five establishments.

II.335

The ten missions in Coahuila and Texas have recently been secularized: the monks have become parish priests. The same change will be made at the missions in the Sierra Gorda, but since the revolution has opened to Mexican youth

ESTABLISHMENT	MONKS	MISSIONS	PROVINCE(S)
Santa Cruz de Querétaro	66	9	Sonora
San Fernando de México	77	20	Alta California
San Francisco de Pachuca	45	9	Nuevo Santander and Coahuila
San José de Gracia de Orizaba	47		
Nuestra Señora de Guadalupe de Zacatecas	94	22	Tarahumara and Texas
Total	329	60	

such a variety of ways of employing their talents, the number of clergy decreases daily.

"The two Californias," writes the minister of the interior, Mr. Alamán in his excellent report to the Congress from 1823, "deserve to be considered from another political point of view than they have been until now. The wide-ranging commerce of which these provinces will one day be the center due to the quantity and richness of their agricultural products; their natural ability to provide a home for a national navy; and the eagerness with which they are coveted by European powers must demand the attention of the Mexican government. If the ministering work of the missions may be considered the most appropriate means of removing from barbarism the savages who roam the woods with no intellectual idea of religion and culture, we must never forget that such ministering can serve to provide only the foundations of society without necessarily leading men toward spiritual perfection. The Indians must be bound to the earth by making them independent proprietors and by distributing land to them; the two Californias must be populated. The thirty-six Fernandist priests who are in charge of the missions of Alta California enjoy annual revenue (*Sinodo*) of only four hundred piasters, which has been rather inadequately paid of late. In Baja California, the missionaries' income is only three hundred and fifty piasters." The number of clergy has decreased noticeably since I left Mexico. Here is the situation in the thirteen provinces, as published by the minister of religious services.   II.336

Before the Republican government had been established, the monasteries loaned the State these funds in the amount of more than three million piasters. The number of convents throughout Mexico is only fifty-seven, five of which are of [La] *Concepción*, in Mexico City, Puebla, Mérida, San Miguel el Grande, and Oaxaca.   II.337

MONKS OF THE FIVE ORDERS—1822

PROVINCE	LOCATION	MONASTERIES	MONKS	CAPITAL IN PIASTERS
I. Dominicos				
Santiago de Predicadores	Mexico City.	10	130	144,074
San Miguel y Santos Ángeles	Puebla.	6	55	10,900
San Hipólito Mártir	Oaxaca.	6	52	81,687
II. Franciscanos				
Santo Evangelio	Mexico City.	20	310	148,531
San Diego	Idem.	14	232	32,990
San Pedro y San Pablo de Michoacán	Querétaro.	15	133	411,985
San Francisco los Zacatecas	Potosí.	11	157	235,615
Santiago Jalisco	Guadalajara.	7	133	
San José de Campeche	Mérida.	4	79	61,562
III. Augustinos				
San Nicolás de Michoacan	Salamanca.	11	87	562,563
Dulce nombre de Jesús	Mexico City.	11	134	216,532
IV. Caramelitas				
San Alberto	Idem.	15	243	919,709
V. Mercedarios				
San Pedro Nolasco de la Visitación.	Idem.	19	186	224,430
	Total 13.	149	1,931	3,050,578

Four of *Santa Clara*, in Mexico City, Puebla, Atlixco, and Querétaro.
Two of *Encarnación*, in Mexico City and Chiapa.
One of *Santa Rosa María*, in Puebla.
Four of *Santa Catalina*, in Mexico City, Puebla, Oaxaca, and Valladolid.
Five of *Santa Teresa*, in Mexico City, Querétaro, Puebla, and Guadalajara.
One of *Nuestra Señora de Soledad*, in Puebla.
Two of *Jesús María*, in Mexico City and Guadalajara.

Two of *Santa Inés*, in Mexico City and Puebla.

One of *Santa Isabel*, in Mexico City.

One of *Regina*, in Mexico City.

One of *San José de Gracia*, in Mexico City.

One of *San Juan de la Penitencia*, in Mexico City.

One of *Santa Brígida*, in Mexico City.

Two of *la Balvanera*, in Mexico City.

One of the *Santísima Trinidad*, in Puebla.

One of the *Enseñanza*, in Mexico City.

Four of the *Enseñanza de Índias*, in Mexico City, Guadalajara, Aguas Calientes, and Irapuato.

One of *San Lorenzo*, in Mexico City.

Two of *San Gerónimo*, in Mexico City and Puebla.

Three of *Santa Monica*, in Puebla, Oaxaca, and Guadalajara.

One of *Santa María de Gracia*, in Guadalajara.

One of *Nazarenas*, in Celaya.

One of *Capuchinas*, in Mexico City.

Ten of *Capuchinas de Indias*, in Mexico City, Oaxaca, Villa de Guadalupe, Querétaro, Villa de Lagos, Valladolid, Guadalajara, Salvatierra, and Puebla.

The minister of religious orders was aware only of the funds of twenty of these convents and the number of nuns in thirty of them. The capital funds (*capitales impuestos*) were 5,200,000 piasters; the number of nuns was 962. According to this count, it would appear that the population of the convents is only slightly different from the population of the monasteries, and that the total number of monks and nuns is approximately 3,800. I estimated this number to be four or five thousand for 1803. The Mexican clergy today, even including the *donados* and *legos*, probably numbers only 8,000 individuals; in other words, from one and a quarter to one and a half per one thousand members of the population, whereas in Spain, they number more than ten to twelve out of every thousand. According to an official notice that includes the census taken in 1786, there were in Spain at that time 57,533 individuals in monasteries (to wit, 37,520 monks by profession, 7,862 lay brothers, 4,225 *donados*) out of a population estimated at 10,409,877 souls; 33,630 individuals in convents; and 86,546 among the secular clergy. These numbers are slightly different from the figures that ▼ Mr. de Bourgoing and Count Alexandre de Laborde published. Among a population of 3,173,000 in Portugal (1822), there were nearly 27,000 secular and regular ecclesiastics of both sexes, in other words, nine out of every thousand members of the total population.

II.338

The province of *Chiapas* was recently separated from the territory of Guatemala and annexed to the Mexican confederation. The cool and temperate part includes the environs of Ciudad Real, Chiapa, Tuxtla, and Ocosocontla; the warm part includes the Tonalá and Maquilapa seaboard on the Pacific Ocean. The beautiful indigo that takes the commerical name of Guatemalan indigo is grown in the latter region. Ciudad Real, the capital of Chiapas, has 8,000 inhabitants; Tuxtla, where there is a very active cacao and tobacco trade, has almost the same population.

Northeastern borders. Despite Article 4 of the Washington treaty of February 22, 1819, the borders between the Mexican confederation and the United States remain undefined. The population of the United States is advancing only very slowly west of the Ozark Mountains and toward Taos in New Mexico; according to very precise information from Major Long (*[Narrative of an] Expedition*, vol. II, p. 361 and 388), the country between the Rocky Mountains and the *Ozark Mountains* as far as the meridian of Council Bluff (41°25′ latitude, 98°3′ longitude) is highly unsuited for agriculture because of the lack of timber and water. *The Great Desert* (▼ Nuttal, *[A Journal of] Travels*, p. 120) extends east of the granite peaks of the Rocky Mountains—Spanish Peak, James's Peak, Long's Peak—between the rivers of Canada, Arkansas, and Padouca from 36° to 41° latitude. This strip of arid land forms a border between the Protestant States of the United States and the Catholic States of the Mexican Confederation, which is difficult to cross. There is closer contact south of the Ozark Mountains, between Texas and Louisiana, and San Antonio de Bejar and Natchitotches, and the fertility of the soil does not prevent a complete joining of the two countries.

Before there was more specific information about the progress of Mexican society and civilization, one often compared the relative population of this country with that of Russian Asia. The following information, taken from recently published official documents (["Bruchstück aus der neuesten Reise nach China durch die Mongolei,"] *Petersburger Zeitschrift*, 1823, *June*, p. 294), proves the vague and erroneous nature of these comparisons. The size of Russian Asia, taking the Kara, the Ural Mountains, and the Iaik as the western border, is 465,000 nautical square leagues (at twenty to a degree) and has a population of two million inhabitants at most. Its surface area is thus six times that of Mexico, and its relative population is twenty-two times as small. The most populous Siberian province, Tobolsk, has only 570,000 inhabitants, whereas the Mexican State alone (the former intendancy) has 1,777,000. The population of Siberia, however, grows very fast in the western provinces: for example, in Tobolsk where the ratio of births to deaths is twenty to eleven. In the southernmost parts (the governments of Tomsk and Yeniseisk), the same ratio is twenty-six to eleven;

II.339

II.340

in the eastern parts (the governments of Irkutsk and Yakutsk) it is eight to five. The most peopled towns of Russian Asia are now Tobolsk, which has 16,700 inhabitants; Irkutsk, which has 11,100; Tyumen, which has 9,900; and Tomsk, which has 9,700. The capital of the Mexican Confederation has a population of 168,000 and five towns with more than 40,000 inhabitants! In Asiatic Russia, the indigenous peoples, formerly called Felltributpflichtige (those subject to a tribute of pelts), are now officially and even more strangely, called Fremdstämmige (those of foreign *race or origin*), and their ratio to Russians is one to two and three-fifths. In Mexico, the ratio of pure indigenous peoples [de race pure] to other classes, either of Spanish origin or from mixed castes, is one to eight-tenths.

As it is with any politico-economic material, numbers become instructive only by comparison with corresponding facts. Therefore, I have carefully examined, with respect to the present situation of the two continents, what may be considered a small or very middling *relative population* in Europe and a very sizable *relative population* in the Americas. I have chosen examples only from provinces with a continuous surface area of more than 600 square nautical leagues in order to exclude the *incidental increases* in population that one finds around large cities, for instance, on the coastline of Brazil, in the Valley of Mexico, on the plateaus of Bogotá and Cuzco, or in the Archipelago of the Lesser Antilles (Barbados, Martinique, St. Thomas). There, the *relative population* ranges from 3,000 to 4,700 inhabitants per square nautical league and thus equals the population of the most fertile parts of Holland, France, and Lombardy (*Relation historique*, vol. III, chap. 26, p. 95).

LEAST POPULOUS GOVERNMENT DISTRICTS OF EUROPE		MOST POPULOUS GOVERNMENT DISTRICTS OF THE AMERICAS	
The four least populous government districts of *European Russia*:	by square league	*Central part* of the States or old intendancies of *Mexico City* and *Puebla*	by square league 1,300
Archangel	10	In the *United States*, Massachusetts, despite its surface area of only 522 sq. lg	900
Olonets	42		
Vologda and Astrakhan	52	*Massachusetts, Rhode Island,* and *Connecticut* combined	840

(*continued*)

II.341	LEAST POPULOUS GOVERNMENT DISTRICTS OF EUROPE		MOST POPULOUS GOVERNMENT DISTRICTS OF THE AMERICAS	
	Finland	106	The entire intendancy of *Puebla*	540
	The least populous province of Spain, Cuenca	311	The entire intendancy of *Mexico City*. The two states (formerly intendan-	460
	The duchy of *Lune- burg* (because of the heaths)	550	cies) of Puebla and of Mexico City are, together, nearly one-third the size	
	The least populated department of continental *France* (Hautes-Alpes)	758	of France and have popu- lations (in 1823, nearly 2,800,000) that are large enough for the big cities	
	Other departments in France with relatively low populations (Creuze, Var, and Aude)	1,300	of Mexico and of Puebla not to have a notable effect on their relative populations	
			Northern part of the province of Caracas in Colombia (excluding the *Llanos*)	208

When I succinctly described (on p. 383) the pyramid of Cholula as it had been preserved in 1803, I compared its construction to the tiered pyra- mids of Sakkara, according to the very partial information that we have about the Sakkara group. Southeast of the village of Sakkara, there is a five-tiered stone pyramid commonly called Mastaba al Faraun. Near Dahshur, there are three five-tiered pyramids built of bricks that contain broken straw. The great pyramid of Sakkara, called Harem al Kebyreh, has a base measuring 618 feet and a height of 316 feet ([France,] *Description de l'Égypte, Antiquités*, vol. II, p. 4 and 6).

II.342   One cannot recommend enough that travelers to Egypt study the nineteen pyramidal monuments at Sakkara, Dahshur, and Abusheer, whose constructions are all the more worthy of attention for greatly resembling the Mexica pyramids. In the Ghiza group near the fourth pyramid, one even finds a four-tiered building divided into steps (*Loc. cit.*, vol. V, plate 26, no. 14).

     Mr. Bullock, whose ever so praiseworthy work recently drew the attention of European scholars to American antiquities, rightly observes that the view I gave in plate seven of my *Views of the Cordilleras* of the church with *two* towers atop the pyramid of Cholula, is inexact ([Bullock,] *Six months' residence*, p. 115). I drew only a sketch of this plate to remind myself of the circumstances surrounding a trigonometric measurement, and so the lines of the cypress trees and the church itself were outlined only roughly.

The study of ancient Mexica monuments, so closely connected to the study of the earliest human civilization, will profit significantly from a more detailed examination of the monuments of Guatemala. One finds there the most astonishing remains of Toltec and Aztec sculpture, as proven by Captain Antonio del Río's work entitled *Description of the Ruins of an Ancient City Discovered Near Palenque*, published in London in 1822. This flowerdecorated cross (plate eight), before which a child is presented, is nearly identical to another relief work from Palenque, of which I have a drawing, in which there appears to be an actual Adoration of the Cross. Are these the crosses of which the first *conquistadores* spoke, not to be confused with the ones that show the marks (the path) of the sun during the equinoxes, solstices, and through its zenith [?] (See my *Views of the Cordilleras*, plate 37, no. 8). Even if the correct proportion of the human figures in Palenque contrasts markedly with the squat, unrealistic figures in the II.343 paintings found in Mexico, the stylistic resemblance of secondary details—for example, the enormous aquiline noses, the perpendicular bars accompanied by large dots (disks) inserted near the head of an animal (plate 45)—is no less striking (*Views of the Cordilleras*, plates 15, 27, and 37). A very well-read Persian, to whom I showed the last plate from the precious manuscript preserved in the Dresden library, at first believed that he recognized figures of Asian geomancy, ▼ *'ilm al raml* (Wilkins, [*A Dictionary Persian, Arabic and English*] 1806, vol. I, p. 482). But there is no doubt of the Mexica origin of the Dresden manuscript, and one finds traces of the same *raml* among the bas-reliefs on the ruins of Palenque in Guatemala, which are a Toltec construction. Also very recently, I was shown ▼ another geomantic tableau purchased in Madrid, which is nearly identical to the one in Dresden, drawn on Agave Americana paper, as are most of the historical, ritual, or astrological books of the ancient Mexica. The most famous ruins preserved in Guatemala are (1) those of the ancient *Ciudad de Palenque* or Colhuacan (province of the Tzendales, the former intendancy of Chiapas), near Santo Domingo de Palenque, which we discussed above; (2) the ruins of Tulha near Ocosingo; (3) the ruins of Copan, commonly called the *Circus maximus*, with pyramids and remarkably clothed, medieval looking human statues; (4) the ruins in the cavern of Tibulca, with its columns fitted out with capitals; (5) the ruins of Utatlan near Santa Cruz de Quiche (province of Sololá), with an immense Toltec palace; (6) the ruins of Petén island, province of Chiapas (Juarros [y Montúfar], *Compendio de la historia de[la cuidad de] Guatemala*, vol. I, pp. 15, 33, 44, 67). I regret not yet having been able to find the *Preambulo de las Constituciones diocesanas* by the Bishop of Chiapas, ▼ Francisco Nuñez de II.344

la Vega, to verify what that prelate suggests regarding the Votan or Wodan of the people of Chiapas who gave his name to a short day, as Buddha and Odin have done for *Boud-var* and *Wodansdag* (see my *Views of the Cordilleras and Monuments of the Indigenous Peoples of the Americas*, octavo edition, vol. I, p. 386; vol. II, p. 356) [pp. 173–175 in the 2011 English edition].

Volcanoes of Mexico. I have already reminded readers of the first edition of this work (quarto edition, vol. I, p. 80) that in clement weather, the inhabitants of Mexico City and Jalapa never see smoke rising up from the craters of Popocatepetl and Orizaba Peak. Only during a time of great eruptions, like those in 1540 (Gómara, *Historia de Mexico*, 1553, folio XXXVIII), were jets of flames and ashes from Popocatepetl visible from the very shores of Lake Texcoco. If these immense eruptions, *visible from a distance*, are as rare in Mexico as they are in the Andes of Quito, it is no less likely "that the crater of Popocatepetl is constantly aflame and that, as it spews smoke and ashes, its mouth opens wide in the midst of perpetual snows" (see vol. I, p. 166 [p. 113 in this edition] and above, p. 40) I offer here an excerpt from my ▼ *Tetimpa Diary* (near San Nicolás de los Ranchos, on January 24, 1804, the day when I attempted a trigonometric measurement of Popocatepetl): "The snow line is easily reached from the south by traveling via Cuautla de las Amilpas, especially via San Pedro Lliyapa, where there is less shifting sand of pumice stone and more solid rock. Popocatepetl consistently has less snow on its southern and southeastern side where winds of warm air from Xochimilco and Atlixco, *las tierras templadas y calientes*, blow on the mountain, than on its northern side, where it is cooled by the snows of the Sierra Iztaccihuatl. The indigenous peoples claim that one smells the sulfurous vapors of the crater before arriving at the perpetual snow line. Through a telescope one can see very clearly that the crater faces southeast, such that its opening is not easily discernible from Mexico City. The mouth of the volcano is surrounded by a ring of ashes and grayish snow that appears to be moistened by the sulfurous vapors and is frozen here and there into ice. Popocatepetl forecasts the weather for the farmers in the surrounding area. When black smoke arises from the crater at sunset and condenses into thick clouds edging northward, the Indians expect rain. If the smoke faces southward, it will be cold and frosty. A straight column of smoke proclaims wind. Two or three hours before the storm breaks over the Tetimpa plain, fountains of pumice stone are seen bursting forth from the crater. The stones roll like sand over the slope of the Volcano which is covered in thick layers of snow. The earthquake and great storm of January 13, 1804, were preceded by eruptions of ashes and smoke. Rather strong tremors were felt in Mexico City. The Indians claim that smoke is not seen in the morning but usually between four and six in the evening, especially at sunset. Would this not be for optical reasons? The eruptions of ashes and smoke are most frequent in May, when the entire summit of the Volcano sometimes appears to reflect a

II.345

yellowish light probably caused by the sulfurous vapors. No fire has been seen at night in recent times. The moment we arrived at San Nicolás de los Ranchos, a column of rather thick smoke rose up from the mouth of the Volcano around five thirty in the evening. Mr. Bonpland thought he could make out a cloud of ashes moving southeast."

I think that these *supplementary notes* to the Statistics of Mexico will be more interesting if I end them with a description of the ports on the eastern and western coasts written there in 1824 by a ▼ sage traveler who was most able to appreciate the benefits of trade relations between Europe and the Confederation of Mexican States.

II.346

"*Pueblo Viejo*, generally known as *Tampico*, is situated on the shore of the large eponymous lake and built on a hillside that increases the natural heat of the climate in summer due to the reflection of the sun's rays. Nearby, to the south at the bottom of this elevation, is the Laguna de Tamiagua that connects with the Laguna de Tampico via streams that flow into the Panuco River. Pirogues can navigate this lake and make connections with Tuspan possible. Pueblo Viejo has approximately 2,000 inhabitants of which the majority are Indians; with the exception of eight or ten stone houses, all the dwellings are built of wood and covered with palm fronds; most of them are huts open to the winds.

"Several river branches, or *bayous,* that are traversable by the large canoes create small islands at the point of confluence and connect the Laguna de Tampico to the Panuco River. On the banks of one of these branches, there is an elevation that overlooks the sandbar and the town as well as all the surrounding area. This place is called the *Mira*, and a lookout has been established there. Various *bayous* empty into the Panuco River opposite Tampico de Tamaulipas, where the river looks like a proud, broad bay, and where the ships that have crossed the sand bar drop anchor.

"When strong winds from the northwest or west sometimes drive the Ocean back from the coastline, the waters of the Laguna de Tampico retreat almost completely, to the point where a rotting mass of beached fish infects the air and makes any stay in Pueblo Viejo unpleasant and injurious. The insalubrity of the town is also the unfortunate effect of other, more general causes, since besides the *vómito*, to which foreigners and inhabitants of the hinterland who travel to Pueblo Viejo in the summer are also susceptible, almost all inhabitants are attacked by intermittent fevers until well into the fall (*Tercianas* and *Frios*). Consequently, it is not possible to do business in the summertime; traders from the hinterland do not come down from the plateau. There are almost no deliveries, and the upper classes take refuge from diseases in Altamira, which is only twelve leagues away.

II.347

"Tampico de Tamaulipas, formerly quite populous, betrays almost no vestige of the dwellings of its former population, which abandoned the place before settling in Pueblo Viejo. But easier connections with the interior and several other circumstances contributed in favor of rebuilding Tampico de Tamaulipas. The Panuco River forms the natural boundary of the former provinces of Veracruz and Nuevo Santander. Pueblo Viejo is very close (one and a half miles) to the river and adjoins the right bank that corresponds to Veracruz, while Tampico de Tamaulipas is on the left bank, which corresponds to Nuevo Santander. The customs office in Pueblo Viejo, which is supervised by Veracruz, insists upon its exclusive rights to anything that is imported through this point on the coastline. The government of New Santander objects, insofar as the state of Veracruz receives ample benefits from the customs offices in the south, whereas the one in New Santander has no other means of remunerating its expenses. A customs office has been established in Tampico de Tamaulipas, where many houses are being built, and where trade and the population of Pueblo Viejo will soon move.

II.348

"The Panuco River, on which one travels upriver for five miles before reaching a mooring, is nearly one-third as wide as the Mississippi and presents no danger inside the sandbar. It is from five to six fathoms deep everywhere. Buildings that collapse on the riverbanks after an accident only fall onto the mud and are not in any danger, since the force of the current is never more than three or four knots. With the exception of the heights of Mira and Tampico, the land along the Panuco River is low and marshy. The sandbar in the river, like the ones in all the rivers along the Gulf of Mexico, is a great obstacle to navigation. The Alvarado and Panuco sandbars are subject to an extraordinary reflux of the river water against the Ocean waters; despite the sandy bottom, shoals along their banks increase this danger. It is generally said that a ship cannot draw more than eight feet of water before crossing the Panuco or Tampico sandbar. I have seen ships that drew ten feet come in, but the departure of other ships was delayed, although they drew only six feet of water. These variations are not periodical in the least, since they depend as much on the influence of sea winds as on rains which, by increasing the strength of the current, carry away sands that block passage and occasionally carve a deeper canal, if only for a very short while.

"Ships that draw more than eight feet of water remain anchored at twelve fathoms in the harbor two or three miles from the Panuco sandbar. This harbor affords no protection from the northwest winds that blow fiercely from October to March; one must take advantage of these often very brief intervals to unload merchandise by means of longboats (*barges*) that transport cargo to Pueblo Viejo. Although the buffets of northwest wind are most violent, they are nevertheless not dangerous when one is always careful to be ready to raise anchor so as not to be forced to cut one's cables; the winds that dominate these areas at those times keep small craft away from the coastline.

"Ships arriving in Tampico harbor give a signal so that the pilot may assign them an appropriate anchorage in the harbor or allow them to enter. If they approach the harbor at night, they usually find lookouts or fires near the sandbar that serve as points of orientation.

II.349

"Signals are given from an old fort at the sandbar on the right bank of the river. The anchorage at Tampico de Tamaulipas is five miles from the sandbar; a mile and a half separates this anchorage from Pueblo Viejo. Like the coast of Veracruz, the entire coastline is composed of shifting sands pushed back by the sea and piled up by winds that create rather high elevations called *meganos* (dunes).

"The Panuco River is passable far above the eponymous town that is located ten or twelve leagues from Pueblo Viejo; some ships sail up the river to load yellow wood or salted meat (*tasajo*). Although the source of the river or, rather, one of its tributaries (the Río Moctezuma) lies in the mountains near Mexico City, it is of no use for transporting goods intended for the capital city. The Río Panuco itself comes from San Luis Potosi to the west, and where it is navigable, even by large dugouts, it flows away from the direct route rather than closer to it.

"*Tuxpan*. Foreign vessels do not put in at this small port located between Tampico and Veracruz, because it is almost uninhabited, and connections with the hinterland are easier through one or the other of the two nearby ports. It appears that the sandbar in the Río Tuxpan suddenly deepened last year, and that a canal eighteen palms deep opened up. We do not know if this canal still exists, but one may assume that the sea would not have held off before destroying what the currents had accidentally produced.

"*Soto la Marina*, located on the Santander River north of Tampico, presents the same obstacles to large vessels that must anchor some distance away in the open sea and the same security to small boats that find a good port after crossing the sandbar. The two sandbars of Tampico and Soto la Marina are indistinguishable from each other as far as coastal navigation is concerned: the description of the one provides a rather specific idea of the other.

II.350

"*Guasacualco* or *Huasacualco*. The sandbar in this river is located in that part of the Gulf of Mexico which is approximately thirty leagues ESE of Alvarado and thirty-five leagues WSW of the Tabasco sandbar. It is the best port that the rivers emptying into the Gulf of Mexico offer, including the Mississippi, since Pensacola is, as we know, located on an actual bay. Frigates can always enter Coatzacoalcoss because there are always eighteen to twenty feet of water on the sandbar. This superb river presents no danger for navigation. It has a muddy bottom and a low, marshy shoreline. Vessels can sail eighteen leagues upriver, but they stop at Paso de la Fábrica, which is eight leagues from the sandbar, and their cargo, if it is intended for the interior, is transported by dugouts as far as Paso de la Puerta, fifteen leagues upstream, where this river ceases to be passable by pirogue.

"The important port of Guasacualco, which offers the easiest connections[1] with a part of the province of Chiapas, Oaxaca, and the Eastern part of the province of Veracruz, is, nevertheless, uninhabited. It is located in a frightful desert where the wild animals of Mexico have taken refuge because humans rarely disturb their tranquility there. The position of Guascualcos is even more advantageous, especially in the naval context, since it is both an excellent port and a place where it is easy to obtain the best construction lumber. It has already attracted the attention of the Government which has received ▼ Don Tadeo Ortiz's reports; for the past two years, this citizen has generously devoted his time to surveying the river and exploring the surrounding country with the intention of establishing a colony. Don Tadeo's hut is located near the Paso de la Fábrica on high ground formed by several small hills. This is the only recognizable high land after the sandbar, which appears to have all the necessary elements for creating a small town in this locale. It would be easier to drive away the jaguars and other wild animals there than the *mosquitoes* that infest the country.

"What is commonly called the Fábrica is a decrepit shanty covered with palms which serves as a lazar house for travelers and a shelter for merchandise. Half a league upstream, one comes to the farm or Rancho de Tlacosulpan, where there are dugouts available for traveling on the nearby rivers. With the exception of Don Tadeo Ortiz's hut, this is the only inhabited place to this point on the Guascualcos River. A fort located at the sandbar defends the entrance to the river.

"The Río Uspanapa flows into the Guascualcos half a league below the Paso de la Fábrica. This river, like the Guascualcos, is at least half or two-thirds as wide as the Mississippi. Passage up the Uspanapa for approximately fifteen leagues to go toward Tabasco, is assured.

"The inhabited places closest to Guascualcos are, to the west on the Acayucan road which is fifteen leagues away, Chinameca, Cosoleacaque, Jaltipa, and Soconusco, where very industrious Indians live who make all kinds of cotton fabric, toiles, and *Pita* (Agave) ropes, like the ones from Campeche, which are in great demand.

"*Villa Hermosa de Tabasco*. This town, improperly called Villa Hermosa, is built on the left bank of the Tabasco, or Guichula, river, twenty-four leagues above its mouth. It is the governmental seat of the state of Tabasco whose entire

II.351

II.352

---

1. Compare this with Mr. Robert Birks Pitman's decisive observations in his *Succinct View on the [on the practicability of joining] the Atlantic and Pacific*, 1825, pp. 67|91.

population is only 75,000, and whose principal industry is growing cacao of excellent quality, which is indigenous to this province.

"Villa Hermosa has nearly 5,000 inhabitants; there are many stone houses, but most houses are constructed of bamboo and palm leaves. Its position controls the trade of the provinces of Chiapas (part of the Mexican federation) and Guatemala. Despite its distance, it even has ties to the province of Oaxaca through the cacao trade and part of the other products it exports. These various connections sometimes take place via the Tabasco River, which is navigable by dugout canoe as far as Quichula seventy-five or eighty leagues above Villa Hermosa. But this route presents real and great dangers because, for most of its course, it is enclosed between the mountains and resembles a torrent more than a peaceful river. The first French traveler to travel up the river perished upon his return; the nearby mountains are so steep that mules, despite their sure-footedness, are of no use for transporting goods, which are carried only on the Indians' backs.

"The Tabasco River near Villa Hermosa is nearly two-thirds as wide as the Mississippi. Its mouth has two branches: one to the northwest, the other to the northeast. The first is the deepest, measuring between twelve and fourteen palms, which corresponds to ten or eleven French feet; the second is only seven feet deep and only serves as an entrance for small boats. Small craft that arrive there can enter when the *Nortes* blow less violently across the northwest mouth of the river. If the wind is in the east or northeast, they must hug the coastline of Campeche and avoid falling under the wind. When the ship draws only seven feet, it can enter through the small harbor channel with these same winds; but if it draws any more, it will sail along the island that encloses the two branches and, after rounding the point, it will drop anchor inside the sandbar itself without the slightest danger. If the ship wants to sail farther, it will have itself towed by cable as far as the fort where it will again drop anchor. It is possible to sail upriver half of the distance to Villa Hermosa with some wind from the north, because its northeast direction is rather direct to this point. But farther along, the river meanders, and one must sail against a strong current, either being towed by stringing landlines or by tying up to a tree, as is done on the Mississippi. One reaches the Escobas fort eight leagues before arriving at the town. The river has a muddy bottom everywhere. Depending on its high water level, the river is five to six fathoms deep opposite Villa Hermosa, and deeper by up to ten or fifteen fathoms. The fort built at the sandbar is called San Fernando. The eponymous village lies one league upstream and offers very good anchorage.

"Along the coastline of Tabasco, the sounding line extends nearly as far as on the coastline of Campeche. Ships sailing toward one or the other of these points find an anchorage sheltered from winds either from the south or the wind from

II.353

the small islands located twenty-six or thirty leagues off the coastline, for example, near the Arcas. There is a sandy bottom. If the ships cannot rest on their anchors, they will remain under sail, protected by the islands.

II.354   "Alvarado, twelve leagues southeast of Veracruz, is a place more or less like Pueblo Viejo de Tampico: there are, however, considerably more brick houses that have been built mostly since trade with Veracruz was established. Before that time, Alvarado was only a sad village, and even now, its filth, humble thatched cottages, and the droves of donkeys that roam its streets stand in marked contrast to the commercial importance that this place has acquired. About three thousand inhabitants, including many foreigners, are housed in a very tight space. Housing is overpriced, because any place that is at all spacious is used as a storehouse for merchandise. Despite the merchants' hope of regaining their lodgings in Veracruz soon, there has been considerable new construction here.

    "The town of Alvarado is built on the left bank of the eponymous river at a distance of approximately one and a half miles from its sandbar, surrounded by sandy hills that the oldest inhabitants recall having often seen change shape and place. In 1824, diseases began to appear only in the month of August. Before then, there was not one sick person among the foreigners. But although the *vómito negro* was devastating later on, one may believe that the air is slightly less unhealthy in Alvarado than in Havana and Veracruz. Intermittent fevers like the ones in Tampico are also not found there.

    "From its mouth or sandbar as far as the town, the Río Alvarado is slightly less than half a mile wide. Farther upstream, it broadens as several other streams enter it and forms a wide bay that is two miles wide and five or six miles long. As we have seen above, *méganos* [dunes] cover the land on the left bank where the town is; but there is good soil below the layer of sand at the end of the bay. Large trees and excellent vegetation prove that if the inhabitants were not so lazy, they might develop the area around Alvarado.

II.355   "The right bank is a low, marshy plain covered with forests submerged in the rainy season. Beautiful rivers that flow down from the Sierra until they connect near Alvarado crisscross this immense plain in all directions; it extends as far as the mountains of Oaxaca thirty leagues away. The riverbanks, which are always higher than the nearby land, are the only inhabitable areas. The land has immense lagoons that are always filled with aquatic fowl.

    "The rivers that join near Alvarado bay are passable by schooner for fifteen to twenty leagues, when they begin to traverse higher land called *Llanos*, savannahs that extend as far as the mountains, where animals take refuge from flooding that lasts all summer. The current becomes much stronger during the rainy season, but the water level drops during the dry season, so much that the rivers are only navigable by pirogues that sail upriver thirty-five to forty-five leagues from

Alvarado. Thus the San Juan River is traversable as far as the eponymous *Paso* eight leagues from Acayucan on the Guasacualco road; the Tesechoacan River is navigable as far as Playa Vicente at the foot of the Oaxaca mountains; the Cosamaloapa [Papaloapan] River as far as Santuario near the same mountains; the Tuxtla River, as far as the small eponymous town located on the slope of the San Martín mountains which are the only ones on the shores of Mexico—they run between Guascualco and Alvarado. There are places of considerable size in the middle of this desert, like Tlacotalpan, which is eight leagues away, Cosamaloapa, twenty, Tesechoacan, twenty-five, and Acayucan, forty-five leagues from Alvarado. The first two and the last of these places are very well built and have many brick houses. Their white population is comprised of those who, because they are civilized, supposedly were not brought up in these desert regions. The Indian farmers, who are the most numerous group, are honest, industrious, and hospitable. Their character contrasts markedly with the character of the class of peasants called *Jarucos* or *Vaqueros* here, whose principal occupation is raising livestock. Their insensitivity, pride, and bad faith are only equaled by their laziness. The women of the Guajiros Jarucos are active and hard-working. Devoted, like the Indians, to industry, farming, and more agreeable occupations, they are honest and very good-natured. The Jarucos spend their lives on horseback, either moving about or chasing and deftly lassoing the wild bulls that roam on the *Llanos*. The peaceable manners of the Indians from southern Mexico, who support themselves through agriculture and the fruits of their labor, contrast with the fierce, uncontrollable character of the northern Indians whose only nourishment is the animals they hunt and whose sole enjoyment is fighting.

II.356

"After the sandbars of the Guascualco and the Mississippi, the Alvarado sandbar is the deepest one in the Gulf of Mexico; it can accommodate ships that draw ten, twelve, and even fourteen feet of water. The latter, however, are often delayed upon entering or leaving the bar, as they await a tide that will allow them passage.

"Frigates and ships that draw more than fourteen feet of water and those that cannot or do not wish to enter the river with such a high load waterline remain anchored one, two, and even three miles out in the harbor; as in Tampico, they load and unload their cargo by means of *lighters*. This sandbar is even more hazardous for small craft than the one in Tampico, especially if they are full, and life is at risk if one tries to cross it while a strong wind is blowing. The best time to go ashore or unload merchandise is during the morning calm. The sea breaks violently upon the reefs that line the bar. There is no more danger beyond the passageway, and one finds water deep enough for dropping anchor.

II.357

"Ships may anchor opposite and very close to Alvarado. The largest ones unload their cargo by means of dugout canoes; others approach a small jetty (*muelle*) that was recently built and that projects out into the river, allowing ships to unload quite easily; others ultimately approach the riverbanks to unload by means of a bridge. Pilots are very attentive to the signals they receive while in the harbor. Because of its position, this harbor is more dangerous in a strong wind than Tampico harbor. There, gales of northwest wind tear ships away from the coastline and leave them to sail up and down the entire gulf. Even in Veracruz, ships have a certain safe distance; but since the Alvarado harbor is at the bottom of the gulf, northwest winds blow small craft toward the coastline. Only the gulf of Guascualco is slightly farther south than Alvarado. We emphasize these circumstances to show that the harbor of Alvarado can become dangerous in strong winds, although it is safe in ordinary weather; here, much more than in Tampico, one must always be prepared to take to the high seas. The Alvarado sandbar is defended by a battery.

"*The Laguna de Términos.* This large lagoon is located approximately fifteen leagues eastward of the Tabasco sandbar and twenty-five or thirty leagues south-southwest of Campeche. It is fifteen leagues long and ten wide; it connects with the sea and the Tabasco River through several navigable channels.

"One can sail up the Palizada River by dugout canoe before entering the bayous that terminate at the Tabasco River. The two main islands in the Laguna de Términos are Laguna and Puerto Real: they are home to two villages by the same names whose inhabitants work mainly in the dyewood [logwood or bloodwood] trade. The large water passage is twelve to thirteen feet deep, and since the bottom is silt, ships sunk six inches in the mud can emerge without any danger. Leaving the anchorage at Laguna, one travels due east toward the mainland until one finds the silty bottom; then one turns northward, moving away from the mainland, because the Laguna sandbars extend nearly thirty-four miles northward. One takes soundings constantly as one travels westward, and whenever less than three fathoms and a sandy bottom are reached, one turns east again. Conversely, one turns west again as soon as one hits a rocky bottom. One must take care to remain in the silt, which unmistakably marks the truly navigable channel. The passage at Puerto Real is only eight palms deep, which corresponds to five and a half feet; one should not risk attempting to enter on a northern wind, even with wind in the sails, because the sea is very high in this area.

"*Tehuantepec,* located on the eponymous isthmus five leagues from the coastline of the Pacific Ocean and twenty-eight or thirty leagues from the Paso de la Puerta, is the first navigable point on the Guascualcos River that flows into the Atlantic. With a population of 14,000, Tehuantepec is comprised of a considerable number of very respectable white families; the great majority of the population, however, is Indian. Its inhabitants are the most active in New Spain and are more industrious than one would expect in a country known to be among the hottest in America. The Tehuantepec River crosses the town, which is nestled

II.358

against the hills. These are the only hills to be found for several leagues; all the nearby countryside consists of a vast sandy plain, although it is watered by streams and irrigation. The town is composed of five or six different districts that resemble as many villages separated from each other by low rises in the terrain, in such a way that it is not possible to see the entire town at once from any vantage point. Many of the streets where the white population lives are paved with stone; churches and public buildings are well-constructed, whereas the Indian neighborhoods are built of bamboo and palm fronds. The inhabitants overall are admirable for their mild manners and amiability. It is so hot in Tehuantepec that even before daybreak, mass is celebrated under a trellis attached to the church. The north winds bring neither rain nor freshness. Despite the marshes and lagoons that surround this place, the climate is very healthy. The inhabitants are not bothered by any type of venomous insect, not even by mosquitoes. It would be difficult to attribute this salubriousness to the absence of foreigners as its only cause, since Tehuantepec is undoubtedly a healthy place for the inhabitants of the Mexican plateau. One sees many Indians accustomed to the temperate climate in the mountains flocking there. They contract none of the diseases there that kill them on the coastline northeast of Acapulco.

II.359

"The inhabitants of this district cultivate cochineal, but even more especially they grow indigo. There is also a large trade in salt and dry fish. Pearl fishing, so well known earlier, has been abandoned. The Indians seeks out only a shell that produces a deep purple color, which they use to dye cotton. They make all types of local silk cloth.

"The closest point to Tehuantepec on the coastline is the Morro: a lookout has been stationed there. To go to the Morro, one crosses the river and passes through the Hacienda de San Diego, commonly called Soleta, which is three leagues away in a sandy country covered with stagnant pools of water. Only one mile separates San Diego from the Morro; the road winds between low mounds of sand and small rocks. It is also possible to go to Morro through the Indian village of Vilotepec, four leagues from Tehuantepec, by following the left bank side of the river; the ground is very muddy here. The Morro is one of the summits of a small mountain that is three times higher than the Morro in Havana. The bay is open to the south and offers hardly any shelter to ships.

II.360

"Since there are no barges or large dugouts on the coast, merchandise can be unloaded in Morro Bay only by average-sized canoes. One must take advantage of the morning calm and bring merchandise ashore near the Morro, where the sea breaks less in windy conditions and very little during a calm spell.

"It is extremely rare that ships arrive on this coastline and send small craft ashore in the Morro bay to collect supplies. On the shore of the South Sea, shepherds recently discovered enormous pieces of iron-shod wood almost buried in the sand, which drew their attention because of the value of iron in this country. This marine debris most likely came from the wreck of a ship that sank long ago.

"The western coastline of Morro is bordered by mountains; five are identifiable, including the Morro itself, which are separated from each other by from three to six miles. In the recesses, or crescent-shaped open bays, one finds lagoons separated from the sea by narrow beaches. In the lagoon located between the second and third of these rocky points, the sea deposits a large quantity of salt. Forty thousand loads of salt are collected there, at a price of four hundred livres each. This amount is enough for the consumption of the entire province of Oaxaca for four or five years: revenue from the salt goes to the Government.

II.361　　"The eastern coastline of the Morro has numerous lagoons that are visible from the summit of the Cerro de San Francisco. This mountain is approximately thirty-five leagues east of Tehuantepec, on the road to Ciudad-Real. The San Mateo lagoon is the closest to the Morro: it is one and a half leagues away, and at least seven leagues long and three leagues deep. This lagoon is almost completely split in two by a very narrow strip of land, at the end of which the village of San Mateo is located. Eight leagues farther away from the Morro, on the shore of the same lagoon, one finds the village of San Francisco. Its sandbar allows passage only to small fishing boats.

"As for the Tehuantepec River itself, it disappears on the plain that surrounds the Morro. Admittedly, its waters reach the beach at the small bay of Morro, but the river is still there and resembles a lagoon dotted with small islands. The tree trunks found where the river ends lead one to assume that when there is high water, the river opens a channel across the beach and creates a sandbar that, in any other season, becomes impassable. This beach, which is only fifty feet wide, separates the Río de Tehuantepec from the South Sea. The river water and the seawater are approximately on the same level. The river is so shallow here that it can be crossed on foot to go hunting on the small islands which abound in wild game. One is often forced to enter the water up to one's armpits, and one risks simultaneously sinking into the mud and being devoured by the caimans. In Tehuantepec itself, the river is rather swift. During the rainy season, the river is a torrent that often blocks travelers' progress; it is only during the dry season that it is safely navigable by pirogue from the ocean shore to the town. We have gone into detail about Tehuantepec Bay to show the obstacles that will need to be overcome when, by digging the Guasacualco canal sometime in the future, a port will be created on the shores of the Pacific Ocean."

II.362　　From my depictions of the ports of Mexico above, I have eliminated all references to the Veracruz harbor, which is amply described in chapters VIII (vol. II, pp. 209–15 [pp. 413ff in this edition]) and XII. This portrait can serve to adjust the above comments about connections between the two oceans by means of an

oceanic canal (vol. I, pp. 209–37 [pp. 150–67 in this edition]). We have just seen that on the isthmus of the Río Huasacualco, it would not be enough to open up a canal and *make channels* of the rivers; a port would also have to be established in Tehuantepec Bay, since this bay is as unapproachable by large craft as the Gulf of Panama. New information that I had from Mexico (in the summer of 1825) also suggests that, since I have returned to Europe, no steps have been taken to confirm the elevation above sea level of the sources of the Huascualco and Chimalapa rivers. But a note, the receipt of which I owe to the kindness of the well-known geographer and navigator Don Felipe Bauzá, confirmed the suspicions I voiced above (vol. I, p. 215 [p. 152 in this edition]) regarding the high elevation of Lake Nicaragua. An order of the Court of Madrid addressed to Captain-General of Guatemala, ▼ Don Matías de Gálvez, demanded a leveling survey to be carried out from the Gulf of Papagayo, on the coastline of the South Sea, to the Laguna de Nicaragua. By using 336 ascent stations and 339 descent stations (*ascensos*: 604 feet, 8 inches, 8 lines in the Castilian measurement; *descensos*: 470 feet, 1 inch, 7 lines), the engineer ▼ Don Manuel Galisteo found that the elevation of the surface of the lake above the level of the South Sea is 134 feet, 7 inches, 1 line. Today, this lake is eighty-eight feet, six inches deep, which means that its bottom is still forty-six *pies castellanos* higher than the level of the ocean. The Río Panaloya, which connects lake León with Lake Nicaragua, has a weir (*salto*) that measures between twenty-five and thirty *varas* (according to Mr. Ciscar, one *vara castellana* is equivalent to three *pies de Burgos* = 0.429). This measurement does not indicate the direction and the terminal points of the survey line. Since the goal of the measurement was simply to determine the el-   II.363
evation of Lake Nicaragua, it still does not prove to me that the watershed ridge between the lake and the Pacific Ocean runs everywhere at the same high elevation of eighty-five toises, or that there is not a depression in the earth or a side valley between Realexo and León, or between the Gulf of Papagayo, or Nicoya, and Lake Nicaragua, that could accommodate the water for a great oceanic canal (*Relation historique*, vol. III, p. 320). On the scouting mission headed up by the commander of Omoa Castle, ▼ Don Ignacio Maestre, and the engineers Don Joaquim Ysasy and Don José Maria Alexandro, it was noted that Lake Nicaragua has no natural connection with the South Sea; they also observed at that time "that the mountainous terrain (*aspero y montuoso*) between Villa de Granada and the port of Culebra makes any connection via canals in this area difficult if not impossible." Since the free governments in the Americas and several large trading companies in England and the United States watch these places closely, we may hope that we shall soon have the results of surveys done with a view to drawing up canals for large or small navigation. The solution to a problem tied to the greatest interests of human civilization depends on a great number of factors:

the absolute elevation of the watershed; the expanse of the terrain to be crossed; the type of soil; *canalizing* the rivers; the amount of water necessary for uninterrupted interior navigation; the healthiness of the climate; and the condition of the ports at the two ends of the navigable line.

Since a new trade route has been opened between the capital of Mexico and Pueblo Viejo de Tampico, I record here some precise information about the direction of this route and partial distances:

II.364   *The Road from Pueblo Viejo to Mexico City*

> Pueblo Viejo de Tampico.
> Rancho de *Arroyo del Monte*.
>   de la *Tortuga*.
>   de la *Ese*.
>   de *Vichin** fifteen mule-driver leagues.
>   de *Buena Vista*.
>   del *Río de Chicallan*.
>   de los *Alacranes*.
>   de *San Rafael*.
>   del *Pabellón*.
>   de los *Paderones** sixteen leagues.
>   de los *Huevos*.

Pueblo de *Tantoyuca*. The entire country up to this point is savanna bordered by low hills and covered with palm trees. The *Cañada road* forks off south of the large village (*pueblo*) of Tantoyuca. This road runs east through the Flores farm (*hacienda*) and the Tecolulo ranch (*rancho*). Mule drivers report that one crosses the small Cañada River seventy-seven times ([Pointsett,] *Notes on Mexico*, p. 228). The Cañada road and the Sierra road reconnect near the village of Tlacolula, so that on the latter road one crosses only the southernmost part of the Cañada as far as the Cerro de Pinolco.

Río de *Tecolulo*.

Rancho de *Uatipan** twelve leagues. Here, one enters the high mountains commonly called the *Sierra Madre*, or *Serranía Grande*.

Pueblo de *Atlapezco*.

Pueblo de *Yagualica*. A very strong military position on a mountain east of the road.

II.365   Pueblo de *Soquitipan*. In a valley on the bank of a river; the warm springs of Atempa are one league away.

Yatipan* eleven leagues. This village is located on a very high plateau. The Indians have *temascales* (steam baths) here.

> Cerro de *Guayatlapa*.
> Pueblo de *Tlacolula*. Here, one comes upon the Cañada road.
> Cerro del *Pinolco*.
> Pueblo de *Bemuchco*, on the summit of the Pinolco mountain.
> Pueblo de *Matlatengo** ten leagues.
> Pueblo de *Teniztengo*.
> Pueblo de *Zacualtipán*. This village is surrounded by apricot and pear
>   trees and other fruit trees.
> Pueblo de *San Bernardo*.
> Río *Oquilcalco*.
> Hacienda de *Río Grande** twelve leagues.
> Villa de *Atotonilco el Grande*.
> Pueblo de *Umitlan*.
> Villa de Real *del Monte*.
> Rancho de *Azoliatla*.
> Venta de *Jagüey de Teyes** eighteen leagues.
> Village of *San Mateo*.
>   de *San Mateo el Chico*.
>   de *Santa* Anita.
>   de *Tecama*.
>   de *Ozumbillo*.
>   de *Chiconautla*.
>   de *San Cristóbal*.
>   de *Tepetlaque*.
>   de *Campedrito*.
>   de *Nuestra Señora de la Guadalupe*.
> *Mexico* City* seventeen miles.

Total distance in leagues, according to the mule-drivers' calculations: 111.

This journey can be accomplished in seven and a half days, without chang- II.366 ing mules and by spending part of the night resting at the way stations marked with an *asterisk*. The numbers give the distance in local leagues from one of these eight stations to the other as the crow flies (according to the mule-drivers' calculations). The distance from Mexico City to Pueblo Viejo de Tampico as the crow flies is, according to my general Map of New Spain (1804) which Mr. Bauzá will soon replace with another more precise map, sixty nautical leagues (at 2,854 toises). Adding one-quarter to compensate for the twisting roads, one finds that each of the mule-drivers' 111 leagues equals 1,928 toises. In measuring the leagues en route from Mexico City to Acapulco, we found 1,725 toises. It takes from ten to twelve days to travel comfortably with pack mules from Pueblo Viejo to Mexico City. The road from the capital to Veracruz is only one eleventh shorter as the crow flies.

DISTANCE AS THE CROW FLIES:				WITH DETOURS:	
from Mexico City	to Acapulco	152,000 toises		190,000	toises
	to Veracruz	157,000		217,000	
	to Tampico	171,000		214,000	

In this table, contours are measured at one-quarter more than the direct distance (see above vol. I, p. 53 [p. 40 in this edition]) and Arago in the *Annuaire du Bureau des longitudes*, 1825, p.126). Only on the road from Mexico City to Veracruz do we assume that the traveler would head toward Puebla, Perote, and Jalapa instead of toward Orizaba and Cordova. Climbing from Tampico to the Mexican plateau, one skirts along the eastern slope of the Cordillera for a very long time and remains in the hot region nearly half of the way, as far as the Uatipan and Yagualica mountains. On the other hand, as one descends from Mexico City to Veracruz, one follows a line that is almost perpendicular to the axis of the great Anahuac Cordillera, and for three-fifths of the way, one is exposed to the cold temperature of the plateau. The new Tampico road offers mule drivers the advantage of finding fodder easily for the beasts of burden; it is so direct that as an *itinerary distance*, in other words, comparing this distance to the detours of the Veracruz road via Puebla (a town located twelve minutes in arc *south* of Veracruz), it is the shortest of all the routes that lead toward the eastern coastline, with the exception of the route one takes from Mexico City to the Tuspa sandbar.

II.367

BOOK IV

The State of Agriculture
in New Spain—Metal Mines

CHAPTER IX

Vegetable Crops in the Mexican Territory—Progress in
Soil Cultivation—The Influence of Mines on Clearing Land
for Cultivation—Plants for Human Nourishment

We have just traversed the immense expanse of land that is understood to be the Kingdom of New Spain. We have quickly described the borders of each province, the physical aspect of the country, its temperature, natural productiveness, and the progress of an incipient population. It is now time to turn our attention more particularly to the state of agriculture and the richness of the territory.

Since Mexico extends from 16° to 37° latitudes, its geographical position offers all the climatic variations that one would encounter on a journey from the shores of Senegal to Spain, or from the coastline of Malabar to the steppes of greater Bukharia [Uzbekistan]. This variety of climates is increased even more by the geological makeup of the country and the extraordinary mass and shape of the Mexican mountains portrayed in the third chapter. Each plateau has a different temperature on the ridge and slopes of the Cordilleras, depending on its relative elevation. These are not isolated peaks whose summits are covered with pines and oaks near the permanent snow line; entire provinces spontaneously produce alpine plants on the highest plateaus, and the farmer who lives in the Torrid Zone often loses hope because of frost damage or heavy snowfalls. The distribution of heat on the globe is so remarkable that one finds colder strata in the aerial Ocean the higher one goes, while the temperature at the bottom of the sea decreases gradually as one moves away from the surface of the water. In these two elements, the same latitude may bring together, so to speak, all climates. At unequal distances from the surface of the Ocean but on the same vertical plane, one finds layers of air and water at the same temperature.

Consequently, in the tropics, on the slopes of the Cordilleras and in the chasm of the Ocean, the plants of Lapland, like the marine animals near the pole, find the degree of heat they need to develop their organs.

II.370

According to the order of things established by nature, one understands that in a vast, mountainous country such as Mexico, the variety of indigenous plants must be enormous and that there is hardly a plant in the rest of the globe that one would not be able to grow in some part of New Spain. Despite the painstaking research of three distinguished botanists—Mr. Sessé, Mr. Mociño, and Mr. Cervantes—who were dispatched by the Court to examine the natural richness of Mexico, we are far from knowing all the plants that are found either scattered on solitary peaks or crowded together in the vast forests at the foot of the Cordilleras. Although new herbaceous species are discovered daily on the central plateau, even around Mexico City, how many arborescent plants must still be hiding from botanists in the hot, humid region that extends along the eastern slopes from the province of Tabasco and the fertile banks of the Guascualco to Tecolula and Papantla and along the western slopes from the port of San Blas and the Sonora to the Tehuantepec plains in the province of Oaxaca? Until now, no species of quinquina (Cinchona), even in the small group whose stamens are longer than the corolla and which forms the genus Exostema, has been identified in the equinoctial part of New Spain. It is likely, however, that this precious discovery will be made someday on the slopes of the Cordillera where tree ferns abound and the region[1] of true antifebrile quinquina with very short stamens and villous corolla begins.

II.371

1. See my *Géographie des plantes*, pp. 61–66, and a report that I published in German, which contains physical observations of the various species of Cinchona that grow on the two continents (["Über die Chinawälder in Südamerica,"] *Gesellschaft Naturforschender Freunde zu Berlin*, 1807, n. 1 and 2). It is believed in Mexico that Portlandia mexicana, which Mr. Sessé discovered, might replace quinquina in Loxa, just as Portlandia hexandra (Coutarea Aublet) does to a certain extent in Cayenne, and Bonplandia trifoliate Willd. or the Cusparé on the banks of the Orinoco, and Switenia febrifuga Roxb. in the East Indies. One should also examine the medicinal properties of Michaux's Pinkneya pubens (Mussaenda bracteolata Bartram), which grows in Georgia and has many similarities with Cinchonas. When considering the peculiar properties of the genera Portlandia, Coutarea, and Bonplandia, or the natural affinity of true prickly creeping Cinchona—which ▼ Mr. Tafalla discovered in Guayaquil—with the genres Pæderia and Danais, one finds that the antifebrile principle of quinquina is found in many rubiaceae. For this reason, rubber is not only extracted from Hevea but also from Urceola elastica, Commiphora madagascarensis, and a great number of other plants in the euphorbia, nettles (Ficus, Cecropia), cucurbitaceae (Carica), and campanulaceae (Lobelia) families. ▼ Mr. Auguste de St.-Hilaire recently (1824) introduced an Apocynum [Apocynacea/dogbane], the Strychnos pseudoquina of Brazil, which acts on intermittent fevers like true Cinchona does, although it

We do not propose to describe here the countless variety of plants with II.372
which nature has endowed the vast expanse of New Spain, whose useful
properties will become better known as civilization progresses in this country. Nor shall we discuss the various types of crops that an enlightened
government might successfully introduce. We shall limit ourselves to an
examination of the indigenous products that currently provide objects for
exportation and form the fundamental basis of Mexican agriculture.

In the tropics, especially in the West Indies, which have become the
center of European commercial activity, the word agriculture has a very
different meaning from its accepted definition in Europe. When one hears
talk about the flourishing state of agriculture in Jamaica or on the island
of Cuba, these words do not offer the imagination the idea of harvests that
contribute to human nourishment but, rather, suggest land that produces
objects of exchange for trade and raw materials for the manufacturing industry. No matter how rich and fertile the countryside may be, for instance
the Güines Valley southeast of Havana, one of the most delectable sites
in the New World, one only sees plains planted with sugarcane and coffee;
but these plains are watered with the sweat of African slaves! Rural life loses
its charms when it becomes inseparable from the tragic face of our species.

In the interior of Mexico, the word agriculture brings to mind less harsh II.373
and depressing ideas. The Indian farmer is poor but free. His situation is
far preferable to that of peasants in a large part of northern Europe. There is
neither forced labor nor serfdom in New Spain. There are almost no slaves.
Sugar is produced for the most part by free labor. The main objects of agriculture are not those products to which European luxury has attached a
variable and arbitrary value, but cereal grains, nutritive root vegetables, and
agave, which is the Indians' grape. The sight of the fields reminds the traveler that the earth there nourishes the person who cultivates it, and that the
true prosperity of the Mexican people hangs neither on chance opportunities for exterior trade nor on Europe's troublesome politics.

Those who know the interior of the Spanish colonies only through the
vague and imprecise information published to date will find it hard to believe that the main sources of Mexico's wealth are not the mines but, rather,
agriculture that has noticeably improved since the end of the last century. If

---

contains neither brucine nor quinine. ▼ (Mr. Lambert translated Mr. von Humboldt's report
on the types of quinquina found on the two continents, expanding it with very instructive
notes. See the *Illustration of the genus Cinchona*, 1821, pp. 2–59, and Humboldt, *Relation
historique*, vol. I, p. 367.

we do not reflect on the vast expanse of the country and especially the large number of provinces that appear to be entirely devoid of precious metals, we might typically imagine that all the activity of the Mexican population is directed toward working the mines. Agriculture has certainly made very considerable progress in the *capitanía general* of Caracas, the kingdom of Guatemala, on the island of Cuba, and wherever the mountains are supposed to lack mineral deposits. But it was wrong to conclude that the lack of attention given to cultivating the soil in other parts of the Spanish colonies is attributable to mining. Such reasoning might be accurate if it were only applied to small pieces of land. In the provinces of Chocó and Antioquia, and on the coastline of Barbacoas, the inhabitants prefer to look for alluvial gold washed in streams and ravines rather than clear the unspoiled, fertile land for cultivation. At the beginning of the conquest, the Spanish who abandoned the peninsula or the archipelago of the Canaries to settle in Peru and Mexico were only interested in discovering precious metals. "Auri rabida sitis a cultural Hispanos diverti" [A frenzied thirst for gold sidetracked Spanish culture], wrote Petrus Martyr d'Anghiera,[1] a writer of the time, in his work on the discovery of the Yucatán and the colonization of the Antilles. But such reasoning cannot now explain why agriculture is languishing in countries whose surface area is three or four times larger than France. The same physical and moral causes that impede all progress of national industry in the Spanish colonies also obstructed the improvement of soil cultivation. There is no doubt that if social institutions are improved, the regions that are the richest in mineral production will be as well, and perhaps better cultivated than those that appear to be devoid of metals. The natural human impulse to oversimplify the cause of all things has introduced a way of thinking into works of political economy that is perpetuated because it flatters the slothful mind of the masses. The depopulation of Spanish America, the state of abandon of the most fertile land, and the lack of manufacturing industry are all attributed to a wealth of metals and the abundance of gold and silver, as if, by the same logic, all of Spain's troubles were attributable either to the discovery of the Americas, to the nomadic life of the merinos, or to the religious intolerance of the clergy!

One scarcely takes note of the fact that agriculture is more neglected in Peru than it is in the province of Cumaná or in Guiana where there are no working mines. The best cultivated fields in Mexico, which remind

II.374

II.375

---

1. De insulis nuper repertis et de moribus incolarum earum [Of the recently discovered islands and their forgotten inhabitants]. Grynaeus, ed. *De orbe novo*, 1555, p. 511.

travelers of the most beautiful landscape in France, are the plains that extend from Salamanca to near Silao, Guanajuato, and Villa de León, and surround the richest mines in the known world. Wherever veins of metal have been discovered in the least inhabited parts of the Cordilleras, on solitary and deserted plateaus, mining, far from obstructing the cultivation II.376 of the soil, has markedly favored it. Travel on the ridge of the Andes or in the mountainous part of Mexico provides the most striking examples of the beneficial influence mines have had on agriculture. Without the settlements created for exploiting the veins, how many sites would have remained deserted, how much terrain would not have been cleared in the four intendancies of Guanajuato, Zacatecas, San Luis Potosí, and Durango, between the twenty-first and twenty-fifth parallels, where the most considerable wealth of metals in New Spain is concentrated! The foundation of a town immediately follows the discovery of a good-sized mine. If the town is built on the arid side or the ridge of the Cordilleras, new settlers can bring from afar only whatever they need for their subsistence and as fodder for the large number of animals used to draw water, haul, and amalgamate minerals. Need soon awakens industry. People begin to work the soil in the ravines and on the slopes of nearby mountains wherever rock is covered with earth. Farms spring up near the mine. The high cost of provisions and the considerable price at which competition among consumers keep all agricultural products compensate the farmer for the hardships to which life in the mountains exposes him. In this way, the expectation of profit alone and the drive of mutual interest that represent the strong bonds of society, and II.377 without governmental interference in the colonization, a mine that appears at first to be isolated in the midst of wild and deserted mountains can soon become connected to land that has long been cultivated.

Furthermore, the influence of the mines on the gradual clearing of land for farming is more enduring than the mines themselves. When the mines are exhausted and underground work has been abandoned, the local population undoubtedly declines, for the miners seek their fortune elsewhere; but the settler remains because of his attachment to the soil where he was born and which his forefathers worked with their own hands. The more isolated the homestead, the more attractive it is for the mountain dweller. At the beginning of civilization as toward its decline, humans appear to regret the trouble they caused themselves by entering society. They love solitude because it restores their former freedom. This moral tendency, the desire for solitude, shows itself especially among the indigenous people of the coppery race whom long and sad experience has rendered them disgusted with

social life, particularly around whites. Like the Arcadians, the people of the Aztec race enjoy living on the summits and slopes of the steepest mountains. This particular aspect of their customs contributes singularly to the expansion of their population in the mountainous region of Mexico. It is
II.378 ever so interesting for the traveler to follow the peaceful conquests of agriculture and contemplate the many Indian huts scattered among the wildest ravines and the strips of cultivated land that advance into a desert country among outcroppings of naked and arid rock!

The plants that are grown in these high, isolated regions are essentially different from those cultivated on the lower plateaus, on the slopes and at the foot of the Cordilleras. I could discuss agriculture in New Spain by following the large divisions that I presented above and sketching the physical tableau of the Mexican territory. I could follow the *growth lines* that are drawn on my geological *profiles* whose elevations were partially given in the third chapter of this work.[1] But it is important to note that these growth lines, like the perpetual snow lines to which they are parallel, drop toward the north, and that the same cereal grains that, below the latitude of the towns of Oaxaca and Mexico City, produce abundantly only at an elevation of fifteen or sixteen hundred meters, are found in the temperate zone in the *Provincias internas* on the lowest plains. The land elevation required by various kinds of crops depends generally on the distance to the pole and the surface of the ocean; but such is the organizational flexibility of the culti-
II.379 vated plants assisted by human care that they often surpass the borders that the naturalist assigned to them.

Below the equator, meteorological phenomena like those of the geography of plants and animals are subject to immutable laws that are easily grasped. There, only elevation can change the climate, and the temperature is almost constant, despite the different seasons. Moving away from the equator, especially between 15° latitude and the tropics, the climate depends on a greater variety of local circumstances; it varies at the same absolute elevation and on the same geographic latitude. This influence of local conditions, whose study is so important to the farmer, is even more obvious in the northern than in the southern hemisphere. The great breadth of the New Continent, its proximity to Canada, winds blowing from the north, and other causes that have been developed above give the equinoctial region of Mexico and the island of Cuba their special character. One might say that in these regions, the temperate zone, where climates vary, expands

1. See vol. I, p. 273, and vol. II, p. 199 [pp. 184 and 409 in this edition].

to the south and goes beyond the tropic of Cancer. It is enough to state here that in the environs of Havana (latitude 23°8′), at the low elevation of eighty meters above sea level, the thermometer has been seen to descend to the freezing point,[1] and snow fell near Valladolid (latitude 19°42′) at 1,900 meters absolute elevation, whereas on the equator, the latter phenomenon is only seen at elevations twice as high.

II.380

These considerations prove that near the tropics, where the Torrid Zone approaches the temperate zone (I am using the commonly accepted but inappropriate names), cultivated plants are not subject to set and invariable elevations. One might be tempted to distribute them according to the average temperature of the places where they grow. One accurately observes that in Europe the minimum average temperature for growing sugarcane successfully is between nineteen and twenty degrees; for coffee plants, eighteen degrees; for orange trees, seventeen degrees; for olive trees, from thirteen and a half to fourteen degrees; for grapes producing drinkable wine from ten to eleven degrees Centigrade. This thermometric scale for agriculture is rather precise when we reflect on such phenomena broadly. But numerous exceptions appear, when we consider countries whose average annual temperature is the same, while average monthly temperatures may vary widely from each other. As Mr. de Candolle[2] has proven so well, the uneven distribution of heat throughout the different seasons of the year is the primary influence on the type of crop best suited to a particular latitude. Several annuals, especially graminacea with farinaceous seeds, are relatively unaffected by the rigors of winter[3]; but like fruit trees and grapes, they need considerable heat during summer. In part of Maryland and especially in Virginia, the average annual temperature is equal to or perhaps even higher than the same temperature in Lombardy; however, wintry chills make it impossible to grow there the same plants that adorn the plains of the countryside around Milan. In the equinoctial region of Peru or Mexico, rye, and

II.381

---

1. Mr. Robredo saw ice form in a wooden trough in the month of January in the village of Wajay, fifteen miles southwest of Havana at seventy-four meters absolute elevation. On January 4, 1801, at eight o'clock in the morning in Río Blanco, I saw a temperature of 7.5° above zero on a Centigrade thermometer. During the night, an unfortunate black man [nègre] froze to death in prison. Average temperatures in the months of December and January on the plains of the island of Cuba, however, are seventeen and eighteen degrees. All these readings were made with excellent ▼ Nairne thermometers.

2. [de Candolle,] *Flore française*, 3rd ed., vol. II, p. 10.

3. In Umea in Westro-Botnia (latitude 63°49′) in 1801, the extremes of the Centigrade thermometer reached +35° in the summer and -45.7° in winter. ▼ Mr. Acerbi complains much about the great summer heat in the northernmost part of Lapland.

to a much lesser extent, wheat, do not reach maturity on plateaus with elevations of 3,500 to 4,000 meters, although the average heat in these alpine countries is above that of parts of Norway and Siberia where these cereals are successfully grown. But for about thirty days, the obliqueness of the sphere or the shortness of the nights makes for considerable summer heat in the countries nearest the pole, while in the tropics, on the plateau of the

II.382 Cordilleras at the same elevation as [Saint-Martin-du-] Canigou, the average temperature almost never rises above ten or twelve degrees Centigrade.

So as not to mix theoretical ideas that are hardly susceptible to rigorous precision with the statement of proven facts, we shall divide the plants grown in New Spain neither according to the elevation where they grow most abundantly nor according to the degrees of average temperature that they appear to require for their development; instead we shall arrange them according to the usefulness that they provide to society. We shall begin with the plants that form the principal basis of the diet of the Mexican people, before discussing the cultivation of plants that provide material for the manufacturing industry. We conclude this research with a description of the plant crops that are a source of significant trade with the metropole.

For a large part of the inhabitants of equinoctial America, the *banana tree* represents what cereal grasses, wheat, barley, and rye are for western Asia and Europe and what the numerous varieties of rice are for the countries located beyond the Indus, especially for Bengal and China. On both continents and the islands comprised in the vast expanse of the equinoctial seas, wherever the average annual temperature is higher than twenty-four degrees Centigrade, the fruit of the banana tree is an object of the greatest

II.383 interest for human subsistence. ▼ The famous traveler Georg Forster and other naturalists who followed him insisted that this precious plant did not exist in the Americas before the arrival of the Spanish but that it was brought from the Canary Islands at the beginning of the sixteenth century. In fact, Oviedo, who, in his *Natural History of the Indies*, carefully distinguishes native plants from those that were introduced, asserts that ▼ Tomás de Berlangas,[1] a Dominican monk, planted the first banana trees in 1516 on the island of San Domingo. He confirms that he himself saw Musa planted in Spain near the town of Armeria in Granada and at the Franciscan monastery on the island of the *Gran Canaria*, where Berlangas obtained the suckers

---

1. [Forster,] *De plantis esculentis [insularum oceani australis] commentatio botanica,* 1786, p. 28. [Fernández de Oviedo,] *Histoire naturelle et générale des îles et terres fermes de la grande mer océane,* 1556, pp. 112–14.

that were transported to Hispaniola and from there to the other islands and the mainland. In support of Mr. Forster's opinion one might also state that corn, the pawpaw tree, Jatropha manihot, and agave are often mentioned in the earliest accounts of the voyages of Columbus, ▼ Alonzo Negro, Pinzón, Vespucci,[1] and Cortés, but never the banana tree. These early voyagers' silence, however, proves only how little attention they paid to the plant production of the American soil. Hernández [de Toledo], who describes many other Mexican plants besides medicinal ones, does not mention Musa. This botanist lived half a century after Oviedo, and those who consider Musa foreign to the New Continent do not doubt that it was commonly grown in Mexico toward the end of the sixteenth century, at a time when many other plants less useful to humans had already been brought there from Spain, the Canary islands, and Peru. An author's silence is therefore not always adequate proof in favor of Mr. Forster's opinion.

II.384

The question of the true country of the banana tree is perhaps the same as for pear and cherry trees. The wild cherry (*prunus avium*), for example, is indigenous to Germany and France; it has existed in our forests from time immemorial, like the English oak and the linden tree, whereas other species of cherry, which are considered permanent varieties and whose fruit is more savory than the wild cherry, came to us through the Romans from Asia minor,[2] especially from the Kingdom of Pontus. Likewise, under the name of the banana tree, a large number of plants that are essentially different because of the shape of their fruit and perhaps constitute true species, are cultivated in the equinoctial regions as far as the thirty-third or thirty-fourth parallel. If there is scant proof to date that all cultivated pears are descended as common stock from the wild pear, we are even more entitled to doubt that the large number of constant varieties of the banana tree descend from the Musa troglodytarum grown in the Moluccas, which itself, according to ▼ Gärtner, is possibly not a Musa but a species of Adanson's genus Ravenala.

II.385

All the Musa, or *Pisang*, that ▼ Rumphius and Rheede described are not known in the Spanish colonies. Three varieties, however, are identifiable, which botanists have only very imperfectly distinguished until now: the true *Plátano* or *Artón* (Musa paradisiacal Lin?), the *Camburi* (M.

---

1. *Christophori Columbi navigatio. De gentibus ab Alonzo repertis. De navigatione Pinzoni socii admirantis. Navigatio Alberici Vesputii.* See Grynaeus, *Novus orbis*, 1555, pp. 64, 84, 85, 87, and 211.

2. ▼ Desfontaines, *Histoire des arbres et arbrisseaux qui peuvent être cultivés [en pleine terre] sur le sol de la France*, 1809, vol. II, p. 208, a work of learned and notable research on the origin of useful plants and the time when they were first grown in Europe.

sapientum Lin?), and the *Dominico* (M. regia Rumph?). In Peru, I saw a fourth variety grown, which has an exquisite flavor: the South Sea *Meiya*, which is called the *Plátano de Taïti* in the marketplace in Lima because the frigate *Aguila* brought the first seedlings there from the island of Otahiti. Now it is an uninterrupted belief in Mexico and on the entire mainland of South America that the *Plátano artón* and the *Dominico* were grown there long before the Spanish arrived but that a variety of the *Camburi* (the *Guineo*), as its name proves, came from the coasts of Africa. The author who most carefully recorded the different periods when American agriculture was enriched with foreign plants, the Peruvian ▼ Garcilaso de la Vega,[1] states expressly that at the time of the Inca, corn, quinoa, potatoes, and, in the hot and temperate regions, the banana formed the base of the people's diet. ▼ Father Acosta[2] also confirms, although less positively, that Americans grew Musa before the Spanish arrived. According to him, the banana is a fruit found throughout the Indies, although some insist that it originated in Ethiopia and came to the Americas from there. On the banks of the Orinoco, the Casiquiare, or the Bení, between the Esmeralda Mountains and the sources of the Caroni River, in the middle of the thickest forests, practically wherever one finds Indian peoples who have had no connection with European settlements, one finds manioc and banana plantations.

II.386

II.387

Father Thomas de Berlangas could not transport from the Canary Island to San Domingo any other species of Musa than the one grown there, which is the *Camburi* (caule nigrescente striato, fructu minore ovato-elongato), not the *Plátano artón* or the Mexican *zapalote* (caule albo-virescente lævi, fructu longiore apicem versus subarcuato acute trigono). Only the first of these two species will grow in temperate climates on the Canary Islands, in Tunis, Algiers, and on the coastline of Málaga. For this reason, in the Caracas valley below 10°30′ latitude but at nine hundred meters absolute elevation, one finds only the *Camburi* and the *Dominico* (caule albo-virescente, fructu minimo obsolete trigono), not the *Plátano artón*, whose fruit ripens only under the effect of a very high temperature. With such abundant evidence, there seems little

---

1. [Garcilaso de la Vega El Inca,] *Comentarios reales de los Incas*, vol. I, p. 282. The small musk-scented banana, the *Dominico*, whose fruit seemed most delicious to me in the province of Jaén de Bracamoros on the banks of the Amazon and the Chamaya, appears to be identical to Jacquin's Musa maculata (Hortus Schœnbrunnensis, table 446) and Rumphius's Musa regia. The latter species is itself perhaps only a variety of the Musa mensaria. Curiously enough, a wild banana exists in the forests of Amboina [Indonesia] whose fruit has no seeds, the Pisang jacki. (Rumph[ius, *Herbarium amboinense*], vol. V, p. 138.)

2. [Acosta,] *Historia natural [y moral] de [las] Indias,* 1608, p. 250.

doubt that the banana tree, of which several travelers insist that they found stock growing wild in Amboina, Gilolo, and on the Mariana Islands, was cultivated in the Americas long before the arrival of Europeans, who merely increased the number of varieties grown there. Nevertheless, we should not be surprised to know that there was no Musa on the island of San Domingo before 1516. Like some animals, the savages nourished themselves most often from a single species of plant. The forests of Guyana offer numerous examples of tribes whose plantations (*conucos*) include manihot, Arum, or Dioscorea,   II.388 and not a single banana plant.

Despite the vast extent of the Mexican plateau and the height of the mountains near the coastline, the area whose temperature is favorable for growing Musa occupies more than fifty thousand square leagues and is inhabited by one and a half million people. In the hot and humid valleys of the intendancy of Veracruz at the foot of the Orizaba Cordillera, the fruit of the *Plátano artón* is sometimes more than three decimeters wide, and most often twenty to twenty-two centimeters (seven to eight inches) long. In these fertile regions, especially around Acapulco, San Blas, and the Río Coatzacoalcos, a cluster of bananas contains 160 to 180 pieces of fruit and weighs thirty to forty kilograms.

I doubt that there is another plant on the globe that can produce such a sizable amount of food on a small expanse of land. Eight or nine months after the seedling is planted, the banana tree begins to develop its cluster. Its fruit may be picked in the tenth or eleventh month. When the stalk is cut, one always finds a tiller (*pimpollo*) among the many shoots that have grown roots, which is two-thirds the height of the mother plant and bears fruit three months later. This is why a Musa plantation, which is called a *Platanar* (a banana plantation) in the Spanish colonies, can thrive without any human attention besides cutting the stalks when the fruit has ripened and working the soil once or twice a year by digging around the roots. A   II.389 plot of land with a surface area of one hundred square meters can contain at least thirty to forty banana trees. In one year's time, the same land, estimating the weight of a cluster at only fifteen to twenty kilograms, produces over two thousand kilograms (or four thousand livres) of food. What a difference between this product and the cereal graminacea in the most fertile parts of Europe! Wheat, assuming that it is sown and not planted using the Chinese method and basing estimates on a tenfold harvest, yields only fifteen kilograms (or thirty livres) of grain on a plot of land of one hundred square meters. In France, for example, half a hectare (or the official arpent measuring 1,344 and a half square toises) is sown by using 160 livres of seed for excellent land and 200 or 220 livres for mediocre or poor land. Production varies between 1,000 to 2,500 livres per arpent. According to ▼ Mr. Tessier, potatoes produce a harvest of forty-five kilograms (or ninety livres) of roots on one hundred square meters of well-cultivated and well-fertilized land. One estimates four to six thousand livres per official arpent. As a result, the production ratio of bananas to wheat is 133 to one, and the ratio of bananas to potatoes forty-four to one.

II.390   Those who have tasted bananas ripened in hot houses in Europe find it difficult to imagine that a fruit so mild that it tastes something like a dried fig could be one of the nutritive staples of a few million people who live in both Indies. We easily forget that as vegetation grows, the same elements form very different chemical mixtures, depending on whether they combine or separate. In fact, would we recognize in the milky mucilage that the seeds of graminacea contain before the spike of grain ripens the farinaceous perisperm of the cereals that nourish most of the people in the temperate zone? In Musa, the formation of the starchy matter precedes the ripening stage. A distinction must be made between the fruit of the banana tree that is picked green and what is left to yellow on the stalk. The sugar in the latter is completely formed; it is mixed with the pulp and is so abundant that if sugarcane were not grown in the same region as banana trees, one could extract the sugar from their fruit more profitably than it is done in Europe from beets and grapes. When a banana is picked green, it contains the same nutritive elements that one observes in wheat, rice, the tuberous roots, and the sago, namely, amyloid starch [fecula] together with a very small portion of vegetal gluten. By kneading the flour of sun-dried bananas with water, I was able to obtain only a few particles of the ductile, viscous mass that abounds in the perisperm and especially in the embryo of cereals. If glu-

II.391   ten, which is so similar to animal matter and puffs up and swells in heat, is so useful for making bread, its presence, on the other hand, is not vital to making a root or fruit nutritional. ▼ Mr. Proust has found gluten in beans, apples, and quince, but he did not discover it in potato flour. Gums, for example, from the Mimosa nilotica (Acacia vera Willd.), which many African tribes eat as they cross the desert, prove that a vegetable substance can be nutritious without containing either gluten or starch.

It would be difficult to describe the numerous preparations by which Americans transform the fruit of the Musa, before or after ripening, into a healthy and wholesome food. Traveling up the rivers, I often saw locals after a hard day's work make a complete dinner of a very small portion of manioc and three large bananas (*Plátano artón*). In ▼ Alexander's time, if we are to believe the ancients, philosophers in Hindustan were even more austere: "Arbori nomen palœ pomo arienœ, quo sapientes Indorum vivunt. Frucuts admirabilis succ dulcedine utu no quaternos satiet" [There is another tree in India, of still larger size, and even more remarkable for the size and sweetness of its fruit, upon which the sages of India live. . . . It puts forth its fruit from the bark, a fruit remarkable for the sweetness of its juice, a single one sufficient to satisfy four persons. The name of this tree is 'pala,' and of the fruit, 'ariena.'" (▼ Pliny, [Historia Naturalis,] XII, 12 [The Natural History, chap. XII.6). People in hot countries generally consider sweet substances not only a temporarily satisfying food but also one that is truly nourishing. I have often observed that on the coastline of Caracas, the mule drivers who carried our baggage preferred raw sugar (papelón) to fresh meat   II.392 as their dinner.

Physiologists have not yet precisely determined what constitutes a highly nutritive substance. Calming the appetite by stimulating the nerves of the gastric system or providing easily assimilated matter for the body are very different activities. Tobacco, the leaves of Erythroxylon coca mixed with quicklime, or opium, which the indigenous peoples of Bengal often used successfully for months on end during times of famine, appease hunger pangs; but these substances act very differently from wheat bread, Jatropha root, gum Arabic, Icelandic lichen, or rotten fish, which is the main staple of several tribes of black Africans [Nègres]. There is no doubt that in equal bulk, *super-nitrogenous* or animal matter is more nourishing than vegetable matter; it appears that among the latter, gluten is more nutritive than starch, and starch more so than mucilage. But we must be careful not to attribute to these isolated elements that which depends, in the act of eating and in the living body, on the varied mixture of hydrogen, carbon, and oxygen. Matter thus becomes exceptionally nutritious if it contains, like the cocoa bean (Theobroma cacao), an aromatic principle that excites and fortifies the nervous system in addition to starch.

These ideas, which we cannot develop further here, will serve to shed   II.393 light on the comparisons that we made above of the yield from different crops. Although by weight one harvests three times as many potatoes as wheat from the same piece of land, one should not conclude that cultivating tuberous plants can provide sustenance for three times as many people as cultivating cereals[1] on an equal surface. A potato is reduced to one-quarter

---

1. "In Cuba, banana trees are ordinarily planted four varas apart in a square (1 vara = ot, 43): some settlers plant them closer together, but then they yield much less. Each clump of

of its weight when it is dried in moderate heat, and the dry starch that could be separated from 2,400 kilograms harvested on half a hectare of land would scarcely equal the amount that eight hundred kilograms of wheat can provide. The same is true for the fruit of the banana tree that, before ripening and even at the stage when it is very farinaceous, contains much more water and sugary pulp than the seeds of graminacea. We have seen that in a favorable climate, the same piece of land can produce 106,000 kilograms of bananas, 2,400 kilograms of tuberous roots, and 800 kilograms of wheat. These amounts are not proportional to the number of individuals who could be fed by these different crops grown on the same land. The aqueous mucilage that the banana or the tuberous Solanum root contains certainly has nutritive properties. As it is found in nature, farinaceous pulp undoubtedly provides more food than the starch that is manually separated from it. But weight alone is not an indication of absolute amounts of nutritive matter, and to show how much more nourishment the cultivation of Musa on the same surface area can provide than the cultivation of wheat, one ought rather to reckon according to the mass of vegetable substance needed to satisfy the appetite of an adult. According to this principle (and it a very curious fact), half a hectare (or one legal arpent) where large bananas (*Plátano artón*) are grown in a very fertile country can feed more than fifty individuals, whereas the same arpent in Europe (supposing grain of the eighth quality) would yield only 576 kilograms of wheat flour annually, a quantity that is not enough for the subsistence of two individuals.[1] As a result, nothing strikes a European newly arrived in the Torrid Zone more than the extreme smallness of the cultivated land around a farmhouse where a large family of indigenous people live.

When it is exposed to the sun, the ripe fruit of the Musa is preserved like our figs. Its skin blackens and takes on a particular odor like smoked

---

trees is composed of four to five stalks that grow new roots, but one should expect only three bunches of bananas per clump each year because some bunches always die or are damaged before fully ripening. There are twenty-five to thirty bananas of the long variety and fifty to seventy small ones per cluster. A worker with a hearty appetite but with no other source of food would require twelve large or thirty small bananas, since one reckons that five of the latter as an amount of food are equivalent to two of the former. The result of this estimate is the consumption of half a bunch of bananas per day; sixty or so clumps planted four varas apart would be approximately enough to feed one man." ▼ Catineau Laroche, *Notice sur la Guyane française . . . et sur la Mana*, 1822, p. 5; Humboldt, *Relation historique*, vol. II, pp. 614–19.

1. The calculation is based on the following principles: one hundred kilograms of wheat produce seventy-two kilograms of flour, and sixteen kilograms of flour are convertible into twenty-one kilograms of bread. A person's annual consumption is estimated at 547 kilograms of bread per year.

ham. In this state, the fruit is called *Plátano pasado* and becomes a trade commodity in the province of Michoacan. This dried banana is a pleasant tasting and very healthy food. But newly arrived Europeans consider the ripe, freshly picked fruit of the *Plátano artón* highly indigestible. This is a very ancient opinion, since Pliny tells us that Alexander ordered his soldiers not to touch the bananas that grew on the banks of the Hyphasis. Flour is extracted from the Musa by cutting the green fruit into slices and drying it in the sun on slopes, then pounding it when it has become crumbly. This flour is less used in Mexico than in the islands[1] and may be used in the same ways as rice flour or corn flour.

The ease with which the banana tree regenerates itself from its roots gives it an extraordinary advantage over other fruit trees, even the breadfruit tree that is loaded with farinaceous fruit during the eighth month of the year. When tribes wage war against each other and destroy trees, this calamity is felt for a long time, but seedlings can renew a banana plantation in a few months' time.

One often hears it repeated in the Spanish colonies that the inhabitants of the hot region (*tierra caliente*) will emerge from the state of apathy into which they have descended over centuries only when a *royal charter* [cédula] orders the destruction of the banana plantations (*plantares*). This is a violent recourse, and those who propose it so vehemently are generally no more active than the lower class that they would force to work by increasing their overall needs. One hopes that industry will make headway among the Mexicans without recurring destructive methods. Further, when we consider how easy it is for people to find sustenance in a climate where banana trees grow, we should not be astonished that in the equinoctial region of the New Continent, civilization began on less fertile soil in the mountains in an atmosphere less favorable to the development of organized beings, where need prompts industry. At the foot of the Cordilleras, in the humid valleys of the intendancies of Veracruz, Valladolid, or Guadalajara, a man who does only two days of easy labor each week can provide sustenance for an entire family. Such is the mountain dweller's love of his native land, however, that although night frosts often dash any hope of harvest, he refuses to descend to the fertile but unpopulous plains where nature spreads her endowments and riches in vain.[2]

II.397

---

1. See ▼ Mr. de Tussac's interesting paper in his *Flora Antillarum,* p. 60 (Paris, F. Schœll).

2. Since the first edition of this work was published, new doubts have arisen concerning the American origin of the banana trees cultivated by the wild Indians of the Orinoco

II.398   The same region where the banana tree is cultivated also produces the valuable plants whose roots provide *manioc* (or *magnoc*) flour. The green fruit of the Musa is eaten cooked or roasted, like the fruit of the breadfruit tree or the tuberous root of the potato. Manioc and corn flour, on the contrary, are converted into bread that provides what the Spanish settlers call *pan de tierra caliente* for the inhabitants of the hot countries. As we shall soon see, corn has the great advantage of being growable in the tropics from sea level to elevations that equal those of the highest summits of the Pyrenees. It possesses the extraordinary flexibility of organization that characterizes plants in the family of grasses [Poaceae] even to a higher degree
II.399   than the cereal crops of the Old Continent that wilt under a burning sun, while corn grows vigorously in the hottest countries on earth. The plant whose root yields the nutritive *manioc* flour is called *Juca* [yucca], a word in the language of *Haiti* or the island of San Domingo. It is difficult to grow

and the Casiquiare. It is my duty to mention them here. While reminding us that ▼ Marggraf and Piso (*Hist[oire] nat[urelle] du Brésil*, p. 554) considered that Brazilian banana trees were introduced from the coastline of Africa, ▼ Mr. Robert Brown, the illustrious author of the *Observations Systematical and Geographical on the Herbarium Collected [by Professor Christian Smith, in the Vicinity] of the Congo* (1818, p. 51), posits the general principle that *whenever there is doubt*, one may admit the likelihood that a cultivated species is foreign to a country where no other species of the same genus is native. According to this principle, which appears to be very well founded, the different varieties of banana cultivated in the Americas were originally native to Asia; that continent already had five different species of the genus Musa that grew spontaneously, while the Americas did not have even one. According to Mr. Brown, all varieties of Musa that provide human nourishment, of which sixteen are grown in the Indian Archipelago alone (▼ Crawford, *Hist[ory] of the Indian Archipelago*, vol. I, pp. 410–413) descend from the Musa sapientum indigenous to Asia and have nonabortive seeds (▼ Roxburgh, *[Plants of the Coast of] Corom[andel*, vol. 3,] plate 275). On the other hand, I might cite in support of my opinion, which Mr. Robertson shares, and against Mr. Robert Brown and ▼ Mr. Desvaux's opinion ("Essai sur l'histoire botanique," *Journal de botanique*, vol. IV, p. [5]), that in the southern hemisphere on the banks of the Prato, the Puris confirm having grown a small species of banana long before their connections with the Portuguese (▼ Caldcleugh, *Travels in South America*, 1825, vol. I, p. 23), and that one finds *non-imported* words in the American languages that describe the fruit of the Musa, for example: *paruru* in Tamanac; *arata* in Maypur (Humboldt, *Re[lation] [hist]orique*, vol. I, p. 104–587; vol. II, pp. 355–67. Leopold von Buch, *Physic[alische] Beschreibung der Can[arischen] Inseln*, 1825, p. 124). I also believe that the word *pala*, which Pliny uses to designate the Musa (*Mouz* in Arabic) is ascribable to one of the misunderstandings that are so frequent even now among travelers, and that it is derived from the Sanskrit *phalam* which means *fruit* in general. The word *pisang* introduced into the German language is Malay (Crawfurd, *History of the Indian Archipelago*, vol. II, p. 158): *banana* is most likely derived from *barana-busa* which, according to the Amarakosha, is synonymous in Sanskrit with *radala, rambha*, and *mocha,* words which all mean Musa (▼ Ainslie, *Materia medica of Hindoostan*, 1813, p. 234).

beyond the Torrid Zone; in the mountainous part of Mexico, it is generally not grown above an absolute elevation of six or eight hundred meters. This height is far surpassed by the absolute elevation of the Camburi or Banana Tree of the Canary Islands, a plant that is grown nearer the central plateau of the Cordilleras.

Like the indigenous peoples of all equinoctial America, the Mexica have grown two types of *Juca* since time immemorial, which botanists, in their inventory of *species*, have grouped together under the name Jatropha manihot. In the Spanish colony, sweet Juca (*dulce*) is distinguished from sharp or bitter Juca (*amarga*). The root of the first one, which is called *Camagnoc* in Cayenne, can be eaten without danger, while the other root is a rather fast-acting poison. The two may be used in making bread; however, only the root of the bitter Juca is generally used for this purpose, after its poisonous juice is carefully extracted from the flour before making manioc bread, called *Cazavi*, or *Cassave*. The separation is made by pressing the grated root in a *Cibucán*, a type of long sack. It appears that according to a passage in ▼ Oviedo (*L'histoire naturelle et generale des Indes*, book VII, chap. 2), II.400 the Juca dulce which he calls *boniata* (and which the Mexica call *huacamote*), was not originally found in the Antilles but was transplanted there from the nearby continent. "The *boniata*," writes Oviedo, "is similar to the one on the mainland; it is not at all poisonous and may be eaten with its juice, either raw, cooked, or roasted." The locals carefully separate the two kinds of Jatropha in their fields (*conucos*).[1]

It is quite remarkable that plants whose chemical properties are so different should be so difficult to distinguish by their external characteristics. In his *Civil and Natural History of Jamaica*, Browne believed that he discovered these characteristics by dissecting the leaves. He called sweet Juca *sweet cassava* and bitter or sharp Juca *common cassava*, Jatropha foliis palmatis pentadactylibus. But having examined many *manihot* plantations, I have seen that, like all lobed or palmate leaves that are cultivated, the appearance of the two kinds of Jatropha varies prodigiously. I observed that the indigenous peoples distinguish sweet manioc from the poisonous variety less by the greater whiteness of the stem and the reddish color of the leaves than by the taste of the root, which is neither sharp nor bitter. In this respect, cultivated Jatropha is like the sweet orange tree that botanists are not able to distinguish from the bitter orange which, however, II.401

1. [Browne,] *Civil and Natural Hist[ory] of Jamaica*, pp. 349 *and* 350. See also [José de] Acosta, [*Historia natural y moral de las Indias*,] book 4, chap. 17.

according to ▼ Mr. Gallesio's elegant experiments, is a primitive species that is propagated from seed like the bitter orange tree. Some naturalists who follow the example of ▼ Dr. Wright of Jamaica take sweet Juca for Linnaeus's true Jatropha janipha or Löfling's Janipha fructens.[1] But the latter species, which is Jacquin's Jatropha carthaginensis, is essentially different from sweet Juca because of the shape of its leaves (lobis utrinque sinuatis), which resemble the leaves of the Pawpaw tree. I strongly doubt that the Janipha can be transformed into the Jatropha manihot by cultivation. It seems equally unlikely that sweet Juca is a poisonous jatropha that gradually lost the pungency of its juices either through human care or the effect of long cultivation: the *Juca amarga* of American fields has remained the same for centuries, although it is planted and cared for like the *Juca dulce*. Nothing is more mysterious than this difference in internal organization among cultivated plants whose external forms are almost identical.

Raynal[2] believes that manioc was brought from Africa to the Americas to serve as sustenance for blacks [nègres]; he adds that if manioc already existed on Terra Firma [the continent] before the arrival of the Spanish, the indigenous peoples of the Antilles did not know it in Columbus's time. I fear that this well-known author, who, nonetheless, describes objects of natural history rather accurately, has confused manioc with yams, or jatropha with a species of Dioscorea. By what authority can one prove that manioc was cultivated in Guinea from the earliest of times? Several travelers have affirmed that corn grew wild in this part of Africa; but it had certainly transported there by the Portuguese in the sixteenth century. Nothing is more difficult to resolve than the problem of the migration of plants that are useful to humankind, especially since connections among the various continents have become so frequent. Fernández de Oviedo, who had gone to the island of Hispaniola or San Domingo in 1513 and lived for more than twenty years in the continental parts of the New World, describes the manioc as a very ancient crop that is proper to the Americas. If black slaves [nègres esclaves] had brought manioc with them, Oviedo would have seen the beginning of this significant branch of agriculture in the tropics with his own eyes. He would have referred to the period when the first manioc stock was planted, just as he describes in great detail the [first] introduction of sugarcane, the banana tree from the Canary Islands, and olive and

II.402

---

1. [Löfling,] *[Iter Hispanicum, eller]Resa til Spanska Landerna*, 1758, p. 309.
2. [Raynal,] *Histoire philosophique*, vol. III, pp. 212–14.

date trees. In his letter addressed to the ▼ Duke of Lorraine,[1] Amerigo Vespucci relates that he saw manioc bread being baked on the coast of Paria in 1497. "The locals," writes the explorer whose narrative, however, is rather inaccurate, "are unfamiliar with our wheat and farinaceous grains. They derive their main sustenance from a root that they reduce to a flour, which some call *iucha*, others *chambi*, and others *igname* [yam]." The word *iucca* is easily recognizable in the word *iucha*. As for the word *igname*, it now designates the root of the *Dioscarea alata*, which Columbus[2] describes by the name *ages* and which we shall discuss below. The indigenous peoples of Guyana, who do not recognize the Europeans' domination, have grown manioc since the earliest of times. Lacking provisions as we crossed the *rapids* of the Orinoco for the second time during our return from the Río Negro, we approached the tribe of Piaroa Indians who live east of Maypures and asked them to provide us with some Jatropha bread. There can be no further doubt that manioc is a plant whose cultivation precedes the Europeans' and the Africans' arrival in the Americas.

II.403

Manioc bread is very nutritious, perhaps because of its sugar content and a viscous matter that binds the farinaceous molecules of the cassava. Some confirm that this matter is somewhat similar to rubber, which is common to all plants in the Tithymaloid group. The cassava is round in shape; the disks that are called *turtas*, or *xauxuu*, in the ancient language of Haiti are five to six decimeters in diameter and three millimeters thick. The indigenous peoples, who are much more reserved than the whites, usually eat less than half a kilogram of manioc every day. The lack of gluten mixed into the starchy matter and the thinness of the bread make it very brittle and difficult to transport. This inconvenience is particularly noticeable during long voyages. The starchy flour made from grated, dried, and smoked manioc is almost unalterable. Insects and worms do not attack it, and every traveler knows the advantages of *couaque* in equinoctial America.

II.404

Flour from the *Juca amarga* does not solely sustain the Indians' diet; they also use the juice pressed from its roots, which is an active poison in its raw state. Fire causes this juice to break down. When boiled for a long time, it loses its toxic properties as it is gradually skimmed. There is no danger to using it as a sauce, and I myself have often eaten the brownish juice which resembles a very nourishing broth. In Cayenne,[3] it is thickened to make

1. Grynaeus, [*Novus orbis*], p. 215.
2. *Ibid.* p. 66.
3. ▼ Aublet, *Histoire des plantes de la Guiane française*, vol. 2, p. 72.

II.405   *cabiou*, which is similar to the *soy* brought from China for seasoning dishes at table. Serious accidents sometimes occur if the pressed juice has not been heated long enough. It is a well-known fact in the islands that long ago a large number of Haitian indigenous peoples committed suicide using the unboiled juice of the *Juca [yucca] amarga* root. As an eyewitness, Oviedo relates that those poor souls who, like many African tribes, preferred death to forced labor gathered in groups of fifty or so to swallow the poisonous Jatropha juice together. This extraordinary contempt for life characterizes the savage in the most far-flung parts of the globe.

    Reflecting on the compilation of fortuitous circumstances that may have determined a tribe's decision to grow specific kinds of crops, one is astonished to see that Americans, in the midst of such a rich natural environment, would seek in the poisonous root of a euphorbia (a tithymaloid) the same starchy substance that other peoples found in the family of grasses, bananas, asparagus (Dioscorea alata), aroids (Arum macrorrhizon, Dracontium polyphyllum), Solanaceae, bindweed (Convolvulus batatas, C. chrysorhizus), narcissus (Tacca pinnatifida), polygonaceae (P. fagopyrum), nettles (Artocarpus), Leguminosae, and arborescent ferns (Cycas circinnalis). One wonders why the savage who discovered the Jatropha manihot did not reject a root whose poisonous properties an un-

II.406   fortunate experience must have shown him before he could recognize its nutritional value. Perhaps the *Juca dulce*, whose juice is not harmful, was cultivated before the *Juca amarga* from which manioc is derived. Perhaps the same people who were the first to be brave enough to eat the root of Jatropha manihot had also grown plants earlier that were similar to Arum and Dracontium, whose juice is pungent without being poisonous. It was easy enough to note that the starchy flour extracted from the root of an aroid is even more pleasant when it is carefully washed to remove its milky nectar. This very simple observation naturally led to the idea of pressing the starchy flour and preparing it in the same way as manioc. It is conceivable that a people who knew how *to sweeten* the roots of an aroid would endeavor to nourish themselves from a plant in the genus Euphorbia. The transition is easy, although there is an ever-growing danger. In fact, the indigenous peoples of the Society Islands and the Moluccas, who are unfamiliar with Jatropha manihot, grow Arum macrorrhizon and Tacca pinnatifida. The root of the latter plant requires the same precautions as manioc, but tacca bread vies with saga palm bread in the marketplace in Banda.

    The cultivation of manioc requires more care than growing bananas. The crop is more like the potato and is harvested only seven to nine months

after slips have been planted. A people who know how to plant Jatropha II.407
have already taken a significant step toward civilization. There are even
some varieties of manioc, for example, the ones called *manioc bois blanc*
[whitewood] and *manioc mai-pourri-rouge* in Cayenne, whose roots can be
dug up only after fifteen months. A savage in New Zealand would certainly
not have the patience to wait for such a late harvest.

Plantations of Jatropha manihot are now found along the coastline from
the mouth of the Coatzacoalcos River to north of Santander, and from Tehu-
ántepec as far as San Blas and Sinaloa in the low, hot regions of the intendancies of Veracruz, Oaxaca, Puebla, Mexico City, Valladolid, and Guadalajara.
Mr. Aublet, an astute botanist who deigned to study agriculture in the
tropics, wrote for good reason "that the manioc is one of the finest and most
useful products of the American soil and that with this plant, an inhabitant
of the Torrid Zone could make do without rice and every sort of wheat as
well as all the roots and fruits that serve to nourish the human species."

Corn [maize] is found in the same region as bananas and manioc, but its
cultivation is even more important and above all more extensive than that
of the two plants we have just described. Ascending toward the central plateau, one finds cornfields from the coastline to the Toluca Valley, which is II.408
2,800 meters above sea level. A year when there is no harvest means a year
of famine and misery for the inhabitants of Mexico.

Botanists no longer doubt that maize or Turkish wheat are true American
grains and that the New Continent gave them to the Old. It also appears that
this plant was cultivated in Spain long before potatoes. Oviedo, whose first
essay on the natural history of the Indies was published in Toledo in 1525,
relates that he saw corn planted in Andalusia and near the Atocha chapel
around Madrid. This assertion is even more remarkable because a passage in
Hernández[1] might lead one to believe that corn was still unknown in Spain
at the time of Philip II, toward the end of the sixteenth century.

When Europeans discovered the Americas, Zea maize (*tlaolli* in the Aztec language, *mahiz* in Haitian, *cara* in Quechua) was already being cultivated from the southernmost part of Chile as far as Pennsylvania. According
to a tradition of the Aztec nations, the Toltec introduced the cultivation of
corn, cotton, and chili peppers in Mexico in the seventh century of our era.
It is possible, however, that these different branches of agriculture existed
before the Toltec and that this nation, whose great civilization is celebrated II.409

---

1. [Hernández de Toledo], *Rerum medicarum Novæ Hispaniæ Thesaurus*, 1651, book 7,
chap. 40, p. 247. [This footnote was moved from Oviedo above to correct an error.]

by all historians, merely extended them successfully. Hernández informs us that the Otomites themselves, who were but a nomadic and barbarous people, planted corn. As a result, the cultivation of this grain extended as far as beyond the *Río grande de Santiago*, formerly called Tololotlan.[1]

II.410   The corn introduced in northern Europe is not resistant to cold wherever the average temperature does not reach seven or eight degrees Centigrade. Similarly, on the ridge of the Cordilleras, one sees rye and especially barley grow vigorously at elevations that, because of the intemperate climate, are not suitable for growing corn. Barley, however, descends to the warmest regions in the Torrid Zone and as far as the plains where spikes of wheat, barley, and rye cannot develop. Consequently, on the scale of different types of crops, corn now occupies a much larger expanse of the equinoctial part of Mexico than the cereal grains of the Old Continent. Of all the cereal grains useful to humankind, the farinaceous perisperm of corn has the largest volume.

It is commonly believed that this plant is the only species of wheat known to Americans before the Europeans arrived. It appears certain enough, however, that besides Zea corn and Zea curagua, two grains called *magu* and *tuca*—of which, according to Abbot Molina, the first was a type of rye and the second, a type of barley—were cultivated in Chile in the fifteenth century II.411   and long before then. The bread made from this Araucan wheat was called

1. Mr. Robert Brown, whose name is a great authority on all questions of the geography and history of plants, also considers corn, manioc, capsicum, and tobacco plants of American origin (*Botany of Congo*, p. 50), whereas Mr. Crawfurd, in his excellent work on the Archipelago of India ([*History of the Indian Archipelago*,] vol. I, p. 366), believes that corn, which has a *non-imported* name, *jagang*, in Malay and *javanala* in Sanskrit (Ainslie, *Mat[eria] Med[ica] of Hindoostan*, p. 218), was cultivated in this archipelago before the discovery of America. In the earliest times, long before the Europeans arrived, could people of the Malay race or from Greater Polynesia have brought corn and bananas from Asia to America? The botanical isolation of the genus Zea and its isolation from all grains that grow spontaneously are very remarkable facts. "In continental eastern Asia, corn does not have a proper name; it is called *yu-chu-chu* in Chinese, chu or yu (jade) grains, or *yu-my* (rice that resembles jade); in Japanese, *nanban-kibi* or nanban grains, commonly called *foreign wheat*; in Manchu, *aikha-chouchou*, grains of colored glass. The great Chinese herbarium entitled *Bencao Gangmu*, composed toward the middle of the seventeenth century, relates that corn was imported into China from *western* countries" (Mr. Klaproth's handwritten note). One might be astonished to see that wheat, one of the five grains that the Chinese have grown from time immemorial, is called *may-tsee*, which corresponds almost exactly to the pronunciation of "maïs;" but one should not forget that the word "maïs" is a corruption of the word *mahiz*, uniquely used in Haiti or San Domingo, and that on opposite coastlines of Asia, the words for this grain present no analogy with the stem *may*. Also, in Celtic and Livonian, *maise* means bread.

covque, a word that was later applied to bread made from European wheat.[1] Hernández [de Toledo] even insists that he found a species of wheat[2] among the Michoacan Indians, which, according to his very precise description, was similar to the *wheat of abundance* (Triticum compositum) thought to be of Egyptian origin. Despite all the information I gathered during my stay in the intendancy of Valladolid, it was not possible for me to shed any light on this important point in the history of cereal grains. No one there knew of a species of wheat native to the country, and I suspect that what Hernández [de Toledo] called *Triticum michuacanense* was a variety of European wheat that had become wild and grew on very fertile soil.

The productive yield of *tlaolli* or Mexican corn is beyond anything imaginable in Europe. Favored by strong heat and high humidity, the plant reaches a height of two to three meters. On the beautiful plains that extend from San Juan del Río to Querétaro, for example, on the farmland of the Esperanza ranch, one fanega of corn sometimes produces eight hundred more. In an average year, fertile land produces three to four hundred fanegas. Around Valladolid, one considers a harvest to be poor if it produces only 130 to 150 times that of the seed planted. Where the soil is most bar- II.412 ren, one can still count on sixty or eighty grains for each planted one. It is believed that in the equinoctial region of New Spain, the yield of corn can be generally figured at 150 grains for each original one. The Toluca valley alone harvests annually more than 600,000 *fanegas*[3] on an expanse of thirty square leagues, of which a large part is planted with Agave. Between the eighteenth and twenty-second parallels, frosts and cold winds make this crop unprofitable on plateaus whose elevation is higher than three thousand feet. The annual production of corn in the intendancy of Guadalajara is, as we have seen above, more than eighty million kilograms.

In the temperate zone, between 33° and 38° latitude, for example, in New California, corn only produces seventy to eighty grains to one in an average year. Comparing Father Fermin Lasuén's unpublished papers that I possess with the statistical tables published in the historical account of Mr. [Alcalá] Galiano's voyage, I should be able to indicate village by village the amount of corn that was sown and harvested. I found that, in 1791,

---

1. Molina, *[Essai sur l']Histoire naturelle du Chili*, p. 101.

2. Hernández [de Toledo, *Rerum medicarum,*] pp. 7, 43. Clavijero, [*Storia antica del Messico,*] p. 56, note F.

3. One *fanega* weight four arrobas or one hundred livres, and in some provinces, it weighs twenty livres (fifty to sixty kilograms).

twelve missions in New California[1] harvested 7,625 fanegas on land that
had been sown with ninety-six fanegas. In 1801, the harvest of sixteen missions was 4,661 fanegas, whereas the amount that had been sown was only
sixty-six fanegas. This results in a production of seventy-nine grains to one
in the first year, and seventy to one in the second year. Like all cold countries, this coastline generally seems more suitable for growing European
cereal grains. The same tables that I am using, however, prove that in some
parts of New California, for example in the fields belonging to the villages
of San Buenaventura and Capistrano, corn often produces 180 to 200 times
the seed amount.

Although a large amount of wheat is grown in Mexico, one must consider corn the people's principal food. It is also that for most domestic animals. The cost of this commodity affects the cost of all others, of which one
might say that it is the natural measure. When the harvest is poor, either
due to a lack of rain or early frosts, there is general famine with the dir-
est consequences. Chickens, turkeys, and even large livestock all suffer
equally. A traveler who crosses a province where the corn has frozen will
find neither eggs, nor poultry, nor *arepa* bread, nor flour for making *atole*, a
pleasant and nourishing soup. The high cost of provisions is especially felt
around the Mexican mines; for example, in the Guanajuato mines where
the fourteen thousand mules that are required by the amalgamation sites
annually consume an enormous amount of corn. We have mentioned above
the impact that famine has had on the progress of the population of New
Spain. The horrible famine in the year 1784 was the result of a strong frost
that struck at a time when it was the least expected in the Torrid Zone, on
August 28, at the negligible elevation of 1,800 meters above sea level.

Of all the cereal grains that people cultivate, none is as uneven in its
production as corn. Depending on changes in humidity and average annual
temperature, production from the same field may vary from forty to two or
three hundred grains to one. When the harvest is good, a settler can make
his fortune more rapidly with corn than with wheat; one might say that this
crop shares the same advantages and disadvantages as grapes. The price of
corn varies from two livres, ten sous to twenty-five livres per *fanega*. The
average price is five livres in the hinterland, but transport increases it so
much that during my stay in the intendancy of Guanajuato, one *fanega* cost
nine livres in Salamanca, twelve in Querétaro, and twenty-two in San Luis

II.413

II.414

---

1. [Espinosa y Tello, ed.] *Relacion del [viage] hecho por las goletas [Sútil] y Mexicana,*
p. 168.

Potosí. In a country where there are no stores and where the local people live from hand to mouth, they suffer immensely when the price of corn remains at two piasters or ten livres per *fanega*; the indigenous people then eat unripe fruit, cactus berries, and roots. Such bad food leads to illnesses   II.415 among them, and one observes that these famines are usually accompanied by high infant mortality.

In hot and very humid regions, corn can yield two to three harvests annually, but there is generally only one. Corn is planted from mid-June to the end of August. Among the numerous varieties of this nourishing grain, there is one whose cob ripens two months after the grain has been sown. This early variety is well-known in Hungary, and ▼ Mr. Parmentier has tried to propagate it in France. The Mexicans who live on the coastline of the South Sea prefer another variety, which Oviedo[1] confirms having seen in his day in the province of Nicaragua and which can be harvested in less than thirty to forty days. I recall having seen this variety near Tomependa on the banks of the Amazon River; but all these varieties of corn that grow so rapidly appear to have less farinaceous grain and are almost as small as the Zea curagua in Chile.

The usefulness of corn for Americans is too well-known for me to dwell on it here. The use of rice is no more varied in China and the East Indies. The ear of corn is eaten boiled in water or roasted. Crushed grains of corn provide a nutritive bread (*arepa*), although it is unfermented and doughy   II.416 because of the small amount of gluten that is mixed with the starchy fecule. The flour is used as grits to make the soups that the Mexicans call *atole*, into which sugar, honey, and sometimes even ground potatoes are mixed. The botanist Hernández [de Toledo][2] describes sixteen types of *atole* that he saw prepared in his time.

It would be difficult for a chemist to prepare the countless variety of alcoholic, acidic, or sweet drinks that the Indians make with special skill by infusing corn grain in which the sugary matter begins to develop by germination. Some of these drinks, which are commonly called *chicha*, are like beer and others are like cider. Under the monastic government of the Inca, it was forbidden in Peru to make intoxicating liquors, especially those called *vinapu* and *sora*.[3] The Mexica despots were less interested in public

---

1. [Fernández de Oviedo, *L'histoire naturelle et generale des Indes,*] book 7, chap. 1, p. 103.

2. Hernández [de Toledo, *Rerum medicarum,*] book 7, chap. 40, p. 244.

3. Garcilaso [de la Vega El Inca, *Comentarios reales,*] book 8, chap. 9 (vol. I, p. 277). Acosta, [*Historia natural y moral de las Indias,*] book 4, chap. 16, p. 238.

and private behavior; consequently, drunkenness was already quite common among the Indians at the time of the Aztec dynasty. But Europeans have increased the lower class's pleasures by introducing the cultivation of sugarcane. Every elevation now offers the Indians its own drinks. The

II.417 plains near the coastline provide spirits made from sugarcane (*guarapo* or *aguardiente de caña*) and *chicha de manioc*. *Chicha de maïs* abounds on the slopes of the Cordilleras. The central plateau is the country of the Mexican grapes: there one finds the agave plantations that provide the people's favorite drink, *pulque de maguey*. The well-to-do Indian adds to these products of the American soil a liquor that is more expensive and rarer: an eau-de-vie from grapes (*aguardiente de Castilla*) that is partially provided by trade with Europe and partially distilled in the country itself. Such are the many resources for a people with an excessive fondness for strong liquor.

Before the Europeans arrived, Mexica and Peruvians pressed the juice of the corn stalk to make sugar. Not content to concentrate this sugar by evaporation, they also knew how to make raw sugar by cooling the thickened syrup. In his description to emperor Charles V of all the commodities sold at the great market in Tlatelolco at the time of his entry into Tenochtitlan, Cortés specifically mentions Mexica sugar. "Here," he writes, "honey from bees and wax are sold, and *honey made from corn stalks* that are as sweet as sugarcane, and honey from a plant that the people call maguay. The indigenous peoples make sugar from these plants and also sell it." The plant stems of all grains contain a sugary substance, especially near the knots. The amount

II.418 of sugar that corn can provide in the temperate zone, however, appears to be small. In the tropics, on the other hand, its fistulous stalk is so sugary that I have often seen Indians sucking it, just as blacks [nègres] suck on sugarcane. In the Toluca valley, corn culms are crushed between cylinders and an alcoholic liquor called *pulque de mahis* or *de tlaolli* is made from the fermented juice; this liquor is a significant object of trade.

Statistical tables drawn up in the province of Guadalajara, whose population is more than half a million, make it probable that in an average year the present production of corn in all of New Spain is over seventeen million kilograms. In temperate climates in Mexico, this grain can be stored for three years, and in the Toluca valley and on all plateaus whose mean temperature is below fourteen degrees Centigrade even for five or six years, especially if the dry stalks are not cut before the ripe grain has been slightly damaged by frost.

In good years, the Kingdom of New Spain produces much more corn than it can consume. Since the country assembles a great variety of climates

in a small space, and since corn almost never grows simultaneously in the hot region (*tierras calientes*) and in the *tierras frías* [cold regions] on the central plateau, the transportation of this grain markedly stimulates inte-   II.419
rior trade. Compared with European wheat, corn has the disadvantage of containing a smaller amount of nutritious substance in a larger volume. This fact, along with the difficulty of the roads on the mountain slopes, are obstacles to exporting it. Export will occur more frequently when the construction of the beautiful highway leading from Veracruz to Jalapa and Perote has been completed. The islands in general, and especially that of Cuba, consume an enormous amount of corn. It is often in shortage on the islands, because their inhabitants' interest is fixed almost exclusively on growing sugarcane and coffee, although learned agriculturalists have long since observed that in the district between Havana, the port of Batabanó, and Matanzas, corn fields cultivated by free workers yield more revenue than a sugarcane plantation that requires enormous advances for buying and maintaining slaves and building workshops.

Although two other grains with farinaceous seeds that belong to the same genus as our barley and wheat were probably sown in Chile besides corn, it is no less certain that before the Spanish arrived in the Americas, none of the cereal grains of the Old Continent were known there. If we suppose that all human beings are descended from the same stock, one might be tempted to think that Americans, like the Atlantes,[1] had already sepa-   II.420
rated from the rest of humanity before wheat was cultivated on the central plateau of Asia. But must we lose ourselves in mythic times to explain the ancient connections that appear to have existed between the two continents? In Herodotus's time, the only agricultural people in the entire northern part of Africa were the Egyptians and the Carthaginians.[2] In the interior of Asia, the tribes of the Mongol race—the Hiong-nu, the Burattes, the Kalkas, and the Sifanes—always lived as wandering shepherds. If these tribes of central Asia or Libyans from Africa had been able to cross over to the New Continent, neither of them could have introduced the cultivation of cereal grains there. The absence of these grains thus does not disprove the Asian origin of the American tribes or the possibility of a fairly recent transmigration.

Since the introduction of European wheat has had the most benefi-cial influence on the well-being of the indigenous peoples of Mexico, it is

---

1. See the opinion expressed by Diodorus of Sicily, *Bibl[iothecae historicae]*, book 3, p. 186 (Rhodom[annus ed.]).

2. ▼ Heeren, ["Bildung und Zustand des Carthagischen Gebiets in] Africa," pp. [39-]41.

interesting to relate when the new branch of agriculture began. One of Cortés's black slaves had found three or four grains of wheat in the rice that provided sustenance for the Spanish army. It appears that these grains were sown before 1530. As a result, the cultivation of wheat is slightly older in Mexico than in Peru. History has preserved the name of a Spanish lady, María de Escobar, wife of Diego de Chaves, who brought the first grains of wheat to the town of Lima, then called Rimac. The yield from the harvests that she reaped from these grains was distributed among the new settlers for three years, such that each farmer received twenty or thirty grains. Garcilaso already complained of the ingratitude of his compatriots who hardly knew the name of María de Escobar. We do not know the specific period when the cultivation of cereal grains began in Peru, but it is certain that in 1547 wheat bread was not yet known in the town of Cuzco.[1] In Quito, ▼ Father [Jodoco] Rixi, a native of Ghent in Flanders, sowed the first European wheat near the monastery of Saint Francis. The monks still avidly display the earthenware vase in which the first wheat came from Europe, which they look upon as a precious relic.[2] Why have the names of those who, instead of ravaging the earth, were the first to enrich it with plants useful to humankind not been preserved everywhere?

II.422   The temperate region, especially those climates in which the average annual temperature does not exceed eighteen to twenty-nine degrees Centigrade, appears to be the most favorable for growing cereal grains, if we include in this name only the nutritious grains known to the ancients, such as wheat, spelt, barley, oats, and rye.[3] In fact, in the equinoctial part of Mexico, European cereal grains are not grown anywhere on the plateaus whose elevation is below eight to nine hundred meters. We have seen above that on the slope of the Cordilleras between Veracruz and Acapulco, such cultivation typically begins only at an elevation of twelve to thirteen hundred meters. Long experience has shown the inhabitants of Jalapa that wheat planted around their town grows vigorously but does not produce

---

1. [Garcilaso de la Vega El Inca,] *Commentatios reales* IX, 24. vol. 2, p. 332. "*Maria de Escobar, digna de un gran estado llevó el trigo al Perú. Por otro tanto adorarón los gentiles á Ceres por Diosa, y de esta matrona no hicieron cuenta los de mi tierra.*"

2. See my *Tableaux de la nature*, vol. 2, p. 166.

3. Triticum (πυρος), Spelta (ξεα), Hordeum (χριθη), Avena (Dioscorides's Βρομος and not Theophrastus's Βρομος), and Secale (τιφη). I shall not discuss here whether the Romans actually grew wheat and rye and whether Theophrastus and Pliny knew our *Secale cereale* [*cereal rye*]. Compare ▼ Dioscorides [*Opera . . . omnia,*] II, 116; IV, 140, Saracen. [Sarassin ed.], pp. 126 and 294; with Columella [*Opera . . . omnia,*] II.10, Theophrastus [*De historia plantarum,*] VIII, 1–04; and Pliny, [*Historia Naturalis,*] I.126.

spikes. They grow it because its stalks and succulent leaves provide feed for livestock (*zacate*). It is quite certain, however, that in the kingdom of Guatemala and thus closer to the equator, wheat matures at much lower elevations than those around the town of Jalapa. A special exposure, cool winds that blow toward the North, and other local factors can modify the influence of climate. In the province of Caracas, I have seen the finest wheat harvests near Victoria (latitude 10°13') at five or six hundred meters absolute elevation, and it appears that the wheat fields around *Quatro Villas* on the island of Cuba (latitude 21°58') are at an even lower elevation. In Île-de-France (latitude 20°10'), wheat is grown on land that is almost at sea level.   II.423

European settlers did not experiment widely enough to determine the *minimum* elevation at which cereal grains can grow in the equinoctial region of Mexico. The absolute lack of rain during the summer months is even more unfavorable to wheat because the heat of the climate is greater. It is true that drought and heat are also very considerable in Syria and Egypt. But Egypt, which is so rich in wheat, has a climate that is essentially different from the climate in the Torrid Zone, and its soil always maintains a certain degree of moisture that is caused by the beneficial flooding of the Nile. Furthermore, the plants that belong to the same genus as our cereal grains only grow wild in temperate climates, even in those of the Old Continent. With the exception of a few giant arundinacea that are *social plants*, gramina generally appear to be infinitely rarer in the Torrid Zone than in the temperate zone where they dominate, so to speak, other plants. Despite the great organizational *flexibility* that is attributed to cereal grains   II.424 (which they share with domestic animals), we should not be surprised to know that they grow better in the hilly part of the central plateau of Mexico, where they find the same climate as Rome and Milan, than on the plains near the equinoctial Ocean!

If the soil of New Spain were watered by more frequent rains, it would be one of the most fertile terrains that humanity has ever cultivated in either hemisphere. Fernando Cortés, the hero[1] whose eyes were set on all branches of the national industry in the midst of a bloody war, wrote to his sovereign shortly after the siege of Tenochtitlan: "All Spanish plants grow admirably well in this land. We shall not do here what we did in the islands, where we neglected cultivation and destroyed the inhabitants. An unfortunate experience must make us more cautious. I beg your majesty to give

---

1. [Cortés,] *Letter to the Emperor Charles V, dated from the great city of Temixtitlan,* October 15, 1524.

orders to the *Casa de Contratación* of Seville that no ship shall set sail for that country without taking on a certain amount of plants and grains." The great productivity of the Mexican soil is incontestable, but the scarcity of water of which we spoke in the third chapter often decreases the abundance of harvests.

II.425   There are only two recognizable seasons in the equinoctial region of Mexico even as far as the twenty-eighth degree of northern latitude: the rainy season (*estación de las aguas*) that begins in the month of June or July and ends in September or October, and the dry season (*el estío*) that lasts for eight months, from October until the end of May. The first rains usually fall on the eastern slope of the Cordilleras. Cloud formations and the precipitation dissolved in the air begin on the coastline of Veracruz. Strong electrical explosions that take place successively in Mexico City, Guadalajara, and on the western coastline accompany these phenomena. Chemical action spreads from east to west in the direction of the trade winds, and rains fall fifteen or twenty days earlier in Veracruz than on the central plateau. During the months of November, December, and January, one sometimes sees rain mixed with sleet and snow on the mountains and even below an absolute elevation of two thousand meters; but these rains are very short-lived and only last for four or five days. However cold they may be, they are considered very useful for the growth of wheat and pastureland. In Mexico as in Europe generally, rain is more frequent in the mountainous region, especially in that part of the Cordilleras that extends from the Orizaba Peak

II.426   via Guanajuato, Sierra de Pinos, Zacatecas, and Bolaños as far as the Guarisamey and Rosario mines.

    The prosperity of New Spain depends on the proportion that is established between the length of the two seasons of rain and drought. The farmer has seldom to complain of too much humidity, and if corn and European cereal grains are exposed to partial floods on the plateaus, several of which form round basins surrounded by mountains, wheat sown on the slopes of the hills grows there all the more vigorously. Rains are rarer and very brief from the twenty-fourth to the thirtieth parallel. Snow that is considerably abundant from 26° latitude fortunately compensates for this lack of rain.

    The extreme dryness to which New Spain is exposed from June to September forces the inhabitants of a large part of this vast country to irrigate artificially. Rich harvests occur only in proportion to water drained from the rivers and brought from a great distance through irrigation canals. Such a system of drainage channels is used especially on the beautiful plains that border the Santiago River, called the *Río Grande*, and on the plains between Salamanca, Irapuato, and Villa de León. Irrigation canals (*acequias*), water reservoirs (*presas*), and water wheels (*norias*) are objects of the greatest importance to Mexican agriculture. Like Persia and the lower part of Peru, the interior of New Spain produces an infinite amount of nutritive cereal grains wherever the industry of humankind has decreased the natural aridity of the soil and the air.[1]

II.427

Nowhere else does the proprietor of a large farm more often feel the need to engage engineers skilled in surveying land and familiar with the principles of hydraulic constructions. In Mexico City, however, like everywhere else, arts that appeal to the imagination are privileged over those that are indispensable to the needs of domestic life. Mexicans manage to turn out architects who learnedly judge the beauty and ordonnance of a building, but nothing is still rarer there than a person capable of building machinery, dikes, and canals. Fortunately, the awareness of their needs has stimulated national industry and a certain wisdom peculiar to all mountain people compensates for the lack of professional education.

In places that are not artificially irrigated, Mexican soil provides pasturage only until March and April. At the time when the southwest wind (*viento de la Misteca*), which is dry and hot, blows frequently, all greenery disappears, and gramina and other herbaceous plants gradually shrivel up. This change is even more deeply felt when the rains from the preceding year have been less abundant and the summer is warmer. Then, and especially in the month of May, wheat suffers greatly if it is not artificially irrigated. Rains only awaken plants in June; with the first showers, the fields are covered with verdure, the leaves on the trees reappear, and the joy of the European who can never forget the climate of his homeland is twofold in the rainy season because it presents him with the image of spring.

II.428

By indicating the months of drought and rain, we have described the pattern that meteorological phenomena commonly follow. In the past few years, however, these phenomena have appeared to deviate from general law, and the exceptions have unfortunately been to the detriment of agriculture. The year I visited the Jorullo volcano, the rainy season was three full months late; it began in September and lasted only until mid-November. In Mexico, one observes that corn, which suffers from autumnal frosts much more than wheat, has the advantage of recovering more easily after long droughts. In the intendancy of Valladolid, between Salamanca and Lake

1. See above, pp. 328 and 377.

Cuizeo, I have seen cornfields that were believed to be ruined grow with astonishing vigor after two or three days of rain. The broad width of the leaves certainly contributes greatly to the nutrition and productive force of this American grain.

II.429

On farms (*haciendas de trigo*) where the irrigation system is well-established, for example near León, Silao, and Irapuato, corn is watered twice—the first time as soon as the young plants sprout in January and the second time in early March when the spike is about to develop. Sometimes an entire field is flooded before sowing begins. By allowing the water to remain for several weeks, the ground observably absorbs so much moisture that the wheat is more easily resistant to long periods of drought. Seed is cast at the same moment as the water is released by opening the irrigation ditches. This method recalls the cultivation of wheat in Lower Egypt, and the prolonged flooding simultaneously decreases the prevalence of parasitic grasses that are mixed in with the harvest as it is reaped, a portion of which has unfortunately been introduced into the Americas with European wheat.

The abundance of harvests from carefully tended land is surprising, especially where the soil has been watered or mellowed several times. The most fertile part of the plateau extends from Querétaro as far as the town of León. These elevated plains are thirty leagues long and eight to ten leagues wide. Thirty-five to forty times more wheat is harvested there than what is sown, and many large farms can count on fifty to sixty grains to one. I found the same productivity in the fields that extend from the village of Santiago to Yurirapundaro in the intendancy of Valladolid. In a large part of the dioceses of Michoacan and Guadalajara, around Puebla, Atlisco, and Zelaya, twenty to thirty grains to one are produced. A field is considered unproductive there when one fanega of wheat sown yields only sixteen fanegas in an average year. The average harvest in Cholula is from thirty to forty grains, but it often exceeds seventy to eighty. In the Valley of Mexico City, one reckons two hundred grains for corn and eighteen to twenty for wheat. I point out that the numbers given here have all the accuracy that can be required of a matter so important to knowledge of natural territorial wealth. Eagerly wishing to familiarize myself with agricultural products in the tropics, I have collected all my information on site; I have compared the data that intelligent settlers living in provinces quite remote from each other provided for me. I have contributed even more accuracy to this work because, as a person born in a country where wheat hardly yields four or five grains to one, I was more apt than anyone else to mistrust agronomists' exaggerations, which are the same in Mexico as in China and

II.430

everywhere where the inhabitants' vanity seeks to take advantage of travelers' credulity.

I am aware that because of the great inconsistency with which different countries sow grain, it would have been better to compare the yield of a harvest with the extent of sown land. But agrarian methods are so inexact, and there are so few farms in Mexico where one accurately knows the number of toises or square varas that they contain, that I had to restrict myself to the simple comparison between wheat harvested and wheat sown. The result of the research I conducted during my stay in Mexico was that in an average year, the mean yield of the entire country was from twenty-two to twenty-five grains to one. Having returned to Europe, I had new doubts as to the accuracy of this significant result, and I might have been reluctant to publish it if I had not been able to consult recently with someone on this subject in Paris, a respectable and learned person who has lived in the Spanish colonies for thirty years and who has had great agricultural success there. Mr. Abad [y Queipo], the deacon of the metropolitan church of *Valladolid de Michoacan*, confirmed that according to his calculations, the average yield of Mexican wheat, far from being below twenty-two grains, is probably from twenty-five to thirty, which, according to Lavoisier's and ▼ Necker's calculations, is five to six times more than the average yield of France.

II.431

Farmers near Zelaya showed me the enormous difference in yield between land that is artificially watered and land that is not. The former, which receives water from the Río Grande via drains into several ponds, yields forty to fifty times the grain sown, whereas the fields that do not enjoy the benefit of irrigation only yield fifteen to twenty times what is sown. The same problem of which agronomists complain in almost all parts of Europe also exists here: too much seed is used, so that the grain is wasted and suffocates. If this were not done, the yield from harvests would appear to be even greater than what we have just indicated.

II.432

It may be useful here to mention an observation made near Zelaya by a trustworthy person who is very familiar with this kind of research.[1] Mr. Abad [y Queipo] randomly collected forty wheat plants (Triticum hybernum) from a healthy wheat field several arpents in size. He submerged their roots in water to remove all the earth from them and found that each grain had sprouted forty, sixty, and even seventy stems. The spikes were almost all uniformly well-endowed. When the number of grains each one

---

1. *Sobre la fertilidad de las tierras en la Nueva España, por Don Manuel Abad y Queipo,* from the bishop of Michoacán (manuscript written in 1808).

contained was counted, one found that this number often exceeded one hundred and even one hundred twenty. Their average number was ninety. A few spikes even had up to 160 grains. What a striking example of productivity! It is generally remarked that wheat tillers well in Mexican fields, that a single grain will produce several culms there, and that each plant has extremely long, tufted roots. Spanish settlers call the effect of such vigorous growth *el macollar del trigo* [the clustering of the wheat].

II.433

The country north of the eminently fertile district of Zelaya, Salamanca, and León is extremely arid, with no rivers or springs, and presents vast extents of hardened clay crusts (*tepetate*) that farmers call *hard* and *cold* terrain, which the roots of herbaceous plants have difficulty penetrating. From a distance, these layers of clay, which I also found in the kingdom of Quito, resemble banks of rocks devoid of any vegetation. They belong to the *trappean formation* and always accompany basalt and grünstein, amygdaloids, and amphibolic porphyry on the ridge of the Peruvian and Mexican Andes. In other parts of New Spain, however, in the beautiful Santiago valley and south of the town of Valladolid, basalts and decomposed amygdaloids have formed a black and very fertile loam after many centuries. The fertile fields around Alberca de Santiago thus recall the basaltic terrain of the Mittelgebirge in Bohemia.

II.434

In our discussion of the specific statistics of the provinces,[1] we have described the waterless deserts that separate Nueva Vizcaya from New Mexico. The entire plateau that extends from Sombrerete to Saltillo and from there toward Punta de Lampazos is a barren, arid plain where only cacti and other spiny plants grow. There is no trace of agriculture, except for a few points where, as around the town of Saltillo, human industriousness has collected a little water for irrigating the fields. We have also drawn a portrait of Old California,[2] whose rocky ground is devoid of loam and natural springs. All these considerations come together to prove that, because of its extreme dryness, a significant part of New Spain located north of the Tropics is not suitable for a large population. What a striking contrast indeed between the physiognomy of the two neighboring countries, between Mexico and the United States! The soil in the latter is but a vast forest scored by a great number of rivers that empty into spacious gulfs. Mexico, by contrast, is a wooded seaboard from east to west and in its center an enormous mass of colossal mountains from whose ridges extend treeless plains that are even

1. Chap. 8, p. 249 [p. 434 in this edition].
2. Chap. 8, p. 264 [p. 443 in this edition].

more arid because the reflection of the sun's rays increases the atmospheric
temperature. In the northern part of New Spain, as in Tibet, Persia, and all   II.435
mountainous regions, a part of the country cannot be adapted to the cultivation of cereal grains until a dense population that has reached a high level
of civilization has conquered the obstacles that nature puts in the way of
the progress of rural economy. But this aridity, I repeat, is not typical: it is
compensated for by the extreme fertility that one observes in the southern
regions, even in that part of the *Provincias internas* near the rivers in the
basins of the Río del Norte, the Gila, the Hiaqui, the Mayo, the Culiacan,
the Río del Rosario, the Río de Conchos, the Río de Santander, the Tigre,
and many torrents in the province of Texas.

Taking the average size of the harvests from eighteen villages over two
years, corn yields from sixteen to seventeen grains for each original one in
the northernmost extremity of the kingdom on the coastline of New California. I believe that agronomists will view with interest the description of
these harvests in a country located on the same parallel as Algiers, Tunis,
and Palestine between 32°39′ and 37°48′ latitude.

NAME OF VILLAGE IN NEW CALIFORNIA	1791 FANEGAS OF WHEAT		1802 FANEGAS OF WHEAT		HARVEST AS MULTIPLE OF SEEDS SOWN		II.436
	SOWN	HARVESTED	SOWN	HARVESTED	1791	1802	
SAN DIEGO	60	3,021	...	...	50 3/10	...	
SAN LUIS REY DE FRANCIA	...	...	100	1,200	...	12	
SAN JUAN CAPISTRANO	80	1,586	103	2,908	19 8/10	28 2/10	
SAN GABRIEL	178	3,700	282	3,800	20 7/10	13 4/10	
SAN FERNANDO	...	...	100	2,800	...	28	
SAN BUENAVEN-TURA	44	259	96	3,500	5 8/10	36 4/10	
SANTA BARBARA	65	1,500	113	2,876	23	25 4/10	
LA PURÍSIMA CONCEPCIÓN	76	800	96	3,500	10 5/10	36 4/10	
SAN LUIS OBISPO	86	1,078	161	4,000	12 5/10	25 4/10	

(continued)

NAME OF VILLAGE IN NEW CALIFORNIA	1791		1802		HARVEST		II.436
	FANEGAS OF WHEAT		FANEGAS OF WHEAT		AS MULTIPLE OF SEEDS SOWN		
	SOWN	HARVESTED	SOWN	HARVESTED	1791	1802	
SAN MIGUEL	...	...	70	1,600	...	22 8/10	
SOLEDAD	...	...	78	500	...	6 4/10	
SAN ANTONIO DE PADUA	90	952	139	1,200	10 5/10	8 7/10	
SAN CARLOS	71	221	60	240	3 1/10	4	
SAN JUAN BAUTISTA	...	...	52	1,200	...	23 1/10	
SANTA CRUZ	...	...	60	550	...	9 1/10	
SANTA CLARA	64	1,400	129	2,000	21 8/10	15 5/10	
SAN JOSÉ	...	...	84	1,200	...	14 3/10	
SAN FRANCISCO	60	680	233	2,322	11 3/10	9 9/10	
	874	15,197	1,956	35,396	17 4/10	17 2/10	

It appears that the northernmost part of this coastline is less favorable to growing corn than the part than extends from San Diego to San Miguel. Moreover, produce from recently planted land is more irregular than in countries that have long been cultivated, although nowhere in New Spain does one observe the gradual decrease in fertility that distresses new settlers wherever forests have been cut down to transform them into arable land.

II.437   Those who have given serious thought to the natural wealth of the Mexican soil know that by cultivating more carefully, and without supposing the extraordinary effort of irrigating the fields, the already planted land could feed a population eight or ten times larger. If the fertile plains of Atlixco, Cholula, and Puebla do not produce more abundant harvests, the main cause must be sought in the lack of consumers and the obstacles that the unevenness of the soil poses to grain trade in the interior, especially to its transportation toward the coastline of the Antillean Sea. We shall return to this interesting topic when we discuss exports from Veracruz below.

What is the actual grain harvest in all of New Spain? One realizes how difficult this problem is to solve in a country where the government has been so unsupportive of statistical research since the death of Count de Revillagigedo. In France even ▼ Quesnay's, Lavoisier's, and Arthur Young's calculations vary by forty-five and fifty to seventy-five million setiers weighing 117 kilograms each. I have no positive information about the amounts of rye and barley harvested in Mexico, but I believe I can calculate approximately the average yield of wheat. The most certain calculation in Europe is based on the estimated consumption of each individual. Mr. Lavoisier and Mr. Arnould used this method successfully, but it cannot be adopted when the population is composed of very heterogeneous elements. The Indians and the Mestizos who live in the countryside eat only cornbread and manioc bread. The white Creoles who live in large towns eat much more wheat bread than those who mainly live on farms. The capital city, where more than 33,000 Indians live, requires annually more than nineteen million kilograms of flour. This consumption is almost the same as in equally populous European cities, and if one wanted to calculate the consumption of the entire Kingdom of New Spain on this basis, one would arrive at a result that would be more than five times too high.

II.438

In light of these considerations, I prefer the method based on partial estimates. According to the statistical table that the intendant of the province of Guadalajara sent to the Chamber of Commerce of Veracruz, the amount of wheat harvested in 1802 was 43,000 *cargas* or 6,450,000 kilograms. Now, the population of the intendancy of Guadalajara is about one-ninth of the total population. There are many Indians in this part of Mexico who eat cornbread, and there are few populous towns inhabited by well-to-do whites. According to the analogy of this partial harvest, the general harvest in New Spain would be only fifty-nine million kilograms.

By adding thirty-six million kilograms, however, because of the beneficial influence of consumption in the towns[1] of Mexico City, Puebla, and

II.439

---

1. Chap. 8, pp. 87 and 158 [pp. 349 and 386 in this edition]. According to the accurate material in my possession, I have drawn up the following table that compares the consumption of flour with the number of inhabitants.

CITY	FLOUR CONSUMPTION	POPULATION
MEXICO CITY	19,100,000 kg	150,000
PUEBLA	7,790,000	67,300
HAVANA	5,230,000	130,000
PARIS	111,300,000	714,000

Consumption in Paris is figured according to information sent by the Count de ▼ Chabrol [*Statistique*] in 1825. See also Peuchet, *Statistique élémentaire de la France*, p. 372. The

Guanajuato on cultivation in surrounding districts, and because of the *Provincias internas* whose inhabitants live almost exclusively on wheat bread, one finds nearly ten million myriagrams or more than 800,000 setiers for the entire kingdom. The result of this estimate is too small, because the northern provinces have not been appropriately separated from the equinoctial regions in the calculation that we have just presented. The very nature of the population itself, however, dictates this separation.

II.440   The largest number of inhabitants in the *Provincias internas* are white or reputed to be so; they number 400,000. If we assume that their wheat consumption is proportional to that of the town of Puebla, we reckon six million myriagrams. Basing our calculation on the annual harvest in the intendancy of Guadalajara, we may allow that in the southern regions of New Spain, whose mixed population is estimated to be 5,437,000, the consumption of wheat in the countryside is 5,800,000 myriagrams. By adding 3,600,000 myriagrams for the consumption in the large interior towns of Mexico City, Puebla, and Guanajuato, we find the total consumption of New Spain to be more than fifteen million myriagrams, or 1,280,000 setiers weighing 240 livres each.

According to this calculation, we might be surprised to find that the *Provincias internas*, whose population is only one-fourteenth of the total population, consume more than one-third of the harvest of Mexico. But we must bear in mind that in these northern provinces, the proportion of the number of whites to the total mass of Spanish (Creoles and Europeans) is one to three, and that this caste is the main consumer of wheat flour. Almost 150,000 of the 800,000 whites in the equinoctial region of New Spain live in an excessively hot climate on the plains near the coasts and eat manioc and bananas.[1]

II.441   These results, I insist, are merely approximations. But it appeared even more interesting to me to publish them since they had already drawn the attention of the government during my stay in Mexico City. One is certain to stimulate the spirit of research when advancing a fact that interests the entire nation and about which one has not yet ventured any calculations.

According to Lavoisier, the total grain harvest in France—that is, the wheat, rye, and barley harvest—before the Revolution and thus at a time when the population of the kingdom had risen to twenty-five million

---

lower classes in Havana eat large amounts of casaba and arepa. Taking the average figure over four years, annual consumption in Havana is 427,018 arrobas or 58,899 *barriles* (*Papel periódico de la Habana*, 1801, n. 12, p. 46).

1. See above, p. 515.

inhabitants, was fifty-eight million setiers or 6,786 million kilograms. According to the authors of the *Feuille du cultivateur* [▼ Jean-Baptiste Dubois de Jancigny, Jean-Laurent Lefebvre, Antoine Augustin Parmentier, and Adrien-Joseph Marchand], the ratio of harvested wheat to entire mass of grains in France is five to seventeen. As a result, the yield of wheat alone before 1789 was seventeen million setiers, which is, in terms of absolute quantities and without considering the populations of the two empires, almost thirteen times more than the wheat harvested in Mexico. This comparison concurs with the basis of my earlier estimate, since the number of inhabitants in New Spain who eat wheat bread regularly does not exceed 1,300,000; it is also a known fact that the French consume more bread than the people of the Spanish race, especially those who live in the Americas.

Because of the extreme fertility of the soil, however, the fifteen million myriagrams of wheat that New Spain now produces are harvested over an expanse of terrain four to five times smaller than that which the same harvest would require in France. It is true that as the Mexican population increases, one must expect to see this *fertility*, which may be called *average* and indicates twenty-four grains to one as the total product of all harvests, decrease. Men begin everywhere by cultivating the least arid land, and the average yield must decrease naturally as agriculture embraces a larger expanse and consequently a greater variety of land. But in an empire as vast as Mexico, this effect is apparent only much later, and its inhabitants' industriousness increases with the population and the extent of its needs.

We are going to collect in one table the information we have acquired about the average yield of cereal grains on the two continents. This is neither a question of examples of extraordinary fertility observed on a small plot of land nor of wheat planted according to the Chinese method. The yield would be relatively the same in all zones if in choosing the terrain, one cultivated cereal grains with the same care as one gives to a vegetable garden. But when dealing with agriculture in general, it can only be a question of broad results, calculations for which the total harvest of a country is considered a *multiple* of the quantity of wheat sown. This multiple, which one may consider one of the primary factors of a people's prosperity, varies in the following manner:

II.442

II.443

Five to six grains to one in *France*, according to Lavoisier and
    Necker. According to Mr. Peuchet, one reckons that 4,400,000
    arpents of wheat sown yield 5,280 million livres annually, which
    amounts to 1,173 kilograms per hectare. This is also the average

yield in northern Germany, Poland, and in Sweden, according to ▼ Mr. Rühs. In France, one reckons fifteen to one for a few highly fertile districts in the département of the Escaut and Nord, eight to ten to one for the good land in Picardy and Île-de-France, and four to five grains for less fertile land.[1]

Eight to ten grains to one in *Hungary*, *Croatia*, and *Slavonia*, according to ▼ Mr. Swartner's research.

Twelve grains to one in the kingdom of La Plata, especially around Montevideo, according to ▼ Don Felix Azara. Near the town of Buenos Aires, one reckons up to sixteen grains. In Paraguay, the cultivation of cereal grains does not extend northward toward the equator beyond the twenty-fourth parallel.[2]

II.444

Seventeen grains to one in *the northern part of Mexico*, and at the same distance from the equator as Paraguay and Buenos Aires.

Twenty-four grains to one in the *equinoctial part of Mexico*, at two or three thousand meters above sea level, reckoning five thousand kilograms per hectare. In the province of *Pasto*, which I crossed in November 1801, and which is part of the kingdom of Santa Fe, the plateaus of the Vega de San Lorenao, Pansitara, and Almaguer[3] generally yield twenty-five grains to one, and in very fertile years thirty-five to one, and twelve to one in cold, dry years. On the beautiful Cajamarca[4] plain in Peru, irrigated by the Mascón and Utusco Rivers and made famous by the defeat of the Inca Atahualpa, wheat yields eighteen to twenty grains to one.

Mexican flour becomes competitive with flour from the United States in the Havana market. When the road being built from the Perote plateau to Veracruz will be completely finished, wheat from New Spain will be exported to Bordeaux, Hamburg, and Bremen. Mexicans will then have a double advantage over the inhabitants of the United States: greater fertility of the land and less expensive labor. In this respect, it would be very interesting to compare here the *average production* of different provinces of the American confederation with the results for Mexico. But the fertility of the soil and the industriousness of the inhabitants vary so greatly from province to province

II.445

1. Peuchet, *Statistique [élémentaire de la France]*, p. 290.
2. Azara, *Voyage[s dans l'Amérique Méridionale,]* vol. I, p. 140.
3. Latitude 1°54′ northern absolute elevation, 2,300 meters.
4. Latitude 7°8′ aust. [south] absolute elevation, 2,860 meters. See my *Recueil d'observations astronomiques*, vol. I, p. 316.

that it is difficult to find the average figure that corresponds to the total harvest. Wht a difference between the beautiful cultivation around Lancaster and that in many parts of New England and North Carolina! "An English farmer," writes the immortal Washington in one of his letters to Arthur Young, "must entertain an extremely comtemptible idea of our husbandry, or *a horrid idea* of our lands, when he shall be informed that not more than eight or ten *bushels* of wheat is the yield of *an acre*; but this low produce may be ascribed, and principally too, to [the fact] . . . that the aim of the farmers in this country (. . .) is not to make the most of the land which is, or has been, cheap but the most of the labour, which is dear; the consequence of which has been that much ground has been *scratched* over, and none cultivated or improved as it ought to have been."[1] According to Mr. Blodget's recent research, which we may consider to be accurate, we find the following results:

II.446

IN THE ATLANTIC PROVINCES EAST OF THE ALLEGHENY MOUNTAINS	PER ACRE	PER HECTARE
In fertile soil	32 bushels	2,372 kg
In poor soil	9	667
In the western territory between the Alleghenies and the Mississippi		
In fertile soil	40	2,965
In poor soil	25	1,853

From this data, one can see that in the Mexican intendancies of Puebla and Guanajuato, where the climate of Rome and Naples prevails on the ridge of the Cordillera, the land is richer and more productive than in the most fertile parts of the United States.

As agricultural progress since the time of General Washington's death has been quite considerable in the *western region*, especially in Kentucky, Tennessee, and Louisiana, I believe that we make take thirteen to fourteen *bushels* as the average number for present harvests, which would still only be 1,000 kilograms per hectare, or less than four grains to one. One generally reckons the wheat harvest in England to be from nineteen to twenty *bushels* per acre, which yields 1,450 kilograms per hectare. This comparison, I repeat, does not signify greater fertility of the soil in Great Britain. Far

II.447

1. This interesting letter was published in the *Statistical Manual for the United States*, 1806 [by Blodget], p. 96. One acre is 4,029 square meters. A *bushel* of wheat weighs thirty kilograms.

from giving us *a frightful idea* of the sterility of the Atlantic provinces of the United States, it proves only that wherever the settler has control of a vast expanse of terrain, the art of farming reaches perfection only very slowly. For this reason, the proceedings of the Agricultural Society of Philadelphia offer various examples of harvests that exceeded thirty-eight to forty bushels per acre whenever fields in Pennsylvania were tilled with the same care as in Ireland and Flanders.[1]

II.448   After comparing the average yield of the land in Mexico, Buenos Aires, the United States, and France, let us now take a rapid glance at the cost of a day's labor in these different countries. In Mexico, it is reckoned at two *reales de plata* (twenty-six soles) in cold regions, and at two and a half reales (thirty-two soles) in warm regions lacking manpower and where the inhabitants are generally very lazy. This cost of labor must appear to be very modest, when we consider the metallic wealth of the country and the amount of money that is constantly in circulation. In the United States, where the whites have pushed back the Indian population beyond the Ohio and the Mississippi, a day's labor is from three francs, ten soles to four francs. In France it can be estimated at from thirty to forty soles,[2] and in Bengal, according to ▼ Mr. Titsingh, at six soles. Thus, despite the enormous difference in freight costs, sugar from East India is cheaper in Philadelphia than

1. According to Mr. Tessier, whose work has contributed so much to the advancement of agriculture, the *legal arpent* or *royal arpent of water and forests* (half a hectare) is 1,344 square toises, and is sown by casting seeds, 160 livres of wheat grain, over good soil; from 200 to 220 livres are cast over ordinary soil. The yield per legal arpent from very fertile soil is from 2,400 to 2,500 livres; from ordinary soil, it is 1,100 to 1,200 livres, and from poor soil, from 900 to 1,000 livres. The average yield per arpent in France is 1,000 livres. In amended and fertilized soil, the potato yields 5,107 square meters per legal arpent on the average, and 3,000 livres of roots; in excellent soil, from 5,000 to 6,000. According to ▼ Mr. Dandolo, the same *pertica* in Lombardy yields 208 livres of wheat and 1,800 livres of potatoes. (*Bibliothèque universelle*, 1807. August, p. 189.) According to Mr. Albert Gallatin, they now sow one or one and a half bushels per acre in Pennsylvania in the United States and harvest an average yield of sixteen bushels of wheat, rarely less than twelve, and from extremely fertile soil, twenty-five to twenty-eight bushels. The harvest that I have seen in the Aragua valleys (in the Republic of Colombia) was 3,200 livres of wheat per *legal arpent* or *arpent des eaux et forets* in France, an amount equivalent to forty-four English bushels per English acre. (*Relation historique*, vol. II, p. 53.

2. [Humboldt made the following correction in his errata section in IV.312: "In the discussion of agricultural products, the cost of a day's labor in France was overestimated, if one considers the total surface area of the kingdom. The cost of a day laborer in Paris is between forty and fifty soles; in the area surrounding Paris, it is eighteen to thirty soles. In 1827, the average cost for France appears to be from twenty to thirty soles."]

sugar from Jamaica. The result of this data is that the ratio of the cost of a day's labor in Mexico to the same cost elsewhere is

IN FRANCE	=	5:6
in the United States	=	5:12
in Bengal	=	5:1

The average price of wheat in New Spain is from four to five piasters or from twenty to twenty-five francs per load (*carga*) weighing 150 kilograms. This is the purchase price from the farmer himself in the countryside. For several years, 150 kilograms of wheat in Paris has cost thirty francs. The high cost of transportation in Mexico City increases the cost of wheat so much that its ordinary price is from nine to ten piasters per load. In times of the greatest or least fertility, the extremes are eight and fourteen piasters. It is easy to foresee that the price of Mexican wheat will drop considerably when roads are built on the slope of the Cordilleras and greater freedom of trade will favor the advancement of agriculture.

II.449

Mexican wheat is of the best quality and is comparable to the finest wheat from Andalusia; it is superior to wheat from Montevideo, whose grain, according to Mr. Azara, is half the size of Spanish wheat. In Mexico, the grain is very large, very white, and very nutritious, especially if it comes from farms where irrigation is used. One observes that mountain wheat (*trigo de sierra*), in other words, the grain that grows at very high levels on the ridge of the Cordilleras, is covered with a thicker husk, while wheat from temperate regions abounds in glutinous matter. The quality of flour depends mainly on the proportion of gluten to starch that is present, and it appears normal that in a climate that is favorable to growing cereal grains, the embryo and the cellular network[1] of the albumen, which physiologists consider the principal location of gluten, would be more voluminous.

II.450

It is difficult to store wheat in Mexico longer than two or three years, especially in temperate climates, and not enough thought has been given to the causes of this phenomenon. It would be wise to set up storehouses in the coldest parts of the country. Furthermore, one finds a prejudice prevalent in several ports in Spanish America that flour from the Cordillera cannot be

---

1. ▼ Mirbel, on the germination of graminaceae. *Annales du Muséum d'hist[oire] nat[urelle]*, vol. 13, p. 147.

stored as long as flour from the United States. It is easy to guess the cause of this prejudice, which has been particularly detrimental to the agriculture of New Granada. It is highly advantageous for the merchants who live on the coastline facing the Antilles, for example, those in Cartagena, who are hampered by trade restrictions, to maintain connections with the United States. Customs officials are sometimes indulgent enough to take a ship from Philadelphia for a ship from the island of Cuba.

Rye, and especially barley, is more resistant to cold than wheat and is grown on the highest plateaus. Barley still produces abundant harvests at elevations where the thermometer rarely rises above fourteen degrees. If we take the average of the harvests in thirteen villages in New California, barley produced twenty-four grains to one in 1791 and eighteen grains to one in 1802.

II.451 Oats are little grown in Mexico. They are seldom seen even in Spain, where horses are fed barley, as in Greek and Roman times. A disease that the Mexicans call *chaquistle*, which often destroys the best wheat harvests when spring and the beginning of summer have been very hot and storms are frequent, rarely attack rye and barley. It is commonly believed that this disease is caused by small insects that fill the interior of the stalk and prevent the life-giving juice from rising to the sprig.

The potato (*solanum tuberosum*), a plant with a nutritious root, which belongs originally to the Americas, appears to have been introduced in Mexico at about the same time as cereal grains from the Old Continent. I shall not decide whether *papas* (the ancient Peruvian name by which potatoes are now known in all the Spanish colonies) were brought to Mexico together with the Schinus molle [Peruvian pepper tree][1] from Peru and consequently by way of the South Sea, or if the first conquerors brought them from the mountains of New Granada. Whatever the case may be, it is certain that the potato was not known in Montezuma's time, and this fact is all the more important, since it is one of those for which the history of the migrations of a plant are linked to the history of people's migrations.

II.452 Certain tribes' predilection for cultivating specific plants indicates most often either a racial correspondence or ancient connections between peoples who live in different climates. In this sense, plants, like languages and the facial characteristics of nations, can become historical monuments. Not only pastoral peoples or those who live exclusively by hunting, driven by restlessness and animosity, undertake long journeys. The Germanic

1. Hernández [de Toledo, *Rerum medicarum*,] book 3, chap.15, p. 54.

hordes, that swarm of people who migrated from the interior of Asia to the banks of the Borysthenes and the Danube, or the savages from Guyana, provide numerous examples of tribes who settled in one place for a few years, cultivated small plots of land, sowed the seeds they had harvested, and abandoned these newly begun fields as soon as a bad year or some other accident disgusted them with the recently occupied site. This is how the tribes of the Mongol race reached the wall that separates China from Tartary as far as the center of Europe, and how American tribes swept back from northern California and the banks of the Gila River to the southern hemisphere. We see torrents of bellicose nomadic hordes everywhere carving a path amidst peaceable agricultural tribes. As stationary as the riverbank, these peoples gather and carefully preserve the nutritious plants and domestic animals that have accompanied the wandering tribes on their far- II.453 flung journeys. The cultivation of a few plants, like foreign words mixed into languages of a different origin, serves to designate the route over which a nation crossed from one extremity of the continent to the other.

Such considerations, which I have developed further in my *Essai sur la géographie des plantes*, suffice to prove the importance for the history of our species of accurately knowing how far the domain of certain plants extended originally before the Europeans' colonizing spirit managed to bring together the most distant climates. If cereal grains and rice[1] from the East Indies were unknown to the earliest inhabitants of the Americas, corn, the potato, and quinoa were not cultivated either in eastern Asia or on the islands in the South Sea. The Chinese, who, some authors insist, must have known corn since time immemorial, introduced it to Japan.[2] If it were founded, this claim would shed light on the ancient connections that are supposed to have II.454 existed between the inhabitants of the two continents. But where are the monuments that testify that corn was cultivated in Asia before the sixteenth century? According to ▼ Father Gaubil's[3] learned research, it even appears doubtful that the Chinese had visited the western coastline of the Americas one thousand years earlier, as a rightly famous historian, Mr. de Guignes,

---

1. What is the wild rice of which Mr. Mackenzie speaks, a gramina that does not grow beyond 50° latitude and which the indigenous people of Canada eat during winter? Mackenzie, *Voyage*, [vol.] 1, p. 156.

2. ▼ Thunberg, *Flora Japonica*, p. 37. In Japanese, corn is called *Sjo Kuso* and *Too Kibbi*. The word *kuso* designates an herbaceous plant and the word *too* suggests an exotic creation.

3. Unpublished astronomical papers by Jesuit Fathers preserved in the archives of the Office of Longitudes in Paris.

had suggested. We persist in believing that corn was not transplanted from the plateau of Tartary to the Mexican plateau and that it is equally improbable that before the Europeans discovered the Americas, this valuable cereal grain was brought from the New Continent to Asia.

The potato presents another very interesting problem, if we consider it in a historical context. It appears certain, as we have related above, that this plant, whose cultivation has had the greatest influence on the advancement of the population in Europe, was unknown in Mexico before the Spanish arrived there. At that time, it was grown in Chile and Peru, in Quito, in the kingdom of New Granada, on the entire Andean Cordillera, from 40° southern latitude nearly to 50° northern latitude. Botanists assume that the potato grows spontaneously in the mountainous part of Peru. By contrast, scholars who have investigated the introduction of potatoes in Europe also confirm that the first settlers whom ▼ Sir Walter Raleigh sent there in 1584 also found them there. So how are we to imagine that a plant said to belong originally in the southern hemisphere was also cultivated at the foot of the Allegheny Mountains, while it was completely unknown in Mexico and in the mountainous and temperate regions of the Antilles? Is it probable that Peruvian tribes had penetrated northward as far as the banks of the Rapahannoc River in Virginia or that potatoes came from north to south like the tribes that since the seventh century have successively appeared on the Anahuac plateau? In either of these hypotheses, how is it possible that this crop was not introduced or preserved in Mexico? These questions have been little discussed until now; however, they well deserve the attention of the naturalist who, by taking a rapid glance that encompasses humankind's influence on nature and the reaction of the physical world to humankind, believes he can read the history of the earliest migrations of our species in the distribution of plants.

II.455

I observe first of all that the potato does not appear to me to be indigenous to Peru and that it is not found wild anywhere in the part of the Cordillera located in the tropics. Mr. Bonpland and I have gathered plants on the ridge and slopes of the Andes from 5° north to 12° south. We have collected information from persons who have examined this chain of colossal mountains as far as La Paz and Orur, and we are sure that no species of Solanaceae with nutritious roots grows spontaneously on this vast expanse of land. It is true that there are very cold and inaccessible places that the locals call *Páramos de las Papas* (deserted plateaus of potatoes); but this name, whose origin is difficult to infer, hardly demonstrates that these high elevations produce the plant whose name they bear.

II.456

Traveling farther south below the tropics, the potato, according to Molina,[1] is found throughout the Chilean countryside. The locals distinguish the wild potato, whose tubers are small and slightly bitter, from the potato that has been grown there for centuries. The first of these plants is called *maglia*, and the second, *pogny*. Another type of Solanum that belongs to the same group with pinnate, non-prickly leaves, and whose root is very sweet and cylindrical shaped, is also cultivated in Chile: the *Solanum cari* is still unknown, not only in Europe but even in Quito and in Mexico.

One might well inquire if these useful plants truly originate in Chile or if, as the result of long cultivation, they have become wild there. The same question has been asked of travelers who found cereal grains growing spontaneously in the mountains of India and the Caucasus. Mr. Ruiz and Mr. Pavón, whose authority is weighty, confirm having found the potato in cultivated ground, *in cultis*, and not in forests and on mountain ridges. But we must note that in Europe, Solanum and different types of wheat do not self-propagate in a lasting way when birds carry their seeds to the prairies and woods. Wherever these plants appear to go wild before our very eyes, far from multiplying like Erigeron canadense, Ocnothera biennis, and other offshoots of the plant kingdom, they disappear in a short period of time. Could Chilean *maglia*, wheat from the banks of the Terek,[2] and the mountain grain (*Hill-wheat*) from Bhutan that ▼ Mr. Banks[3] has recently described be the prototype of Solanum and other cultivated cereal grains?

It is probable that the cultivation of potatoes spread little by little from the mountains of Chile northward through Peru and the kingdom of Quito to the plateau of Bogotá, the ancient Cundinamarca. This was also the path that the Inca took after their conquests. It is easy to imagine why long before ▼ Manco Capac's arrival, in those distant times when the province of Collao and the plains of Tiahuanacu were the center of the first civilization of humankind,[4] tribal migrations in South America occurred from South to North rather than in the opposite direction. Everywhere in the two hemispheres, mountain people showed the desire to move closer to

II.457

II.458

1. [Molina, *Essai sur l'] hist[oire] nat[urelle] du Chili*, p. 102.

2. ▼ Marschall von Bieberstein, *[Tableau des provinces situées] sur la côte occid[entale] de la mer Caspienne*, 1798, pp. 65 and 105.

3. [Banks, "An Attempt to Ascertain the Time,"] *Bibl[iothèque] Brit[annique]*, 1809, *n.* 322, p. 86.

4. Pedro Cieza de León, [*La crónica del Perú*], chap. 105; Garcilaso [de la Vega El Inca, *Comentarios reales,*] 3, I.

the equator or at least closer to the Torrid Zone, which offers at high elevations a mild climate and the other advantages of the temperate zone. By following the Cordilleras either from the banks of the Gila River to the center
of Mexico or from Chile to the beautiful valleys of Quito, the indigenous
people found more vigorous plant life, less premature frosts, and less abundant snow at the same elevations without descending toward the plains. The
plains of Tiahuanacu (latitude 17°10′ south) that are covered with the ruins
of majestic grandeur and the shores of Lake Chucuito, a basin that resembles
a small inland sea, are the Himalayas and the Tibet of South America. In
those places, people governed by laws who had gathered on unfertile land
were the first to devote themselves to agriculture. Many powerful tribes descended from this remarkable plateau, located between the towns of Cuzco
and La Paz, bearing arms, their language, and their skills as far as the northern hemisphere.

II.459   The plants that were the source of agriculture in the Andes spread
northward in two ways, either through the Inca's conquests, which were
followed by the establishment of a few Peruvian settlements in the occupied country, or by the slow but peaceable connections that always take
place among neighboring tribes. The sovereigns of Cuzco did not push
their conquests beyond the Mayo River (latitude 1°34′ north), which flows
north of the town of Pasto. The potatoes that the Spanish found cultivated
among the Muysca people in the kingdom of the Zaque of Bogotá (latitude 4°–6° north) can only have come there from Peru as the result of these
relationships that are gradually established, even among mountain tribes
separated from each other by badlands covered with snow or impassable
valleys. After maintaining their majestic height from Chile to the province
of Antioquia, the Cordilleras drop suddenly toward the sources of the
great Río Atrato. Chocó and Darién are simply a group of hills on the Isthmus of Panama that are only a few hundred toises high. The cultivation
of potatoes thrives in the tropics only on very elevated plateaus in a cold
and foggy climate. The Indian in hot countries prefers corn, manioc, and
bananas. Furthermore, hordes of savages and hunters, the enemies of any
II.460   and all civilization, have inhabited Chocó, the Darién, and the isthmus
covered with thick forests for centuries. We should not be surprised, then,
that these causes have collectively prevented the potato from making its
way to Mexico.

We are not aware of a single fact by which the history of South America
is linked with the history of North America. As we have already observed
several times, the migration of tribes in New Spain moved from North to

South. A great similarity of customs and civilization is recognizable[1] between the Toltec, whom a plague appears to have chased off the Anahuac plateau in the middle of the twelfth century, and the Peruvians ruled by Manco-Capac. It is possible that peoples who left Aztlan advanced as far as beyond the Isthmus or the Gulf of Panama. But it is improbable that products from Peru, Quito, and New Granada ever reached Mexico and Canada because of migrations from South to North.

The result of all these considerations is that if the settlers sent by Raleigh actually found potatoes among the Indians of Virginia, it is difficult to refute the idea that this plant originally grew wild in some region of the northern hemisphere, as it did in Chile. ▼ Mr. Beckmann's, Mr. Banks's, and Mr. Dryander's interesting research[2] proves that the vessels returning from Albemarle Bay in 1586 brought the first potatoes to Ireland, and that ▼ Thomas Harriot, more famous as a mathematician than a navigator, described this nutritious root by the name *openawk*. In his *Herbal* published in 1597, ▼ Gerard calls it the Virginia potato or *norembega*. One might be tempted to believe that the English settlers had received it from Spanish America. Their settlement existed from July of 1584. The navigators of the time would not set course directly westward to land on the coastline of North America; they were still accustomed to following the route Columbus had indicated, taking advantage of the trade winds in the Torrid Zone. This passage facilitated connections with the Antilles, which were the center of Spanish trade. Sir Francis Drake, who had just sailed among these islands and along the coastline of the mainland, had landed at Roanoke[3] in Virginia. It appears then normal enough to suppose that the English themselves had brought potatoes from South America or Mexico to Virginia. When they were sent from Virginia to England, potatoes were already common in Spain and Italy. It is not surprising then that produce that had passed from one continent to another was able to reach America from the Spanish colonies to the English colonies. The very name by which

II.461

II.462

---

1. I have discussed Chevalier Boturini's interesting hypothesis in my paper on the earliest inhabitants of the Americas. ("Über die Urvölker") *Neue Berlin[ische] Monatsschrift*, 1806, p. 205.

2. Beckmann, *Grundsätze der teutschen Landwirthschaft*, 1806, p. 289; Sir Joseph Banks, "An Attempt to Ascertain the Time When the Potatoes," 1808 [1809]. The potato has been broadly cultivated in Lancashire since 1684; in Saxony since 1717; in Scotland since 1728; in Prussia since 1738.

3. Roanoke and Albermarle, where ▼ Amadas and Barlow had established their first settlement, now belong to the state of North Carolina. On Raleigh's colony, see Marshall, *Life of [George] Washington*, vol. I, p. 12.

Harriot describes the potato would seem to support the hypothesis of a Virginian origin. Would the indigenous peoples have had a word for a foreign plant, and would Harriot not have known the word *Papas*?

The plants that belong to the highest and coldest part of the Andes and the Mexican Cordilleras are the potato, the Tropæolum tuberosum,[1] and the Chenopodium quinoa, whose grain is a healthy and appetizing food. The first of these crops is all the more important and widespread in New Spain because it does not require very moist soil. Both Mexicans and Peruvians know how to preserve potatoes for years on end, by exposing them to frost and by drying them in the sun. The hardened, waterless root is called *chunu*, from a word in the Quechua language. It would probably be very useful to imitate this preparation in Europe where the onset of germination often leads to the loss of winter provisions. But it would be even more important to acquire seeds of the potatoes grown in Quito and on the Santa Fe plateau. I have seen some of them that are spherical, more than three decimeters (twelve to thirteen inches) in diameter, and much tastier than the potatoes of our continent. It is known that some herbaceous plants whose roots have been multiplied for a long time ultimately degenerate, especially when one has the bad habit of cutting the roots into several pieces. Experience has shown that in some parts of Germany, potatoes grown from seed are the best tasting of all. Gathering seeds from their place of origin and selecting the most highly recommended varieties even on the Andean Cordillera by size and the taste of their roots could improve the species. We have had a potato in Europe for a long time that agronomists know as the red Bedfordshire potato whose tubers weigh more than one kilogram. But this variety (*conglomerated potatoe*) has a bland taste and is used mainly as animal fodder, while the *papa de Bogotá*, which contains less water, is very starchy, slightly sweet, and very savory.

Among the great number of useful products that tribal migrations and distant navigations have brought to our knowledge, no plant since the discovery of cereals, that is, since time immemorial, has had such a striking influence on the well-being of humankind as the potato.[2] According to ▼ Sir

II.463

II.464

II.465

---

1. This type of nasturtium, similar to Tropæolum peregrinum, is grown in the provinces of Popayán and Pasto on plateaus at three thousand meters absolute elevation.

2. Since the publication of the first edition of this work, the opinion according to which *Solanum tuberosum* is considered only a plant introduced in Virginia has become more widespread. It has been claimed that long before Drake, ▼ John Hawkins, a slave merchant, had brought the valuable product in 1545 from the coastline of New Granada to Ireland. Gerard was the first to cultivate it in England, having received it from Sir Francis Drake

John Sinclair's calculations, this crop can nourish nine persons per *acre* composed of 5,368 square meters. Potatoes have become common in New Zealand,[1] Japan, on the island of Java, in Bhutan, and in Bengal, where according to ▼ Mr. Beckford's testimony, they are considered more useful than the breadfruit tree introduced in Madras. They are cultivated from the extremity of Africa to Labrador, in Iceland, and in Lapland. It is an interesting spectacle to see a plant descended from mountains below the equator advance toward the [north] pole and resist the wintry coldness of the North more than cereal gramina.

We have just examined successively the plant products that form the basic sustenance of the Mexican people: the *banana, manioc, corn,* and *cereals*. We have tried to spread some interest on this subject by comparing agriculture in the equinoctial regions with that in the temperate climates of Europe and by linking the history of the migrations of plants to the events that led mankind to sweep from one part of the globe to another. Without

II.466

---

himself. Cultivation spread to Belgium in 1590, but it was neglected in Ireland until Raleigh reintroduced the potato there at the beginning of the seventeenth century, bringing it from Virginia. ▼ Putsche und Bertuch, [*Versuch einer*] *Monographie der Kartoffeln,* 1822 [1819]. I have examined the very rare book entitled *General History of Virginia, New-England and the Summer Isles, from* 1584 *to* 1626, *by Capt. John Smith, Governor in These Countries and Admiral of New-England* (London 1632); but in the part that contains the observations of *Thomas Harriot, a learned mathematician* (p. 9), I was unable to find the description of the potato. This tuberous plant was introduced to the Bermudas not from Virginia but from Europe. The names *Norembega* and *Openawk* (see above, p. 553) given to the *Solanum tuberosum* by the first English writers were also not the names of the indigenous plants: they owe their origin, as far as I know, to one of those so common misunderstandings among travelers who do not know the local language. I have recently discovered that *Norembega* is the former name of New England (Smith, *General Hist[ory],* p. 203). The word *Openawk*, possibly derived from the name of the Lenni-lenape [or Delaware] Indians with whom the first settlers had frequent relations, and which by corruption they called *Openagi* and *Apenagi* instead of *Wapanachki* ([Heckewelder, "An Account of the History, Manners, and Customs, of the Indian Natives,"] *Transactions of the Hist[orical] Committee of the American Philos[ophical] Society*, 1819, vol. I, p. 25. (Could the navigators who brought the plant to England have given it the name of the country and its inhabitants where the settlers had attempted to plant it? Mr. Bonpland and I have never found *Solanum tuberosum* growing wild in any part of America, but Mr. Caldcleugh and ▼ Mr. Baldwin have recently made an important discovery, in Chile and near Montevideo and Maldonado, respectively: this is perhaps ▼ Mr. Dunal's *Solanum Commersonii*, but Mr. Lambert considers this species as simply a variety of the common potato (*Journal of Science [and the] Arts*, [vol. X,] No. 19, p. 28). ▼ Sabine, ["On the Native Country of the Wild Potato,"] in *Trans[actions] of the Horticultur[al] Society*, vol. 5, II, p. 137 [249–59]. Long, [*Narrative of an] Exped[ition]*, vol. I, p. 94. Lambert, "On the Native Country of the Potatoe," in his great work on Pines [*A description of the genus Pinus*], p. 41.)

1. ▼ John Savage, [*Some] Account of New Zealand*, 1807, p.18.

entering into botanical details, which would not be germane to the principal goal of this work, we shall conclude this chapter by indicating briefly the other food plants that are grown in Mexico.

Many of these plants have been introduced since the sixteenth century. The inhabitants of Western Europe brought to the Americas what they had received over the past two thousand years by way of their connections with the Greeks and Romans, the irruption of the hordes of central Asia, the Arab conquests, the crusades, and the Portuguese navigations. All of these botanical treasures, accumulated in one extremity of the ancient continent through the constant westward movement of tribes and preserved by the beneficial influence of an ever-growing civilization, have become almost simultaneously the heritage of Mexico and Peru. We see them later, augmented by products from the Americas, crossing even farther to the islands of the South Sea, to those settlements that a powerful people had recently built on the coastline of New Holland. In this way, if a small corner of the earth becomes the domain of European colonists and presents a wide variety of tiered climates, like the mountainous part of equinoctial America, it can exhibit the prodigious activity that humankind has exerted for centuries. One colony can collect in a small area the most valuable discoveries on the surface of the globe that tribes have made during their long migrations.

II.467

The Americas are extremely rich in plants with nutritious roots. After *manioc* and *papas* or potatoes, there is nothing more useful to a people's sustenance than the *oca* (Oxalis tuberosa), the *batate*, and the *yam*. The first of these products only grows in cold and temperate countries, on the summit and slope of the Cordilleras; the two others belong to the hot region of Mexico. The Spanish historians who have described the discovery of the Americas confuse[1] the words *axes* and *batates*, although one designates a plant in the asparagus group, and the other a convovulus.

The *yam* or *Dioscorea alata*, like the banana tree, appears to be indigenous to the equinoctial region of the globe. The account of Aloysio Cadamusto's[2] voyage informs us that the Arabs were familiar with this root. Its American name may even shed some light on a very important fact for the history of geographical discoveries, one that until now does not appear to

1. *Gómara,* [*La Conquista de México,*] book III, chap. 21.
2. *Cadamusti Navigatio ad terras incognitas* (*Grynæus Orb. Nov.*, pp. 47, 67, 215, ▼ Herrera [y Tordesillas, *Historia general,*] Dec. I, book IV, chap. 7.) Cadamusto describes the famous Admiral Cabral in these words: "Petrus quidam Aliares ac Abrilus Fidalcus" [Petrus something Aliares and Abrilus Fidalcus]. The unfortunate ▼ captain Tuckey found the bitter yam (Dioscorea) growing wild on the banks of the Congo (Brown, *Botany of Congo*, p. 54).

have drawn any scientific attention. ▼ Cadamusto relates that in 1500 the   II.468
King of Portugal had sent a fleet of twelve ships around the Cape of Good
Hope to Calcutta, under ▼ Pedro Alvarez Cabral's orders. After having seen
the Cape Verde islands, this admiral discovered a great unknown land that
he mistook for a continent. He found naked brown men there, painted red,
with very long hair, who plucked their beards and pierced their chins, slept
in hammocks, and were completely ignorant of the use of metals. These
are easily recognizable traits of the indigenous people of the Americas. Cabral landed on the coast of Brazil (Terre de Sainte-Croix or *Insula Psit-
tacorum*). He found a type of millet (corn) grown there and a root *that bore
the name igname* [yam] used to make bread. Three years before Cabral,
Vespucci had heard the inhabitants of the seacoast of Paria pronounce this
same word. The Haitian name Dioscorea alata is *axes* or *ajes*. Columbus
describes the yam by this name in the account of his first voyage; it had also
born this name in the time of Garcilaso, Acosta, and Oviedo,[1] who have
well described the characteristics by which *axes* are distinguishable from
batates.

The first roots of Dioscorea were brought to Portugal in 1596, from the   II.469
small island of Saint Thomas, located near the seacoast of Africa, almost
below the equator.[2] A ship bringing slaves to Lisbon had taken on these
yams to feed the blacks [nègres] during the crossing. Under similar circumstances, several food plants from Guinea have been introduced in the West
Indies. They have been carefully propagated to provide nourishment for
the slaves who were accustomed to them from their native country. It has
been observed that these unfortunate beings' melancholy decreases noticeably when, after disembarking in a new land, they recognize the plants that
surrounded them at birth.

In the warm regions of the Spanish colonies, the inhabitants distinguish
between *axe* and *ñames de Guinea*. The latter came from the coastline of
Africa to the Antilles, and the name *yam* gradually replaced the name *axe*.
These two plants are perhaps only varieties of Dioscorea alata, although
Brown tried to raise them to the rank of species, forgetting that the shape
of the *yam* leaf undergoes a singular change as it grows. Nowhere have we

---

1. *Christophori Columbi navigatio*, chap. 89. [Garcilaso de la Vega El Inca,] *Comentarios Reales*, vol. I, p. 278. *Historia natural de Indias*, p. 242. [Fernández de] Oviedo,
[*L'histoire naturelle et generale des Indes*,] book VII, chap. 3.

2. ▼ Clusius, *Rariorum plantarum hist[oria]*, book IV, p. 77.

II.470   found the plant that Linnaeus calls Dioscorea sativa;[1] nor does it exist on the islands in the South Sea where the root of Dioscorea alata, mixed with coconut milk and banana pulp, is the favorite dish of the Tahitians. The root of the yam grows to an enormous size when planted in fertile ground. We have seen roots weighing from twenty-five to thirty kilograms in the Aragua valleys west of Caracas.

    Batates are called *apichu* in Peru and *camotes* in Mexico, a name that is a corruption of the Aztec word *cacamotic*.[2] Several varieties with white and yellow roots are grown; the ones in Querétaro, which are grown in a climate similar to that of Andalusia, are the most sought after. I doubt that Spanish navigators ever found wild *batates,* although Clusius assumes so. Besides *Convolvulus batatas*, I have seen ▼ Vahl's *C. platanifolius* cultivated in the colonies, and I am inclined to believe that these two plants, Tahitian *Umara* (▼ Solander's *C. chrysorrizus*[3]) and Thunberg's *C. edulis*, which the Portuguese have introduced in Japan, have become constant varieties descended from the same species. It would be interesting to know if the batates grown in Peru and those Cook found on Easter Island are one and the same, since

II.471   the position of that island and its monuments have led many scholars to suspect that ancient relations may have existed between Peruvians and the inhabitants of the land ▼ Roggeveen discovered.

    Gómara relates that when Columbus made his first appearance before Queen Isabella after returning to Spain, he offered her grain of corn, yam roots, and *batates*. Consequently, the cultivation of *batates* was already common in the southern part of Spain toward the middle of the sixteenth century. In 1591, they were even sold in the marketplace in London.[4] It is commonly believed that the famous Drake or Sir John Hawkins introduced them to England where they were long thought to be endowed with the mysterious properties, for which the Greeks recommended onions from Megara. *Batates* grow quite well in southern France. They require less heat than yams, which, furthermore, because of the enormous amount of nutrition that their roots provide, would be much preferable to the batate and

---

    1. Thunberg confirms, however, having seen it cultivated in Japan. Much confusion reigns over the genus Dioscorea, and one hopes that a monograph will be written on it. We have brought back several new species that are described briefly in the *Species plantarum* [by Linné] published by ▼ Mr. Willdenow, vol. IV, part I, pp. 794–96.

    2. The *cacamotic-tlanoquiloni* or *caxtlatlapan*, represented in Hernández [de Toledo, *Rerum medicarum*,] chap. 54, appears to be the Convolvulus jalapa.

    3. Forster, *[De] plantis esculentis*, p. 56.

    4. Clusius, [*Rariorum plantarum historia*,] III, chap. 51.

even the potato, if they could be grown successfully in countries whose average temperature is below eighteen degrees Centigrade.

We must also include among the useful plants native to Mexico the *cacomite* or *oceloxochitl,* a type of Tigridia whose root provided a nutritious flour to the inhabitants of the Valley of Mexico City; the numerous varieties of love-apple or *Tomatl* (Solanum lycopersicum) that used to be sown and mixed with corn; the earth pistachio or *mani*[1] (Arachis hypogea), whose fruit lies beneath the soil and appears to have existed in Africa and Asia, especially in Cochinchina[2] long before the discovery of the Americas; and finally, the different types of pimento (Capsicum baccatum, C. annuum, and C. fructens), which the Mexicans call *chilli* and the Peruvians *uchu,* whose fruit is as necessary to the indigenous people as salt is to whites. The Spanish call the pimento *chile* or *axi* (ahi). The first word is derived from *quauh-chili,* the second is a Haitian word that must not be confused with *axe* which, as we have seen above, designates the English *yam* (Dioscorea alata).

II.472

I cannot recall having seen *topinambours* [Jerusalem artichokes] or what the French call *Canadian truffles* (Helianthus tuberosus) cultivated anywhere in the Spanish colony. According to ▼ Mr. Correia [da Serra], they are not even found in Brazil, although in all our botanical works they are considered native to the country of the Brazilians called Topinambas. The *chamalitl* or sun with large flowers (Helianthus) came from Peru to New Spain. It was formerly sown in many parts of Spanish America, not only to extract the oil from its seeds but also to roast them and make very nutritious bread.

II.473

Rice (Oryza sativa) was unknown to the peoples of the New Continent as it was to the inhabitants of the South Sea Islands. Whenever the first historians use the expression small Peruvian rice (*arroz pequeño*), they mean to designate *Chenopodium quinoa,* which I found very common in Peru and in the beautiful valley of Bogotá. The cultivation of rice, which the Arabs introduced in Europe,[3] is of little importance in New Spain. The

---

1. Like most of the words that the Spanish settlers give to cultivated plants, *mani* comes from the language of Haiti, which is now a dead language. In Peru, Arachis is called *inchic.* Mr. Brown also believes that Arachis is common to both continents. ([*Botany of*] *Congo,* p. 54.)

2. ▼ Loureiro, *Flora cochinchinensis,* p. 522.

3. The Greeks knew rice but did not cultivate it. ▼ Aristobulus in Strabo, [*Rerum geographicarum*] book XV, Casaub[on], p. 1014. Theophr[astes], [*De historia plantarum,*] book. IV, chap. 5. Dioscor[ides, *Opera . . . omnia,*] book II, chap. 116, Sara[ssin ed.,] p. 127.

extreme aridity that prevails in the interior of the country seems contrary to its cultivation. There is divided opinion in Mexico as to the usefulness that might be derived from introducing *mountain rice*, which is common in China and Japan, and known to all the Spanish who have lived in the Philippine Islands. There is no doubt that this *mountain rice*, so extolled of late, only grows on the slope of hills that are watered either by natural tor-

II.474   rents or irrigation canals[1] dug at high elevations. On the seacoast of Mexico, especially southeast of Veracruz on the fertile, marshy land located between the mouths of the Alvarado and the Goascualco Rivers, the cultivation of ordinary rice may one day become as important as it has been for a long time to the province of Guayaquil, Louisiana, and the southern part of the United States.

There is even greater reason to hope that this branch of agriculture will be practiced vigorously, for great droughts and early frosts often thwart the wheat and corn harvests in the mountainous regions, and the Mexican people suffer periodically from the dire consequences of a general famine. Rice has a high nutritional content in a very small volume. In Bengal, where one can buy forty kilograms of rice for three francs, a family of five will consume four kilograms of rice, two of peas, and two ounces of salt.[2] The frugality of the Aztec is almost as great as that of the Hindu; one could avoid the frequent food shortages in Mexico by multiplying the types of crops and encouraging industry to plant crops that are more easily stored and

II.475   transported than corn and farinaceous root. Furthermore, and I make this proposal without touching on the significant problem of the population of China, there is no doubt that land planted with rice can feed a larger number of families than the same expanse planted with wheat. In Louisiana in the Mississippi basin,[3] one arpent of land generally produces eighteen barrels of *rice*, eight of *wheat* and *oats*, twenty of *corn*, and twenty-six of *potatoes*. According to Mr. Blodget, one reckons in Virginia that one arpent (*acre*) yields twenty to thirty *bushels* of *rice*, while wheat yields only fifteen to sixteen. I am aware that rice fields in Europe are considered to be very

1. Crescit oryza Japonica in Collibus et montibus *artificio singulari*. Thunberg, *Flora Japon[ica]*, p. 147. Mr. Titzing, who lived in Japan for a long time, confirms that *mountain rice* near Nagasaki in Japan is irrigated but requires less water than rice grown on the plains. Mr. Crawford informs us, on the contrary, that *mountain* or *dry land* rice is grown without any irrigation. (*Hist[ory] of the Ind[ian] Archipelago,* vol. I, p. 361.)

2. Beckford, *Indian Recreations, Calcutta*, 1807, p. 18.

3. A handwritten note *on the value of land in Louisiana* sent to me by General Wilkinson.

detrimental to the inhabitants' health; but long experience in eastern Asia seems to prove that their effect is not the same in all climates. Whatever the case may be, one must not fear that irrigating rice fields may contribute to the unhealthiness of a country that is already filled with marshlands and mangroves (Rhizophora mangle) and forms a true Delta between the Alvarado, San Juan, and Guasacualco Rivers.

Mexicans now have all the European vegetables and fruit trees. It is not easy to indicate which of the former existed on the New Continent before the Spanish arrived. The same uncertainty prevails among botanists concerning the species of turnips, lettuces, and cabbages grown by the Greeks and Romans. We know with certainty that the Americans have always known onions (*xonacatl* in Mexican), beans (*ayacotl* in Mexican [Nahuatl], *purutu* in Peruvian or the Quechua language), gourds (*capallu* in Peruvian), and some varieties of chickpeas (Cicer, Lin[naeus]). Describing the food products that were sold daily in the market of ancient Tenochtitlan, Cortés[1] writes specifically that all types of vegetables were found there, especially onions, leeks, garlic, garden cress and watercress (*mastuerzo y berro*), borage, sorrel, and artichokes (*cardo y tagarninas*). It appears that no types of cabbage or turnips (Brassica and Raphanus) were cultivated in the Americas, although the indigenous peoples were very fond of cooked herbs. They would mix together all sorts of leaves and even flowers, and this dish was called *iraca*. It appears that the Mexica did not originally have peas, a fact that is all the more remarkable, since we believe that our *Pisum sativum* grows wild on the northwest coast of the Americas.[2]

II.476

In general, if we look at the vegetables of the Aztecs and the large number of farinaceous and sweet roots that were grown in Mexico and Peru, we see that the Americas were not nearly as poor in nutritious plants as scholars who only know the New Continent through ▼ Herrera's and de Solís's works have systematically suggested. There is no relationship between a people's level of civilization and the variety of their produce from agriculture or gardening. This variety is greater or smaller depending on the

II.477

1. [Cortés, *Historia de Nueva-España*,] Lorenzana ed., p. 103; Garcilaso [de la Vega El Inca, *Primera parte de los Comentarios reales*,] pp. 278 and 336; Acosta, [*De natura novi orbis*,] p. 245. Onions were unknown in Peru, and the *chocos* of America were not garbanzos (Cicer arietinum). I do not know if the famous *frijolitos de Veracruz*, which are now exported, are descended from a Spanish *Phasolus* or if they are a variety of Mexican *ayacotli*.

2. On the Queen Charlotte Islands in Norfolk Bay or Tchinkitané. Marchand, *Voyage*, vol. I, pp. 226 and 360. Could a European navigator not have sown these peas? We know that cabbages have recently become will in New Zealand.

frequency of connections between distant regions or whether nations separated from the rest of humankind in distant times were more or less isolated in their local situation. One should not be surprised not to find among the Mexica in the sixteenth century the wealth of plant life that our European gardens now contain. Even Greeks and the Romans did not know spinach, nor cauliflower, nor black salsify, nor artichokes, nor a large number of other vegetables.

II.478   The central plateau of New Spain produces cherries, plums, peaches, apricots, figs, grapes, melons, apples, and pears in the greatest abundance. Around Mexico City, in the villages of San Augustín de las Cuevas and Tacubaya, the famous garden of the Carmelite convent in San Angel and the Fagoaga family garden in Tanepantla produce an incalculable amount of fruit, most of it quite delicious, in the months of June, July, and August, although the trees are in general poorly taken care of. The traveler in Mexico as well as in Peru and New Granada is astonished to see the tables of the wealthy laden at once with fruits from southern Europe, pineapple,[1] small pomegranates (different kinds of *Passiflora* and *Tacsonia*), sapotes, mameyes, guavas, anonas, cherimoyas, and other valuable produce of the Torrid Zone. This variety of fruits is found in almost the entire country, from Guatemala as far as New California.

    In studying the history of the conquest, one admires the extraordinary
II.479 activity with which the Spanish in the sixteenth century spread the cultivation of European vegetable plants on the ridge of the Cordilleras, from one end of the continent to the other. Ecclesiastics and especially the religious missionaries have contributed to this rapid advancement of industry. The gardens of the convents and parish priests have been like so many plant nurseries from which recently issued useful plants have issued. Even the *conquistadores*, who must not all be considered barbarous warriors, became devoted to country life in their old age. Surrounded by Indians whose language they did not speak, these simple men enjoyed growing the plants that reminded them of the soil of the Estremadura and Castille to console themselves in their isolation. The season when a fruit from Europe ripened for the first time was marked by a family celebration. In his moving description

---

1. During their first navigations, the Spanish would take on pineapples, which, when the crossing was short, provided a pleasant repast in Spain. They were presented to Emperor Charles V, who found the fruit very beautiful but would not taste it. We have found the most delicious wild pineapple at the foot of the great mountain of Duida on the banks of the Alto Orinoco. All the seeds are often not perfectly developed. Pineapple was grown in China as early as 1594, where it was brought from Peru. ▼ Kircher, *China [. . .] illustrata*, p. 188.

of the earliest settlers' way of life, the Inca Garcilaso relates with touching naïveté how his father, the courageous *Andrès de la Vega,* gathered his old companions in arms to share with them the first three asparagus stalks that grew on the Cuzco plateau.

Before the arrival of the Spanish, Mexico and the Cordillera of South America produced several fruits that are very similar to those found in the temperate climates on the Old Continent. The physiognomy of plants presents similar characteristics wherever the temperature and humidity are the same. The mountainous part of equinoctial America has cherry (capuli), II.480 walnut, and apple trees, blackberries, strawberries, Rubus, and gooseberries that are indigenous to it, which I shall describe in the botanical section of my *Voyage aux régions équinoxiales.*[1] When he arrived in Mexico, Cortés relates that beside the local cherries, which are very tart, he saw plums, *ciruelas.* He adds that they resembled perfectly the ones in Spain. I doubt that these Mexican plums exist, although Abbot Clavijero also mentions them. The first Spaniards perhaps mistook the fruit of the *Spondias,* which is an ovoid drupa, for European plums.

Although the Great Ocean washes the western coastline of New Spain, and although Mendaña, Gaetano, Quiros, and other Spanish navigators were the first to visit the islands located between the Americas and Asia, the most useful products of these countries—the breadfruit tree, flax from New Zealand (Phormium tenax), and sugar cane from Tahiti—remained II.481 unknown to the inhabitants of Mexico. After traveling almost around the world, these plants would gradually reach them from the Antilles. After ▼ Captain Bligh left them in Jamaica, they have propagated rapidly on the island of Cuba, in Trinidad, and on the seacoast of Caracas. The breadfruit tree (Artocarpus incisa) of which I have seen large plantations in Spanish Guyana, would grow vigorously on the warm and humid coastline of Tabasco, Tustla, and San Blas. It is improbable, however, that this crop could ever persuade the indigenous people to abandon cultivating banana trees which provide more nutritious food on the same expanse of land. It is true that for eight months of the year, the Artocarpus is continuously laden

---

1. ▼ Mr. Kunth, a famous botanist, has described these species in his *Nova genera et Spec. Plant. æquin. Orbis novi,* by the names Mespilus rubescens (Moran in Mexico), M. Stipulosa (Chillo near Quito), Cerasus salicifolius (New Granada), Morus celtidifolia and M. corylifolia (Mexico), Ribes multiflorum, R. alline, R. microphyllum and R. jorullense (Mexico), R. frigidum (Quito), Rubus floribundus (Loxa), R. bogotensis, R. glabratus and R. nubigena (Andes of New Granada). The wild strawberry plant that we found as we crossed the Cordillera de Quindin is the true Fragaria vesca.

with fruit and that three trees are enough to nourish one adult.[1] But one arpent or half a hectare of land can also contain only thirty-five to forty breadfruit trees[2] because they bear less fruit when they are planted too densely
and their roots meet.

The extreme slowness of the crossing from the Philippine Islands and
the Mariannas to Acapulco, and the fact that galleons from Manila must sail
to higher latitudes to catch the northwest winds, make it very difficult to
introduce plants from eastern Asia. For these reasons, no plants from China
or the Philippine Islands are found on the western coastline of Mexico,
except perhaps the *Triphasia aurantiola* (*Limonia trifoliata*), an elegant
bush whose fruit is preserved, and which according to Loureiro is identical
to the Citrus trifoliate or ▼ Kämpfer's *Karatats-banna*. As for orange and
lemon trees, which in southern Europe can withstand cold to five or six
degrees above zero without suffering, they are now grown throughout New
Spain, even on the central plateau. It has often been debated whether these
trees existed in the Spanish colonies before the discovery of the Americas,
or if Europeans brought them from the Canary Islands, the island of Saint
Thomas, or the seacoast of Africa. It is certain that an orange tree with
small, bitter fruit and a prickly lemon tree that produces round, green fruit
with a particularly oily rind scarcely the size of a walnut grow wild on the
island of Cuba and the coastline of the continent (Terra Firma). But in all
my research, I have never found a single one of them in the interior of the
forests of Guyana between the Orinoco, the Casiquiare, and the Brazilian
borders. Perhaps the citrus tree with small green fruit (*Limoncito verde*) was
formerly grown by the indigenous people, and perhaps it has gone wild only
where the population and the area of cultivated land were the greatest. I am
inclined to believe that the Portuguese and the Spanish[3] introduced only
the citrus tree with large yellow fruit (*Limón sútil*) and the sweet orange
tree. We saw them on the banks of the Orinoco only where the Jesuits had
established their missions. At the time of the discovery of the Americas,
the orange tree had existed even in Europe for only a few centuries. If there
had been ancient connections between the New Continent and the islands
in the South Sea, the true *Citrus aurantium* could have arrived in Peru or

II.482

II.483

---

1. Georg Forster, "Vom Brodbaume," 1784, p. 23.
2. Compare with what we have said above about the yield of bananas, wheat, and potatoes, pp. 515ff.
3. [Fernández de] Oviedo, [*L'histoire naturelle et generale des Indes,*] book VIII, chap. I.

Mexico from the West, since Mr. Forster found this tree in the Hebrides where Quiros had seen it long before him.[1]

The great similarity that the climate of the plateau of New Spain provides with Italy, Greece, and southern France should encourage the Mexican to grow olives. This crop was successfully attempted at the beginning of the conquest, but through its unfair policy, the government tried to prevent it indirectly rather than support it. There was no formal prohibition, but the settlers did not dare take up a branch of national industry that would   II.484
have soon aroused the jealousy of the homeland. As a rule, the court of Madrid looked askance on the cultivation of olives, mulberries or blackberries, hemp, flax, and grapevines in the New Continent. If it tolerated trade in wine and local oils in Chile and Peru, this is only because those colonies located beyond Cape Horn are often poorly supplied from Europe, and one fears the repercussion of irksome measures in such remote provinces. The most offensive set of bans was carried out strictly in all the colonies on the Atlantic seaboard. During my stay in Mexico, the viceroy received the order from the court to have the vines uprooted (*arancar las cepas*) in the northern provinces of Mexico because the trade in Cádiz had complained of a decrease in the consumption of Spanish wine. Fortunately, this order, like many others handed down by ministers, was not executed. It was felt that although the Mexican people were extremely patient, it could prove dangerous to drive them to despair by devastating their property and forcing them to buy from European monopolists what Mother Nature produces on Mexican soil.

The olive tree is still not very common in New Spain; there is, however, a very beautiful plantation that belongs to the archbishop of Mexico   II.485
in Tacubaya, two leagues southeast of the capital. This *olivar del Arzobispo* produces annually two hundred arrobas (nearly 2,500 kilograms) of oil of very good quality. Many olives are also grown around Tacubaya on the Hacienda de los Morales near Chapultepec, in Tuyagualco near Lake Chalco and in the district of Celaya. We have also already discussed above (on p. 448) the olives grown by missionaries in New California, especially near the village of San Diego. Freely involved in cultivating his land, the Mexican will in time be able to do without oil, wine, hemp, and flax from

1. [Forster, *De*] *plantis esculentis insularum [oceani] australis*, p. 35. The common orange of the island in the Great Ocean is the Citrus decumana. The mango tree (*Garcinia mangostana*) of which countless varieties are carefully grown in the East Indies and in the Archipelago of the Asian Seas has become very widespread in the Antilles in the last ten years. During my time in Mexico, it did not yet exist.

Europe. The Andalusian olive that Cortés introduced suffers occasionally from the cold on the central plateau; the frosts there, while not hard, are frequent and very protracted. It would be more useful to plant the Corsican olive in Mexico, which more than any other is resistant to changes in climate.

To end our list of nutritious plants, we shall briefly discuss the plants that provide beverages for the Mexican people. We shall see that in this context, the history of Aztec agriculture offers an attribute that is all the more interesting, as one finds nothing like it among other much more civilized nations than the ancient inhabitants of Anahuac.

There hardly exists a tribe of indigenous peoples in the world who does not know how to prepare a beverage from the plant kingdom. The miserable hordes that roam in the forests of Guyana make blends of various fruits of the palm tree that are as refreshing as the barley water that Europeans make. Relegated to a mass of dry rocks with no springs, the inhabitants of Easter Island drink the juice extracted from sugar cane besides seawater. Most civilized people derive their beverages from the same plants that form the base of the diet, whose roots or seeds contain sweet property together with the amylaceous substance. Rice in southern and eastern Asia, the roots of yams and some arums in Africa, and cereal grains in northern Europe provide fermented liquors. Few nations grow certain plants simply to make beverages from them. Only west of the Indus does one find vineyards on the Old Continent. When Greece flourished, this cultivation was even restricted to the countries located between the Oxus and the Euphrates, Asia Minor, and Western Europe. In the rest of the world, nature provides wild species of *Vitis*, but nowhere else has humankind attempted to assemble them around him to improve them by cultivation.

The New Continent offers the example of a people who derived beverages not only from the sweet, amylaceous substance[1] of *corn, manioc,* and *bananas,* or from the pulp of some species of *Mimosa,* but who also grew a plant in the family of the Bromelaceae solely to convert its juice into spirits. On the interior plateau, in the intendancies of Puebla and Mexico City, one crosses great expanses of country where one sees only fields planted with *pittes* or *maguey.* Since the sixteenth century, this plant with its leathery and prickly leaves together with the *Cactus Opuntia* has become wild throughout southern Europe, in the Canary Islands, and on the seacoast of Africa and gives a special character to the Mexican landscape. The traveler is struck by the contrast among the plant forms in a wheat field, an agave

II.486

II.487

1. See above, p. 530.

plantation, or a stand of banana trees whose glossy leaves are always a soft, subtle green color. By multiplying certain crops, humankind can change the appearance of the country under cultivation on every latitude!

In the Spanish colonies, there are several species of maguey that deserve to be examined carefully, of which some appear to belong to different genuses because of the division of their corolla, the length of the stamens, and the shape of their stigmata. The *maguey* or *metl* grown in Mexico represent several varieties of the Agave americana that has become so common in our gardens, with its fibrous yellow flowers and straight leaves and stamens twice as long as the contours of the corolla. This *metl* must not be confused with the *maguey de ocui*[1] (floribus ex albo virentibus, longè paniculatis, endulis, staminibus corolla duplo brevioribus) or with the Agave cubensis Jaq. that ▼ Mr. Lamarck has called A. Mexicana and which some botanists, I know not why, have believed to be the mainstay of Mexican agriculture.

II.488

The plantations of *maguey de pulque* are as widespread as the Aztec language. The people of the Tomite, Totonaque, and Mistec races are not addicted to *octli*, which the Spanish call *pulque*. North of Salamanca, maguey is scarcely found growing on the central plateau. The best fields that I ever saw were in the Toluca valley and on the Cholula plains. Agave plants there are arranged in rows, fifteen decimeters apart from each other. The plants begin to produce what is called their *honey*, because of its abundant sweet property, only when the stalk is about to develop. For this reason, it is very important for the grower to know the exact flowering time. The direction of the radical (/) leaves, which the Indian watches attentively, indicates that it is imminent. Until that point, the leaves that were drooping toward the earth suddenly straighten up and begin to cluster, as if to cover the stalk that is about to form. The bundle of central leaves (*el corazón*) turns a lighter green at the same time and lengthens noticeably. The locals have informed me that it is difficult to mistake these signs, but that there are other no less important ones that cannot be described precisely because they refer simply to the aspect of the plant. The grower inspects his agave plantation daily and marks the plants that are about to flower. If he has any doubt, he asks old Indian *experts* in the village, whose judgment, or rather tact, is more reliable.

II.489

1. In the provinces of Caracas and Cumana, one distinguishes the *maguey de cocuyza* from the *maguey de cocuy*: the first one is Agave americana, the second is the Yucca acaulis a port d'Agave. I have seen stalks of the latter from twelve to fourteen meters tall laden with flowers (See our *Nov[a] Gen[era] et Spec[ies]*, vol. I, pp. 289 and 197).

Near Cholula and between Toluca and Cacanumacan, an eight-year-old maguey will already show signs of developing its stalk. Then the harvest of the juice from which *pulque* is made begins. The *corazón* or central bundle of leaves is cut, the wound is insensitively enlarged and covered with lateral leaves that are lifted up by gathering them together and tying their edges. The vessels appear to deposit into this wound all the juice that would have formed the colossal stalk laden with flowers. It is truly a botanical fountain that keeps running two or three months, which the Indians drain three times daily. The amount of *honey* drawn off from the *maguey* at different times of the day tells us how fast or slowly the sap is moving. In twenty-four hours, a plant ordinarily produces four cubic decimeters or two hundred cubic inches, which equal eight *quartillos*. Three quartillos of this total amount are obtained at dawn, two at noon, and three more at six o'clock in the evening. A robust plant provides occasionally up to fifteen quartillos or 375 cubic inches daily for four to five months, which represents the enormous volume of more than 1,100 cubic decimeters. Such a prolific amount of juice produced by a *maguey* that is scarcely one and a half meters high is all the more astonishing because the agave are planted in the driest terrain, often on shelves of rock barely covered by topsoil. A maguey plant that is about to flower is worth five piasters or twenty-five francs in Pachuca. From unforgiving soil, the Indian can reckon on only 150 bottles from each maguey, and ten to twelve sous for the *pulque,* provided in a day. The yield is as inconsistent as from grapevines, which have sometimes more, sometimes fewer clusters of grapes. In chapter six, I have already given the example of an Indian woman from Cholula who bequeathed fields planted with maguey valued at seventy to eight thousand piasters to her children.

II.490

Growing agave has real advantages over growing corn, wheat, and potatoes. With its firm and vigorous leaves, it can withstand drought, hail, and the excessive winter cold in the high Cordilleras of Mexico. After flowering, the stem dies. If the central bundle has been removed, the stem withers after the juice that nature appears to have intended for the growth of the stem has been completely exhausted. An infinite number of shoots then spring from the root of the decayed plant, since there is no plant that multiplies itself more easily. One arpent of land contains twelve to thirteen hundred *maguey* plants. If it is an old field, one can reckon annually that one-twelfth to one-fourteenth of these plants will produce *honey*. A proprietor who plants from thirty to forty thousand *magueys* is sure to provide wealth or his children, but it takes patience and courage to devote oneself to a crop that only begins to be lucrative after fifteen years. In good soil, the agave begins to flower

II.491

after five years; in poor soil, one can only expect to harvest after eighteen years. Although rapid growth is of the greatest interest to Mexican farmers, they do not, however, attempt to accelerate the development of the stalk artificially by mutilating the roots or by watering them with warm water. It has been found that when these methods are adopted, the flow of juice toward the center of the plant is noticeably reduced. A *maguey* plant is destroyed if the Indian, deceived by false appearances, makes the cut long before the time when the blossoms would have developed naturally.

The honey or agave sugar is pleasantly bittersweet. It ferments easily because of the sugar and mucilage that it contains. To accelerate fermentation, however, some old, acidic *pulque* is added. The process is complete in three to four days. The vinous beverage, which is like cider, smells like decayed meat and is very unpleasant. The Europeans who have been able to overcome the disgust produced by this fetid smell prefer *pulque* to all other beverages, finding it to be stomachic, fortifying, and above all very nutritious. It is recommended to very thin persons. I have seen whites, like Mexican Indians, abstain completely from water, beer, and wine, and drink no other liquid except agave juice. Connoisseurs speak enthusiastically of the *pulque* that is made in the village of Hocotitlan, located north of the town of Toluca, at the foot of a mountain almost as high as the eponymous Nevado. They confirm that the excellent quality of this *pulque* depends not only on the art with which the beverage is prepared but also on the taste of the *terroir* reflected in the juice, according to the fields where the plant is grown. There are *maguey* plantations near Hocotitlan (*haciendas de pulque*) with annual revenue of over 40,000 livres. The inhabitants' opinions are much divided as to the true cause of the fetid smell that comes from pulque. They confirm generally that this smell, which is similar to that of animal matter, is attributable to the skins into which the fresh agave juice is poured, but several well-informed persons insist that pulque prepared in these vessels has the same odor, and that if it is absent from the pulque made in Toluca, this is because the fierce cold on the plateau modifies the fermentation process. I became aware of the latter opinion only at the time I was leaving Mexico City, so I must regret not having been able to clarify this interesting point of organic chemistry by direct experiments. The odor may come from the decomposition of a vegetal-animal matter, similar to gluten, in the agave juice itself.

The cultivation of maguey is so significant for tax purposes that the entry duties paid in the three towns of Mexico City, Toluca, and Puebla amounted to 817,739 piasters in 1793. Collection fees were then 56,608

II.492

II.493

piasters, so the government reaped a net profit from agave juice of 761,131 piasters or more than 3,800,000 francs. The hope of increasing the crown's revenue has led lately to a surcharge on making pulque that is as irritating as it is inconsiderate. It is time to change the system in this respect. Otherwise, it is to be expected that this crop, one of the oldest and most lucrative, will decline insensibly despite the people's obvious predilection for the fermented juice of the *maguey.*

II.494 A very intoxicating liquor called *mexical* or *aguardiente de maguey* is distilled from pulque. I have been assured that the plant that is grown for the distillation of its juice is essentially different from the *common maguey* or *maguey de pulque.* It appeared smaller to me, with less milky colored leaves. Not having seen it in bloom, I cannot judge the difference between the two species. There is also a specific variety of sugar cane with a purple stem that comes from the seacoast of Africa (*Caña de Guinea*), which is preferred to Tahitian sugar cane for making rum. The Spanish government, and especially the *Real Hacienda*, has long since clamped down on *mexical* [mezcal], which is strictly prohibited because its consumption is detrimental to the trade of Spanish brandy. An enormous amount of this liquor is made, however, in the intendancies of Valladolid, Mexico City, and Durango, especially in the new kingdom of León. We can estimate the value of this illegal traffic by considering the disproportion that exists between the population of Mexico and the importation of liquors from Europe, which takes place annually through Veracruz. The entire importation amounts to only 32,000 barrels! In some parts of the kingdom, for example in the *Provincias internas* and in the district of Tuxpan, which belongs to the intendancy of Guadalajara, the public sale of *mexical* has been allowed for some time, by imposing a slight tax on this liquor. This measure, which should be generalized, has been both profitable for the tax office and has brought an end to the inhabitants' complaints.

II.495 But *maguey* is not only the vine of the Aztec people; it may also replace hemp from Asia and the Egyptians' papyrus (Cyperus papyrus). The paper on which the ancient Egyptians drew their hieroglyphic figures was made of fibers of agave leaves macerated in water and pasted together in layers, like the fibers of Cyperus in Egypt and the mulberry bush (Broussonetia) from the islands of the South Sea. I have brought back several fragments of Aztec manuscripts[1] written on maguey paper of such different thicknesses that some resemble cardboard and others, Chinese paper. These fragments

---

1. See chapter VI, vol. I, p. 373 [p. 245 in this edition].

are all the more interesting because the only hieroglyphs that exist in Vienna, Rome, and Velletri are written on Mexican deerskin. The fiber that is extracted from maguey leaves is known as pita fiber in Europe; naturalists prefer it to all others because it is less likely to become twisted. When it is still far from budding season, agave juice (*xugo* [jugo] *de cocuyza*) is very bitter and used successfully as a caustic to clean wounds. The Indians used to use the thorns, like those of the cactus, at the end of the leaves as pins and nails. Mexican priests pierced their arms and chests with it in acts of atonement similar to those of Buddhists in Hindustan.

One may conclude from what we have just related about the use of different parts of the maguey that this plant, after corn and potatoes, is the most useful of all the products that nature has conferred on the mountain people of equinoctial America.

When the obstacles that the government has imposed on different branches of national industry have been removed, when Mexican agriculture shall no longer be chained to an administrative system that impoverishes the colonies without contributing to the wealth of the homeland, vineyards will gradually replace maguey plantations. Grape growing will increase above all with the number of whites who drink a large amount of wine from Spain, France, Madeira, and the Canary Islands. But in the present situation, vineyards can hardly be counted among the territorial wealth of Mexico, since the harvest is so negligible. The best grapes come from Zapotitlan in the intendancy of Oaxaca. There are also vineyards near Dolores and San Luís de la Paz, north of Guanajuato, and in the *Provincias internas* near Parras and Paso del Norte. Paso wine is highly valued, especially the wines from the Marquis de San Miguel's estate. This wine can be cellared for several years, although it is not carefully made. The locals complain that the must harvested on the plateau does not ferment easily. They usually add *arope*, a small amount of wine mixed with sugar that has been reduced by cooking it down to syrup, to the grape juice. This process gives a slight taste of must to Mexican wines that they would lose if the art of winemaking were studied in more depth. If in the course of the centuries, the New Continent, jealous of its independence, were to choose to do without the products of the Old World, the mountainous and temperate parts of Mexico, Guatemala, New Granada, and Caracas might provide wine for all of North America. These locales would become for that country what France, Italy, and Spain have long been for Northern Europe!

II.496

II.497

END OF VOLUME TWO

Chronology

1769. Friedrich Wilhelm Heinrich Alexander von Humboldt is born in Berlin (September 14).

1779. Death of Humboldt's father, Major Alexander Georg von Humboldt (January 6).

1787–88. Studies at the University of Frankfurt/Oder; returns to Berlin for private tutoring in botany under the tutelage of Karl Ludwig Willdenow.

1790. Travels with Georg Forster, from Mayence via Cologne, Brussels, and Amsterdam to England. Returns via revolutionary Paris, where Humboldt carts sand for the "temple of freedom that is still under construction" (March to July).

1791–92. Studies at the School of Mines at Freiberg in Saxony.

1792. Returns to Berlin. Begins his career as inspector of mining operations in Prussia.

1793. *Florae Fribergensis specimen plantas cryptogamicus praesertim subterraneas exhibens* (Examples of the Flora of Freiburg, especially displaying cryptogamic underground plants).

1794. First visit with Johann Wolfgang von Goethe in Jena.

1795. Scientific travels to northern Italy, Switzerland, and the French Alps.

1796. Death of his mother Elisabeth, née Colomb (November 19). Decides to retire from his public office to focus on the preparations for his planned voyage.

1797. Humboldt's relations with Goethe and Friedrich Schiller deepen. Conducts astronomical studies based on Franz Xavier von Zach's method of geographically determining locations. Travels to Vienna at the end of May, where he prepares intensively for traveling to the tropics (August to October). Excursion to Ödenburg (Sopron) in Hungary.

1797–98. Humboldt in Salzburg: research with Leopold von Buch; numerous excursions.

1798. Leaves Salzburg for Paris. Meets the young physician and botanist Aimé Bonpland. They become friends and travel together to Marseille, where they wait in vain for transport to North Africa. Finally depart Marseille for Spain in December.

1799. Travels through Spain, from Barcelona via Valencia to Madrid. Audience with the Spanish King in Aranjuez and approval of the expedition to the Spanish colonies. Departure for the New World from La Coruña (June 5). Stopover in the Canary Islands (June 25), where Humboldt climbs the volcano Teide. Humboldt and Bonpland arrive in Cumaná on July 16 after a forty-one-day sea voyage.

1800. Travels on the Orinoco (February to July). Travels across the cataracts of Atures and Maipures up to the border of the Portuguese colonies. Return voyage up the Orinoco to Angostura (Ciudad Bolívar); through the Llanos to Nueva Barcelona and Cumaná. Crossing from Nueva Barcelona to Cuba (November 24 to December 19).

1801. First stay in Cuba (until early March), then departure from the port of Trinidad for Cartagena de Indias. Travels to Santa Fe de Bogotá, where the botanist José Celestino Mutis welcomes Humboldt and Bonpland.

1802. In Quito and the Ecuadorian Andes, where Humboldt climbs several volcanoes, among them the Pichincha and the Chimborazo (January 6 to October 21). Proceeds to Lima and environs; then sea voyage from Callao via Guayaquil to Acapulco.

1803. Arrives in New Spain (Mexico) on March 23. Travels from Acapulco to Mexico City.

1804. Travels from Mexico City to Veracruz, then again to Havana, where Humboldt and Bonpland arrive in April. Second stay in Cuba (April 29 to May 19). Sails from Havana to Philadelphia; meets with President Thomas Jefferson in Monticello. Leaves Philadelphia on July 9 to return to France; arrives in Bordeaux on August 3. Humboldt returns to Paris on August 27 and stays there until March of the following year.

1805. Appointed a member of the Berlin Academy of Sciences. Travels to Italy (March to October) to visit his brother Wilhelm in Rome. Travels to Naples with Joseph-Louis Gay-Lussac and Leopold von Buch; climbs to Vesuvius and conducts comparative studies in volcanology. Returns to Berlin, where he stays until November of 1807. Commences work on the *Voyage aux régions équinoxiales du Nouveau Continent.*

1807. *Ideen zu einer Geographie der Pflanzen, nebst einem Naturgemälde der Tropenländer.* English: *Essay on Plant Geography* (2009).

1807-27. Humboldt returns to Paris, where he concentrates on his *opus americanum* and on other publications. He also engages in collaborations with French scientists and extends his international network further.

1808. *Ansichten der Natur.* English: *Aspects of Nature* (1849) and *Views of Nature* (1850).

1808-11. *Essai politique sur le royaume de la Nouvelle-Espagne.* English: *Political Essay on the Kingdom of New Spain* (1811).

1810-13. *Vues des Cordillères et Monumens des Peuples Indigènes de l'Amérique.* English: *Researches, Concerning the Institutions & Monuments of the Ancient Inhabitants of America, with Descriptions and Views of some of the most Striking Scenes in the Cordilleras* (1813).

1814-31. *Relation historique,* the actual travelogue of the voyage to the Americas. English: *Personal Narrative of Travels to the Equinoctial Regions of the New Continent* (1814-31).

1814-38. *Atlas géographique et physique des régions équinoxiales du Nouveau Continent.* (Geographical and physical Atlas of the equinoctial regions of the New Continent).

1817. *Des lignes isothermes et la distribution de la chaleur sur le globe.* English: "On Isothermal Lines, and the Distribution of Heat over the Globe" (1820-21).

1823. *Essai géognostique sur le gisement des roches dans les deux hémisphères.* English: *Geognostical Essay on the Superposition of Rocks in both Hemispheres* (1823).

1827. Humboldt returns to Berlin, which will remain his permanent residence until the end of his life.

1829. Travels in Russia, to Siberia, and to the largely unknown Central Asia up to the Chinese frontier.

1834-38. *Examen critique de l'histoire de la géographie du Nouveau Continent* (Critical Examination of the historical development of the geographical knowledge about the New World).

1835. Death of his brother Wilhelm von Humboldt (April 8).

1843. *Asie Centrale* (Central Asia).

1845. First volume of *Kosmos* (Vol. 2: 1847; Vol. 3: 1850; Vol. 4: 1858; Vol. 5: 1862 posthumously). English: *Kosmos: A General Survey of the Physical Phenomena of the Universe* (1845–58).

1848. Humboldt joins the funeral procession for the dead revolutionaries in Berlin.

1859. Alexander von Humboldt dies at his residence in Berlin (May 6).

Editorial Note

From today's perspective, it is astounding that an early-nineteenth-century text, to which scholars in the Spanish-, French-, and German-speaking parts of the world have long attributed great significance and influence, has not been available in a complete and reliable English translation until now. Humboldt's texts were often radically abridged in their English translations, and his language was homogenized. Translators tacitly converted currencies and other units of measure; they removed italics and foreign-language words, and excised sentences, footnotes, and even entire passages; and they also added formal features, such as chapter titles and subsections. The only prior English version, also entitled *Political Essay on the Kingdom of New Spain*, dates back to the early nineteenth century; it abounds with transcriptional and translational errors.

About the French Editions

Originally written in French, Alexander von Humboldt's *Essai politique sur le royaume de la Nouvelle-Espagne* appeared in two different versions between 1808 and 1827. It was first published in two volumes in quarto format as part of Humboldt's thirty-volume *Voyage aux régions équinoxiales du Nouveau Continent* (1805–1839): *Essai politique sur le royaume de la Nouvelle Espagne. Par Alexandre de Humboldt. Avec un Atlas physique et géographique, fondé sur des observations astronomiques, des mesures trigonométriques et des nivellemens barométriques* (Paris: Schoell, 1808–1811). The five-volume octavo edition Schoell also released in 1811 was based on the same typeset as the first edition. Depending on the quality of the paper used, the two volumes of the quarto edition (25×34 cm) would have cost either 306 or 308 francs, whereas the five-volume octavo version (18×15 cm) sold for only forty-two francs. A second octavo edition of the *Essai*, revised and enlarged under Humboldt's direction, was published as four volumes without the subtitle between 1825 and 1827 (the last install-ment was most likely delivered at the end of 1826). The revisions added about 60,000 words, the equivalent of nearly 200 octavo pages. The second edition, entitled *Essai politique sur le royaume de la Nouvelle Espagne par Alexandre de Humboldt. Deuxième Edition* (Paris: Antoine-Augustin Ren-ouard), encompasses four volumes of text—about 442,000 words on 1,768 pages—and an atlas. We based our annotated translation on this second edi-tion, which is the most complete edition of Humboldt's *Essai* on New Spain.

Unlike Humboldt's other political essay, the *Essai politique sur l'île de Cuba* (*Political Essay on the Island of Cuba*), the *Essai politique sur le roy-aume de la Nouvelle Espagne* was not part of his travelogue known as the *Relation historique* (*Personal Narrative*). His volumes on New Spain are, in a way, a substitute for the part of Humboldt's travels that is missing from his travel narrative.

The seed for Alexander von Humboldt's *Essai politique sur le royaume de la Nouvelle-Espagne* was a summary document known as the *Tablas geográfico-políticas del reino de Nueva Espana*, which Humboldt also called his *Statistique du Mexique*. Humboldt offered a copy of this document, which he himself had written in Spanish, to Viceroy Jose de Iturrigaray in January of 1804. Excerpts from the *Tablas* were printed in the Mexican journal *Diario de Mexico* between May 1 and May 31. Humboldt's original manuscript is at the Biblioteka Jagiellonska in Cracow, Poland. (For further information, see Horst Fiedler and Ulrike Leitner's comprehensive bibliog-raphy of Alexander von Humboldt's separately published writings, *Alex-ander von Humboldts Schriften: Bibliographie der selbständig erschienenen Werke*, Berlin, 2000).

Translations

Still hoping to write and publish his entire American œuvre almost simul-taneously in both French and German, Alexander von Humboldt himself contributed substantially to the first German edition, *Versuch über den poli-tischen Zustand des Königreichs Neu-Spanien* (five vols., 1809–1814). In ad-dition to working on almost the entire first volume by himself, he provided a new introduction and multiple other additions. As the French manuscript drafts continued to grow, however, Humboldt was forced to end the bilingual experiment. He handed the production rights over to his editor Johann Friedrich Cotta, who contracted Philipp Joseph von Rehfues for the translation, with which Humboldt was very pleased. Cotta published the remaining four German volumes together with Humboldt's in a slightly abbreviated octavo edition (five vols, Tübingen, 1809–1814).

The first English edition, John Black's *Political Essay on the Kingdom of New Spain* (four vols, London: Longman, Hurst, Rees, Orme, and Brown, 1811), which followed hard on the heels of Humboldt's first French edition, did not enjoy any editorial supervision by Humboldt. Black's translation was used in 1811, 1814, and 1822 for a variety of reprints; it appeared as a facsimile in 1966 in New York (AMS Press) and on microfilm in 1967 (Mexican travel literature on microfilm). Several abbreviated editions appeared during the course of the twentieth century: in 1813 "by a citizen of Maryland" (Baltimore: Wane and O'Reilly); in 1824 by John Taylor (*Selections from the Work of Baron von Humboldt, relating to the Climate, Inhabitants, Productions, and Mines of Mexico*, London: Longman et al.); in 1957 by the librarian Hensley C. Woodbridge of Murray State College, who translated and annotated three chapters of book one (*Scripta Humanistica Kentuckiensia* I, Lexington, Kentucky, i–72; his introduction indicates that he also translated most of books two, three, and six, but that those remain unpublished); and in 1972 by Mary Maples Dunn, who abridged Black's versions (New York: Knopf). While the 1824 and 1972 editions include notes and emendations by the editors, only the 1813 and 1957 editions are actually based on new translations.

The first Spanish version, Vicente González Arnao's *Ensayo político sobre la Nueva España* (four vols., Paris: Rosa, 1822), has been the foundation for all further Spanish editions. The text was at least in part revised by Humboldt himself, who had become acquainted with González Arnao in Paris. While omitting Humboldt's dedication to the King, the original editor's preface, the entire "Analyse raisonnée," and the index, Gonzalez Arnao added some information about the mining industry in New Spain to chapter XI, including a variety of tables. González Arnao's translation was reprinted five times before 1918, and new critical editions of it appeared in 1941 and 1966, one by Vito Alessio Robles (Mexico: Robredo), the other by Juan Antonio Ortega y Medina (Mexico: Porrúa). Ortega y Medina's edition is considered the standard Spanish reference for Humboldt's text and has been reprinted in 1973, 1978, and 1984. (For additional information about the translations, see Fiedler and Leitner's *Alexander von Humboldts Schriften*, 193–208.)

About This Edition

COMPLETENESS: This edition provides the first unabridged English translation of the four-volume *Essai politique sur le royaume de la Nouvelle Espagne* from 1827, in the exact order of the French original. We have restored Humboldt's text, adding neither chapters nor chapter titles, nor subheadings. Nothing was subtracted. Any editorial changes have been marked in [square brackets].

ACCURACY AND RELIABILITY: It is well worth asking what *accuracy* and *reliability* might mean, given that Humboldt revised and amended his writings almost obsessively and that his texts were issued by different publishers in different editions, often simultaneously, and over a period of almost thirty years. We decided to translate the second edition of the *Essai politique sur le royaume de la Nouvelle-Espagne*, because Humboldt stopped revising the essay with this edition, which has never before appeared in English in its entirety. Our goal was not to make this chronologically last edition *the* authoritative text—the one original—to which all other versions must defer. Rather than trying to immobilize Humboldt's unstable textual creations to satisfy some readers' desire for an original, we have tried to account for the historical existence of multiple originals by rendering transparent the exact nature and extent of his revisions. Technology has greatly facilitated this detailed textual comparison, which will eventually be made available online.

Being sensitive to the historical dimensions that retrospectively render a text multiple and unstable has affected our translation in various ways. One instance in which we have not simply modernized Humboldt's language, however cautiously, is nineteenth-century racial terminology. While we have translated "nègre" as "black" throughout, we have rendered Humboldt's "gens de couleur" as "people of color" and his "mulâtres" and "mulâtresses" as men and women of "mixed race." For these and any other racial categories, we have always preserved the specific French terms Humboldt employed in [square brackets].

READABILITY: In addition to completeness and accuracy, readability has been our third major concern in this edition. Our goal was not to create a full-scale critical edition—something that would require a team far larger and more disciplinarily diverse than ours—but to provide a readable scholarly edition unencumbered by an extensive critical apparatus. In keeping with this aim, we have limited our own apparatus to the online annotations, a bibliography of Humboldt's directly and indirectly referenced sources (Humboldt's Library), and an index. Annotations are marked in the text with a ▼ symbol so as not to interfere with Humboldt's own footnote numbers. Although intended primarily for nonspecialist readers, the annotations will (we hope) also be informative for readers already familiar with Humboldt's cast of characters and the historical contexts of his writings. While they include historical and select scientific information, our annotations focus particularly on giving readers a sense of the extensive global network that Humboldt created and carefully nurtured during his lifetime. We have done so by emphasizing relevant connections between the historical personalities that populate his essay's pages.

Above all, we wanted to create a readable English version that also remains very close to Humboldt's French text. Since French has changed far less, and far less quickly, during the nineteenth and twentieth century than has English, careful modernization seemed to us the best approach to rendering Humboldt's voice. We have cautiously modernized Humboldt's language and updated his spellings, avoiding archaisms that might sound precious, without, however, collapsing entirely the inevitable distance between the twenty-first century and the time of Humboldt's writing. We have marked this distance by retaining in English Humboldt's at times elaborate linguistic formalities, especially in accounts of his collaborations, which have an endearing ring in modern English. We have also tried to retain other linguistic characteristics at the level of adjectives and adverbs. In this way, we hope to have kept Humboldt's French, which carries the unmistakable cadences of his native German, from sounding too much like colloquial American English.

ORTHOGRAPHY: In the *Essai politique sur le royaume de la Nouvelle-Espagne*, Alexander von Humboldt used a plethora of terms and phrases in Spanish and Nahuatl either in their contemporary or historical spellings. As was typical for his time, Humboldt often eschewed accents in the indigenous names of places, persons, and idioms. While we have tacitly modernized Humboldt's Spanish orthography in places where we did not have to interfere with his original quotations in other languages, we also decided not to do so in the case of originally pre-Columbian appellations. At times, Humboldt went back and forth between different spellings of the same term. To increase the text's readability, we have, in such cases, created consistencies, but never at the expense of possibly important nuances or different levels of signification. Place names have been updated to current use, except in cases where a new name is noticeably different. In those cases, the old name was kept, followed by the new name in [square brackets]. To give the reader a strong sense of the texture of Humboldt's writing, we have kept intact Humboldt's seemingly capricious uses of capitalization, lower case, and italics. Our sole interference has been with numbers: to minimize the potential for future errors, which are easy to make when all numbers are represented in Arabic numerals, we have spelled out all fractions and numbers below one hundred—even though some may consider the results of this practice awkward.

CORRECTIONS: We have silently corrected errors that we could determine to be merely typographical. Most of those are numbers in the tables or in the pagination of references. Questionable instances that went beyond purely formal aspects into the realm of content have only been corrected when it was clear that they were indeed the result of an error. All other corrections, including in Humboldt's references, appear in [square brackets]. We have integrated Humboldt's own corrections into the text of the translations and eliminated his errata section from book IV.

USE OF FOREIGN LANGUAGES: In the *Essai politique sur le royaume de la Nouvelle-Espagne*, as in many of his writings, Humboldt used many foreign-language terms and even quotations without offering his readers translations. To enhance our translation's readability, we have provided English translations in [square brackets] immediately after each word or citation in all cases where context does not render their meanings self-evident. In each case, we were concerned with creating linguistic and conceptual transparency, while retaining the characteristic polyvocality of Humboldt's writing.

REFERENCES: Humboldt's bibliographical references in both the footnotes and the textual parentheses (at times even directly in the text) follow his four-volume French edition, except that we have formally unified his references to volumes and other sources and have replaced his sometimes odd abbreviations of titles with fuller information. We have provided additional information in [square brackets] to enable readers to locate specific items in our bibliography of Alexander von Humboldt's sources (see Humboldt's Library), which, in addition to identifying the editions Humboldt actually used (insofar as that was possible), also contains additional information, for instance, about other editions and translations.

TABLES: This edition includes all of Humboldt's tables exactly where he placed them in the four-volume 1827 edition. We have tried to approximate the visual appearance of these tables as much as was sensible.

INDEX: Our indexes follow Humboldt's own. His index, like our own, consists of proper names of persons and places, scientific instruments and concepts, historical events, and other subject categories that will help the reader better to navigate the text of the translation and the annotations.

MAPS: All maps and drawings from Humboldt's *Atlas géographique et physique du royaume de la Nouvelle Espagne* (Paris: Schoell, 1811), which are part of this edition, are accessible online at www.press.uchicago.edu/hie. Even though Humboldt discusses the maps in his "Reasoned Analysis" and not in the *Atlas*, he thought of image and text as belonging together. In the translations of the *Essai* into other languages (see above), the maps were typically omitted; at best, they were only partially reproduced.

WEBSITE: This edition, like future volumes in this series, has an accompanying website, which makes available the annotations, Humboldt's Library, and additional materials to interested readers and researchers. The site at press.uchicago.edu/books/humboldt is free and open to the public.

ACKNOWLEDGMENTS: The members of the Alexander von Humboldt in English team wish to thank the staff of the former Center for the Americas at Vanderbilt and at the University of Potsdam for photocopying, scanning, and all other forms of logistical support. Many people at Vanderbilt's Heard Library have assisted us in our research in so many invaluable ways. For patiently and efficiently managing the flood of questions and requests for books and articles, for assisting in background research, and for digitalizing images, we heartily thank Peter Brush, Paula Covington, Henry Shipman, and Jim Toplon. The staff of various libraries in Berlin and Mexico City is no less deserving of our gratitude, as are friends, and family members, who have encouraged and helped each of us in a variety of ways: F. David T. Arens, Detlev Eggers, Marion Humbert-Oyss, Helmut and Margarete Kutzinski, Dana Nelson, Anna Potter, and Santiago Sevilla. We especially thank R. J. Boutelle and Thea Autry for proofing significant parts of this hefty manuscript; Angel Matos and Julian Drews for their assistance with the annotations; and Christine Richter-Nillson, Magnus Nillson, and Brigitte Anschuetz for helping prepare the indexes. Ingo Schwarz and Ulrike Leitner of the Berlin-Brandenburg Academy of Sciences have been invaluable resources throughout the long process of completing these two volumes. Christie Henry, our first editor at Chicago, was a fount of steady encouragement and good cheer. Karen Merikangas Darling, who took over from Christie, continued in this vein. We cannot thank both of them enough for their engagement with and support for HiE.

The work of the HiE team began in the fall of 2007. Various aspects of the work on *The Political Essay on the Kingdom of New Spain* were supported by grants from the National Endowment for the Humanities (for the annotations and for Humboldt's Library), the former Center for the Americas at Vanderbilt, and the Martha Rivers Ingram Chair of English at Vanderbilt. We are deeply grateful to all of them for making these volumes possible.

Contributors

Vera M. Kutzinski is the Martha Rivers Ingram Professor of English and professor of comparative literature at Vanderbilt University.

Ottmar Ette is professor of Romance literatures at University of Potsdam, Germany.

J. Ryan Poynter is Associate Vice Provost for Undergraduate Academic Affairs at New York University.

Kenneth Berri is an independent translator.

Giorleny Altamirano Rayo is an independent scholar of political science.

Tobias Kraft is project director of Alexander von Humboldt auf Reisen–Wissenschaft aus der Bewegung at the Berlin-Brandenburg Academy of Sciences in Berlin.